N-Heterocyclic Carbenes in Transition Metal Catalysis and Organocatalysis

CATALYSIS BY METAL COMPLEXES

This book series covers topics of interest to a wide range of academic and industrial chemists, and biochemists. Catalysis by metal complexes plays a prominent role in many processes. Developments in analytical and synthetic techniques and instrumentation, particularly over the last 30 years, have resulted in an increasingly sophisticated understanding of catalytic processes.

Industrial applications include the production of petrochemicals, fine chemicals and pharmaceuticals (particularly through asymmetric catalysis), hydrometallurgy, and waste-treatment processes. Many life processes are based on metallo-enzyme systems that catalyse redox and acid-base reactions.

Catalysis by metal complexes is an exciting, fast developing and challenging interdisciplinary topic which spans and embraces the three areas of catalysis: heterogeneous, homogeneous, and metallo-enzyme.

Catalysis by Metal Complexes deals with all aspects of catalysis which involve metal complexes and seeks to publish authoritative, state-of-the-art volumes which serve to document the progress being made in this interdisciplinary area of science.

VOLUME 32:
***N*-HETEROCYCLIC CARBENES IN TRANSITION METAL CATALYSIS AND ORGANOCATALYSIS**

Volume Editor

Catherine S. J. Cazin
EaStCHEM School of Chemistry
University of St Andrews
St Andrews, KY16 9ST
United Kingdom

For other titles published in this series, go to www.springer.com/series/5763

Catherine S. J. Cazin
Editor

N-Heterocyclic Carbenes in Transition Metal Catalysis and Organocatalysis

Springer

Editor
Catherine S. J. Cazin
EaStCHEM School of Chemistry
University of St Andrews
St Andrews, KY16 9ST
United Kingdom
cc111@st-andrews.ac.uk

ISBN 978-90-481-2865-5 e-ISBN 978-90-481-2866-2
DOI 10.1007/978-90-481-2866-2
Springer Dordrecht Heidelberg London New York

Library of Congress Control Number: 2010932004

Printed on acid-free paper

Springer is part of Springer Science+Business Media (www.springer.com)

Preface

The origins of amino carbenes as ligands can be traced back almost a century to the complex first synthesised by Tschugajeff (Chugaev). Interestingly, *N*-heterocyclic carbenes (NHCs) remained a lab curiosity until the mid-1990s. A few years later, this new class of ligands exploded in the literature, so much so that NHCs have become a ubiquitous class of ligands.

During the past decade, NHCs have been coordinated to virtually all transition metals (TM) and studied in numerous catalytic transformations, pushing back the frontiers of catalysis. In this regard, the most salient examples are found in olefin metathesis and cross coupling reactions, and more recently in organocatalysis.

The monograph commences with an introductory overview of NHCs, including a complete description of their steric and electronic properties, that shatters long-standing dogmas such as "phosphine mimicry" and "inexistent pi-acidity". This sets the stage for catalytic applications that are thoroughly discussed throughout eleven chapters. The penultimate chapter is devoted to decomposition pathways of TM-NHC systems. The closing chapter brings a unique industrial context to this book by describing applications of NHCs in industrial processes, a first of its kind.

In order to provide the reader with a *fresh* perspective on NHCs, the book has been assembled mainly by young emerging researchers, most of whom studied NHCs in undergraduate classes. This is therefore a perspective from a new generation of researchers that never considered NHCs as laboratory curiosities. A complementary perspective is brought by prominent, well-established academic researchers and an industrialist.

Believe it or not, I have been associated with NHCs in one form or another for the past eleven years. I went through it all, from the frustrations of tar-making to the distress of being *scooped* past tar-stage. I have even been told to give it all up. For some reason NHCs keep crossing my path, and I find them so intriguing that I keep coming back to them. This book has been an exciting project and I hope it will trigger activity from novices and provide inspiration to researchers already in the field.

St Andrews, UK
March 2010

Catherine S. J. Cazin

Contents

Contributors

Manuel Alcarazo
Max Planck Institut für Kohlenforschung, Kaiser Wilhelm Platz 1,
45470 Mülheim an der Ruhr, Germany
alcarazo@mpi-muelheim.mpg.de

Siegfried Blechert
Technische Universität Berlin, Institut für Chemie, Sekretariat C3,
Straße des 17, Juni 135 10623 Berlin, Germany
blechert@chem.tu-berlin.de

Oliver Briel
Umicore AG & Co. KG, Rodenbacher Chaussee 4, P.O. Box 1351,
63403 Hanau-Wolfgang, Germany
oliver.briel@eu.umicore.com

Luigi Cavallo
Dipartimento di Chimica, Università di Salerno, Via ponte don Melillo,
84084 Fisciano SA, Italy
lcavallo@unisa.it

Kingsley J. Cavell
School of Chemistry, Cardiff University, Main Building, Park Place,
CF10 3AT Cardiff, UK
Cavellkj@cf.ac.uk

Catherine S. J. Cazin
EaStCHEM School of Chemistry, University of St Andrews,
St Andrews, KY16 9ST, UK
cc111@st-andrews.ac.uk

Andreas A. Danopoulos
School of Chemistry, University of Southampton, Highfield SO17 1BJ, UK
ad1@soton.ac.uk

Reto Dorta
Organic Chemistry Institute, University of Zurich, Winterthurerstrasse 190, 8057 Zurich, Switzerland
dorta@oci.uzh.ch

Janis Louie
Department of Chemistry, University of Utah, Salt Lake City, Utah 84112-0850, USA
louie@chem.utah.edu

David S. McGuinness
School of Chemistry, University of Tasmania, Private Bag 75, Hobart, Tasmania 7000, Australia
David.McGuinness@utas.edu.au

Oscar Navarro
Department of Chemistry, University of Hawaii at Manoa, Honolulu, Hawaii 96822, USA
oscarnf@hawaii.edu

Christian Slugovc
Institute for Chemistry and Technology of Materials (ICTM), Graz University of Technology, 8010 Graz, Austria
slugovc@tugraz.at

Andrew D. Smith
EaStCHEM School of Chemistry, University of St Andrews, St Andrews, KY16 9ST, UK
ads10@st-andrews.ac.uk

Adam Veige
Department of Chemistry, Center for Catalysis, University of Florida, P.O. Box 117200, Gainesville, FL, USA
veige@chem.ufl.edu

Michael K. Whittlesey
Department of Chemistry, University of Bath, Claverton Down, Bath BA2 7AY, UK
m.k.whittlesey@bath.ac.uk

Jonathan M. J. Williams
Department of Chemistry, University of Bath, Claverton Down, Bath BA2 7AY, UK
j.m.j.williams@bath.ac.uk

Abbreviations

Å	angstrom(s)
AAA	Asymmetric Allylic Alkylation
acac	acetylacetonate
ACM	Asymmetric Cross-Metathesis
ADMET	Acyclic Diene Metathesis
AE	allyl ether
aNHC	*abnormally* bound NHC
ALTMET	Alternating Diene Metathesis Polycondensation
AM3	Amphidinol 3
Ar	aryl
AROCM	Asymmetric Ring Opening Cross Metathesis
AT	Advanced Technology
atm	atmosphere(s)
ATRP	Atom-Transfer Radical Polymerisation
av	average
BAr_4^F	tetrakis[3,5-bis(trifluoromethyl)phenyl]borate
BBN	borabicyclo[3.3.1]nonyl
$(Bcat)_2$	bis(catecholato)diboron
BDE	Bond Dissociation Energy
β-elim.	β-elimination
binap	2,2′-bis(diphenylphosphino)-1,1′-binaphthyl
bipy	2,2′-bipyridine
b:l	branched:linear ratio
$bmim^+$	1-*n*-butyl-3-methylimidazolium
bmiy	1-*n*-butyl-3-methylimidazolin-2-ylidene
Bn	benzyl
Boc	*tert*-butoxycarbonyl
$(Bpin)_2$	bis(pinacolato)diborane
Bu	butyl
^{t}Bu	*tert*-butyl
Bz	benzoyl

°C	degrees Celsius
ca.	*circa*
CAAC	cyclic (alkyl)(amino)carbenes
cal	calorie(s)
cat.	catalyst
cf	*confer*
CM	Cross Metathesis
cm^{-1}	wavenumber(s)
COD	1,5-cyclooctadiene
COE	cyclooctene
Conv.	conversion
Cp	cyclopentadienyl
Cp*	1,2,3,4,5-pentamethylcyclopentadienyl
Cy	cyclohexyl
d	days
Δ	heat
dr	diastereomeric ratio
dba	dibenzylidene acetone
DBM	dibenzoylmethane
DBU	1,8-diazabicyclo[5.4.0]undec-7-ene
DBP	phenyldibenzophosphole
DCE	1,2-dichloroethane
de	diastereomeric excess
Dec	decyl
DFT	Density Functional Theory
DIBAL-H	di-*iso*-butylaluminium hydride
DIPEA	diisopropylethylamine (Hunig's base)
DMA	dimethylacetamide
DME	dimethoxyethane
DMF	dimethylformamide
DMI	1,3-dimethyl-2-imidazolidinone
dmpe	dimethylphosphinoethane
DMSO	dimethylsulphoxide
DPC	diphenyl carbonate
dppe	(diphenylphosphino)ethane
dppf	(diphenylphosphino)ferrocene
dppp	1,3-bis(diphenylphosphino)propane
dr	diastereomeric ratio
dvds	1,3-divinyltetramethyldisiloxane
dvtms	divinyltetramethylsiloxane
$E_{1/2}$	half-wave potential
E_{act}	activation energy
ECM	Enyne Cross Metathesis
EDA	ethyldiazoacetate

ee	enantiomeric excess
e.g.	for example
$emim^+$	1-ethyl-3-methylimidazolium
emiy	1- ethyl-3-methylimidazolin-2-ylidene
Ent	enantiomeric
equiv	equivalent
Et	ethyl
et al.	*et alii*
EWG	Electron Withdrawing Group
Fc	ferrocenyl
FDA	US Food and Drug Administration
g	gram(s)
GC	Gas Chromatography
gem	geminal
GSK	Glaxo Smith Kline
h	hour(s)
Hbbtm	*bis*-{benzothiazol-2-yl}methane
HBpin	pinacolborane
HCV	Hepatitis C Virus
Hex	hexyl
HMDS	hexamethyldisilazane
HNBR	Hydrogenated Nitrile Butadiene Rubbers
HOAt	1-hydroxy-7-azabenzotriazole
HOBt	hydroxybenzotriazole
Hz	Hertz
IAd	1,3-diadamantylimidazolium
ICy	*N,N'*-(dicyclohexyl)imidazol-2-ylidene
IDTB	*N,N'*-bis-[2,5-(di-*tert*-butyl)phenyl]imidazol-2-ylidene
i.e.	*id est*
IiPr	*N,N'*-(di-*iso*-propyl)imidazol-2-ylidene
IL	Ionic Liquid
Im	Imidazolium
IMe	*N,N'*-(dimethyl)imidazol-2-ylidene
IMes	*N,N'*-bis-[2,4,6-(trimethyl)phenyl]imidazol-2-ylidene
Inc.	Incorporated
IND	Investigational New Drug
IP	Intellectual Property
IPr	*N,N'*-bis-[2,6-(di-*iso*-propyl)phenyl]imidazol-2-ylidene
IR	Infrared
I*t*Bu	*N,N'*-(di-*tert*-butyl)imidazol-2-ylidene
ITM	1,3,4,5-tetramethylimidazol-2-ylidene
ITmt	*N,N'*-bis(2,2",6,6"-tetramethyl-*m*-terphenyl-5'-yl)imidazole-2-ylidene

ITol	*N,N'*-bis-(4-methylphenyl)imidazol-2-ylidene
J	coupling constant
J	joule(s)
k	kilo
L	litre(s)
Load.	loading
μ	micro
M	molar
M	metal
m	milli
m	*meta*
MAH	maleic anhydride
MALDI-TOF MS	Matrix-Assisted Laser Desorption Ionisation Time-Of-Flight Mass Spectrometry
MAO	methylaluminoxane
Me	methyl
Mes	mesityl (2,4,6-trimethylphenyl)
min.	minutes
MMA	methyl methacrylate
MMAO	modified methylaluminoxane
MOM	methoxymethylether
MS	Molecular Sieves
mV	millivolt
n	*normal*
nbd	norbornadiene
NHC	*N*-Heterocyclic Carbene
NMO	*N*-methylmorpholine-*N*-oxide
NMP	*N*-methyl-2-pyrrolidone
NMR	Nuclear Magnetic Resonnance
o	*ortho*
OAc	acetate
OTf	trifluoromethanesulfonate
p	*para*
PCBs	polychlorinated biphenyls
PCs	polycarbonates
PG	Protecting Group
Ph	phenyl
Piv	pivaloyl
POPs	Persistent Organic Pollutants
Pr	propyl
^{i}Pr	*iso*-propyl
Ph	phenyl

pin	pinacol
Piv	pivaloyl
PMP	*para*-methoxyphenyl
ppm	parts per million
psi	pound per square inch
quant.	quantitative
rac	racemic
RCM	Ring Closing Metathesis
R. E.	Reductive Elimination (Red. Elim.)
Ref	reference
REMP	Ring Expansion Metathesis Polymerisation
ROCM	Ring Opening Cross Metathesis
ROIMP	Ring Opening Insertion Metathesis Polymerisation
ROMP	Ring Opening Metathesis Polymerisation
RRM	Ring Rearrangement Metathesis
rt	room temperature
SEM	2-(trimethylsilyl)ethoxy-methyl
SICy	*N,N'*-(dicyclohexyl)imidazolidin-2-ylidene
SIEt	*N,N'*-(diethyl)imidazolidin-2-ylidene
SIMes	*N,N'*-bis[2,4,6-(trimethyl)phenyl]imidazolidin-2-ylidene
SIPr	*N,N'*-bis[2,6-(diisopropyl)phenyl]imidazolidin-2-ylidene
Substr.	substrate
syngas	synthesis gas $CO:H_2$ mixture
T	temperature
TASF	tris(dimethylamino)sulfonium difluorotrimethylsilicate
TBAF	tetra-*n*-butylammonium fluoride
TBDPS	*tert*-butyldiphenylsilyl
TBHP	*tert*-butyl hydroperoxide
TBS	*tert*-butylsilyl
TEP	Tolman Electronic Parameter
TES	triethylsilyl
Tf	trifluoromethanesulfonyl (triflyl)
TFA	trifluoroacetic acid
TFAA	trifluoroacetic anhydride
thf	tetrahydrofuran
TIMEN	tris[2-(3-alkylimidazol-2-ylidene)ethyl]amine
TM	Transition Metal
TMHD	2,2,6,6-tetramethyl-3,5-heptanedionate
tmiy	1,3,4,5-tetramethylimidazol-2-ylidene
TMPDA	tetramethylpropylenediamine
TMS	trimethylsilyl
TOF	Turnover Frequency
Tol	tolyl

TON	Turnover Number
TS	Transition State
Ts	*para*-toluenesulfonyl (tosyl)
ν_{av}	average value of stretching frequencies in IR
$\%V_{Bur}$	Percent Volume Buried

Chapter 1
N-Heterocyclic Carbenes: An Introductory Overview

Luigi Cavallo and Catherine S. J. Cazin

Abstract *N*-heterocyclic carbenes (NHCs) are probably the class of ligands that not only has attracted the most attention during the past decade, but also for which the greatest advances have been made. These include a wider availability, applicability and understanding. In this chapter, an overview of all aspects of NHCs is given, starting with an historical discussion that begins almost a century ago. An inventory of the structural diversity of NHCs found in the literature is given, followed by the nomenclature of NHCs and the trivial names used. A section is devoted to the synthetic strategies developed for the formation of NHC-precursors, NHC ligands and NHC-complexes. The most diagnostic spectroscopic features of NHCs and NHC complexes are listed as well as the manner in which NHCs are usually represented. NHCs have become indubitably one of the most important and unique class of ligands as they have very distinctive and interesting electronic and steric features. A large section of this chapter is hence devoted to the discussion of these features and presents the recent advances made for determination of NHC properties and their understanding.

1.1 Generalities

1.1.1 Historical Aspects

N-Heterocyclic carbenes (NHCs) have become an incontrovertible class of molecules for transition-metal- (TM) and organo-catalysis. Inception of the field dates back almost a century ago when Tschugajeff (Chugaev) and co-workers reacted

L. Cavallo (✉)
Dipartimento di Chimica, Università di Salerno, Via ponte don Melillo,
84084 Fisciano, SA, Italy
e-mail: lcavallo@unisa.it

C. S. J. Cazin (✉)
EaStCHEM School of Chemistry, University of St Andrews,
St Andrews, KY16 9ST, UK
e-mail: cc111@st-andrews.ac.uk

C. S. J. Cazin (ed.), *N-Heterocyclic Carbenes in Transition Metal Catalysis and Organocatalysis*, Catalysis by Metal Complexes 32,
DOI 10.1007/978-90-481-2866-2_1, © Springer Science+Business Media B.V. 2011

potassium tetrachloroplatinate with methyl isocyanide, followed by the addition of hydrazine [1, 2]. Contrarily to the authors' expectations, this reaction does not lead to a dimeric species composed of tetracyanide platinum moieties bridged by hydrazine molecules, but leads to a compound that is probably the first diamino carbene complex isolated in pure form. The structure of this salt **1**, and the one of its biscarbene derivative **2**, were only solved decades later (Scheme 1.1) [3–6]. It was later shown that this methodology is applicable to the synthesis of NHC complexes when functionalised isocyanides are used [7].

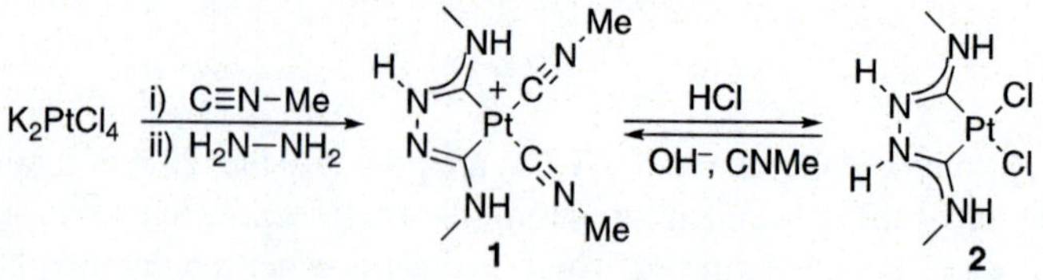

Scheme 1.1 Tschugajeff's (Chugaev) carbene complexes [1, 2]

In the early 1960s, Wanzlick pioneered investigations on NHCs [8]. This was followed by the first description of NHCs as ligands for metal complexes [9, 10]. However, it is only in the 1990s that this new class of ligands was brought under the spotlight. Firstly by Arduengo's seminal isolation and characterisation of a free NHC [11], secondly by the recognition that NHCs could act as promising ligands for homogeneous catalysts [12]. These reports ignited the curiosity of researchers as the possible role of NHCs as ligands was revealed, an interest that has seen an incredible research activity (Fig. 1.1).

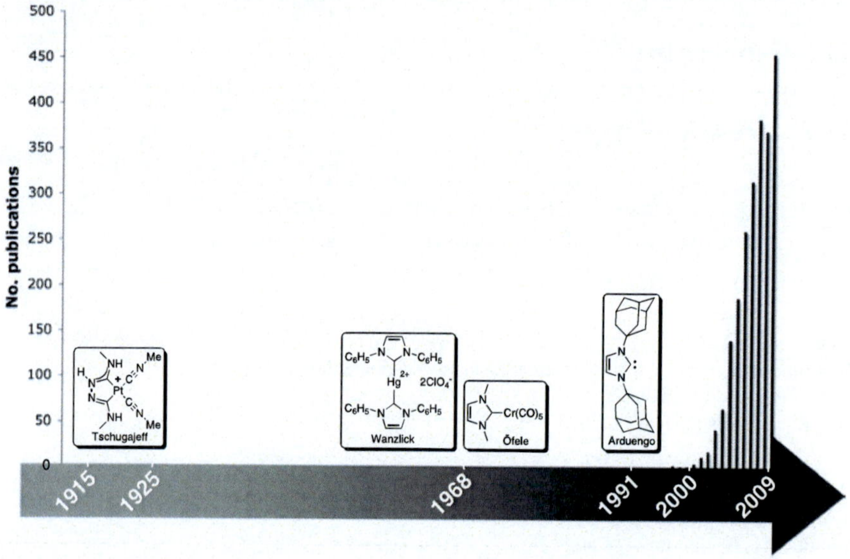

Fig. 1.1 Number of publications (*N-heterocyclic carbene* as research topic)

1.1.2 Structural Diversity of NHCs

At the origin of this abundance of publications on *N*-heterocyclic carbenes is the structural ligand diversity now available (Fig. 1.2). This developing area is noteworthy as most early developments were mainly focused on imidazolylidene and imidazolidinylidene NHC-types.

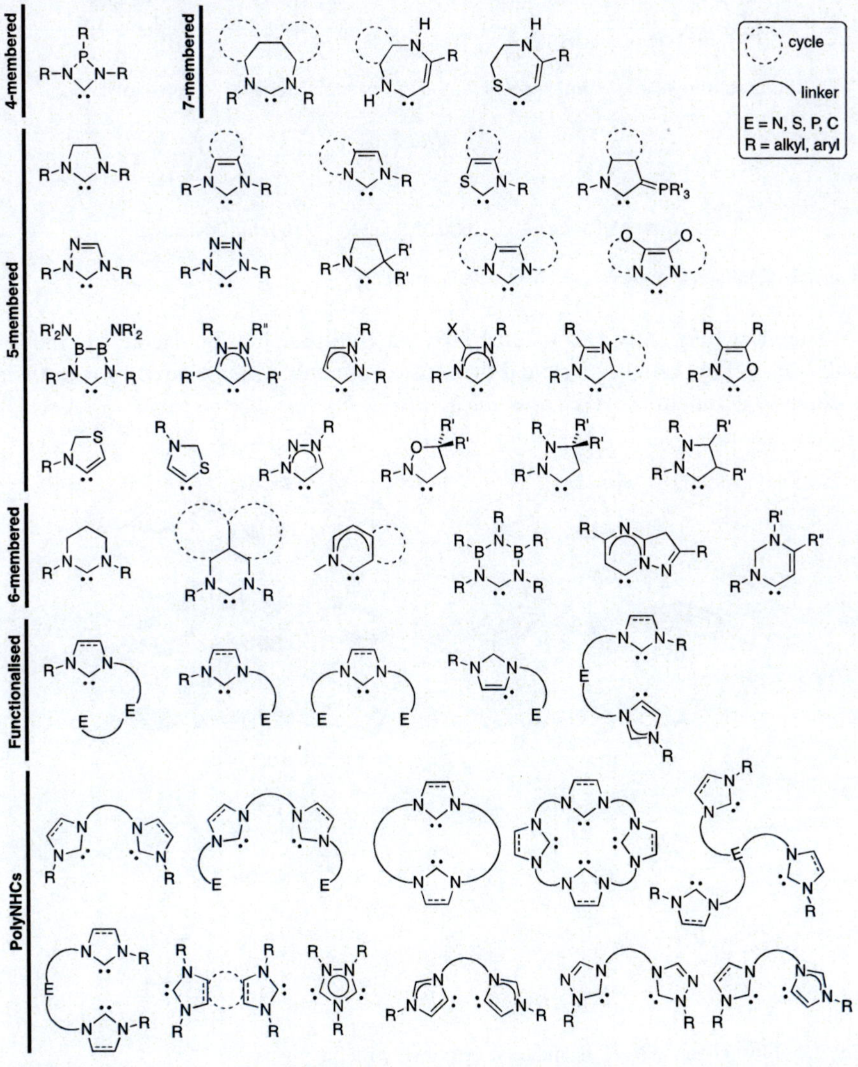

Fig. 1.2 Structural diversity of NHC ligands

This schematic overview of NHC ligands found in the literature shows that most are based on five-membered heterocyclic cores. The most common are listed in Section 1.1.3.

1.1.3 NHC Ligands: Nomenclature and Trivial Names

The most commonly encountered NHCs are those based on five-membered heterocycles. Figure 1.3 summarises these heterocycles and the associated name of the corresponding NHC.

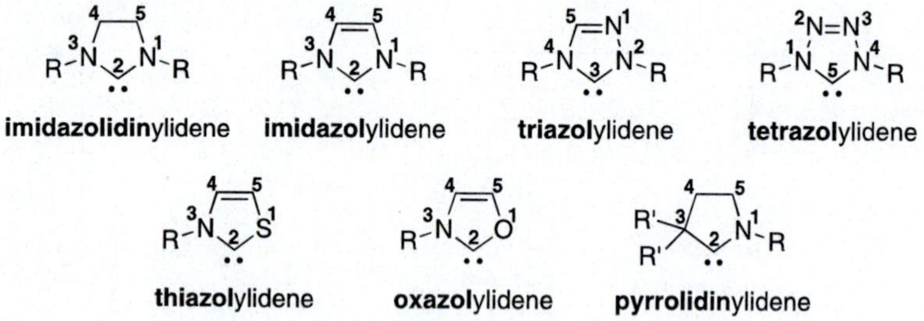

Fig. 1.3 Common five-membered heterocyclic carbenes

The most frequently encountered NHCs are all based on imidazole and imidazolidine. In Fig. 1.4 are presented the most commonly found examples with their associated acronym.

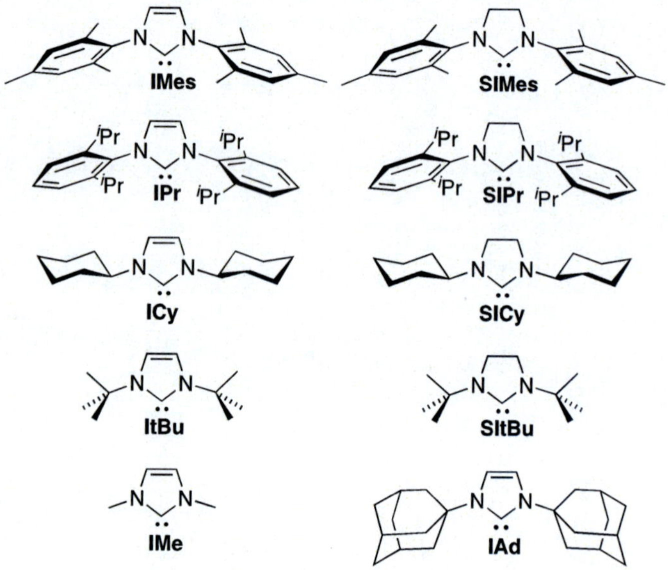

Fig. 1.4 Most common NHC ligands and their respective acronyms

Since Arduengo's first isolation of a free NHC (IAd) [11], a few others have been isolated and characterised. Despite the early assumption that bulky substituents on the nitrogen atoms were necessary in order to stabilise free NHCs, compounds such as IMe have been found stable. However, depending on the class of NHCs, the isolation of the free ligand can be difficult or impossible. This is, for example, the case of NHCs based on imidazolidinylidene bearing small *N*-substituents as they dimerise to the corresponding tetraaminoethylene [13]. In order to overcome this limitation, alternate synthetic strategies have been developed for the formation of NHC complexes.

1.1.4 Synthesis of NHC Precursors, NHCs and Complexes

The main synthetic routes leading to the formation of NHC complexes are depicted in Scheme 1.2. The methodologies given are shown with imidazolidinylidene and imidazolylidene ligands, however, they are applicable to other NHCs [16, 17].

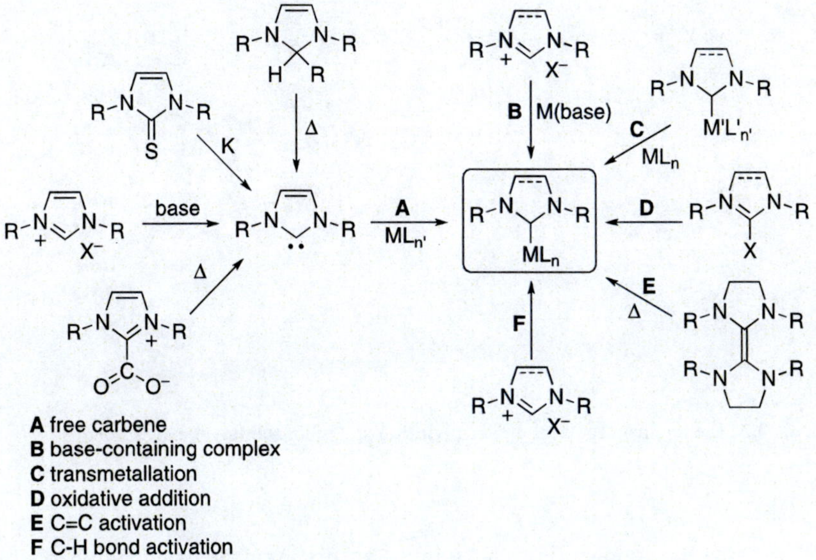

Scheme 1.2 Main synthetic strategies for the formation of NHC-complexes

The most often encountered routes are **A**, **B** and **C**. Route **A** consists of generating the free carbene (by deprotonation of the corresponding salt, by reductive desulfurisation or by thermal α-elimination from appropriate NHC precursors) followed by coordination to a metal centre (often with concomitant ligand displacement). Method **B** consists of using a metal precursor containing a base as ligand. The base deprotonates the imidazol(idin)ium salt, leading to the coordination of the NHC and of the counter-anion of the salt (if X is a coordinating anion). Method **C** employs a carbene transfer reagent (often a Ag-complex) that, by transmetallation, delivers the NHC to the second metal. Methods **D** and **F** are less frequently encountered, they consist of a C–X or a C–H bond activation *via* oxidative addition. Route **E** is a possibility for the formation of imidazolidinylidene complexes by C=C bond activation of the dimerised imidazolidinylidene. Other routes that have not been depicted in Scheme 1.2 (because they are much less versatile) include the cycloaddition to Fischer carbenes and the intramolecular addition to a C≡N bond of coordinated isocyanides. The latter methodology is the intramolecular version of the Tschugajeff synthesis (Scheme 1.1) [7, 14, 15]. The most frequently encountered NHC-complexes are undoubtedly imidazolylidene and imidazolidinylidene due to the easy access of their precursors (imidazol(idin)ium salts). The most common routes for their synthesis are described in Scheme 1.3 [16, 17].

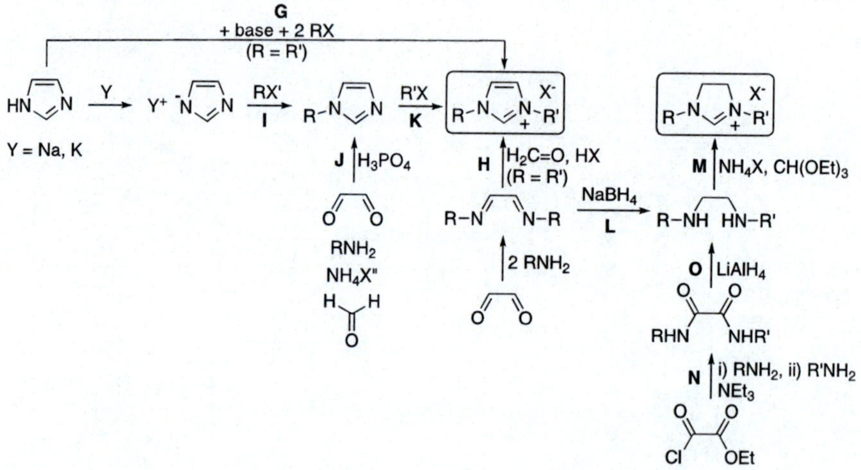

Scheme 1.3 Predominant routes to imidazolylidene and imidazolidinylidene precursors

Two types of imidazolium salts must be distinguished depending on the *N,N'*-substitution: symmetrically (R = R') and unsymmetrically substituted (R ≠ R') versions. For the first type (which is also the predominant one, see Fig. 1.4), two main strategies are viable: reaction of imidazole with RX in the presence of a base (**G**) or cyclisation of an α-diimine or diazobutadiene (obtained by the condensation of the amine with glyoxal) with formaldehyde in the presence of a Brønsted acid (**H**). On the other hand, the synthesis of unsymmetrically *N,N'*-substituted congeners is less straightforward as a functionalised imidazole has to be isolated prior to alkylation or arylation. Two main methods are available for imidazole functionalisation: deprotonation with metallic Na or K leading to an imidazolide (**I**) followed by reaction with RX; or reaction of glyoxal with a primary amine, an ammonium salt and formaldehyde (**J**). *N*-functionalised imidazole can then be alkylated or arylated (**K**) to the imidazolium salt. Symmetrically *N,N'*-substituted imidazolidinium salts can be easily obtained by reduction of an α-diimine (**L**) followed by cyclisation using triethyl orthoformate in the presence of an ammonium salt (**M**). The same strategy is operative for the synthesis of unsymmetrically *N,N'*-substituted imidazolidinium salts (**M**, R ≠ R'). In such cases, the diamine is synthesised in two steps in order to introduce different *N,N'*-substituents: stepwise reaction of ethyl 2-chloro-2-oxoacetate with two primary amines leading to the corresponding oxalamide (**N**) followed by reduction to the diamine (**O**). Cyclisation (**M**) leads to the unsymmetrically *N,N'*-substituted imidazolidinium salt.

The methodologies described above lead to NHC precursors rather limited in terms of substitution at the four- and five-positions as their access is restricted to the accessibility of the appropriate diimine. As such substitutions are of great interest in particular for the design of asymmetric catalysts, routes to the synthesis of the NHC precursors have more recently been developed. Some of these approaches are described in Scheme 1.4.

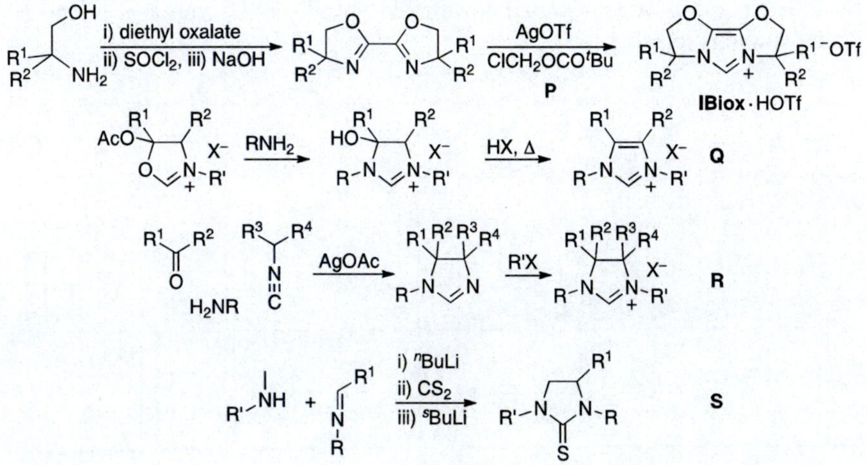

Scheme 1.4 Synthetic pathways to NHC precursors with a substituted backbone (C^4 C^5)

The cyclisation of α-diimines can be efficiently performed using chloromethyl pivalate in the presence of silver triflate (**P**) [18]. This method is a good alternative to the use of formaldehyde (pathway **H**, Scheme 1.3) as it overcomes the problem of ring-closing of sterically hindered substrates encountered when using **H**. This method allowed Glorius and co-workers to introduce a new class of sterically demanding NHCs: the tricyclic Biox ligands. Imidazolium salts with substituted backbones can also be obtained by reaction of oxazolinium acetals with a primary amine, leading to hydroxylated imidazolidinium salts that lead to the imidazolium salt after elimination of water (**Q**) [19]. Imidazolidinium salts with a substituted backbone can be obtained by alkylation of the parent 2-imidazolidine [20]. The latter can be obtained by the reaction of an aldehyde with an amine and an isocyanide (**R**)

[21]. Imidazolidine-2-thiones functionalised on the four-position can be obtained by reaction of $HN(CH_3)R'$ with *n*-butyllithium, followed by addition of carbon disulfide. The lithium thicarbamate can then by further lithiated and cyclisation occurs upon reaction of this species with an imine (**S**) [22].

1.1.5 Spectroscopic Features of NHCs and Complexes

The most convenient tool for the characterisation of NHCs is NMR spectroscopy, in particular $^{13}C\{^1H\}$ NMR. As a case study, the carbenes IPr and SIPr, and their corresponding salts IPr·HCl and SIPr·HCl were chosen. As described above (Scheme 1.2), free carbenes are often obtained by deprotonation of the corresponding salt. The best diagnostic tool to observe the salt deprotonation, and thus indirectly monitor the carbene formation, is ^1H NMR spectroscopy, by means of the disappearance of the characteristic acidic proton resonance. The signal corresponding to the latter (H^2) is largely shifted downfield (typically 8–12 ppm) and disappears upon deprotonation (Fig. 1.5).

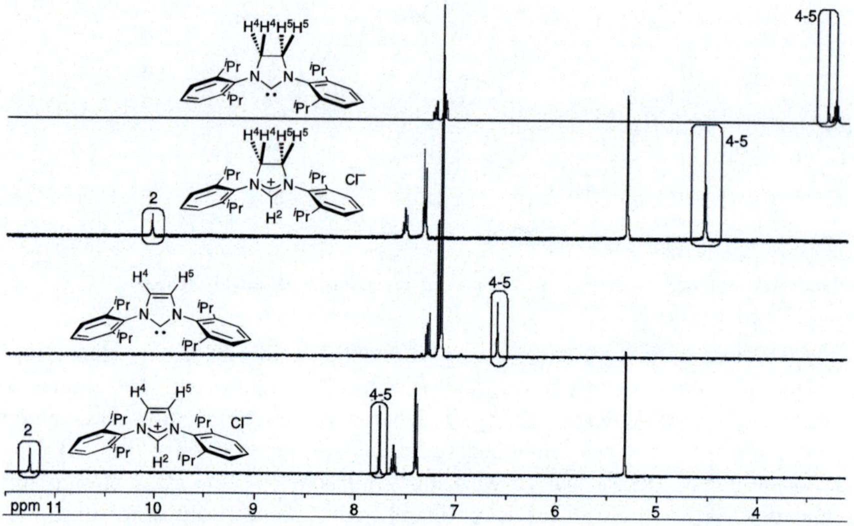

Fig. 1.5 ^1H NMR (CD_2Cl_2 – salt; C_6D_6 – free NHC) spectra of IPr, IPr·HCl, SIPr and SIPr·HCl

The carbene formation can be monitored by $^{13}C\{^1H\}$ NMR, as the carbene carbon atom of free NHCs has a signal significantly shifted downfield. Typically, the signal for the C^2 atom is found between 200 and 250 ppm for the free carbene, and between 130 and 160 ppm for the corresponding salt (Fig. 1.6) [23].

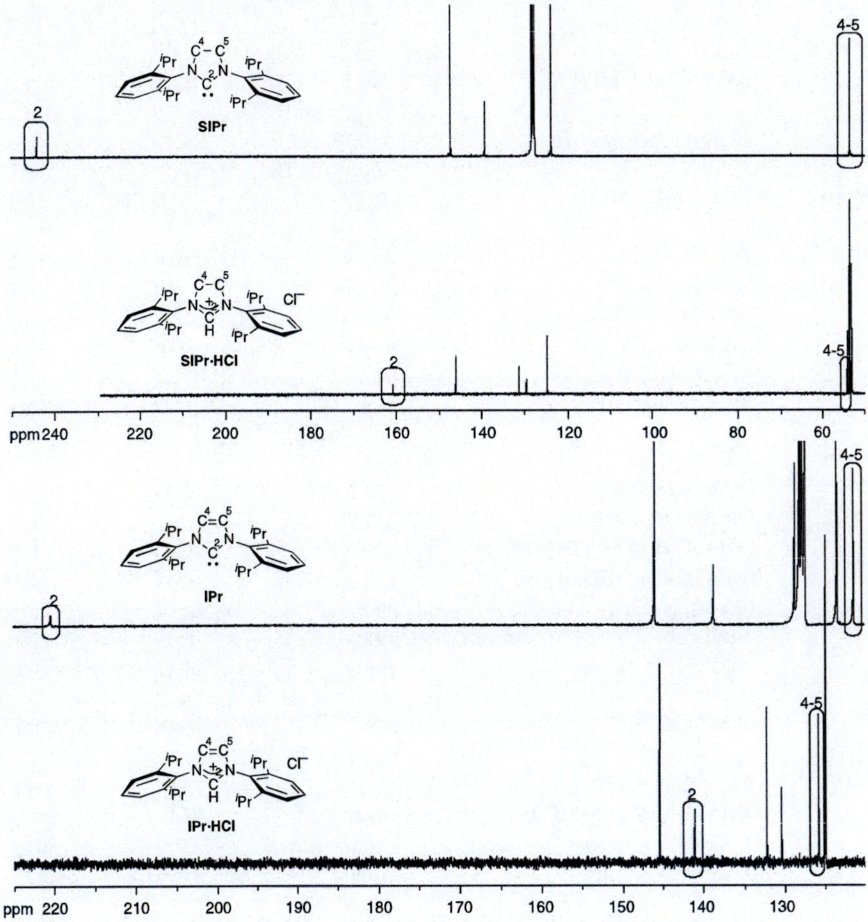

Fig. 1.6 $^{13}C\{^1H\}$ NMR (CD_2Cl_2 – salt; C_6D_6 – free NHC) spectra of IPr, IPr·HCl, SIPr and SIPr·HCl

Once coordinated to a metal centre, the signal corresponding to the carbene carbon atom is usually shifted upfield. The chemical shift of the carbene carbon atom (C^2) for a given metal in a given oxidation state is usually characteristic (Table 1.1).

Table 1.1 $^{13}C\{^1H\}$ NMR chemical shifts of the carbenic carbon atom C^2

Metal	Complex	δC^2 (ppm)	Solvent	Ref.
Ru(II)	$[RuHCl(CO)_2(IPr)_2]$	185.9	thf-d_8	[24]
	$[RuHCl(CO)(IPr)_2]$	195.6	thf-d_8	[24]
	$[RuHCl(CO)_2(SIPr)_2]$	193.0	thf-d_8	[24]
	$[RuHCl(CO)(SIPr)_2]$	201.0	thf-d_8	[24]
Rh(I)	[RhCl(COD)(IPr)]	187.7 $^1J_{Rh–C}$ = 53 Hz	C_6D_6	[25]
	$[RhCl(COE)(IPr)]_2$	181.7 $^1J_{Rh–C}$ = 64 Hz	C_6D_6	[26]
	$[Rh(OAc)(CO)(IPr)_2]$	190.6 $^1J_{Rh–C}$ = 44 Hz	$CDCl_3$	[27]
	[Rh(acac)(CO)(SIPr)]	210.75 $^1J_{Rh–C}$ = 54 Hz	$CDCl_3$	[28]
Rh(III)	$[RhCl(\eta^2\text{-}O_2)(IPr)_2]$	180.8 $^1J_{Rh–C}$ = 40 Hz	$C_6D_5CD_3$	[29]
Ir(I)	[IrCl(COD)(IPr)]	182.6	$CDCl_3$	[30]
	$[IrCl(CO)_2(IPr)]$	178.6	$CDCl_3$	[30]
	[IrCl(COD)(SIPr)]	209.5	$CDCl_3$	[30]
	$[IrCl(CO)_2(SIPr)]$	204.9	$CDCl_3$	[30]
Ir(III)	$[IrCl(\eta^2\text{-}O_2)(IPr)_2]$	167.4	C_6D_6	[31]
	$[IrCl(H)_2(IPr)(a\text{IPr})]$[a]	187.5 (IPr) 166.8 (*a*IPr)	C_6D_6	[32]
Ni(0)	$[Ni(CO)_3(IPr)]$	198.2	C_6D_6	[33]
	$[Ni(CO)_3(SIPr)]$	223.2	C_6D_6	[33]
Ni(II)	$[Ni(\eta^3\text{-}C_3H_5)Cl(IPr)]$	187.8	C_6D_6	[34]
	$[Ni(\eta^3\text{-}C_3H_5)(OH_2)(IPr)]^+[BAr_4^F]^-$	176.4	CD_2Cl_2	[35]
	$[Ni(\eta^3\text{-}C_3H_5)Cl(SIPr)]$	218.4	C_6D_6	[36]
Pd(0)	[Pd(dvtms)(IPr)]	200.8	thf-d_8	[37]
	$[Pd(IPr)_2]$	199.0	C_6D_6	[38]
	$[Pd(IPr)(PPh_3)]$	198.0 $^2J_{P–C}$ = 94 Hz	C_6D_6	[38]
	$[Pd(SIPr)(PPh_3)]$	218.1 $^2J_{P–C}$ = 86 Hz	C_6D_6	[38]
Pd(II)	$[Pd(\eta^3\text{-}C_3H_5)Cl(IPr)]$	188.5	C_6D_6	[39]
	$[Pd(\eta^3\text{-}2\text{-}MeC_3H_4)Cl(IPr)]$	189.6	C_6D_6	[40]
	$[PdCl_2(IPr)_2]$	172.5	$CDCl_3$	[41]
	$[PdCl_2(IPr)(PPh_3)]$	170.9 $^2J_{P–C}$ = 198 Hz	$CDCl_3$	[42]
	$[PdH_2(IPr)(PCy_3)]$	200.5 $^2J_{P–C}$ = 138 Hz	C_6D_6	[43]
	$[Pd(\eta^2\text{-}O_2)(IPr)(PCy_3)]$	192.0 $^2J_{P–C}$ = 16 Hz	C_6D_6	[43]
	$[Pd(\eta^3\text{-}C_3H_5)Cl(SIPr)]$	215.4	C_6D_6	[39]
Pt(0)	[Pt(AE)(IPr)]	187.5[b]	$CDCl_3$	[44]
	[Pt(dvtms)(IPr)]	186.4[b]	$CDCl_3$	[45]
	[Pt(dvtms)(SIPr)]	213.3[b]	$CDCl_3$	[45]

(continued)

Table 1.1 (continued)

Metal	Complex	δC² (ppm)	Solvent	Ref.
Pt(II)	[Pt(η^3-2-MeC_3H_4)Cl(IPr)]	180.8[b]	$CDCl_3$	[46]
	[PtCl_2(dmso)(IPr)]	147.4 $^1J_{Pt-C}$ = 1,479 Hz	$CDCl_3$	[47]
	[PtCl_2(dmso)(SIPr)]	174.4 $^1J_{Pt-C}$ = 1,373 Hz	$CDCl_3$	[47]
Cu(I)	[CuCl(IPr)]	182.3	$(CD_3)_2CO$	[48]
	[Cu$(IPr)_2$][BF_4]	177.4	$CDCl_3$	[49]
	[Cu$(IPr)_2$][PF_6]	178.4	$CDCl_3$	[49]
	[CuCl(SIPr)]	204.3	$CDCl_3$	[50]
	[Cu$(SIPr)_2$][BF_4]	201.4	$CDCl_3$	[51]
	[Cu$(SIPr)_2$][PF_6]	199.8	$CDCl_3$	[51]
Ag(I)	[AgCl(IPr)]	184.6 $^1J_{Ag-C}$ = 271 Hz $^1J_{Ag-C}$ = 235 Hz	$CDCl_3$	[52]
	[Ag$(IPr)_2$][PF_6]	183.6 $^1J_{Ag-C}$ = 211 Hz $^1J_{Ag-C}$ = 183 Hz	$CDCl_3$	[25]
	[AgCl(SIPr)]	207.7 $^1J_{Ag-C}$ = 253 Hz $^1J_{Ag-C}$ = 219 Hz	CD_2Cl_2	[52]
Au(I)	[AuCl(IPr)]	175.1	CD_2Cl_2	[53]
	[AuBr(IPr)]	179.0	$CDCl_3$	[54]
	[AuCl(SIPr)]	196.1	$CDCl_3$	[53]
	[AuBr(SIPr)]	199.0	$CDCl_3$	[54]
Au(III)	[AuBr_3(IPr)]	146.2	$CDCl_3$	[54]
	[AuBr_3(SIPr)]	174.1	$CDCl_3$	[54]

[a] *a*IPr = abnormal IPr (*i.e.* IPr bound through C⁴)
[b] $^1J_{Pt-C}$ not observed

1.1.6 NHC Complexes: Representation and Convention

The standardised representation of NHC metal complexes is still not fully established, and different representations are used. Figure 1.7 summarises the representations found in the literature.

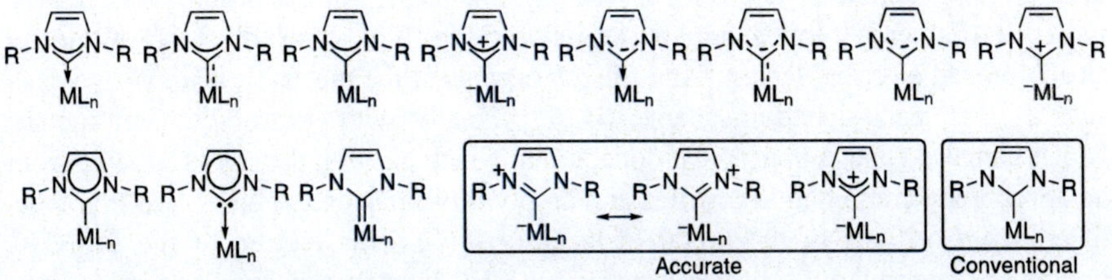

Fig. 1.7 Representations of NHCs found in the literature

Early representations exhibited a double bond between the metal centre and the carbene carbon atom. This was soon recognised as being a misleading representation as NHCs are two-electron donor ligands [55]. With respect to the disparity of representation still currently found in the literature, none is very accurate. The most accurate representation is the one displaying the charges [56], however, presumably for the sake of clarity, the most often encountered representation (conventional) does not contain any charge. It will also be the representation used throughout this book.

By convention, the carbenes displayed in Fig. 1.7 are "normal" NHCs as they are coordinated to the metal centre through the C^2 atom. By contrast, abnormal (also named non-classical or unusual) are those bound through the C^4 atom. Abnormal is also a term used for NHCs having a valence representation requiring additional charges. Remote is a term used to describe a carbene which does not have any heteroatom on the α-position to the carbenic carbon (Fig. 1.8) [57].

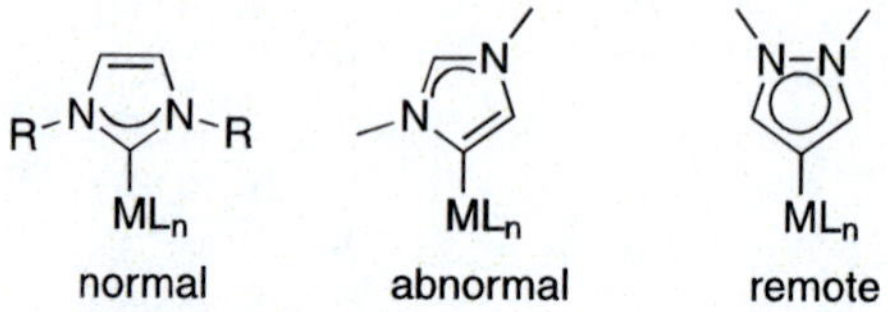

Fig. 1.8 Normal, abnormal and remote NHCs

As depicted above, there is a large variety of NHCs, and their access is relatively easy. This is a veritable advantage in the use of NHCs for homogeneous catalytic systems. However, it is probably their unique electronic and steric properties that make NHCs an exceptional class of molecules. These unique features are described in the following sections.

1.2 Electronic Properties of NHCs

NHCs can be classified as typical σ-basic/π*-acid ligands [58–61], whose electronics can be rationalised considering the Molecular Orbitals diagram presented in Fig. 1.9. The diagrams show the interaction of the basic imidazolidinylidene skeleton with a transition metal. The NHC presents a lone electrons pair in a high energy σ orbital, see Fig. 1.9a, which confers to NHCs a σ-donicity (basicity) clearly-stronger than that of even basic phosphines, such as PCy_3 [62, 63]. This key feature is accompanied by the presence of an empty low energy π* orbital, see Fig. 1.9b, which allows NHCs to act as acceptor of electron density from filled *d* orbitals of the metals (π-acidity) in a classical $d \rightarrow \pi^*$ back-donation scheme [64, 65]. Finally, with electron-deficient metals, NHCs can also engage in a $\pi \rightarrow d$ donation, in which electron density is donated from an appropriate combination of filled and empty π orbitals on the NHC, see Fig. 1.9c, to empty *d* orbitals of the metal (π-basicity) [66]. This picture of the M–NHC bonding is the result of years of research, since NHC ligands were initially considered to be pure σ-donors with insignificant π-acidity capability. Almost a decade later, seminal reports [64, 65, 67, 68] clearly showed that NHCs can accept electron density from the metal into π* orbitals, and this contribution cannot be neglected. Finally, the π-donor capability of NHC ligands towards electron poor metals, which completes the picture, has been recognised only in the past few years [59, 60, 66, 69].

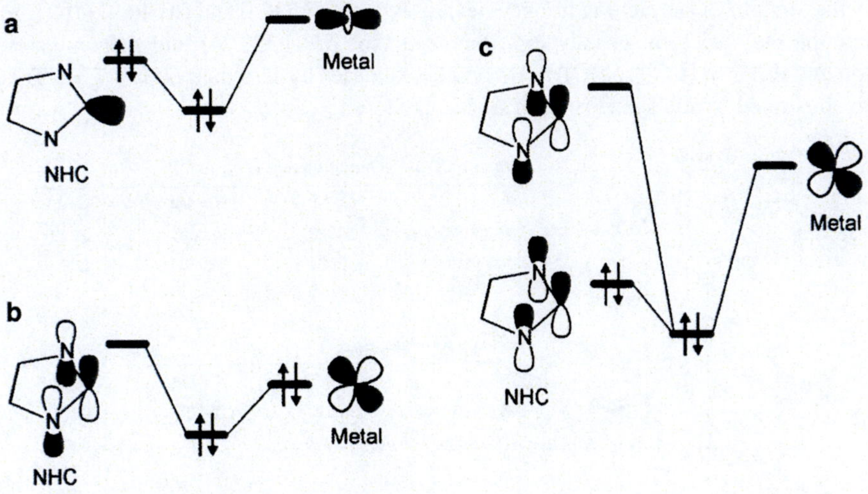

Fig. 1.9 Diagram illustrating the $\sigma \rightarrow d$ (**a**), the $d \rightarrow \pi^*$ (**b**), and the $\pi \rightarrow d$ (**c**) bonding modes occurring between NHCs and transition metals

Different electronic properties of the NHC ligand can result in drastic consequences on the catalytic efficiency of the corresponding metal complexes. The key structural parameters that can be modified to tune the electronic properties in five-membered NHCs, shown in Fig. 1.10, are: (a) the NHC skeleton; (b) the nature of the substituents on the C^4 and C^5 carbon atoms of the NHC skeleton; (c) the *N*-substituents. To discuss these points, we will privilege studies focused on the $[IrCl(CO)_2(NHC)]$ and [IrCl(cod)(NHC)] complexes (cod = 1,5-cyclooctadiene), since these two classes of compounds are gaining the status of model systems to investigate the stereoelectronic properties of NHC ligands [30, 70–75]. The former through the measurement of the average CO stretching frequency, ν_{av}, by IR spectroscopy [30, 70–74], the latter through the measurement of the redox potential, $E_{1/2}$, by cyclic voltammetry [18].

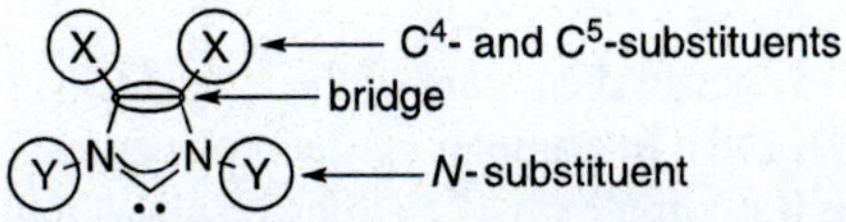

Fig. 1.10 Schematic illustration of the structural points whose modification can be used to tune the electronic properties of NHC ligands

With regard to the nature of the bridge of the NHC skeleton, no strong effects have been noticed by comparing transition metal complexes bearing strictly related imidazolidin-, imidazol- and benzimidazol-2-ylidene based NHCs. Indeed, these studies have indicated that the NHC bond dissociation energy (BDE) in [Ni(CO)$_3$(NHC)] and [RuClCp*(NHC)] complexes (Cp* = η^5-C$_5$Me$_5$) usually varies by less than 1 kcal/mol [33, 76], the average CO stretching frequency ν_{av} (average value of both CO stretching frequencies) in [Ni(CO)$_3$(NHC)] and [IrCl(CO)$_2$(NHC)] complexes varies by usually less than 2–3 cm^{-1} [30, 33, 74] and the oxidation potential $E_{1/2}$ in [IrCl(cod)(NHC)] complexes varies by less than 60 mV [30]. This is illustrated by the systems reported in Fig. 1.11.

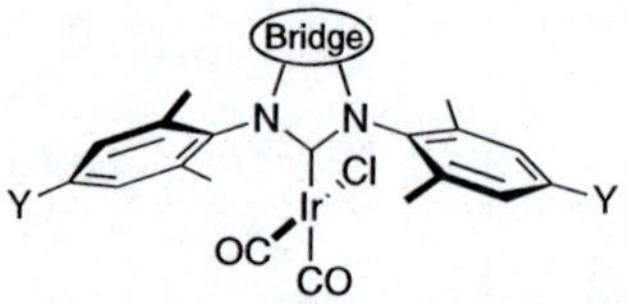

Average CO stretching frequency, ν_{av} (cm^{-1})

	Y = NEt$_2$	Y = Me	Y = SO$_2$-(*p*-tolyl)
Saturated bridge	2022	2025	2030
Unsaturated bridge	2021	2023	2029

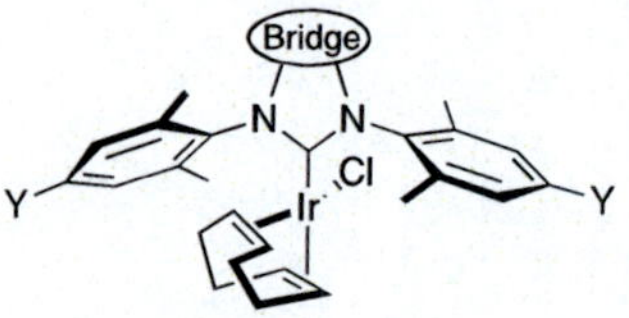

Redox potentials, $E_{1/2}$ (V)

	Y = NEt$_2$	Y = Me	Y = SO$_2$-(*p*-tolyl)
Saturated bridge	0.591	0.735	0.910
Unsaturated bridge	0.648	0.765	0.920

Fig. 1.11 Variation in the average CO stretching frequency and in the redox potential from [74]

These results suggest that imidazolidin- and imidazol-based skeletons transfer similar amounts of electron density to the metal. The conclusion that changes in the bridge of the NHC skeleton have such a small effect on the electronic properties of the NHC is quite surprising, considering that SIMes- and IMes-based catalysts often show remarkably different catalytic behaviour. It is still unclear if these small changes in the electronic properties of the NHC ligand confer such different catalytic behaviours, or other effects (steric, flexibility, etc.) should be invoked.

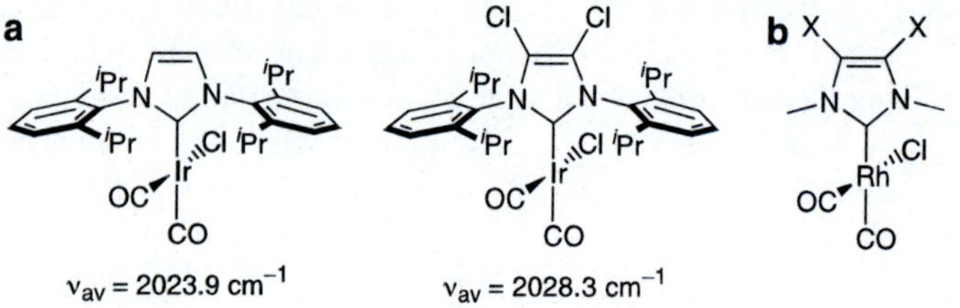

X	ν_{av} in cm^{-1}
H	2046
Cl	2051
NO$_2$	2053
CN	2058

Fig. 1.12 Effect of the C^4 and C^5 substituents on the average CO stretching frequency [77]

More significant is to change the nature of the C^4 and C^5 substituents. Remaining in the class of [IrCl(CO)$_2$(NHC)] complexes, the replacement of the H atoms on the C^4 and C^5 atoms of the IPr complex by Cl atoms resulted in the rather high shift of the average CO stretching frequency from 2023.9 to 2028.3 cm^{-1}, see Fig. 1.12a [30]. Similar effects have been found in the closely related [RhCl(CO)$_2$(NHC)] systems, see Fig. 1.12b, with the electron-withdrawing CN group able to shift the average CO stretching frequency by 12 cm^{-1}, which is a quite remarkable result for NHC ligands [77].

Strong effects have also been noticed by varying the *N*-substituents. The bond dissociation energy of NHC ligands bearing alkyl *N*-substituents, such as ICy, in [RuClCp*(NHC)] complexes can differ by 3-5 kcal/mol from that of NHC ligands bearing aryl *N*-susbtituents, such as ITol and IMes [76], although this comparison is influenced by the different steric demand of the different NHCs (*vide infra*). Greater insights come from the IR spectroscopic study of a series of [IrCl$(CO)_2$(NHC)] model systems, which indicated that NHCs with alkyl *N*-substituents are better electron donors than NHCs with aromatic substituents, but the range spanned by the average CO stretching frequency remains relatively small (less than 5 cm^{-1}), see Fig. 1.13 [30, 74]. Similar effects have been measured for the saturated analogues.

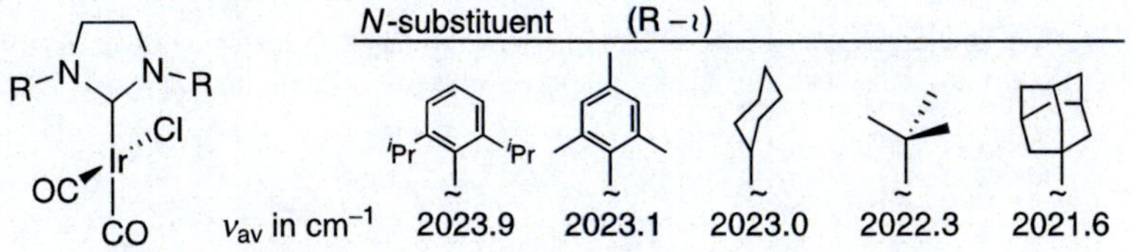

Fig. 1.13 Effect of the *N*-substituent on the average CO stretching frequency [30]

Stronger effects can be obtained by placing an electron-withdrawing or electron-donating group in the *para*-position of an aromatic *N*-substituent of the NHC ligand. As shown in Fig. 1.11, on going from a strongly electron-donating group, such as NEt_2, to a strongly electron-withdrawing group, such as SO_2-(*p*-tolyl), the average CO stretching frequency of [IrCl$(CO)_2$(NHC)] complexes increases by 8 cm^{-1} [74]. Similar effect is on the redox potential of the analogues [IrCl(cod)(NHC)] complexes [74], which increases by roughly 300 mV (Fig. 1.11).

Finally, a comparison between NHCs and tertiary phosphines is next presented. The ν_{av} of a series of NHC ligands discussed above is plotted *versus* the Tolman electronic parameter (TEP) in Fig. 1.14, together with some of the most typical tertiary phosphines. The TEP is the measured frequency of the A_1 carbonyl mode in [Ni$(CO)_3$L] complexes whilst ν_{av} is the average of both stretching frequencies found for *trans*-[IrCl$(CO)_2$L] (L = PR_3 or NHC). Visual inspection clearly shows that quite lower TEP values correspond to standard NHCs, and that they are clustered in the relatively small range of 3–5 cm^{-1}. Phosphines, instead, span over the much larger range of roughly 12 cm^{-1}, and consistently at higher TEP values.

Nevertheless, the placement of two Cl atoms on the C^4 and C^5 carbon atoms of the NHC skeleton, the presence of the strong SO_2Ar electron-withdrawing group on the aryl *N*-substituents, or the modification of the NHC skeleton from an imidazol-to a triazol-type, have allowed to fill the gap between the TEP of phosphines and NHCs, see Fig. 1.14.

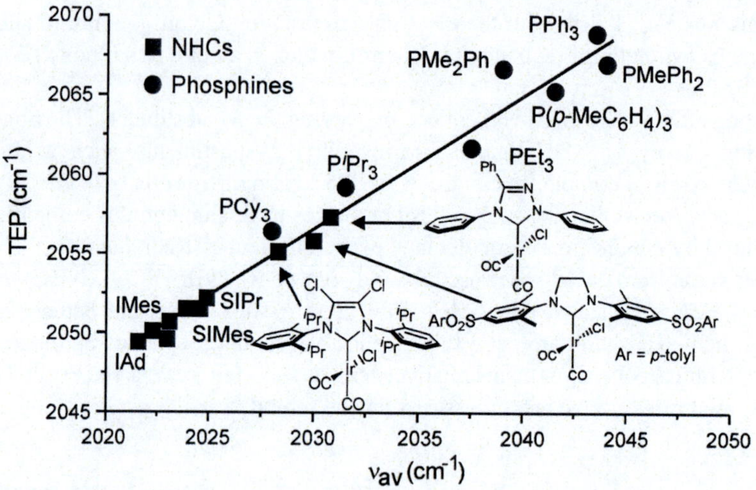

Fig. 1.14 Tolman electronic parameter (TEP) versus the average CO stretching frequency (ν_{av}) for NHCs (■) and phosphines (•). For clarity purposes, only data for the most popular NHCs are labelled

Finally, it is also worth mentioning the abnormal NHC ligand of Fig. 1.15, which extends the ν_{av} range to the low value of 2003 cm^{-1} [71].

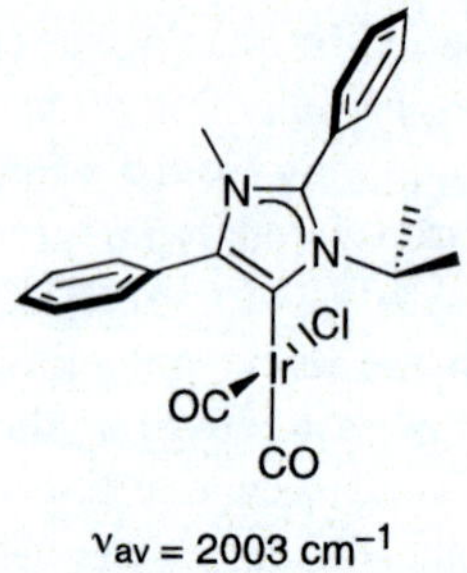

Fig. 1.15 Ir complex containing an abnormal NHC ligand

1.3 Steric Properties of NHCs

Based on the experience with tertiary phosphines, the importance of the steric properties of NHC ligands in determining chemical behaviour has been immediately recognised. The main practical problem, however, is that NHC ligands substantially present a local C_2 symmetry axis, whereas phosphines present a local C_3 symmetry axis. This implies that the well-accepted molecular descriptor used to quantify steric properties in phosphines, the Tolman cone angle [78], cannot be applied to NHC ligands.

This difference between NHCs and phosphines was acknowledged early, and the so-called fence model, see Fig. 1.16, was proposed [79]. In this model the NHC is seen as a "fence", with a "length" and a "height". These quantities were initially used to quantify the steric effect of NHCs. However, this model showed some limitations in properly capturing the behaviour of some common NHCs, such as ICy, and was thus abandoned [79].

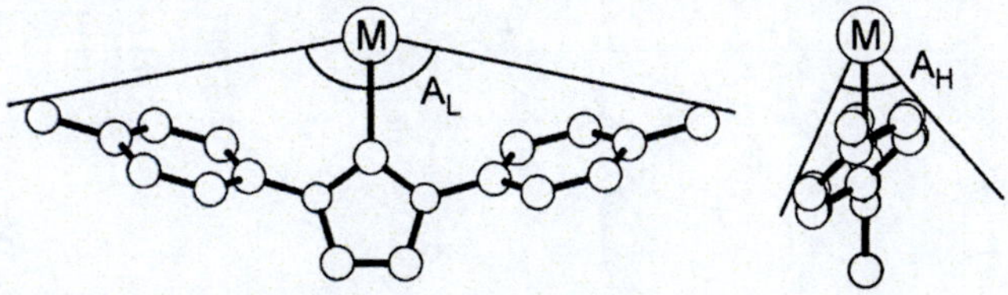

Fig. 1.16 Graphical representation of the fence model, with the key "length" (A_L) and "height" (A_H) parameters

In the following years, the percent buried volume, $\%V_{Bur}$, has emerged as a standard and intuitive parameter able to describe reasonably well the steric properties of NHCs [75]. The $\%V_{Bur}$ is the fraction of the volume of the first coordination sphere around the metal occupied by a given ligand, see Fig. 1.17 [75].

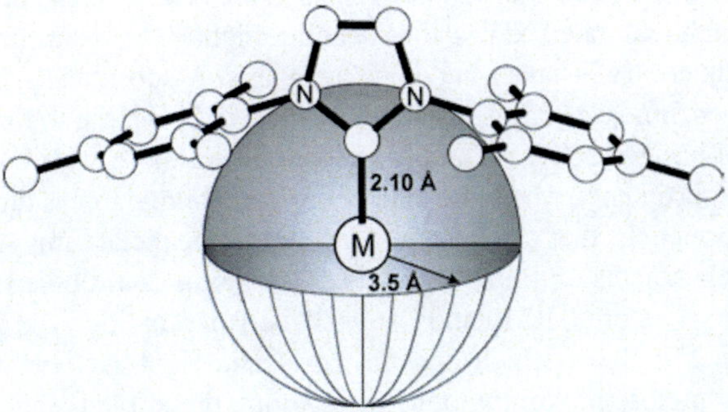

Fig. 1.17 Graphical representation of the sphere used to calculate the $\%V_{Bur}$ values

The $\%V_{Bur}$ has been initially used to rationalise the trend in the bond dissociation energy of NHC ligands in [RuClCp*(NHC)] complexes [76]. The almost linear correlation between the BDEs and the $\%V_{Bur}$ indicated that the steric properties of the NHC is the key parameter that controls their binding behaviour in these systems. This result is in agreement with the small variation in the electronic properties of NHC ligands discussed in the previous section. Similar conclusions were obtained when the BDE of the NHC ligand in a series of [$Ni(CO)_3$(NHC)] complexes was investigated [33]. The $\%V_{Bur}$ has also been used to rationalise the energetics of NHC dimerisation [80]. An extended set of $\%V_{Bur}$ values, obtained by analysing the steric bulk of the NHC ligand in the quantum mechanically optimised structure of the model [$IrCl(CO)_2$(NHC)] complexes is reported in Fig. 1.18.

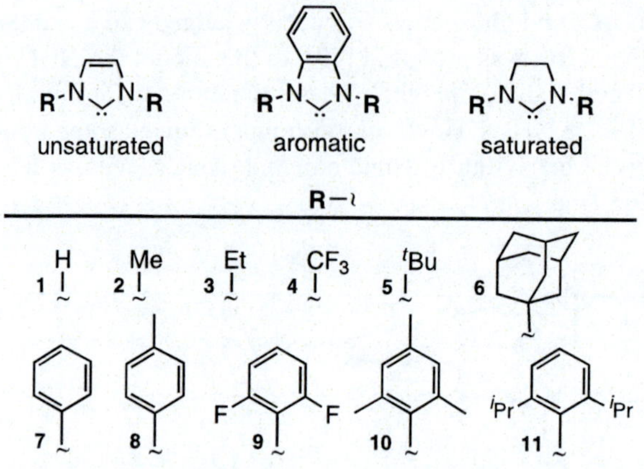

NHC	Uns.	Arom.	Sat.
1	18.8	18.9	19.0
2	24.9	25.1	25.4
3	26.0	26.4	25.9
4	31.1	30.4	31.8
5	35.5	38.9	36.2
6	36.1	40.8	36.6
7	30.5	30.2	31.6
8	30.5	30.2	32.4
9	31.3	30.9	32.3
10	31.6	31.2	32.7
11	33.6	31.9	35.7

(Uns., Arom. and Sat. are all $\%V_{Bur}$ values.)

PPh_3 = 30.5 PCy_3 = 35.3

Fig. 1.18 $\%V_{Bur}$ of frequently encountered NHC ligands from the quantum mechanically optimised structure of $[IrCl(CO)_2(NHC)]$ complexes from [75]. For comparison, the $\%V_{Bur}$ of two popular phosphines, PPh_3 and PCy_3, as obtained from optimisation of $[Ni(CO)_3(PR_3)]$ complexes, is also reported

The data reported in Fig. 1.18 indicate that the saturated NHCs show a $\%V_{Bur}$ slightly greater than that of the unsaturated analogues. This is a consequence of the different nature of the NHC skeleton that results in the N–C–N angle being consistently 4–5° greater in the saturated NHCs, thus bending slightly the *N*-substituent toward the metal. The bulkier NHCs have *t*-butyl and adamantyl *N*-substituents. The NHCs with the aromatic benzimidazol-bridge usually present a $\%V_{Bur}$ similar to that of the unsaturated imidazol-bridged NHC. The only exceptions are those with bulky *t*-butyl and adamantyl *N*-substituents, where the bulkier aromatic bridge forces the *N*-substituents to adopt conformations that place them more toward the metal. This suggests that the presence of substituents on the C^4 and C^5 carbon atoms could be a key to fine-tune the steric properties of NHC ligands. Indeed, bioxazoline derived NHCs, such as that shown in Fig. 1.19, which contain a tricyclic rigid skeleton that places the *N*-substituents in close proximity to the metal atom, present the highest $\%V_{Bur}$ calculated so far for systems containing an unsaturated imidazol-based NHC ring [81].

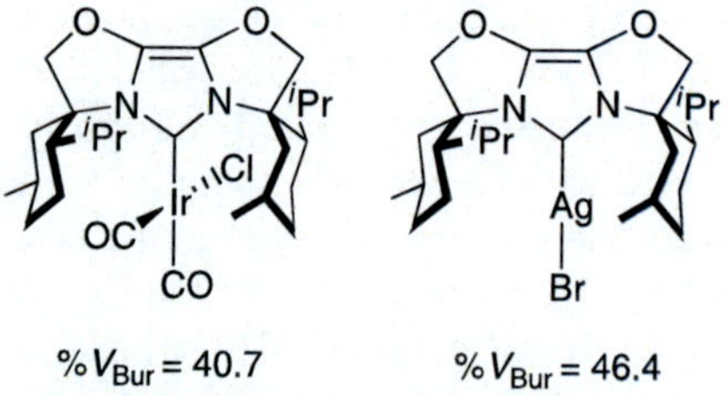

Fig. 1.19 $\%V_{Bur}$ from the quantum mechanically optimised structure of the $[IrCl(CO)_2(NHC)]$ and the $[AgBr(NHC)]$ complexes

One of the advantages of the $\%V_{Bur}$ as molecular descriptor to characterise the steric properties of ligands is its generality. This has allowed the placement of tertiary phosphines and NHCs on the same scale. The values reported in Fig. 1.18 for two of the most classical phosphines indicate that the less bulky PPh_3 compares with NHCs of intermediate bulkiness, such as those presenting *p*-tolyl *N*-substituents, while the bulkier PCy_3 compares with the bulky IPr and SIPr NHCs. Finally, further refinement of the model was recently disclosed in the form of the dihedral angles ϕ_1 and ϕ_2 (Fig. 1.20).

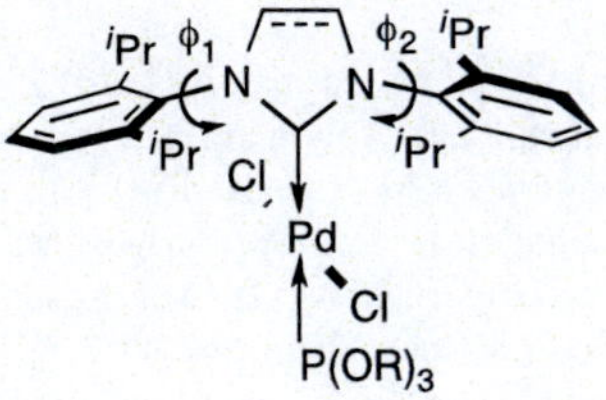

Fig. 1.20 Schematic representation of the dihedral angles ϕ_1 and ϕ_2 [82]

The measurement of these angles for a series of $[PdCl_2(NHC)\{P(OR)_3\}]$ complexes permitted to evidence the remarkable flexibility of NHCs due to rotations around the *N*-substituent bonds [82]. This flexibility, captured by the $\%V_{Bur}$, allows NHCs to respond actively to the steric requirements of co-ligands. This is further confirmed by *ab initio* molecular dynamics simulations aimed at understanding the variability of ϕ_1 and ϕ_2 in a series of NHCs containing Ru-complexes relevant to olefin metathesis [83].

1.4 Conclusion

NHCs are entities relatively easy to synthesise, characterise and coordinate. They have unique steric and electronic properties that make them one of the most interesting class of ligands and organocatalysts. Fundamental studies still have to be carried out to allow for better understanding of their properties. However, great advances have recently been made, leading to better models for NHC description, understanding of NHCs inherent properties and an inception of explanation of their catalytic superior behaviour. Progresses in NHC understanding also allowed ruling out of longstanding dogmas. NHCs are now considered as an inimitable class of ligands, an idea far from the "phosphine mimic" term that they have been abusively attributed. The profusion of work published every year on NHCs attests to the exceptional character of this class of ligands that is now very far from "laboratory curiosity". NHCs have rapidly become a standard class of ligands, as attested by their commercial availability (precursors, free NHCs, NHC complexes of Ru, Pd, Cu, Au and Ni are commercialised). This chapter provided an overview of the unique features that make NHCs so interesting, it also provided methods to synthesise, characterise and understand these ligands. The remainder of the Book presents an extensive treatise discussing numerous applications of NHCs in organometallic and organo-catalysis.

Acknowledgments The authors gratefully thank Thibault Schmid for recording the NMR spectra shown in Section 1.1.5.

References

1. Tschugajeff L, Skanawy-Grigorjewa M, Posnjak A (1925) Z Anorg Chem 148:37–42
2. In reference [1], the authors mention the following reference as a preliminary communication. Tschugajeff L, Skanawy-Grigorjewa M (1915) J Russ Chem Ges 47:776. This article was subsequently cited as Tschugajeff L, Skanawy-Grigorjewa M (1915) J Russ Chem Soc 47:776; Chugaev L, Skanavy-Grigorieva M (1915) J Russ Chem Soc 47:776; Chugaev L, Skanavy-Grigorizeva M (1915) J Russ Chem Soc 47:776. Despite intensive research, this manuscript could not be located
3. Rouschias G, Shaw BL (1970) Chem Commun 183
4. Burke A, Balch AL, Enemark JH (1970) J Am Chem Soc 92:2555–2557
5. Butler WM, Enemark JH, Parks J, Balch AL (1973) Inorg Chem 12:451–457
6. Stork JR, Olmstead MM, Balch AL (2004) Inorg Chem 43:7508–7515
7. Fehlhammer WP, Fritz M (1993) Chem Rev 93:1243–1280
8. Wanzlick HW (1962) Angew Chem Int Ed 1:75–80
9. Wanzlick HW, Schönherr HJ (1968) Angew Chem Int Ed 7:141–142
10. Öfele K (1968) J Organomet Chem 12:P42–P43
11. Arduengo AJ III, Harlow RL, Kline M (1991) J Am Chem Soc 113:361–363
12. Herrmann WA, Elison M, Fischer J, Kocher C, Artus GRJ (1995) Angew Chem Int Ed 34:2371–2374
13. Alder RW, Blake ME, Chaker L, Harvey JN, Paolini F, Schütz J (2004) Angew Chem Int Ed 43:5896–5911
14. Tamm M, Hahn FE (1999) Coord Chem Rev 182:175–209
15. Basato M, Michelin RA, Mozzon M, Sgarbossa P, Tassan A (2005) J Organomet Chem 690:5414–5420
16. Hahn FE, Jahnke MC (2008) Angew Chem Int Ed 47:3122–3172
17. de Frémont P, Marion N, Nolan SP (2009) Coord Chem Rev 253:862–892
18. Würtz S, Glorius F (2008) Acc Chem Res 41:1523–1533
19. Fürstner A, Alcarazo M, César V, Lehmann CW (2006) Chem Commun 2176–2178
20. Bon RS, de Kanter FJJ, Lutz M, Spek AL, Jahnke MC, Hahn FE, Groen MB, Orru RVA (2007) Organometallics 26:3639–3650
21. Bon RS, van Vliet B, Sprenkels NE, Schmitz RF, de Kanter FJJ, Stevens CV, Swart M, Bickelhaupt FM, Groen MB, Orru RVA (2005) J Org Chem 70:3542–3553
22. Hahn FE, Paas M, Le Van D, Lügger T (2003) Angew Chem Int Ed 42:5243–5246
23. Tapu D, Dixon DA, Roe C (2009) Chem Rev 109:3385–3407
24. Chantler VL, Chatwin SL, Jazzar RFR, Mahon MF, Saker O, Whittlesey MK (2008) Dalton Trans 19:2603–2614
25. Yu X-Y, Patrick BO, James BR (2006) Organometallics 25:2359–2363
26. Yu X-Y, Patrick BO, James BR (2006) Organometallics 25:4870–4877
27. Praetorius JM, Kotyk MW, Webb JD, Wang R, Crudden CM (2007) Organometallics 26:1057–1061
28. Datt MS, Nair JJ, Otto S (2005) J Organomet Chem 690:3422–3426
29. Praetorius JM, Allen DP, Wang R, Webb JD, Grein F, Kennepohl P, Crudden CM (2008) J Am Chem Soc 130:3724–3725

30. Kelly RA III, Clavier H, Giudice S, Scott NM, Stevens ED, Bordner J, Samardjiev I, Hoff CD, Cavallo L, Nolan SP (2008) Organometallics 27:202–210
31. DeBoef B, Pastine SJ, Sames D (2004) J Am Chem Soc 126:6556–6557
32. Tang CY, Smith W, Vidovic D, Thompson AL, Chaplin AB, Aldridge S (2009) Organometallics 28:3059–3066
33. Dorta R, Stevens ED, Scott NM, Costabile C, Cavallo L, Hoff CD, Nolan SP (2005) J Am Chem Soc 127:2485–2495
34. Dible BR, Sigman MS (2003) J Am Chem Soc 125:872–873
35. Cámpora J, Ortiz de la Tabla L, Palma P, Álvarez E, Lahoz F, Mereiter K (2006) Organometallics 25:3314–3316
36. Dible BR, Sigman MS (2006) Inorg Chem 45:8430–8441
37. Jackstell R, Harkal S, Jiao H, Spannenberg A, Borgmann C, Röttger D, Nierlich F, Elliot M, Niven S, Cavell K, Navarro O, Viciu MS, Nolan SP, Beller M (2004) Chem Eur J 10:3891–3900
38. Fantasia S, Nolan SP (2008) Chem Eur J 14:6987–6993
39. Viciu MS, Navarro O, Germaneau RF, Kelly RA III, Sommer W, Marion N, Stevens ED, Cavallo L, Nolan SP (2004) Organometallics 23:1629–1635
40. Navarro O, Oonishi Y, Kelly RA, Stevens ED, Briel O, Nolan SP (2004) J Organomet Chem 689:3722–3727
41. Campeau L-C, Thansandote P, Fagnou K (2005) Org Lett 7:1857–1860
42. Flahaut A, Toutah K, Mangeney P, Roland S (2009) Eur J Inorg Chem 5422–5432
43. Fantasia S, Egbert JD, Jurčík V, Cazin CSJ, Jacobsen H, Cavallo L, Heinekey DM, Nolan SP (2009) Angew Chem Int Ed 48:5182–5186
44. Berthon-Gelloz G, Schumers J-M, De Bo G, Markó IE (2008) J Org Chem 73:4190–4197
45. Berthon-Gelloz G, Buisine O, Brière J-F, Michaud G, Stérin S, Mignani G, Tinant B, Declercq J-P, Chapon D, Markó IE (2005) J Organomet Chem 690:6156–6168
46. Jung IG, Seo J, Lee SI, Choi SY, Chung YK (2006) Organometallics 25:4240–4242
47. Fantasia S, Petersen JL, Jacobsen H, Cavallo L, Nolan SP (2007) Organometallics 26:5880–5889
48. Kaur H, Kauer Zinn F, Stevens ED, Nolan SP (2004) Organometallics 23:1157–1160
49. Díez-González S, Scott NM, Nolan SP (2006) Organometallics 25:2355–2358
50. Goj LA, Blue ED, Delp SA, Gunnoe TB, Cundari TR, Pierpont AW, Petersen JL, Boyle PD (2006) Inorg Chem 45:9032–9045
51. Díez-González S, Stevens ED, Scott NM, Petersen JL, Nolan SP (2008) Chem Eur J 14:158–168
52. de Frémont P, Scott NM, Stevens ED, Ramnial T, Lightbody OC, Macdonald CLB, Clyburne JAC, Abernethy CD, Nolan SP (2005) Organometallics 24:6301–6309
53. de Frémont P, Scott NM, Stevens ED, Nolan SP (2005) Organometallics 24:2411–2418
54. de Frémont P, Singh R, Stevens ED, Petersen JL, Nolan SP (2007) Organometallics 26:1376–1385
55. Herrmann WA, Elison M, Fischer J, Kocher C, Artus GRJ (1995) Angew Chem Int Ed 34:2371–2374
56. Perry MC, Burgess K (2003) Tetrahedron: Asymmetry 14:951–961
57. Schuster O, Yang L, Raubenheimer HG, Albrecht M (2009) Chem Rev 109:3445–3478
58. Frenking G, Solà M, Vyboishchikov SF (2005) J Organomet Chem 690:6178–6204
59. Jacobsen H, Correa A, Costabile C, Cavallo L (2006) J Organomet Chem 691:4350–4358
60. Tonner R, Heydenrych G, Frenking G (2007) Chem Asian J 2:1555–1567
61. Jacobsen H, Correa A, Poater A, Costabile C, Cavallo L (2009) Coord Chem Rev 253:687–703
62. Fröhlich N, Pidun U, Stahl M, Frenking G (1997) Organometallics 16:442–448
63. Herrmann WA (2002) Angew Chem Int Ed 41:1291–1309
64. Hu XL, Tang YJ, Gantzel P, Meyer K (2003) Organometallics 22:612–614
65. Hu XL, Castro-Rodriguez I, Olsen K, Meyer K (2004) Organometallics 23:755–764
66. Scott NM, Dorta R, Stevens ED, Correa A, Cavallo L, Nolan SP (2005) J Am Chem Soc 127:3516–3526

67. Garrison JC, Simons RS, Kofron WG, Tessier CA, Youngs WJ (2001) Chem Commun1780–1781
68. Tulloch AAD, Danopoulos AA, Kleinhenz S, Light ME, Hursthouse MB, Eastham G (2001) Organometallics 20:2027–2031
69. Dorta R, Stevens ED, Nolan SP (2004) J Am Chem Soc 126:5054–5055
70. Chianese AR, Li X, Janzen MC, Faller JW, Crabtree RH (2003) Organometallics 22:1663–1667
71. Chianese AR, Kovacevic A, Zeglis BM, Faller JW, Crabtree RH (2004) Organometallics 23:2461–2468
72. Altenhoff G, Goddard R, Lehmann CW, Glorius F (2004) J Am Chem Soc 126:15195–15201
73. Frey GD, Rentzsch CF, von Preysing D, Scherg T, Muehlhofer M, Herdtweck E, Herrmann WA (2006) J Organomet Chem 691:5725–5738
74. Leuthäußer S, Schwarz D, Plenio H (2007) Chem Eur J 13:7195–7203
75. Poater A, Cosenza B, Correa A, Giudice S, Ragone F, Scarano V, Cavallo L (2009) Eur J Inorg Chem 1759–1766
76. Hillier AC, Sommer WJ, Yong BS, Petersen JL, Cavallo L, Nolan SP (2003) Organometallics 22:4322–4326
77. Khramov DM, Lynch VM, Bielawski CW (2007) Organometallics 26:6042–6049
78. Tolman CA (1977) Chem Rev 77:313–348
79. Huang J, Schanz HJ, Stevens ED, Nolan SP (1999) Organometallics 18:2370–2375
80. Poater A, Ragone F, Giudice S, Costabile C, Dorta R, Nolan SP, Cavallo L (2008) Organometallics 27:2679–2681
81. Würtz S, Lohre C, Fröhlich R, Bergander K, Glorius F (2009) J Am Chem Soc 131:8344–8345
82. Diebolt O, Jurčík V, Correa da Costa R, Braunstein P, Cavallo L, Nolan SP, Slawin AMZ, Cazin CSJ (2010) Organometallics. 29:1443–1450
83. Ragone F, Poater A, Cavallo L (2010) J Am Chem Soc 132:4249–4258

Chapter 2
N-Heterocyclic Carbene Complexes in Additions to Multiple Bonds

Andreas A. Danopoulos

Abstract The use of *N*-heterocyclic carbene (NHC) complexes as homogeneous catalysts in addition reactions across carbon-carbon double and triple bonds and carbon-heteroatom double bonds is described. The discussion is focused on the description of the catalytic systems, their current mechanistic understanding and occasionally the relevant organometallic chemistry. The reaction types covered include hydrogenation, transfer hydrogenation, hydrosilylation, hydroboration and diboration, hydroamination, hydrothiolation, hydration, hydroarylation, allylic substitution, addition, chloroesterification and chloroacylation.

2.1 Introduction

The first isolation of stable cyclic and acyclic nucleophilic carbenes by Arduengo and Bertrand, and the development of synthetic methods for their introduction to metals, either by direct reactions with metal precursors of the preformed, or *in situ* generated carbenes, or by the silver transmetallation methodology, initiated detailed studies aiming at a better understanding and use of these ligands in homogeneous catalysis. The current refined description of the metal-NHC bonding across the Periodic Table and the factors affecting it, the semi-quantitative parametrisation of the ligand topology and the space it occupies around the metal, the dynamic processes in which *N*-heterocyclic carbene (NHC) ligands are involved, and the invention of ways to introduce ligand chirality, provide the essential tools for further catalyst discovery, tuning and development. In this chapter an overview of the use of NHC complexes as catalysts in addition reactions across C–C and C-heteroatom multiple bonds is presented. The emerging picture is that NHC ligands are unique in some catalytic applications and complementary to the well-established tertiary phosphines, which we initially thought they were mimicking.

A. A. Danopoulos (✉)
School of Chemistry, University of Southampton, Highfield SO17 1BJ, UK
e-mail: ad1@soton.ac.uk

C. S. J. Cazin (ed.), *N-Heterocyclic Carbenes in Transition Metal Catalysis and Organocatalysis*, Catalysis by Metal Complexes 32,
DOI 10.1007/978-90-481-2866-2_2, © Springer Science+Business Media B.V. 2011

2.2 Hydrogenation of Alkenes and Alkynes

The Rh–NHC complexes, with or without phosphine co-ligands, have been studied as hydrogenation catalysts of alkenes with molecular hydrogen, with the aim to develop more active, selective (and/or enantioselective) and thermally stable catalysts.

Rh(III)(NHC) hydrides have been studied as catalysts for this type of hydrogenation. The products from the reaction of Rh(I) complexes with H_2 are dependent on the nature of the NHC. The reaction of $[RhCl(IPr)_2(N_2)]$ **1** (IPr = *N,N'*-bis-[2,6-(di-*iso*-propyl)phenyl]imidazol-2-ylidene) with H_2 gave the monomeric complex **3** [1], which was also obtained from the reaction of $[RhCl(COE)(IPr)]_2$ **2** with H_2 and excess IPr, while the reaction of $[RhCl(COE)(IMes)]_2$ with H_2 gave the chloride bridged species **4** (Scheme 2.1) [2].

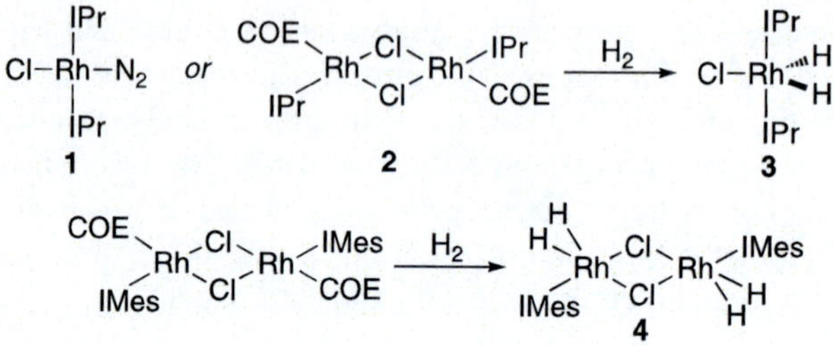

Scheme 2.1 Reactions of Rh(I) NHC complexes with hydrogen

$[Rh(COE)_2Cl]_2$ in the presence of IMes, *N,N'*-bis-[2,4,6-(trimethyl)phenyl]imidazol-2-ylidene, reacted with H_2 at room temperature to give the trigonal bipyramidal $[Rh(H)_2Cl(IMes)_2]$ **6** (which is analogous to **3**) *via* intermediate formation of the isolable cyclometallated **5** (Scheme 2.2) [3].

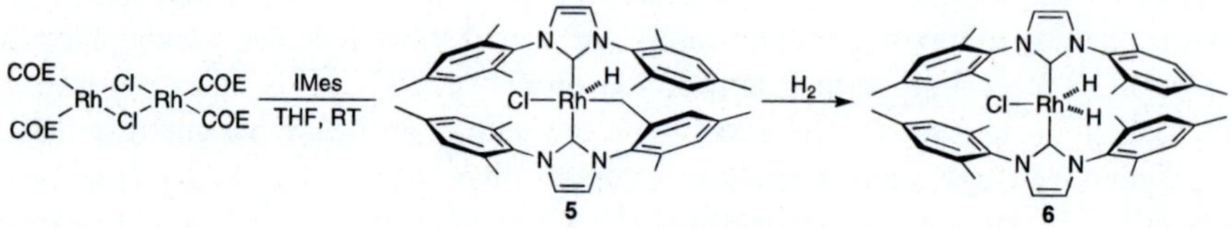

Scheme 2.2 The formation of the rhodium dihydride complex *via* a cyclometallated intermediate

The hydrogenation activity of the isolated hydrides **3** and **6** towards cyclooctene or 1-octene was much lower than the Wilkinson's complex, $[RhCl(PPh_3)_3]$, under the same conditions [2]; furthermore, isomerisation of the terminal to internal alkenes competed with the hydrogenation reaction. The reduced activity may be related to the high stability of the Rh(III) hydrides, while displacement of a coordinated NHC by alkene may lead to decomposition and Rh metal formation.

The catalytic hydrogenation of alkenes by mixed NHC/phosphine complexes of rhodium was also studied. Initial results of the hydrogenation of cyclohexene by *trans*-$[RhCl(ICy)(L)_2]$ (ICy = *N,N'*-(dicyclohexyl)imidazol-2-ylidene, L = PPh_3,

$P(OPh)_3$) and *trans*-[$RhCl(ICy)_2PPh_3$] showed poor activity compared to Crabtree's catalyst [Ir(1,5-COD)(Py)(PCy_3)][PF_6] or its IMes analogue, [Ir(1,5-COD)(Py)(IMes)][PF_6], under the same conditions [4, 5]. Systematic optimisation was conducted by generating the catalysts *in situ* from [RhCl(1,5-COD)(ICy)] and two or four equivalents of PPh_3 under H_2 [5]. In these studies an induction period was observed, while the phosphine was required to prevent catalyst decomposition. The activities were one fifth to one tenth of the original Wilkinson's catalyst. The reduced activities may be ascribed to reduced phosphine pre-dissociation to form the coordinatively and electronically unsaturated active species. In view of the observation that NHCs dissociate from [RhCl(NHC)$(L)_2$] in the presence of L, it may be that the active species of the *in situ* systems contains only phosphine ligands [6].

Chiral monodentate carbene complexes of Rh and Ir of the type [MCl(1,5-COD)(NHC)] (M = Rh, Ir) with the ligands **7–9** (Fig. 2.1) have been studied as catalysts for the enantioselective hydrogenation of methyl-2-acetamido acrylate. Even though the activities were high, the enantiomeric excesses (*ee*) were poor [7, 8].

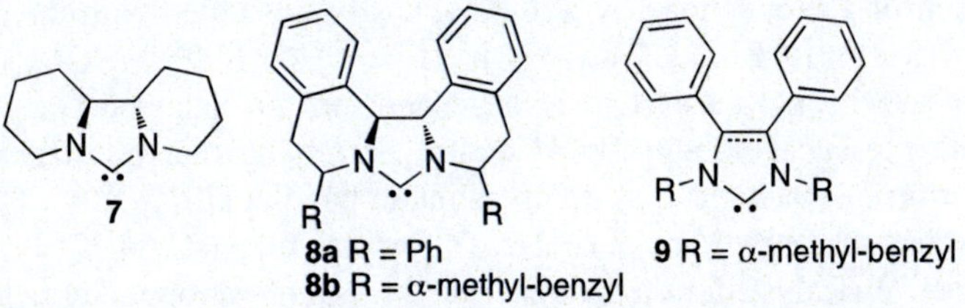

Fig. 2.1 Chiral NHC ligand designs used in the Rh-catalysed enantioselective hydrogenation of methyl-2-acetamido acrylate by dihydrogen

Replacement of one PCy_3 in [RuCl(H)(CO)$(PCy_3)_2$] **10** by IMes gave [RuCl(H)(CO)(PCy_3)(IMes)], **11** (Fig. 2.2) which is an active catalyst for the direct hydrogenation of 1-hexene, allylbenzene and cyclooctene. At higher temperatures (*ca.* 100°C) the activity of **11** is superior to **10**. Furthermore, small (1–2 equiv.) quantities of $HBF_4 \cdot Et_2O$ act as co-catalyst, by enhancing the dissociation of PCy_3 (through the formation of [PCy_3H^+][BF_4^-]) [9]. Similar behaviour was observed in the hydrogenation of 1-octene by [Ru(CO)(Cl)(Ph)(PCy_3)(SIMes)], **12** [10]. [RuCl(H)(CO)(PPh_3)(NHC)] (NHC = IMes, SIMes), **13,** is more active for the hydrogenation of cyclooctene or *cis*- and *trans*-cyclododecene [11] due to the increased lability of PPh_3 compared to PCy_3. In the latter case isomerisation competes with hydrogenation [12].

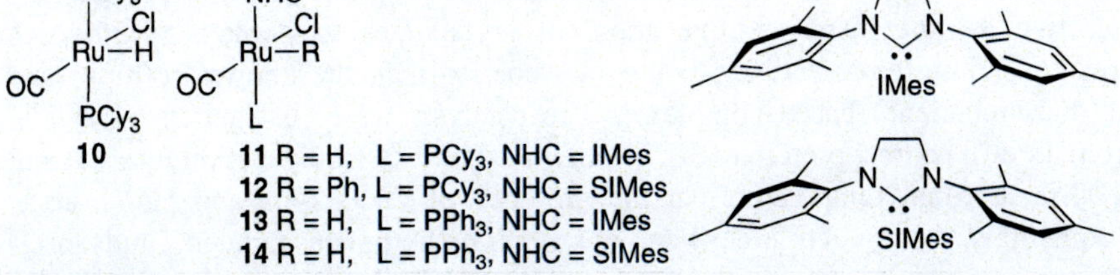

Fig. 2.2 Ruthenium-based hydrogenation catalysts bearing NHC ligands

Reaction of the 18e-species [Ru(H)$_2$(CO)(IMes)(PPh$_3$)$_2$] **15** with trimethyl vinylsilane gave the cyclometallated complex **16** and trimethylethylsilane; **16** can be converted to the original dihydride **15** by reaction with H_2 (1 atm, room temperature) or ^{i}PrOH (50°C) [13]. Catalytically, vinylsilane can be hydrogenated by **16** in ^{i}PrOH in quantitative yield (Fig. 2.3) [14].

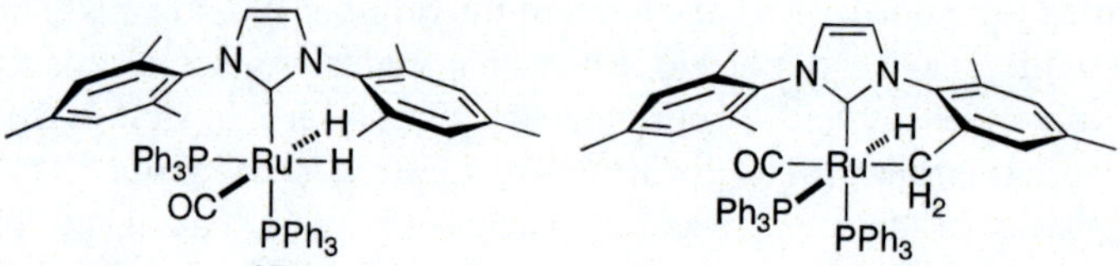

15 **16**

Fig. 2.3 Ru-IMes and cyclometallated Ru-IMes complexes used as catalysts for the hydrogenation of trimethylvinylsilane

Reduction of acetophenone by ^{i}PrOH/H_2 has been studied with the ruthenium complexes [Ru(H)(η^2-BH$_4$)(CO)L(NHC)], (L = NHC, PPh$_3$, NHC = IMes, IPr, SIPr). The activity of the system is dependent on the nature of the NHC and requires the presence of both ^{i}PrOH and H_2, implying that transfer and direct hydrogenation mechanisms may be operating in parallel [15].

Replacement of the pyridine in Crabtree's complex by NHC (NHC = IMes, IMe) gave very active catalysts for the hydrogenation by H_2 of various alkenes to alkanes, including hindered tertiary and quaternary alkenes. For terminal 1-alkenes (1-octene), the IMes analogue gives best conversions, while for tertiary and quaternary alkenes (1-methyl-cyclohexene and 2,3-dimethyl-2-butene) the IMe analogue is more suitable [16]. On the other hand, replacement of PCy$_3$ by NHC (NHC = IMes, ICy, SIMes) in Crabtree's complex [Ir(1,5-COD)(Py)(PCy$_3$)]$^+$, gave a catalyst for the hydrogen transfer reduction of aryl- and alkyl-ketones to alcohols, cyclic alkenes to alkanes, dienes to alkenes, α,β-unsaturated carbonyl compounds to alcohols, and aromatic nitro compounds to anilines. Best activities were observed with NHC = ICy [12].

Initial attempts to develop Pd–NHC complexes as hydrogenation catalysts for alkenes or alkynes focused on the use of [Pd(NHC)$_2$] species. However, these did not withstand the reaction conditions decomposing to metallic Pd and imidazolium salts or imidazolidine, presumably after H_2 oxidative addition to form Pd(NHC) hydrides, followed by imidazolium elimination. Reports by Elsevier on the stability of Pd(NHC) fragment under hydrogenation conditions opened the way for the development of Pd(NHC) hydrogenation catalysts [17]. The catalysts proved very selective for the 'semi-hydrogenation' of arylalkynes to *Z*-alkenes with high chemo- and stereo-selectivity, being far superior than the known Pd(bipy) and Pd(diazabutadiene) based complexes. The catalysts were best generated *in situ* from a Pd(0) source, preferentially **17**, an imidazolium or imidazolidinium salt and KO^{t}Bu, at room temperature, in the presence of the alkyne and H_2 (1 atm). Surprisingly, better activities were observed with the unsaturated imidazol-2-ylidene precursors such as **18**. Preformed Pd(NHC) complexes, for example **19**, were not as active as the *in situ* generated catalysts (Fig. 2.4).

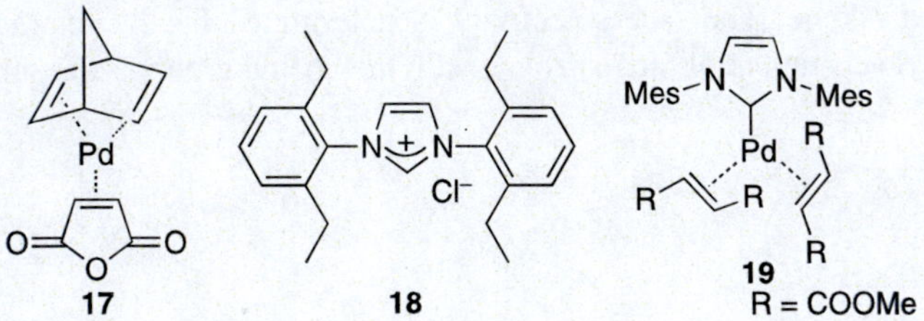

Fig. 2.4 Components of the Pd-based catalytic system for the hydrogenation of arylalkynes. The pre-formed complex **19** shows reduced activity

Recently, the direct hydrogenation of a wide range of alkenes was successfully carried out using [Pd(SIPr)(PCy_3)] **20** in alcohols or THF. Complex **20** in the presence of H_2 forms *trans*-[$Pd(H)_2$(SIPr)(PCy_3)]. The activity of the catalytic system is very high under mild conditions (1 bar, rt, 0.25 mol%Pd). Reduction of the alkene functionality can be selective in α,β-unsaturated carbonyl compounds, while alkynes can be 'semi-hydrogenated' to Z-alkenes with good selectivity as seen above [18].

Bidentate NHC-Pd complexes have been tested as hydrogenation catalysts of cyclooctene under mild conditions (room temperature, 1 atm, ethanol). The complex **22** (Fig. 2.5), featuring 'abnormal' carbene binding from the C^4 carbon of the imidazole heterocycles, has stronger Pd–C_{NHC} bonds and more nucleophilic metal centre than the C^2 bound 'normal' carbene chelate **21**. The different ligand properties are reflected in the superior activity of **22** in the hydrogenation of cyclooctene at 1–2 mol% loadings under mild conditions. The exact reasons for the reactivity difference in terms of elementary reaction steps are not clearly understood [19].

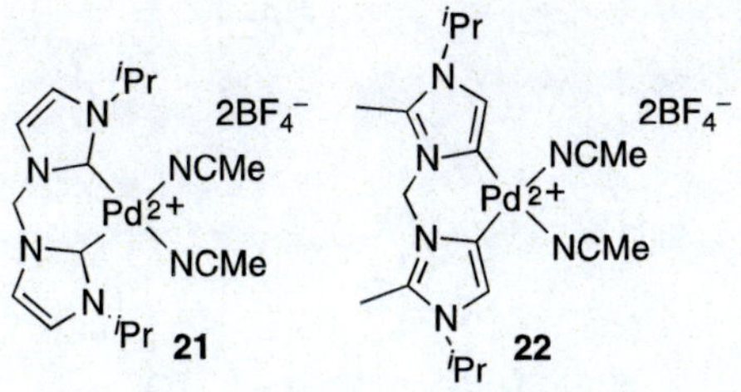

Fig. 2.5 Palladium-based hydrogenation catalysts with bidentate 'normal' and 'abnormal' NHC ligands

The development of functionalised NHC ligands, bearing NHC functional group(s) tethered to classical donors, provides additional ways of tuning the coordination sphere of the metal both sterically and electronically, and opens new ways for catalyst design. The chiral complexes **23**, **24** and **25** (Fig. 2.6) have been used for the enantioselective hydrogenation of substituted aryl alkenes (**23** and **24**), functionalised alkenes **24** and α,β-unsaturated esters **25** with good *ees*. The oxazoline carbene species (**23** and **24**) operate under milder conditions than analogous complexes where NHC is replaced by phosphine [20–23]. Theoretical calculations on the mechanism of hydrogenation of the arylalkenes with **23** and the origin optical induction, support an Ir(III)/(V) cycle with rate determining metathetical insertion of the hydride to a coordinated alkene. The face selectivity is determined by the interaction of the 1-adamantyl substituent of the oxazoline with the phenyl group of the substrate [24].

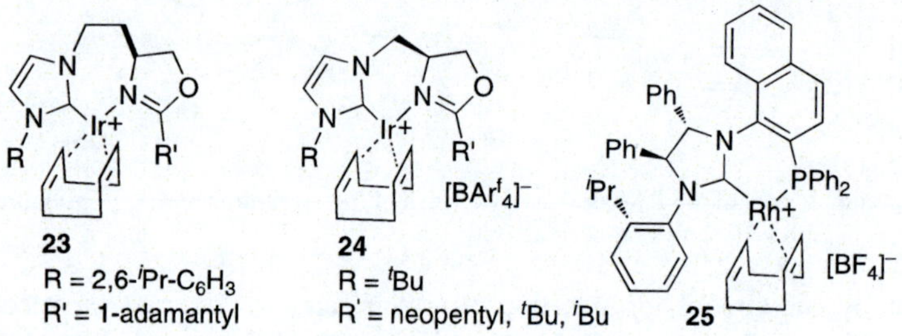

Fig. 2.6 Chiral functionalised NHC complexes as enantioselective hydrogenation catalysts with dihydrogen

The supported Au complexes presented in Scheme 2.3 show increased activity in the hydrogenation by H_2 of diethylcitraconate **26** and diethylbenzylidene succinate **27**, compared to [AuCl(PPh_3)] (40°C, 4 atm, ethanol). The nature of the support has an effect on the TOF (turnover frequency) of the reaction, being highest with ITQ-2 and lowest with silica. The preferred mechanism of hydrogen activation involves heterolytic splitting rather than homolysis or oxidative addition steps [25].

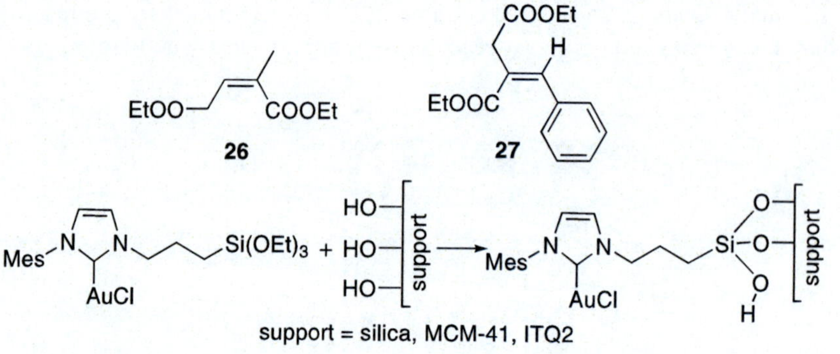

Scheme 2.3 Hydrogenations catalysed by a supported Au-NHC complex

2.3 Transfer Hydrogenation and Related Reactions

The transfer hydrogenations of carbonyl compounds to alcohols catalysed by a variety of NHC complexes have been intensively studied. The strong M–C_{NHC} bond and the excellent σ-donating properties of the NHC ligands are responsible for the stabilisation of hydrides and alkoxides that are intermediates in the postulated catalytic cycles. Rh, Ir, and Ru species, where the metal is coordinated by monodentate NHCs and classical co-ligands (pyridine, phosphines or cyclopentadienyl analogues) [26–28], 'pincer' pyridine di-carbenes [29, 30] bidentate di-carbenes [31] and functionalised NHCs [28, 32–35] are particularly successful catalysts. Steric and electronic tuning of the coordinated NHCs is important for the successful catalysts. Less hindered, alkyl substituted NHCs usually give more active catalysts than their analogues with bulky aryl substituents. The influence of the electronic properties of the NHC on activity and selectivity is more subtle. For example, imidazol-based NHC complexes are generally most active for the reduction of ketones, while in specific cases rhodium complexes with 'abnormal' C^4 bound NHCs are more active compared to C^2 bound analogues [36]; iridium complexes with the weaker σ-donating triazol based NHCs have been developed for the reduction of imines, alkenes and the stepwise one-pot reductive amination of aldehydes [27]. However, the structure and property principles underlying catalyst tuning are still underdeveloped with these ligand systems and the NHC ligands in general. Co-ligands (for example pyridine donors, cyclopentadienyl analogues, etc.) are also important in the catalyst design, presumably by stabilising the metal in the active oxidation state, or providing labile sites for substrate coordination. In addition, in polydentate functionalised NHC complexes, a possible hemilability of the tethered classical donor has been implied during catalysis and supported by spectroscopic studies [34, 37].

In the majority of the known examples, the donor of hydrogen equivalents is isopropanol; $HCOO^-$ or $HCOOH/Et_3N$ azeotrope are less successful. Aromatic ketones (mainly acetophenone and benzophenone) were the easiest to reduce even under mild conditions and low catalyst loadings.

The required bases for the reaction include KOH, KO^{t}Bu and K_2CO_3, the latter usually added in higher base/substrate ratios than KOH. However, K_2CO_3 is the base of choice for the reduction of aldehydes, where aldehyde decarbonylation and subsequent catalyst deactivation or formation of aldol side products interfere with the reduction [38]. The transfer hydrogenation of imines is more challenging due to competing formation of the catalytically inactive complex involving σ-coordination of the imine, and therefore NHC based catalysts for this purpose are less common [27, 29, 39]. The activity of certain NHC complexes in imine reduction has been exploited for the one-pot sequential reductive amination of aldehydes to amines. In the reported system, the added base K_2CO_3 acts also as a drying agent during the imine formation [27]. Selected complexes that have been studied as transfer hydrogenation catalysts and related catalytic data are shown in Figs. 2.7 and 2.8 and Tables 2.1 and 2.2, respectively.

Two versatile hydrogen auto-transfer processes were catalysed by the iridium complexes **41** and **42** [28, 40]. The first, *N*-alkylation of primary amines using alcohols as electrophiles, involves oxidative hydrogen elimination from the alcohol to form an aldehyde, which, in the presence of amine, is converted to imine and the latter reduced to the secondary amine by transfer hydrogenation. The second system involves oxidative hydrogen elimination from one primary and one secondary alcohol to form an aldehyde and ketone, respectively, which undergo aldol coupling and reduction of the α,β-enone to alkylated alcohols. Both transformations are very attractive compared to conventional alkylations with organic halides, due to their reduced waste [41]. The reactions are carried out in refluxing toluene in the presence of base (KOH, KO^{t}Bu, $NaHCO_3$) with high conversions at low catalyst loadings (1.0 mol%).

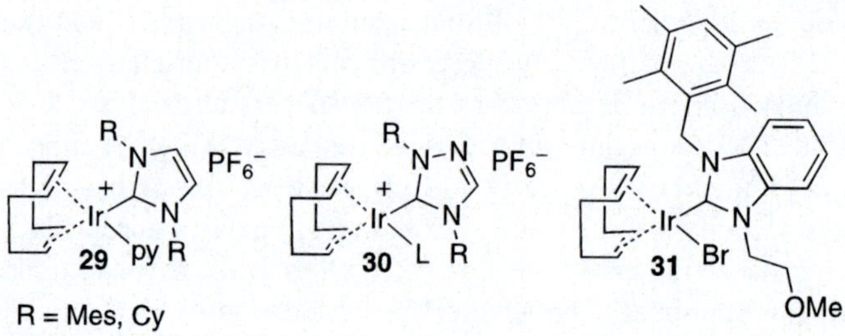

Fig. 2.7 Selected NHC complexes that have been studied as transfer hydrogenation catalysts

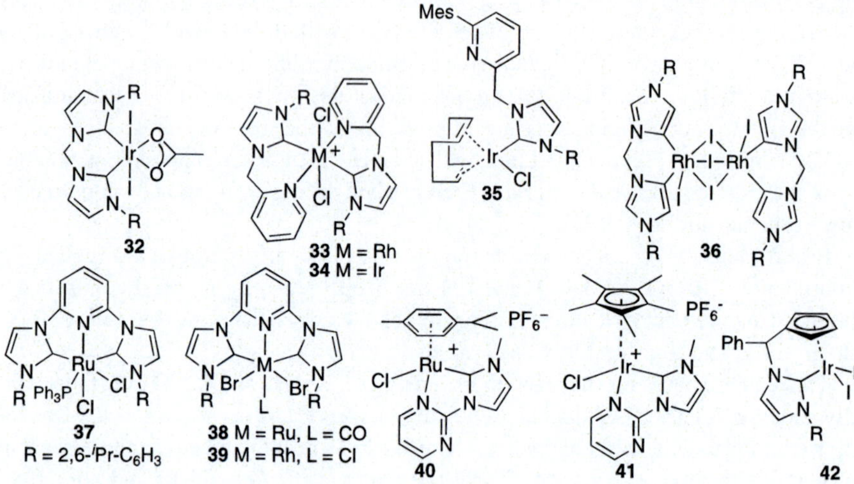

Fig. 2.8 Selected NHC complexes that have been studied as transfer hydrogenation catalysts

Table 2.1 Selected transfer hydrogenation data of $R^1R^2C{=}O$, R^1= alkyl, phenyl, R^2 = alkyl, H or Ph. All reactions were carried out in refluxing ^{i}PrOH (82°C)

R^1	R^2	Complex (mol%)	Base (subst./base)	Time (h)	Conv. (%)	Ref
Ph, Ar	Me	**32** (0.01)	KOH (2000/1)	2	98	[31]
p-tolyl	H	**32** (0.1)	K_2CO_3 (2:1)	0.25	98	[38]
Ph	Me	**40/41** (1.0)	KOH (10:1)	3	98	[28]
Ph	Me	**37** (0.015)	KO^{t}Bu (10:1)	12	80	[29]
Ph	Me	**39** (0.006)	KOH (2:1)	6	30	[30]
Ph	Me	**31** (0.5)	KOH (20:1)	1.5	98	[35]
Me	Bu^n	**32** (0.1)	KOH (200:1)	24	88	[42]
$-(CH_2)_5-$		**38** (0.005)	KOH (2:1)	20	65	[30]
$-(CH_2)_5-$		**41** (0.1)	KOH (1:1)	4.5	99	[28]
$-(CH_2)_5-$		**40** (1.0)	KOH (10:1)	5	99	[28]
$-(CH_2)_5-$		**31** (0.5)	KOH (20:1)	2	98	[28]
Ph	Ph	**36** (0.5)	KOH (10:1)	2	97	[36]
Ph	Ph	**39** (0.006)	KOH (2:1)	24	98	[30]
Ph	Ph	**34** (0.1)	KOH (2:1)	2	99	[32]

Table 2.2 Transfer hydrogenation of *N*-benzylidene aniline. Solvent is ^{i}PrOH

Complex (mol%)	Base (base:substr.)	T (°C)	Time (h)	Conv. (%)	Ref
37 (0.015)	KO^{t}Bu (10:1)	55	20	60–70	[29]
30 (1.0)	K_2CO_3 (2:1)	82	0.5	92–100	[27]
40 (1.0)	KOH (10:1)	82	5	70–80	[28]

Transfer hydrogenation of aldehydes with isopropanol without addition of external base has been achieved using the electronically and coordinatively unsaturated Os complex **43** as catalyst. High turnover frequencies have been observed with aldehyde substrates, however the catalyst was very poor for the hydrogenation of ketones. The stoichiometric conversion of **43** to the spectroscopically identifiable in solution ketone complex **45**, *via* the non-isolable complex **44** (Scheme 2.4), provides evidence for two steps of the operating mechanism (alkoxide exchange, β-hydride elimination to form ketone hydride complex) of the transfer hydrogenation reaction [43].

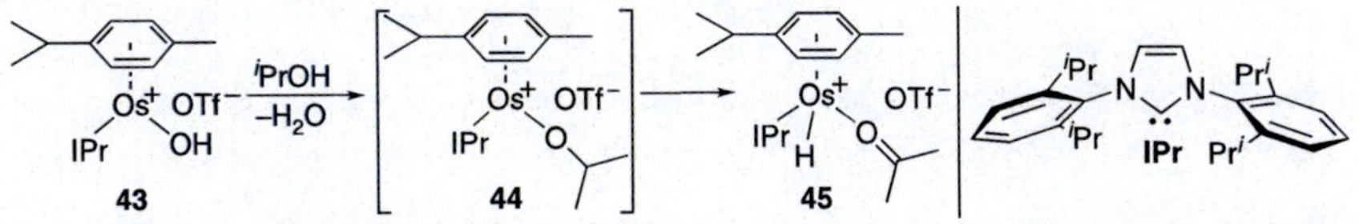

Scheme 2.4 Experimental evidence in support of the mechanism for the base-free transfer hydrogenation of carbonyl compounds catalysed by complex **43**

The iridium complex **35** has been also used as catalyst for the transfer hydrogenation of substituted nitroarenes [34]. Good to very good conversions were observed (2.5 mol%, in refluxing isopropanol, 12 h). A mixture of two products was obtained, the relative ratio of which depends on the concentration of added base (KOH) and catalyst. (Scheme 2.5)

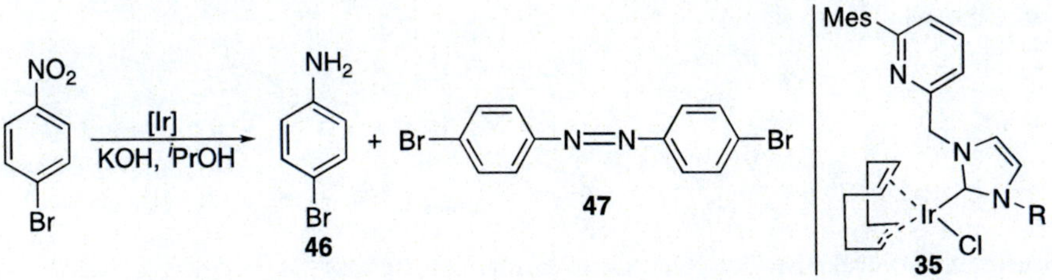

Scheme 2.5 Products formed in the transfer hydrogenation reduction of nitroarenes by complex **35**

This behaviour was rationalised by a stepwise reduction mechanism, in which a high catalyst or KOH concentration gives a high hydride concentration and leads to the aniline formation and suppression of intermolecular reactions to the dimeric azo-compound.

2.4 Hydrosilylation Reactions

The hydrosi(ly)lations of alkenes and alkynes are very important catalytic processes for the synthesis of alkyl- and alkenyl-silanes, respectively, which can be further transformed into aldehydes, ketones or alcohols by established stoichiometric organic transformations, or used as nucleophiles in cross-coupling reactions. Hydrosilylation is also used for the derivatisation of Si containing polymers. The drawbacks of the most widespread hydrosilylation catalysts [the Speier's system, H_2PtCl_6/^{i}PrOH, and Karstedt's complex [Pt_2(divinyl-disiloxane)$_3$] include the formation of side-products, in addition to the desired *anti*-Markovnikov Si–H addition product. In the hydrosilylation of alkynes, formation of di-silanes (by competing further reaction of the product alkenyl-silane) and of geometrical isomers (α-isomer from the Markovnikov addition and *Z*-β and *E*-β from the *anti*-Markovnikov addition, Scheme 2.6) are also possible.

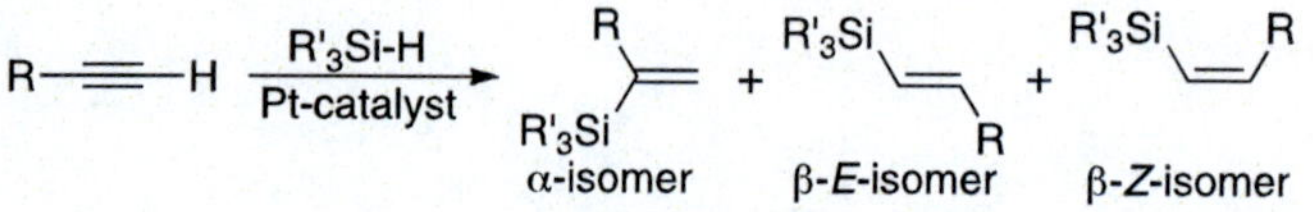

Scheme 2.6 Possible isomeric products in the hydrosilylation of alkynes (disilanes from further hydrosilylation of the alkenyl silanes are excluded)

Therefore, improved chemo- and stereo-selectivities of the catalysts are challenging goals of current research in the area. The reduced chemoselectivity and the formation of side-products with the Speier and Karstedt catalysts is attributed to poor catalyst stability under the reaction conditions, and efforts to improve it have focused on the use of NHC ligand designs, taking advantage of the high strength of the Pt–C_{NHC} bond.

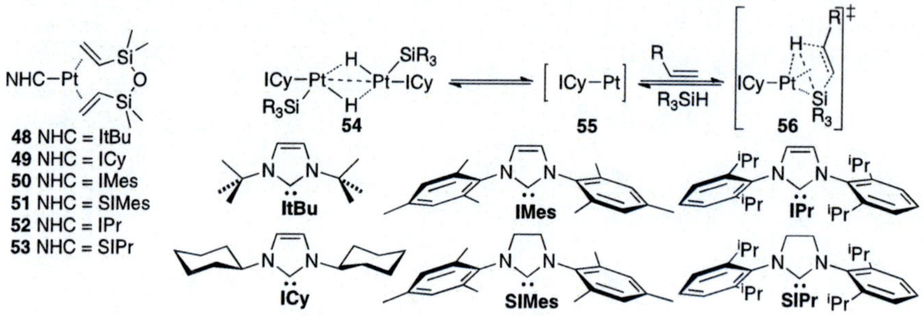

Scheme 2.7 Platinum-NHC complexes as hydrosilylation catalysts

Complexes of the type **48–53** (Scheme 2.7) have been targeted as pre-catalysts for the hydrosilylation of alkenes [44]. For example, in the hydrosilylation of 1-octene with $(Me_3SiO)_2Si(Me)H$, which was studied in detail as a model reaction, the activity of complexes **48–49** with alkyl substituted NHC ligands, is inferior to that of the Karstedt's system. However, selectivity and conversions are dramatically improved due to the suppression of side-product formation. In this reaction only the *anti*-Markovnikov addition product is observed. The complexes **50–52** with bulky aryl substituted NHCs, show activity comparable to that of the Karstedt's complex with excellent chemo- and regio-selectivity, even though they exhibit variable induction periods. The higher activity is attributed to electronic factors and the steric organisation around the Pt imposed by the ligand; the induction period is possibly due to slow diene dissociation from the pre-catalysts to form the 'underligated' 'Pt-NHC' active species. Further insight into the nature of the catalytic species was obtained by the reaction of **49** with excess silane, leading to the isolation of **54**, which was fully characterised; **54** converts to the active monomeric catalyst **55** by cleavage of the weakly bound dimer and loss of silane. Based on kinetic data, the reaction of **54** with alkene and silane is more likely to proceed *via* the concerted transition state **56** rather than two distinct oxidative addition and insertion elementary steps. Hydrosilylation of styrene with Et_3SiH catalysed by **51** (formed *in situ* from $Pt(norbornene)_3$ and SIMes·HCl/KO^{t}Bu) also gave excellent conversions (~99%) with high selectivity (*ca.* 82%) to the *anti*-Markovnikov product [45].

The hydrosilylation of alkynes has also been studied using as catalysts Pt, Rh, Ir and Ni complexes. The improvement of the regioselectivity of the catalyst and the understanding of stereoelectronic factors that control it have been major incentives for the ongoing research. From numerous studies involving non-NHC catalysts, it has been established that there is a complex dependence of the product ratio on the type of metal, the alkyne, the metal coordination sphere, the charge (cationic *versus* neutral) of the catalytic complex and the reaction conditions. In the Speier's and Karstedt's systems, mixtures of the thermodynamically more stable α- and β-*E*-isomers are observed. Bulky phosphine ligands have been used on many occasions in order to obtain selectively β-*E*-isomers.

The Pt complexes **48–53** mentioned previously are pre-catalysts for the hydrosilylation of 1-octyne with bis-(trimethylsilyloxy)methylsilane and other silanes [46]. Complexes **52** and **53** provide good conversions at very low loading (0.005 mol%) with high selectivity (10:1) for the β-*E*-isomer. Further studies have pointed to the steric influence of the *o*-groups of the aryl substituents of the NHC as being responsible for the high stereoselectivity. The linear correlation of the A_H angle (see Chapter 1, Fig. 1.16) of the NHC with the observed β-*E*-/α- ratio (larger A_H angles in more selective catalysts) was explained by the steric interactions between the out-of-the-coordination-plane aryl substituents of the NHC with the alkyne substituents prior to the insertion step. In addition to steric interactions, electronic factors are also responsible for the improved stereoselectivity.

The Rh and Ir complexes **57–62** (Fig. 2.9) with functionalised *N*-heterocyclic carbene ligands are also catalyst precursors for the hydrosilylation of alkynes. For example, terminal alkynes were hydrosilylated with $HSi(Me)_2Ph$ in the presence of **57** or **58**, yielding very good conversions to mixtures of the three possible stereoisomers, with preference for the β-*E*-isomer. A more detailed study of the catalytic reaction profile using **58** showed that the β-*Z*-isomer was formed during the initial stages of the reaction, and later isomerised to the thermodynamically more stable *E*-isomer. Interestingly, hydrosilylations with $HSi(OEt)_3$ showed preference for the β-*Z*-isomer, without any tendency for isomerisation. Rationalisation of these data is further complicated by the ambiguous nuclearity of the active catalysts originating from **58** and the exact coordination sphere of the active species [47]. On the other hand, reactions catalysed by the cationic **57** gave from the outset the β-*E*-isomer.

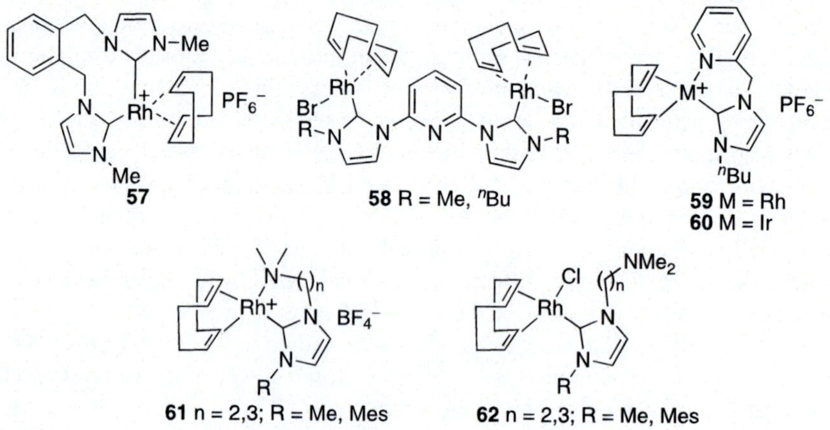

Fig. 2.9 Hydrosilylation catalysts based on Rh- and Ir-NHC complexes

Complexes **59** and **60** catalyse the hydrosilylation of phenylacetylene (but not other terminal alkyl alkynes) with $HSi(Me)_2Ph$. Generally, the Rh analogue is more active than the relative Ir. Both catalysts gave mixtures of all regioisomers, with a preference for the β-*Z*-isomer, in contrast to what has been reported with other non-NHC cationic complexes of Rh, where the β-*E* isomers are predominating. Here also the exact nature of the catalytic species is unclear [48].

Complexes **61** and **62** have been used for the hydrosilylation of a range of terminal alkynes with $HSi(Me)_2Ph$. The product distribution is strongly dependent on the nature of the alkyne and the catalyst. In general, catalysts with less bulky NHC ligands are more active. With terminal alkynes substituted with small linear alkyls (e.g. 1-hexyne), high conversions under mild conditions were obtained, and in some cases very good selectivities for the β-*Z*-vinylsilane. In contrast, the hydrosilylations of $Et_3SiC{\equiv}CH$ gave predominantly the β-*E*-vinylsilane. These data have been rationalised by postulating the classical Chalk-Harrod hydrosilylation mechanism (Si–H oxidative addition followed by alkyne insertion into the M–Si bond to give intermediate **63** and reductive elimination to the product β-*E*-vinylsilane). Isomerisation of **63–64** followed by reductive elimination leads to the β-*Z*-isomer. With small alkyls on the alkyne-C (for example in 1-hexyne), intermediate **64** is favoured due to reduced Rh–SiR_3 interaction leading to product β-*Z*. Larger alkyls on the alkyne-C result in reduced selectivities due to the establishment of an equilibrium between **63** and **64**. On the other hand, in the hydrosilylation of $Et_3SiC{\equiv}CH$, electronic factors are more important in determining the stability of **63** relative to **64** leading to substantial amounts of β-*E*-vinylsilane (Scheme 2.8) [49].

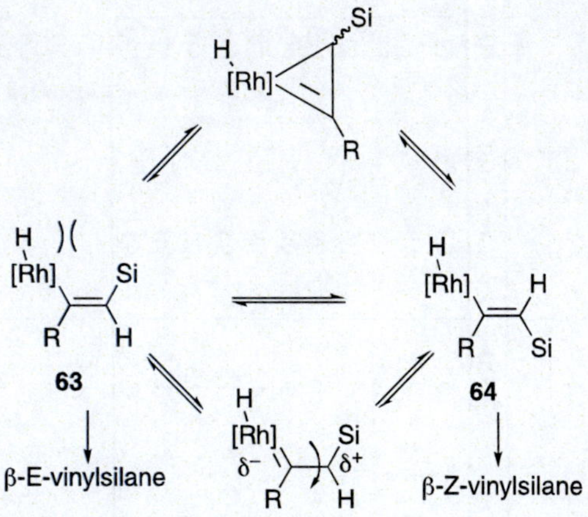

Scheme 2.8 Rationalisation of the selectivity observed in the hydrosilylation of alkynes by the complexes **61** and **62**

Nickel complexes formed *in situ* by the reaction of Ni(1,5-COD)$_2$ with the imidazolium salts IMes·HCl or IPr·HCl in the presence KO^{t}Bu catalyse the hydrosilylation of internal or terminal alkynes with Et$_3$SiH. Interestingly, Ni tri-butylphosphine complexes are inactive in this hydrosilylation reaction. The monosilylated addition products were obtained with slow addition rates of the alkyne in the reaction mixture and were formed with variable degree of stereoselectivity, depending on the type of the alkyne, the silane and the ligand on Ni [50].

The catalytic hydrosi(ly)lations of other C=X functional groups (X = O, NR) constitute alternative routes to the reduction of aldehydes, ketones, imines and other carbonyl compounds (Scheme 2.9), circumventing the use of molecular hydrogen or occasionally harsh transfer hydrogenation conditions.

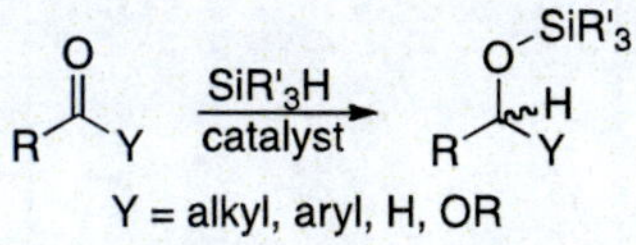

Scheme 2.9 Hydrosilylation of carbonyl compounds to silyl ethers

The hydrosilylation of carbonyl compounds by Et$_3$SiH catalysed by the copper NHC complexes **65** and **66–67** constitutes a convenient method for the direct synthesis of silyl-protected alcohols (silyl ethers). The catalysts can be generated *in situ* from the corresponding imidazolium salts, base and CuCl or [Cu(MeCN)$_4$]X$^-$, respectively. The catalytic reactions usually occur at room temperature in THF with very good conversions and exhibit good functional group tolerance. Complex **66**, which is more active than **65**, allows the reactions to be run under lower silane loadings and is preferred for the hydrosilylation of hindered ketones. The wide scope of application of the copper catalyst [dialkyl-, arylalkyl-ketones, aldehydes (even enolisable) and esters] is evident from some examples compiled in Table 2.3 [51–53].

Table 2.3 Selected data for the hydrosilylation of $R^1YC{=}O$, R^1 = alkyl, phenyl, Y = alkyl, H or OEt

R^1	Y	Catalyst (3 mol%)	Silane (Si:substrate)	Base (mol%)	T (°C)	t (h)	Conv. (%)	Ref
$-(CH_2)_5-$		IPr·HBF_4/CuCl	$HSiEt_3$ (5:1)	NaO^tBu (20%)	rt	2	99	[33]
$-(CH_2)_5-$		**65**	$HSiEt_3$ (3:1)	NaO^tBu (6–12%)	rt	0.75	100	[33]
$-(CH_2)_5-$		**66**	$HSiEt_3$ (2:1)	NaO^tBu (12%)	rt	0.5	95	[32]
Me	Et	IPr·HBF_4/CuCl	$HSiEt_3$ (5:1)	NaO^tBu (20%)	rt	4	96	[33]
Ph	Me	IPr·HBF_4/CuCl	$HSiEt_3$ (5:1)	NaO^tBu (20%)	rt	3	99	[33]
Ph	Me	**65**	$HSiEt_3$ (3:1)	NaO^tBu (6–12%)	rt	3	98	[33]
Ph	Me	**66**	$HSiEt_3$ (2:1)	NaO^tBu (12%)	rt	4	98	[32]
^{i}Pr	^{i}Pr	**66**	$HSiEt_3$ (2:1)	NaO^tBu (12%)	rt	0.3	94	[32]
p-tolyl	H	**66**	$HSiEt_3$ (2:1)	NaO^tBu (12%)	rt	0.3	92	[32]
C_6H_{11}	H	**66**	$HSiEt_3$ (2:1)	NaO^tBu (12%)	rt	0.6	70	[32]
Benzyl	OEt	**66**	$HSiEt_3$ (2:1)	NaO^tBu (12%)	55	6	69	[32]

The proposed mechanism of the catalytic reaction involves the formation of the Cu(I) alkoxide **68** by displacement of either the chloride or the NHC from **65–67**, followed by conversion to the hydride **69** by metathetical exchange of the *tert*-butoxide by the H of the silane (Fig. 2.10).

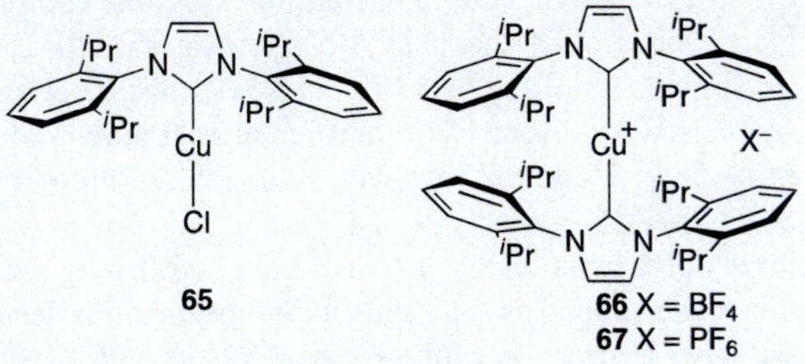

Fig. 2.10 Hydrosilylation catalysts of carbonyl compounds based on Cu-NHC complexes

Transfer of the hydride from the Cu to the electrophilic carbon and cleavage of the copper alkoxide by the silane regenerates **69**. Recent reports point to the influence of the type of the counter ion X^- of the homoleptic **66–67** on the activity, the BF_4^- being superior to the PF_6^- analogue; this effect has been attributed to differences in the rate of active catalyst generation from the homoleptic $[Cu(NHC)_2]^+X^-$ and NaO^{t}Bu due to solubility differences of the inorganic salts formed during the displacement of the NHC by ^{t}BuO$^-$ [54] (Scheme 2.10).

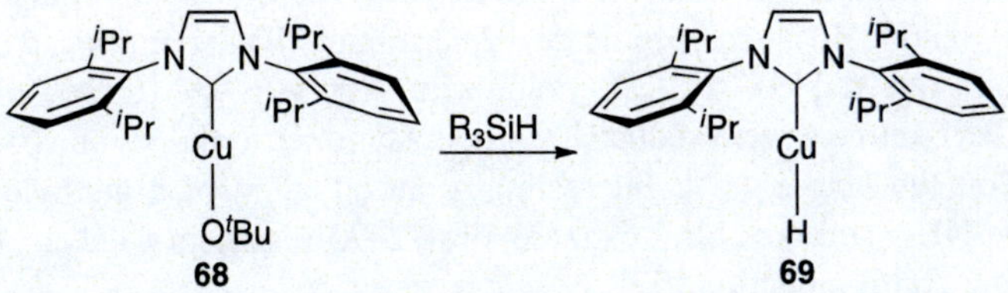

Scheme 2.10 Intermediates in the Cu-catalysed hydrosilylation of carbonyl compounds

The Rh-catalysed asymmetric hydrosilylation of prochiral ketones has been studied with complexes bearing monodentate or heteroatom functionalised NHC ligands. For example, complexes of the type [RhCl(1,5-cod)(NHC)] and [RhL(1,5-cod)(NHC)][SbF_6^-], **70**, where L = isoquinoline, 3,5-lutidine and NHC are the chiral monodentate ligands **71** (Fig. 2.11).

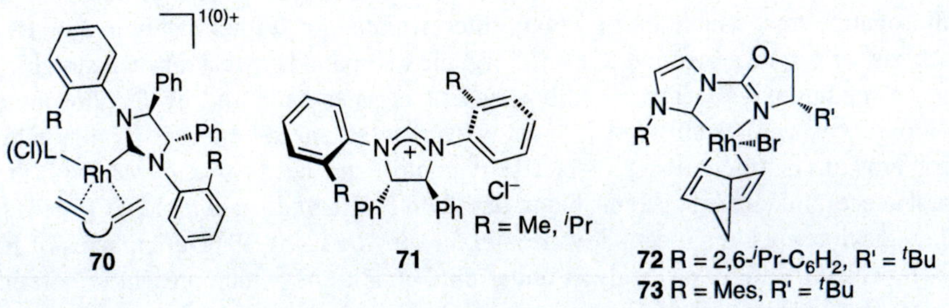

Fig. 2.11 Chiral ligand designs in Rh catalysts for the enantioselective hydrosilylation of carbonyl compounds

X-ray crystallography and variable temperature ^{1}H NMR studies show that the conformation of the coordinated imidazolidin-2-ylidene, in both the neutral and cationic complexes **70**, is *anti*, *anti* with respect to the Ph of the backbone of the NHC, exclusively in the solid state and predominantly in solution at lower temperatures (–75°C). At room temperature in solution, possible conformer interconversion by the rotation around the phenyl-N bond of the NHC substituent is apparent from the broadness of the peaks in the NMR spectra. Hydrosilylation of acetophenone by Ph_2SiH_2 catalysed by **70** at room temperature or at –20°C results in maximum *ee* of 58%. However, at lower temperatures the reaction rates are much slower [55].

NHC complexes of Rh functionalised with chiral oxazoline ligands, for example **72** and **73**, have been developed as pre-catalysts for the enantioselective hydrosilylation of ketones. After abstraction of the Br from **72–73** with silver reagents, the cationic complexes obtained were employed for the hydrosilylation of prochiral ketones with various silanes. The conversions and enantioselectivities observed were dependent on the nature of the R and R' groups of the ligand, the silane and reaction conditions. Higher enantioselectivities for the hydrosilylation of acetophenone to 1-phenyl-ethanol were observed with bulky silanes [Ph_2SiH_2, (p-tolyl)$_2SiH_2$] at the optimum temperature (–60°C); both higher or lower temperatures had a detrimental effect on the *ee* observed. This nonlinear temperature dependence of the *ee* has been interpreted by assuming two enantioselectivity determining steps in the catalytic reaction, one possibly involving the reversible coordination of the prochiral ketone onto Rh and the other the migratory insertion of the coordinated ketone into the Rh–Si bond, that can become rate-determining. Under optimised conditions acetophenone was reduced at 92% yield, with very good *ee* (90%), while other methyl-*n*-alkyl ketones were reduced at good asymmetric induction (65–80% *ee*) depending on the ketone [56]. Interestingly, introduction of a methylene spacer between the NHC and the chiral oxazoline in **72–73** results in catalysts that show disappointing chiral induction [57].

2.5 Hydroboration and Diboration Reactions

The metal catalysed hydroboration and diboration of alkenes and alkynes (addition of H–B and B–B bonds, respectively) gives rise to alkyl- or alkenyl-boronate or diboronate esters, which are important intermediates for further catalytic transformations, or can be converted to useful organic compounds by established stoichiometric methodologies. The *syn*-diboration of alkynes catalysed by Pt phosphine complexes is well-established [58]. However, in alkene diborations, challenging problems of chemo- and stereo-selectivity control still need to be solved, with the most successful current systems being based on Pt, Rh and Au complexes [59–61]. There have been some recent advances in the area by using NHC complexes of Ir, Pd, Pt, Cu, Ag and Au as catalysts under mild conditions, which present important advantages in terms of activity and selectivity over the established catalysts.

The Ag complex **74** catalyses the diboration of internal and terminal alkenes to 1,2-bis-diboronate esters by bis(catecholato)diboron, $(Bcat)_2$, in THF at room temperature. Variable conversions (30–90%) were obtained at 5 mol% loading after 60 h.

Terminal alkenes with electron-rich substituents (Ph, Cy) gave the highest conversions. This reaction represents the first example of a Ag-NHC complex catalysed reaction. The choice of silver was rationalised based on the mechanism of diboration, which involves the insertion of the electron rich metal centre into the B–B bond, followed by the alkene insertion and the release of the mono- or di-boronated organic products. The electron rich Ag^I with low energy d orbitals minimises π-alkene bonding and, therefore, β-H elimination is suppressed, resulting in inhibition of vinyl boronates formation. It is interesting to note that although **74** is chiral, it does not induce any enantioselectivity in the diboration reactions of prochiral substrates [62].

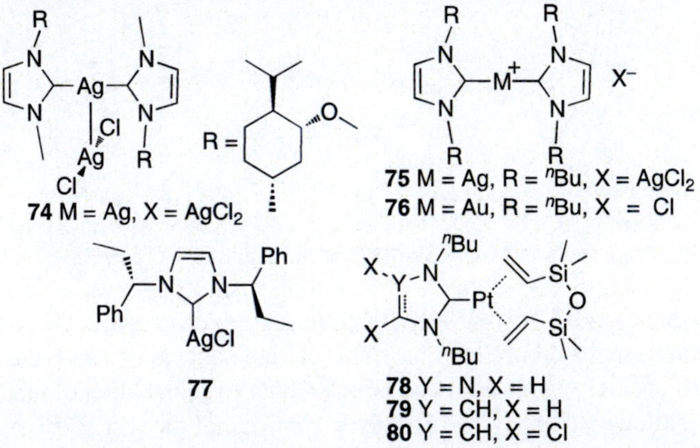

Fig. 2.12 Silver, gold and platinum complexes with monodentate NHC ligands as catalysts for the diboration of alkenes and alkynes

Diboration of terminal alkenes has also been studied with other d^{10} metals (Fig. 2.12) including the Ag^I and Au^I complexes **75–77** and the Pt^0 complexes **78–79**. Styrene is diborylated with 100% selectivity and good conversions in THF (46% for **75** and 94% for **77** at 5 mol%, 60 h) using equimolecular amounts of $(Bcat)_2$. The difference in activity between the Ag and Au complexes has been ascribed to the increased lability of the Ag–NHC bond, which may lead to catalyst decomposition under the reaction conditions. In both catalytic systems it is believed that the active species involves only one coordinated NHC ligand. Complex **77** is less active than **74** and **75**, possibly due to steric reasons. The enantioselectivity of **77** in the diboration of prochiral alkenes is very low [63].

Diboration of terminal and internal alkynes by $(Bcat)_2$ [$B(cat)_2$:alkene = 1:1] was also achieved by using complexes **78** and **79** at room temperature in THF. Best activity was observed with **78** containing the less electron donating triazolylidene carbene ligand. Terminal alkenes can also be diborylated under the same conditions, however the selectivity for the diborylated product was much lower (up to 65%) [64].

The palladium (II) NHC complexes **81** and **82** (Fig. 2.13) have also been used as catalysts in the diboration of styrene. In the presence of $NaOOCCH_3$, (1 equiv.,

5 mol% catalyst, in THF at rt), excellent conversions were obtained when $(Bcat)_2$ was the boron source. Selectivities were also high for the diborylated product when excess of $(Bcat)_2$ was used [$(Bcat)_2$:styrene 3:1]. Other terminal and internal alkenes were also diborylated in high yields and selectivities under the same conditions. A mechanism involving metathetical Pd–B bond formation (without change of the oxidation state) may be operating. However, an involvement of Pd^{II}/Pd^{IV} species, formed by oxidative addition of B–B to a cationic Pd^{II} and formation of an NHC stabilised Pd^{IV}-boryl complex, is also plausible. In the presence of base (NaOAc), the mechanism could involve transmetallation, and in this case interaction of the boron entering group with the base giving borate species may increase its nucleophilicity and facilitate the Pd–B(cat) bond formation. DFT and *in situ* ^{11}B NMR experiments support these mechanistic models [65].

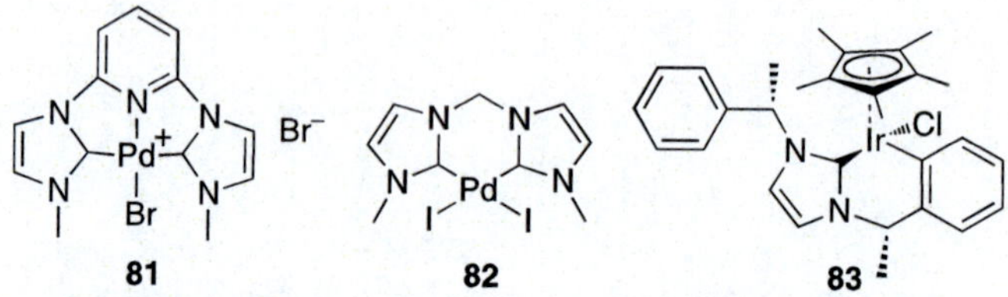

Fig. 2.13 Diboration catalysts based on palladium and iridium NHC complexes

The Ir complexes **83** or [$Ir(IMes)Cl_2Cp^*$], in the presence of NaOAc and excess of $(Bcat)_2$, catalyse the diboration of styrene, at high conversions and selectivities for the diborated species, under mild conditions. Other terminal alkenes react similarly. The base is believed to assist the heterolytic cleavage of the (cat)B–B(cat) bond and the formation of Ir–B(cat) species, without the need of B–B oxidative addition [66].

The stoichiometric insertion of terminal alkenes into the Cu–B bond of the (NHC)Cu–B(cat) complex, and the isolation and full characterisation of the β-boryl-alkyl-copper (I) complex has been reported. The alkyl complex decomposes at higher temperatures by β-H elimination to vinylboronate ester [67]. These data provide experimental evidence for a mechanism involving insertion of alkenes into Cu–boryl bonds, and establish a versatile and inexpensive catalytic system of wide scope for the diboration of alkenes and alkynes based on copper.

The complexes [$Cu(NHC)(MeCN)][BF_4]$, NHC = IPr, SIPr, IMes, catalyse the diboration of styrene with $(Bcat)_2$ in high conversions (5 mol%, THF, rt or reflux). The $(Bcat)_2$/styrene ratio has also an important effect on chemoselectivity (mono-*versus* di-substituted borylated species). Use of equimolecular ratios or excess of $B(cat)_2$ results in the diborylated product, while higher alkene:$B(cat)_2$ ratios lead selectively to mono-borylated species. Alkynes (phenylacetylene, diphenylacetylene) are converted selectively (90–95%) to the *cis*-di-borylated products under the same conditions. The mechanism of the reaction possibly involves σ-bond metathetical reactions, but no oxidative addition at the copper. This mechanistic model was supported by DFT calculations [68].

Platinum mediated regioselective H–B addition of H–B(cat) to vinyl arenes and alkynes has been realised using the complexes **79** and **80** (Fig. 2.12). The reactions with vinyl arenes proceed at rt in THF (0.05 mol%) giving high conversions (80–100%) of mixtures of alkyl or alkenyl boranes, with predominantly branched product (70–90%). Interestingly, the activity of the catalytic system is maintained days after the completion of the reaction, while addition of PPh_3 into the catalytically active system retards or deactivates it. Aryl alkynes are hydroborated selectively to the mono-borylated product, with higher selectivity to the linear alkenyl borane. The addition of aryl iodides into the catalytic system results in the cross coupling of the *in situ* formed alkenyl boranes with the aryliodide in one pot, giving alkenyl arenes [69].

In an attempt to study the electronic effects of the substituents in the four- and five-positions of the coordinated imidazol-2-ylidene on the M–C_{NHC} bonding, and the related catalytic reactivity, the complexes [RhCl(1,5-COD)(NHC)], where NHC is imidazol-2-ylidene substituted in the four- or five-positions with the σ- or π-electron withdrawing groups –CF_3, –Cl, –NO_2 and –CN, were used as catalysts in the hydroboration of alkynes with HB(pin). In the hydroboration of phenylacetylene, it was found that complexes with ligands bearing π-electron withdrawing groups in the four- and/or five-position of the imidazolylidene, were affording lower yields of products than the ones with σ-electron withdrawing groups. These data have been interpreted as an indication that π-interactions between the coordinated NHC with a metal exist, a fact that was disputed in the early era of the NHC ligand development, and may have important implications in catalyst design [70]. However, in the hydroboration of 1-octyne no effect was observed.

Hydroboration of acyclic and cyclic aryl alkenes with $(Bpin)_2$ (1.1 equiv.) in the presence of NaO^{t}Bu (1–100 mol%) and MeOH (2 equiv.) is catalysed by [CuCl(NHC)], (0.5–5 mol%) NHC = IMes, SIMes and ICy, and proceeds with very good conversions and regioselectivity (Scheme 2.11).

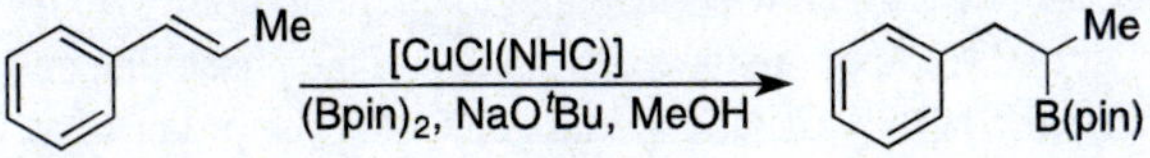

Scheme 2.11 Hydroboration of arylalkenes catalysed by copper NHC complexes

This regioselectivity is opposite to the one observed by the non-catalysed additions of BH_3·THF or 9–BBN to the same alkene, or those catalysed by Rh and Ir catalysts. Chiral NHC ligands (generated from **84**) on Cu under the same conditions proceed with high enantioselectivity (enantiomeric ratio 99:1) [71] (Scheme 2.12).

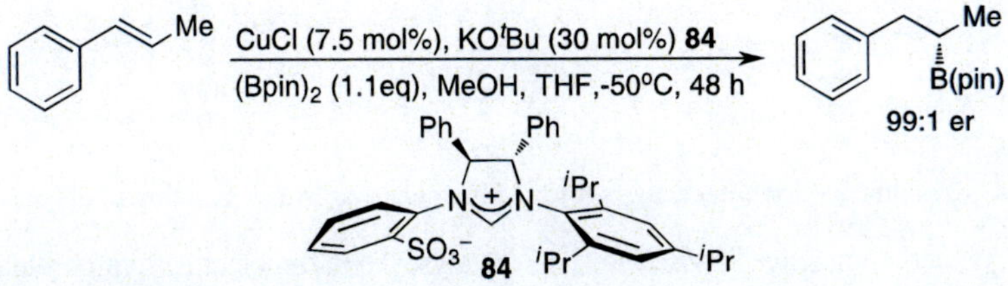

Scheme 2.12 Enantioselective hydroboration of arylalkenes catalysed by copper NHC complexes

2.6 Hydroamination Reactions

Intermolecular and intramolecular hydroamination of alkenes and alkynes is an atom economical method for the synthesis of a range of acyclic and cyclic alk(en)ylamines from simpler amine precursors. The reaction can be catalysed by either electropositive (main group metals, early transition metals and lanthanides-actinides) or late transition metals, under different mechanistic regimes. The relatively facile intramolecular hydroamination of alkynes and allenes is more ubiquitous and commonly studied. Intermolecular hydroamination of alkenes is more challenging and examples with activated, electron-deficient alkenes have recently appeared. Control of regioselectivity (Markovnikov, *anti*-Markovnikov) and suppression of competing side-reactions (alkene isomerisation, oligomerisation, etc.) in addition to high activity, especially when expensive late transition metals are used as catalysts, are important features of the catalyst development [72, 73]. NHC complexes with both electropositive and late transition metals have been studied as alkene and alkyne hydroamination catalysts.

A catalytic system comprising $Ti(NMe_2)_4$, $LiN(SiMe_3)_2$ and IMes has been developed for the intermolecular hydroamination of terminal aliphatic alkynes (1-hexyne, 1-octyne, etc.) with anilines [toluene, 100°C, 10 mol% $Ti(NMe_2)_4$]. Markovnikov products were dominant. Substituted anilines reacted similarly. High conversions (85–95%) were observed with specific anilines. The optimum Ti/IMes/$LiN(SiMe_3)_2$ ratio was 1:2:1. However, the nature of the active species and especially the role of $LiN(SiMe_3)_2$ are unclear [74].

The Rh and Ir complexes **85–88** (Fig. 2.14) have been tested for the intramolecular hydroamination/cyclisation of 4-pentyn-1-amine to 2-methyl-1-pyrroline (n = 1). The reactions were carried out at 60°C (1–1.5 mol%) in THF or $CDCl_3$. The analogous rhodium systems were more active. Furthermore, the activity of **87** is higher than **85** under the same conditions, which was attributed to the hemilability of the P donor in the former complex, or to differences in the *trans*-effects of the phosphine and NHC ligands, which may increase the lability of the coordinated CO in the pre-catalyst [75, 76].

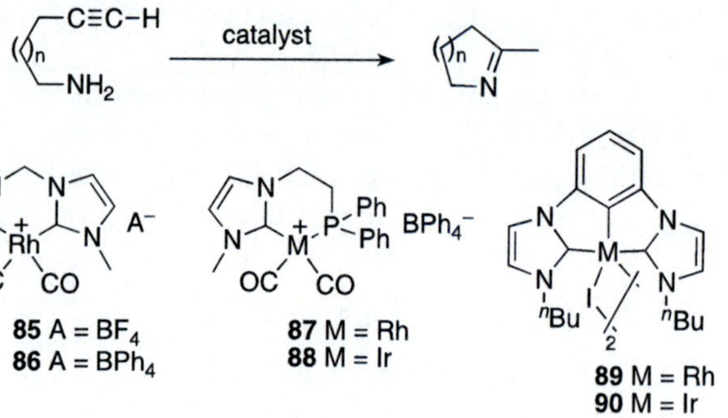

Fig. 2.14 Rhodium and iridium catalysts for the intramolecular hydroamination of alkynes

The 'pincer' complexes **89–90** (Fig. 2.14) catalyse the intramolecular hydroamination/cyclisation of unactivated alkenes, yielding pyrrolidines and piperidines (n = 1, 2, respectively). The reactions can be carried out in benzene or water with high conversions (>95% at 2.5 mol% at 110°C, overnight). Rh and Ir catalysts show very similar activity without competing alkene isomerisation. Interestingly, no reaction was observed with primary amines. The reaction mechanism may involve metal-amido bond formation followed by alkene insertion and reductive elimination, or π-coordination of the alkene followed by nucleophilic attack on the activated coordinated alkene [77, 78].

Hydroamination of activated alkenes has been reported with complexes **91–93** (Fig. 2.15). For example, **91** catalyses the hydroamination of methacrylonitrile (X = CN in Scheme 2.13) by a range of secondary amines (morpholine, thiomorpholine, piperidine, *N*-methylpiperazine or aniline) in good to excellent conversions (67–99%) and *anti*-Markovnikov regioselectivity (5 mol%, –80°C or rt, 24–72 h). Low enantioselectivies were induced (*ee* 30–50%) depending on the amine used and the reaction temperature [79].

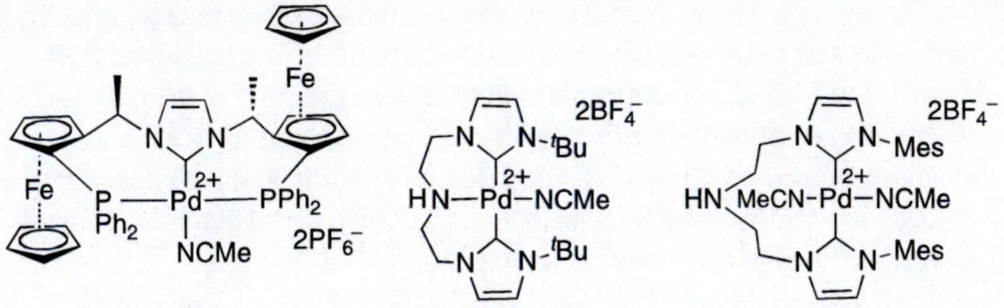

91 **92** **93**

Fig. 2.15 Hydroamination catalysts based on palladium NHC complexes

Complexes **92** and **93** also show good activity for the hydroamination of methacrylonitrile with morpholine, piperidine or *N*-methylpiperazine (70–93% conversion at 2.5 mol%, 90°C in 24 h) [80].

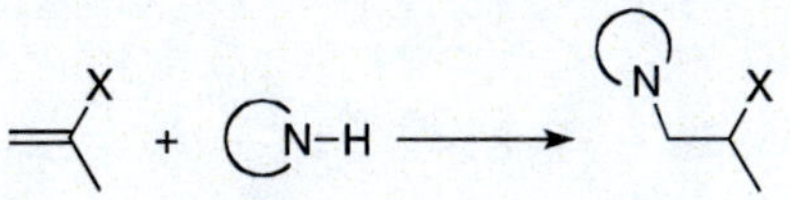

Scheme 2.13 Intermolecular hydroamination of activated alkenes

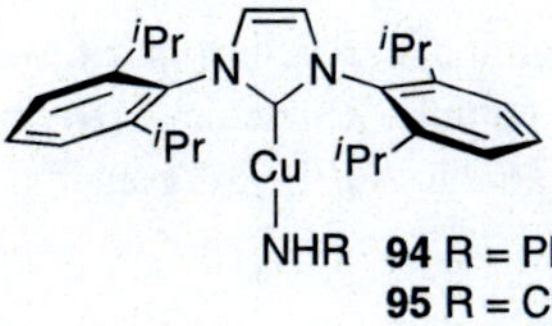

94 R = Ph
95 R = CH_2Ph

Fig. 2.16 Copper-amido complexes as catalysts for the intermolecular hydroamination of electron-deficient alkenes

The well-defined copper complexes **94** and **95** (Fig. 2.16) have been used as catalysts for the intermolecular hydroamination of electron-deficient alkenes [Michael acceptors, X = CN, C(=O)Me, C(=O)(OMe)] and vinyl arenes substituted by the electron-withdrawing groups NO_2 CN, CF_3 but not Cl, Br or H (Schemes 2.13 and 2.14).

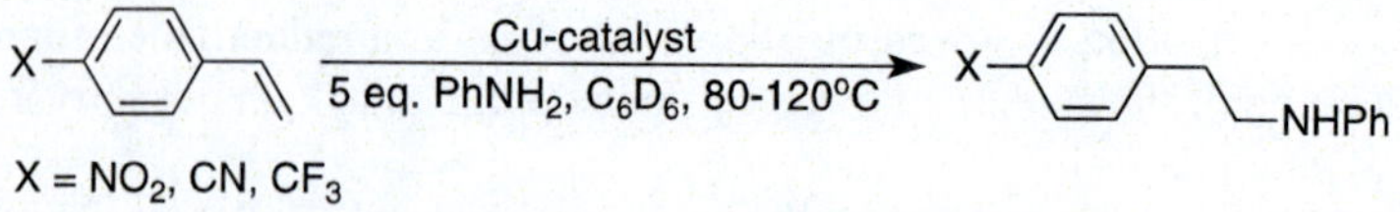

Scheme 2.14 Copper-catalysed intermolecular hydroamination of electron-deficient aryl alkenes

The hydroaminations of electron-deficient alkenes with aniline or small primary alkylamines proceed at high conversions (85–95%, under mild conditions, 5 mol%, rt), giving exclusively the *anti*-Markovnikov addition product. Secondary dialkyl or bulky primary amines require longer reaction times. With amines containing β-hydrogens, no imine side-products were observed.

The proposed reaction mechanism involves intermolecular nucleophilic addition of the amido ligand to the olefin to produce a zwitterionic intermediate, followed by proton transfer to form a new copper amido complex. Reaction with additional amine (presumably *via* coordination to Cu) yields the hydroamination product and regenerates the original copper catalyst (Scheme 2.15). In addition to the NHC complexes **94** and **95**, copper amido complexes with the chelating diphosphine 1,2-bis-(di-*tert*-butylphosphino)-ethane also catalyse the reaction [81, 82].

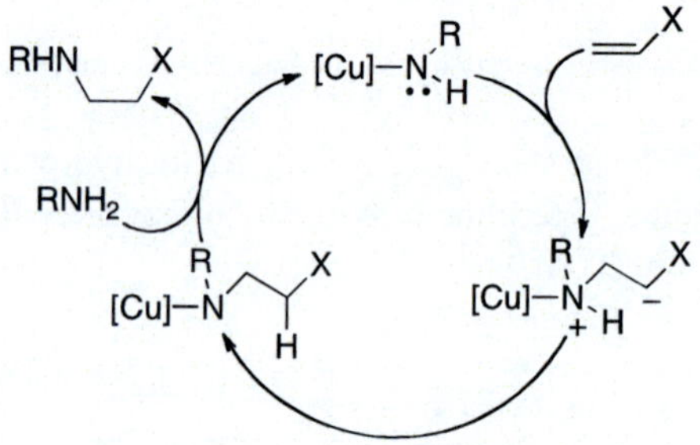

Scheme 2.15 Postulated mechanism for the copper-catalysed hydroamination of electron-deficient alkenes

Finally, intramolecular hydroamination/cyclisation of *N*-alkenyl ureas was catalysed by the well-defined [AuCl(IPr)] complex (Scheme 2.16), in the presence of AgOTf (5 mol%, rt, methanol, 22 h). The cationic $Au(IPr)^+$ is presumably the active species [83].

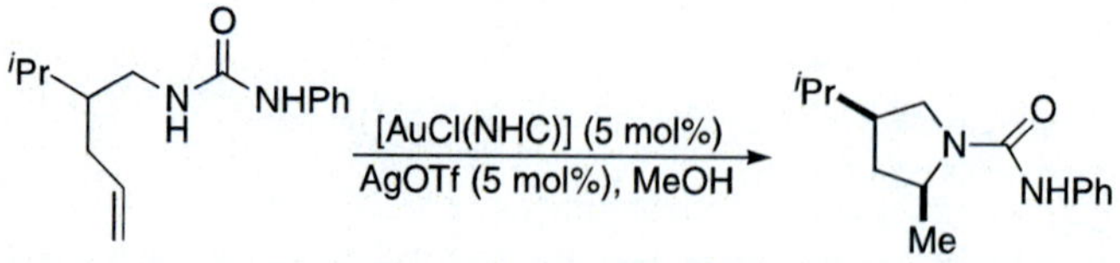

Scheme 2.16 Gold-catalysed intramolecular hydroamination of alkenes

2.7 Hydrothiolation, Hydroalkoxylation and Hydroaryloxylation Reactions

Hydrothiolations (addition of H–SR across the CC multiple bond) of alkynes, electron-deficient alkenes and electron-deficient vinyl arenes have been catalysed by NHC complexes of Ni and Cu, respectively [Scheme 2.17a-c].

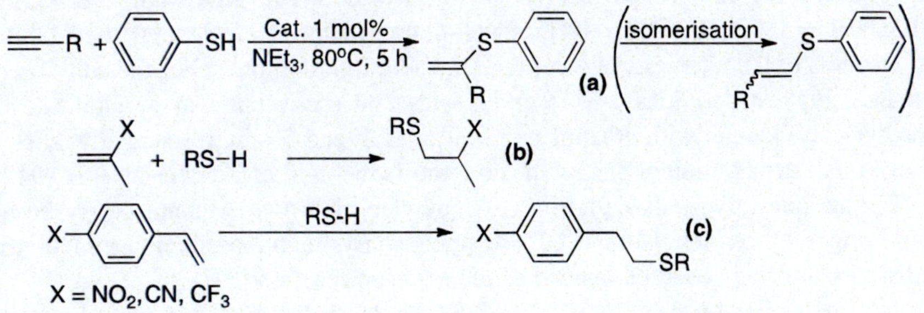

Scheme 2.17 Product formation in the hydrothiolation of alkynes and alkenes

The hydrothiolation of terminal alkyl alkynes with **96** (Fig. 2.17) proceeds with good degree of regio- and chemo-selectivity, especially with thiophenol and *p*-methoxy-thiophenol as substrates. Isomerisation to the internal alkenyl thiolates accounts for less than 9% of the thiolated products under the reaction conditions. In addition, further hydrothiolation of the vinyl thioether product is not observed. Typical conversions of 70–85% at 1 mol% loading at 80°C within 5 h are observed. Arylthiols substituted with electron-withdrawing groups afford lower conversions.

The thiolato complex **97** that was postulated as the active catalytic species in the reaction was prepared from **96** and the thiol in the presence of NEt_3. Certain analogues of **97** (NHC = IMes, SIMes, IPr, SIPr; R = Ph) have also been independently synthesised, isolated and fully characterised. A plausible mechanism for the hydrothiolation involves insertion of the alkyne into the Ni–SR bond forming the (non-isolable) β-thioalkenyl complex, from which the product can be released *via* alkanolysis of the Ni–C bond by the thiol and regeneration of the active catalyst **97** [84].

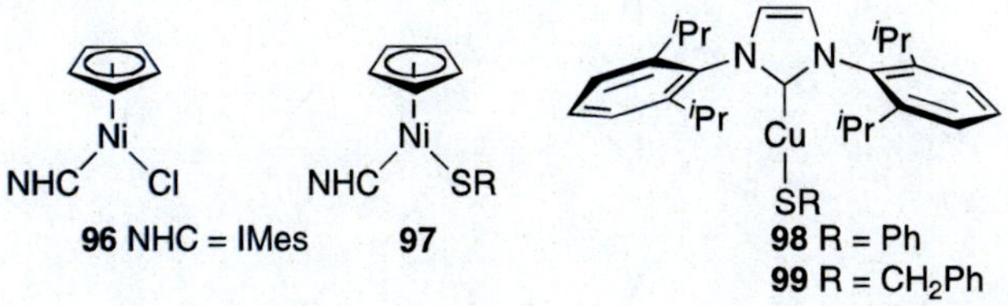

Fig. 2.17 Nickel and copper complexes as catalysts for the hydrothiolation of alkynes and activated alkenes

The hydrothiolation of electron-deficient alkenes [X = CN, C(=O)(OMe)] and *p*-nitro-styrene was catalysed by the Cu^I complexes **98** and **99**. The reactions with phenyl- or benzyl-thiol proceed with high conversions (>90%, rt, 5 mol%).

Anti-Markovnikov products are only observed. The postulated mechanism for these reactions is analogous to the previously discussed for the copper-catalysed hydroamination (Scheme 2.15) with the coordinated thiolate (rather than the amide) acting as nucleophile [82, 85].

Hydroalkoxylation and hydroaryloxylation (addition of H–OR and H–OAr, respectively, across the CC multiple bonds) of electron-deficient alkenes, by alcohols or phenols was studied in the presence of [Cu(OR)(NHC)], NHC = IPr, R = OEt, **100**; NHC = IPr, R = OPh, **101**; NHC = SIPr, R = OEt, **102**; NHC = SIPr, R = OPh, **103**; NHC = IMes, R = OPh, **104**], as well-defined catalysts. In general, the addition of alcohols to the alkenes is slower than the addition of amines. For the addition of ethanol, highest activities are observed with **100** (conversion of acrylonitrile or methyl vinyl ketone with ethanol >95% after 12 h and 5 min, respectively, at rt, 5 mol%). For the addition of phenol to acrylonitrile the most efficient catalyst is **104**.

In attempted hydroalkoxylations of methylacrylate with ethanol catalysed by the copper ethoxides **100** or **102**, copper-catalysed transesterification to the ethylacrylate was observed instead of the addition reaction [81].

Intermolecular hydroalkoxylation of 1,1- and 1,3-di-substituted, tri-substituted and tetra-substituted allenes with a range of primary and secondary alcohols, methanol, phenol and propionic acid was catalysed by the system [AuCl(IPr)]/AgOTf (1:1, 5 mol% each component) at room temperature in toluene, giving excellent conversions to the allylic ethers. Hydroalkoxylation of monosubstituted or trisubstituted allenes led to the selective addition of the alcohol to the less hindered allene terminus and the formation of allylic ethers. A plausible mechanism involves the reaction of the *in situ* formed cationic (IPr)Au$^+$ with the substituted allene to form the π-allenyl complex **105**, which after nucleophilic attack of the alcohol gives the σ-alkenyl complex **106**, which, in turn, is converted to the product by protonolysis and concomitant regeneration of the cationic active species (IPr)-Au$^+$ (Scheme 2.18) [86].

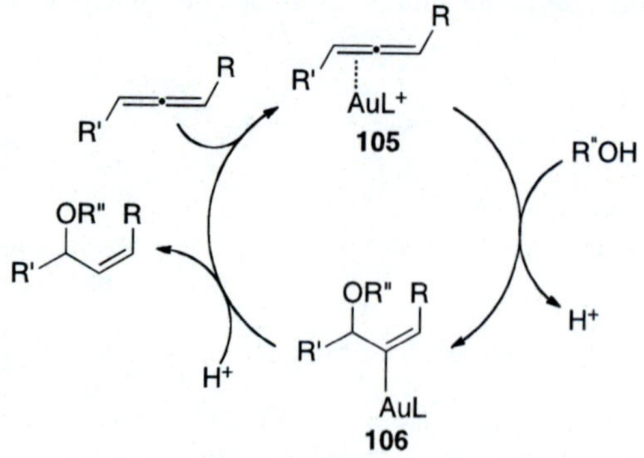

Scheme 2.18 Postulated mechanism for the Au-NHC catalysed hydroalkoxylation of allenes

2.8 Hydration Reactions

The hydration of alkynes to ketones, catalysed by [AuCl(IPr)]/$AgSbF_6$ in 1,4--dioxane/H_2O (2:1) or methanol has been studied in detail. The reaction proceeds at surprisingly low catalyst loadings (50–1,000 ppm), especially, for terminal alkynes in methanol, and is extremely sensitive to the nature of the catalyst, the silver activator and the reaction conditions. The best activity is obtained with the IPr ligand on Au, while silver salts with anions other than $[SbF_6]^-$, either reduce the conversion or result in catalytically inactive species. Interestingly, diphenylacetylene is hydrated faster in 1,4-dioxane than methanol, the reverse behaviour being observed for the terminal phenylacetylene. This feature may imply that two different reaction mechanisms may be operating in dioxane and methanol, with methanol being a non-innocent solvent. This proposition is also supported by the detection of vinyl methylether intermediates, resulting from the direct addition of methanol to the alkyne, which in turn converts to the final ketone product. The nature of the catalytic species involved in this reaction is not known, but may involve solvated, hydroxo- or alkoxo-gold complexes $(NHC)Au(solvent)^+$, (NHC)Au(OH), (NHC)Au(OR). The Au–NHC based catalytic system described here is much more active than the previously known phosphine analogue, $[AuMe(PPh_3)]$ which requires the use of strong acids (H_2SO_4, CF_3SO_3H, etc.) for catalyst activation [87, 88].

The intermolecular hydration of allenes catalysed by [AuCl(IPr)]/AgOTf (1:1, 5 mol%) in dioxane/water at room temperature, has also been studied. In most cases, low to modest yields (25–65%) of *E*-allylic alcohols were obtained by selective addition of the water to the terminal C atom of the allene group [89].

2.9 Hydroarylation Reactions

Hydroarylation, (addition of H–Ar, Ar = aryl), of alkynes, catalysed by $Pd(OOCCH_3)_2$ or $Pd(OOCCF_3)_2$ in acetic acid, is an atom-economic reaction, giving rise to substituted *cis*-stilbenes (Fujiwara reaction). Catalytic conversions and improved chemoselectivity to the mono-coupled product under mild conditions can be achieved by modification of the metal coordination sphere with NHC ligands. Hydroarylation of mesitylene by ethylpropiolate (Scheme 2.19) catalysed by complex **107** (Fig. 2.18) proceeds in good conversions (80–99%, 1 mol%) under mild conditions at room temperature.

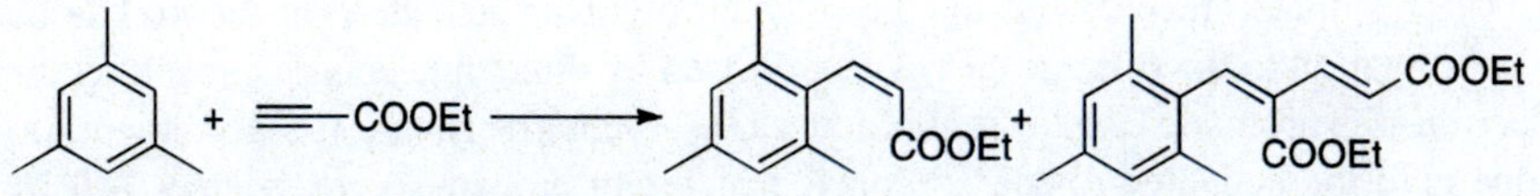

Scheme 2.19 Products from the hydroarylation reactions of activated alkynes

Surprisingly, the organometallic catalyst shows good stability under the reaction conditions (CF_3COOH/CH_2Cl_2). In the absence of **107** (Fig. 2.18), $Pd(OAc)_2$ under the same conditions catalyses the same reaction with reduced activity (*ca.* 50% conversion in 24 h) and different chemoselectivity. Arenes, substituted by electron-withdrawing substituents react slower. Both internal and terminal alkynes undergo the reaction, however, the former require more forcing conditions [90].

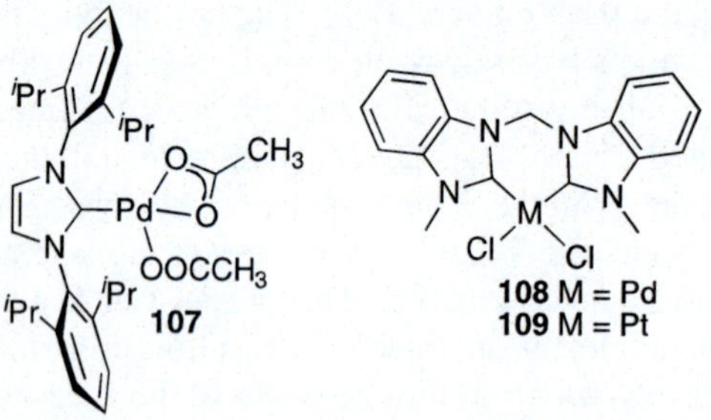

107

108 M = Pd
109 M = Pt

Fig. 2.18 Palladium and platinum NHC complexes as catalysts in hydroarylations of alkynes

High activities and selectivities for the *Z*-monoarylated stilbene were also obtained by the intermolecular hydroarylation of pentamethylbenezene by ethylpropiolate in $CF_3COOH/1,2\text{-}C_2H_4Cl_2$ (1:4) catalysed by **108** (0.1 mol%, 80°C, 95% in 7 h). NHC ligands with bulkier substituents on the NHC show higher activity. The mechanism operating in this transformation has not been fully elucidated, but there are indications that during the reaction the NHC ligand of **108** remains bidentate. Encouraging results were also obtained when using the Pt analogue **109** as catalyst in hydroarylation reactions [91].

The intramolecular hydroarylation/cyclisation of aryl propargylic acetates catalysed by the system [AuCl(IPr)]/$AgBF_4$ (1:1, 2 mol%, 72–92%, rt, 5 min) was developed as a versatile and efficient method leading to indene derivatives **110** (Scheme 2.20). Analogous catalytic systems, where the IPr was substituted by PPh_3, gave lower conversions and chemo-/regio-selectivity.

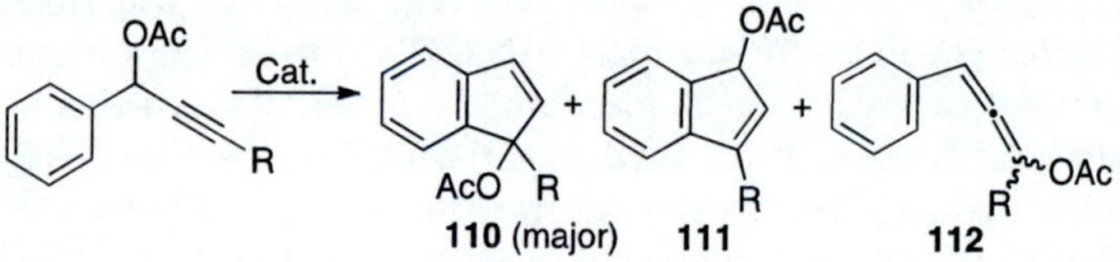

110 (major) **111** **112**

Scheme 2.20 The synthesis of substituted indenes by intramolecular hydroarylation of aryl propargylic acetates

The postulated mechanism for the reaction involves activation of the alkyne by π-coordination to the cationic $(IPr)Au^+$, followed by direct nucleophilic attack by the electron-rich aromatic ring to form product **111**. Alternatively, two 1, 2-acetate migrations give the activated allene complex, which can be cyclised to product **110** by nucleophilic attack of the aromatic ring on the activated allene (Scheme 2.21) [92].

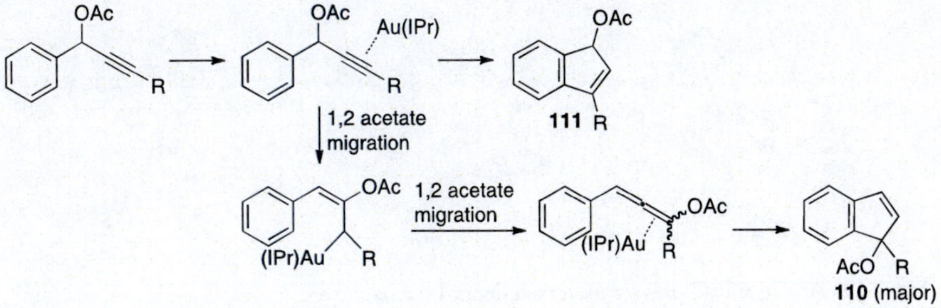

Scheme 2.21 Postulated mechanism for the synthesis of substituted indenes by intramolecular hydroarylation of propargylic acetates

2.10 Allylic Alkylations and Other Substitution Reactions

Allylic substitutions catalysed by palladium NHC complexes have been studied and the activity and selectivity of the catalysts compared to analogous Pd phosphine complexes. A simple catalytic system involves the generation of a Pd(NHC) catalyst *in situ* in THF, from $Pd_2(dba)_3$, imidazolium salt and Cs_2CO_3. This system showed very good activities for the substitution of the allylic acetates by the soft nucleophilic sodium dimethyl malonate (2.5 mol% $Pd_2(dba)_3$, 5 mol% IPr·HCl, 0.1 equiv. $Cs_2(CO_3)_2$, THF, 50°C) (Scheme 2.22). Generation of the malonate nucleophile can also be carried out *in situ* from the dimethylmalonate pro-nucleophile, in which case excess (2.1 equivalents) of Cs_2CO_3 was used. The nature of the catalytic species, especially the number of IPr ligands on the metal is not clear.

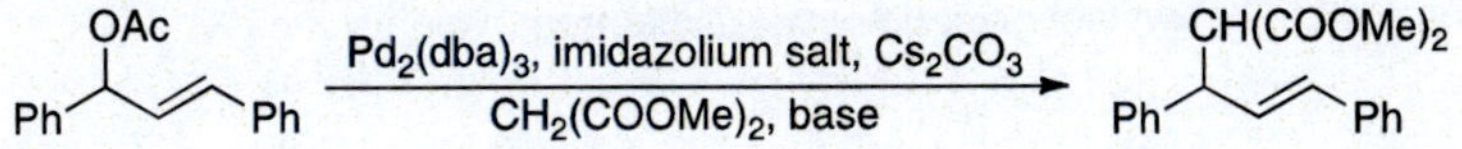

Scheme 2.22 Palladium-NHC complex catalysed substitution of allylic acetate by malonate

The substitution proceeds with retention of configuration at the substituted C atom as confirmed by the reaction of suitable isomeric acetates. Retention of configuration has also been observed in analogous Pd-phosphine catalysed reactions [93].

Compared to the Pd phosphine catalysts, the Pd/IPr system showed reduced reactivity, operating at higher reaction temperatures. Allylic carbonates, as well as harder nitrogen nucleophiles, such as amine or sulfonamides, were unreactive [94].

A detailed comparative study of the allylic substitution in allylacetates by amines has been carried out for the two related mixed donor ligand systems **113** and **114** shown in Fig. 2.19. The P–C system shows dramatically lower activity than N–P, which has been attributed to the decreased electrophilicity of the coordinated allyl group, originating from the strongly σ-donating NHC and P donors, compared to N and P donors.

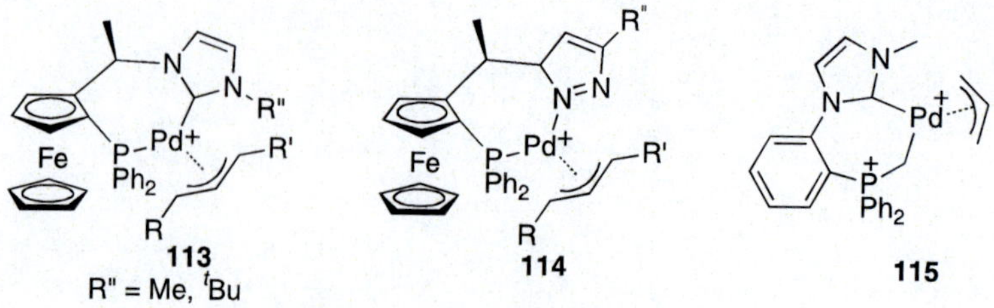

Fig. 2.19 Allylic substitutions with mixed donor Pd complexes

Enantioselectivity (which is linked to the regioselectivity of the attack of the nucleophile to the coordinated allyl) in the allylic amination of 1,3-diphenyl-allyl ethyl carbonate was also very low compared to the P–N system. This was attributed to the comparable *trans*-influence of P and NHC functionalities, leading to poor regioselection of the two allyl termini *trans* to the P and NHC ligands by the nucleophile [95].

Allylic alkylation of 3-acetoxy-1,3-diphenylpropene by sodium dimethylmalonate, catalysed by the Pd-allyl complex **115**, bearing the non-symmetric phosphonium ylide NHC ligand (5 mol%), proceeds to completion with 100% regioselectivity.

Asymmetric allylic alkylation (AAA) has been studied using a variety of novel ligand designs containing one NHC functionality and, usually, a classical heteroatom donor. In addition to the work mentioned earlier with ferrocenylphosphine-functionalised NHC complexes of Pd (**113**), the complexes **116–119** (Fig. 2.20) promote enantioselectivity in AAA reactions. A family of complexes **116** was found to induce up to 90% enantioselectivity in AAA, especially with the more rigid analogues. The nature of the NHC group has an important influence on the stereochemical outcome of the reaction [96]. The AAA of *E*-1,3-diphenylprop-3-en-1-yl acetate by sodium malonate catalysed by a class of imine functionalised NHC complexes **117** gave excellent conversions and enantioselectivities, with *ee* up to 92% in one case. The synthesis of the chiral bidentate ligand uses commercially available chiral *trans*-1,2-diamino-cyclohexane [97].

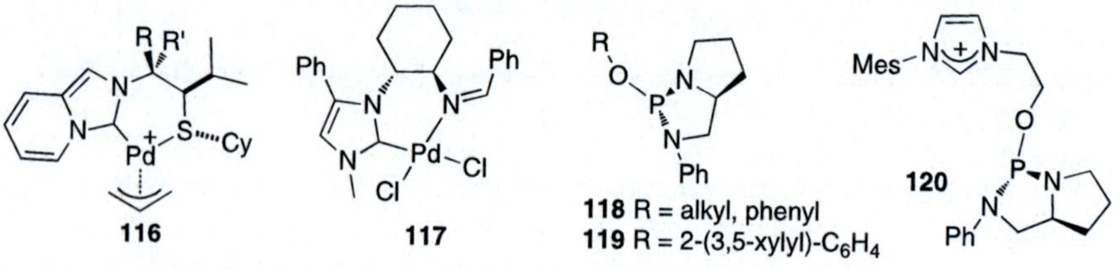

Fig. 2.20 Chiral Pd-NHC catalysts and ligands used for the asymmetric allylic alkylation

A comparative study of the AAA of *E*-1,3-diphenylprop-3-en-1-yl acetate by sodium malonate using the chiral phosphine ligands **118–119** showed that better *ee* were observed if the phosphine ligands were combined with a monodentate NHC. Best enantioselectivities were obtained in a catalytic system generated from [Pd(allyl)(μ-Cl)]$_2$ and **119**/IMes, indicating a synergistic effect of the NHC and phosphine on the enantioselectivity of the reaction. The bidentate **120** showed also this synergism but not as pronounced as the combination of the two monodentate ligands [98].

The Ag complex **121** in the presence of $CuCl_2{\cdot}2H_2O$ or $Cu(OTf)_2{\cdot}C_6H_6$ catalyses the allylic alkylations of allyl phosphates by dialkylzinc reagents with high enantioselectivity (Scheme 2.23). A copper complex **122** which is the precursor to the catalytic species was also isolated and structurally characterised (Figs. 2.21 and 2.22) [99].

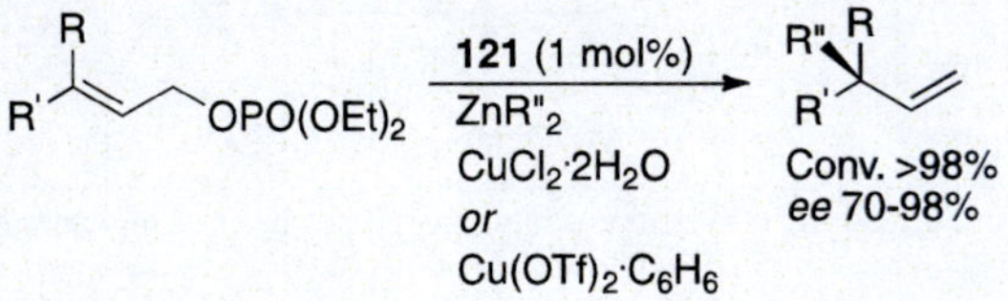

Scheme 2.23 Copper-catalysed enantioselective allylic alkylation of allylphosphates

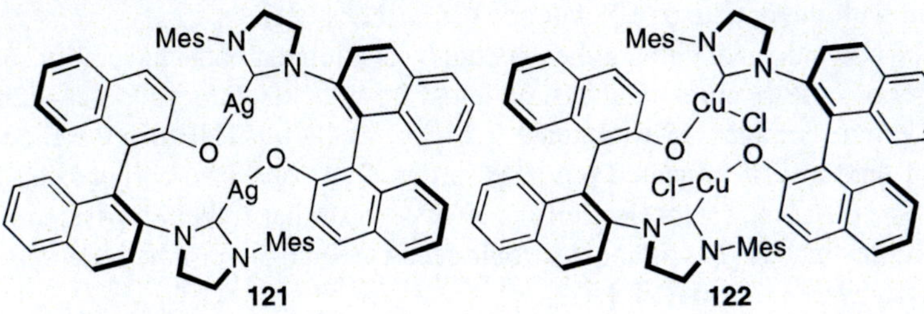

Fig. 2.21 Chiral binap-based catalysts or catalyst precursors for the enantioselective allylic alkylation of allylphosphates

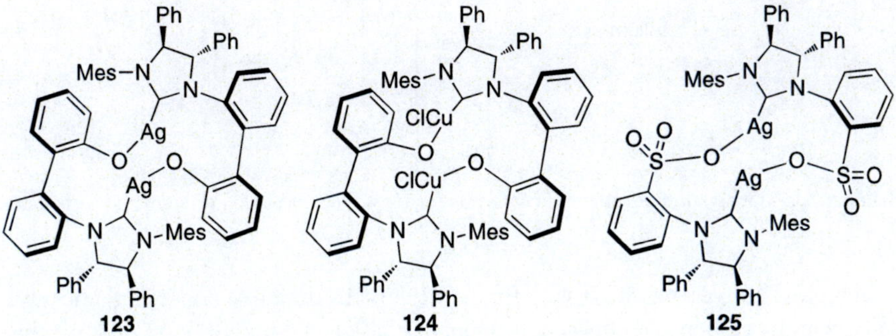

Fig. 2.22 Combination of chiral imidazolidin-2-ylidenes and biphenyl linkers in the chiral catalysts or catalyst precursors for the asymmetric allylic alkylations

The disadvantage of using optically pure binaphthyl building blocks for the synthesis of the NHC ligands in **121** and **122** was addressed by introducing on the imidazolidin-2-ylidene backbone chirality originating from the easily available in optically pure forms 1,2-diphenyl-1,2-diamino-ethane, in combination with racemic biphenyl- rather than optically pure binapthyl-NHC wingtips. The analogous silver and copper complexes **123** and **124** (Fig. 2.22) were formed as single atropoisomers, and catalysed the alkylation of allyl phosphates by dialkylzincs giving excellent chemical yields and enantioselectivities [100].

The AAA of allylphosphates by vinyl aluminium reagents generated *in situ* by hydroalumination of terminal alkynes with DIBAL-H *in situ* has recently been reported (Scheme 2.24).

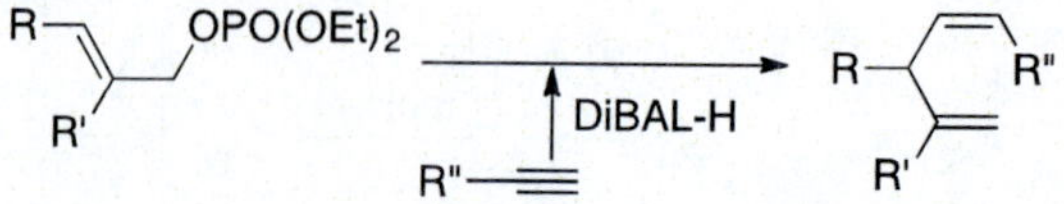

Scheme 2.24 Asymmetric allylic alkylation of allylphosphates by *in situ* generated vinyl Al reagents

The most active catalyst for this reaction is formed *in situ* from **125** (fig. 2.22), in the presence of copper salts. Interestingly, high conversions and *ee* are only observed with the precursor **125** but not **123** or **121** [101].

Allylic alkylations of cinnamyl carbonate by sodium malonate have been studied with a series of ruthenium catalysts, obtained from the azolium salts **126–128** and the ruthenium complex **129** (Scheme 2.25) in MeCN or THF to give moderate yields of mixtures of alkylated products in the allylic and *ipso*-carbons (90:10 to 65:35). The observed regioselectivity is inferior to similar ruthenium systems with non-NHC co-ligands. The stereoelectronic factors which govern the observed regioselectivity were not apparent [102].

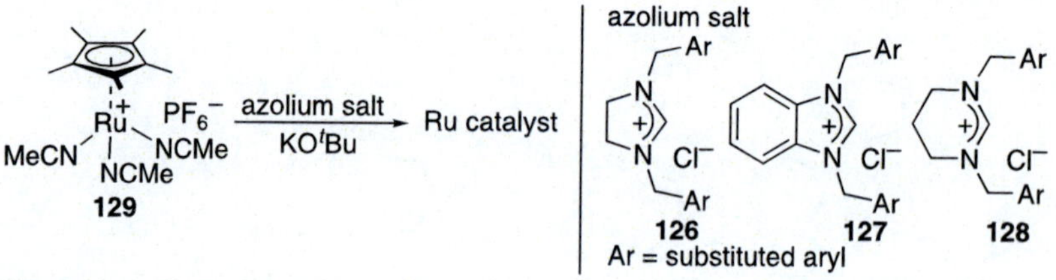

Scheme 2.25 Generation of ruthenium catalysts for the allylic alkylation of cinnamyl carbonate

A regioselective catalytic system for the allylic substitution of non-symmetric allyl carbonates by carbon and nitrogen nucleophiles based on $[^{n}Bu_4N][Fe(NO)(CO)_3]$ and PPh_3 was developed (Scheme 2.26). The high regioselectivity was ascribed to the slow σ-allyl- to π-allyl-isomerisation relative to the rate of substitution. However, the use of high excess of the pro-nucleophile and DMF solvent are drawbacks on the atom efficiency and functional group tolerance of the system.

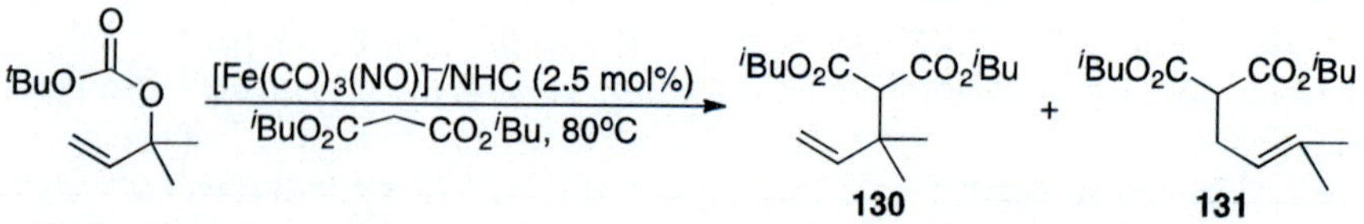

Scheme 2.26 Regioselective iron-catalysed allylic substitution of nonsymmetric allyl carbonates

Replacement of PPh_3 by various NHC ligands, generated *in situ*, resulted in higher conversions by using stoichiometric quantities of the C nucleophile. However, a dependence of the ratio of the isomeric products **130** and **131** on the nature of the NHC ligand was noticed. In general, the bulky SItBu favoured the formation of **130**, while with SIPr variable ratios of both **130** and **131** were obtained, depending on the substituents of the allyl source and the nucleophilicity of the nucleophile; stronger nucleophiles favoured the formation of **130**. The experimental results were rationalised by the assumption that substitution from the initially formed σ-allyl complex (leading to **130**) is possible, either when bulky SI^{t}Bu is present, which disfavours σ-allyl isomerisation *via* a π-allyl intermediate, and/or when the high nucleophilicity entering group is present, resulting in a fast substitution reaction. When SIPr ligands are used, a relatively fast isomerisation process (σ-, π-, σ-allyl) competes with the direct substitution giving rise to substantial amounts of **131** (Scheme 2.27) [103].

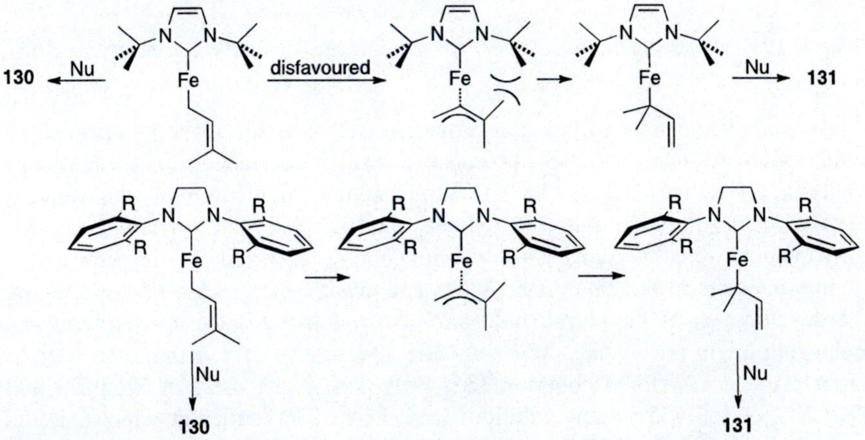

Scheme 2.27 Proposed mechanism to account for the observed regioselectivity in the allylic alkylations catalysed by Fe-NHC complexes. Other co-ligands on Fe are omitted for clarity

In an interesting system, the reduced electrophilicity of the Pd-allyl complexes with NHC ligands, mentioned previously, was exploited in order to promote stoichiometric and catalytic '*umpolung*' allylation reactions of benzaldehyde, by nucleophilic attack of the coordinated allyl group (Scheme 2.28). The best results were obtained with the family of phosphine- or amine-functionalised NHC complexes [104]. Interestingly, [Pd(η^3-C_3H_5)Cl(IMes)] complex does not react with aldehydes. It is plausible that the nucleophilic η^1-allyl Pd intermediate attacks the carbonyl carbon.

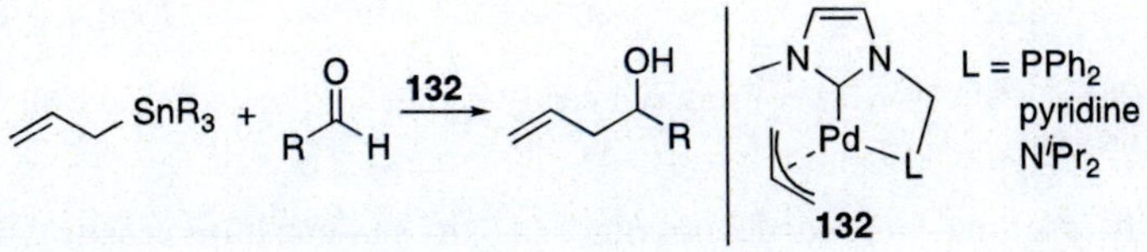

Scheme 2.28 Allylation of benzaldehyde mediated by Pd-NHC complexes

2.11 Additions to Conjugated Enones and Related Reactions

The Cu^{II} (for example $Cu(OTf)_2$) catalysed addition of dialkylzincs or Grignards to α,β-unsaturated carbonyl compounds was found to be accelerated in the presence of *N*-heterocyclic carbenes, generated *in situ* (especially the strongly σ-donating SIMes), or formed by transmetallation from Ag-carbene complexes to copper (Scheme 2.29). For the addition to cyclohex-2-enone, high conversions (>90%) at low catalyst loading (2–5 mol%) were obtained at temperatures –20°C to rt within 5–10 min [105]. The ligand acceleration was attributed to the stabilisation of a putative Cu^{III} transition state by the strongly σ-donating NHC.

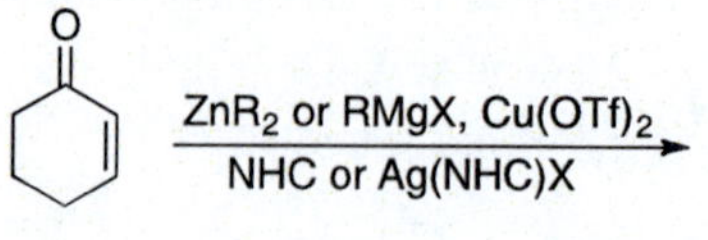

Scheme 2.29 Addition to α,β-unsaturated carbonyls mediated by Cu-NHC complexes

Asymmetric versions of this transformation were also developed by using chiral imidazolium pro-ligands as NHC precursors, or silver transmetallation methodology with chiral NHC ligands (Fig. 2.23) [106]. Imidazolium salts with chiral *N*-substituents (**132**) or imidazolidinium salts with chirality at the backbone of the heterocycle (**133**) gave quantitative conversions at –78°C with good *ee* (58% and 70% respectively).

Improvement in the catalyst activities and enantioselectivities was realised by the development of the chiral, bidentate alkoxy-functionalised imidazolium and imidazolidinium pro-ligands (**134** and **136**). **134**, after deprotonation, was used to prepare the well-defined complex **135**. Both **136** in the presence of BuLi and $Cu(OTf)_2$ or **135** without any additional co-reagents were efficient catalysts in the asymmetric 1,4 addition of dialkylzincs and Grignards to cyclohexen-2-one giving higher *ee* (83% at rt and 51% at –30°C, respectively) [107, 108].

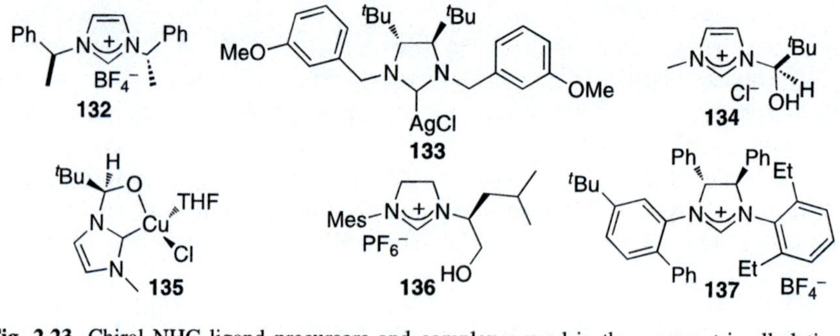

Fig. 2.23 Chiral NHC ligand precursors and complexes used in the asymmetric alkylation of conjugated enones

Recently, attempts were made to replace the air- and moisture-sensitive zinc and magnesium reagents in the copper-catalysed asymmetric conjugate addition, with easier to handle air-stable carbosilanes. Aryl- or alkenyl-trifluorosilanes, in the presence of the fluoride source tris(dimethylamino)sulfonium difluorotrimethylsilicate (TASF, generating *in situ* the nucleophilic aryl- or alkenyl-tetrafluorosilicates) gave good to excellent yields (63–97%) and *ee* (47–97%) of the 1,4 addition products to cyclic enones. The catalysts were generated *in situ* from the imidazolium salt, NaO^{t}Bu and CuBr at rt. In this study, evaluation of the efficiency of various chiral (*C2* and non-*C2* symmetric) NHC-pro-ligands showed that the non-*C2* analogue **137** gave the best yields and *ee* [109].

Asymmetric conjugate addition of dialkyl or diaryl zincs for the formation of all carbon quaternary chiral centres was demonstrated by the combination of the chiral **123** and $Cu(OTf)_2{\cdot}C_6H_6$ (2.5 mol% each component). Yields of 94–98% and *ee* of up to 93% were observed in some cases. Interestingly, the reactions with dialkyl zincs proceed in the opposite enantioselective sense to the ones with diaryl zincs, which has been rationalised by coordination of the opposite enantiofaces of the prochiral enone in the alkyl- and aryl-cuprate intermediates, which precedes the C–C bond formation, and determines the configuration of the product. The copper enolate intermediates can also be trapped by TMS triflate or triflic anhydride giving directly the versatile chiral enolsilanes or enoltriflates that can be used in further transformations (Scheme 2.30) [110].

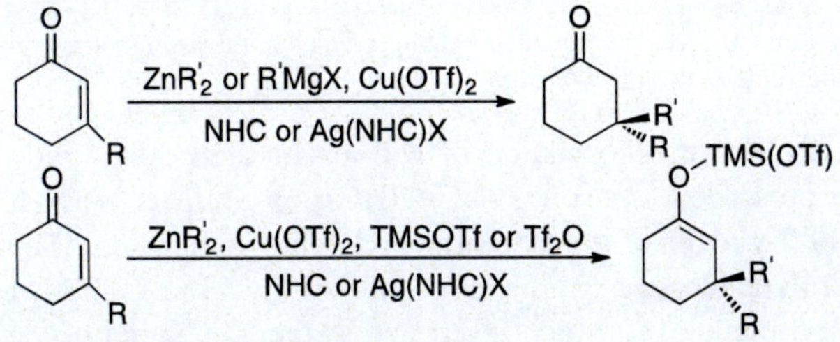

Scheme 2.30 Copper-catalysed asymmetric conjugate addition of organozincs to enones

Similarly, highly enantioselective transformations were reported by using other chiral functionalised non-symmetric (**138**) or *C2* symmetric NHC pro-ligands (**139–140**) (Fig. 2.24) in the presence of organometallic bases and copper salts [111, 112].

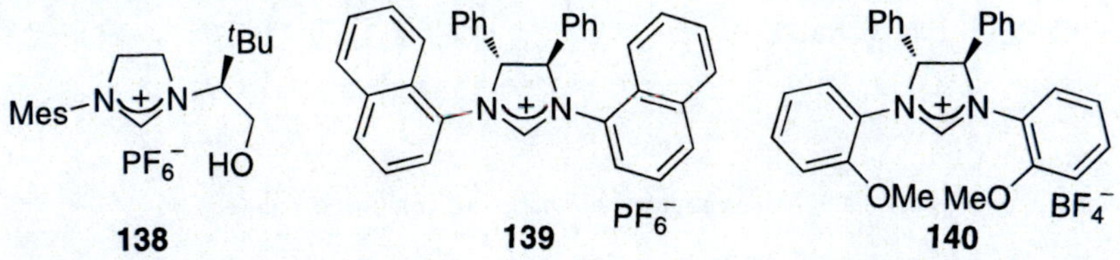

Fig. 2.24 Chiral NHC pro-ligands used in copper-catalysed asymmetric conjugate additions

The asymmetric 1,4-conjugate addition of phenyl boronic acids to cyclohex-2-enone was catalysed by the Pd complex **141** (Fig. 2.25). Good to excellent yields and high *ee* (90–97%) were obtained under mild conditions and low catalyst loadings (rt, 3 mol%)

in THF/H_2O in the presence of base (KOH or K_2CO_3). The labile acetates in **141** are crucial for the success of the reaction as demonstrated by the inactivity of the corresponding iodide complexes. Interestingly, the configuration of the isolated addition product is the opposite to that obtained by the analogous Rh-(binap) catalysed reaction with ligand of the same configuration [113]. Other cyclic enones gave similar results. A plausible catalytic cycle involves reaction of the catalyst with base to form the corresponding hydroxo palladium complex which undergoes transmetallation with the phenyl boronic acids. Insertion of the C=C bond of the enone into the Pd–C bond gives the oxa-allyl species **142** or the enolate **143** which are releasing the product after hydrolysis.

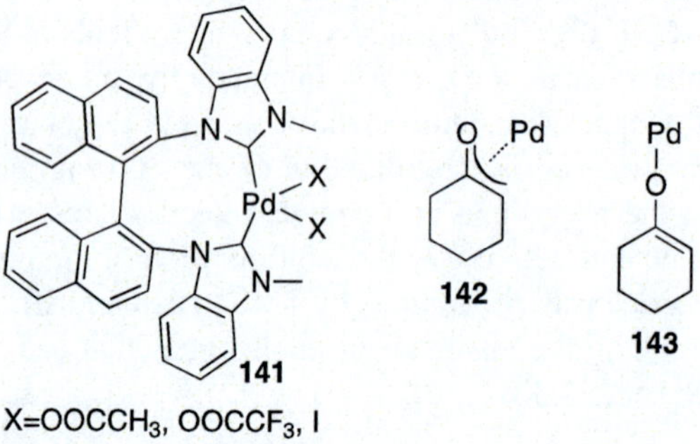

Fig. 2.25 Palladium catalysts and postulated intermediates in the asymmetric conjugate addition of phenyl boronic acids to cyclohex-2-enone

The enantioselective β-borylation of α,β-unsaturated esters with $(Bpin)_2$ was studied in the presence of various [CuCl(NHC)] or $[Cu(MeCN)(NHC)]^+$ (NHC = chiral imidazol-2-ylidene or imidazolidin-2-ylidene) complexes. The reaction proceeds by heterolytic cleavage of the B–B bond of the $(Bpin)_2$, followed by formation of Cu-boryl complexes which insert across the C=C bond of the unsaturated ester. Best yields and *ee* were observed with complex **144**, featuring a non-*C2* symmetric NHC ligand (Scheme 2.31) [114].

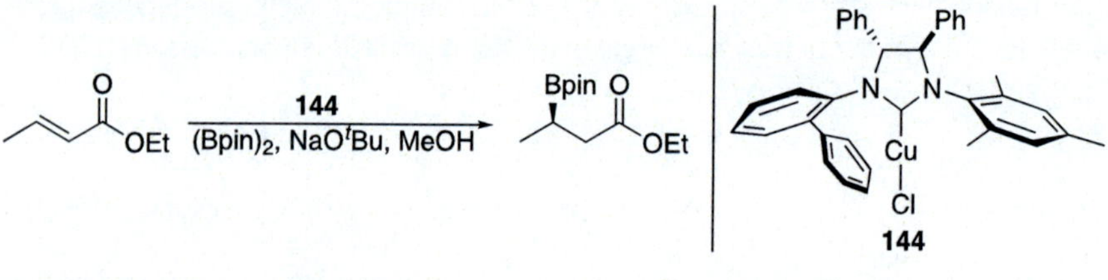

Scheme 2.31 Copper-catalysed asymmetric borylation of conjugated enones

Further insight into the β-borylation reaction of α,β-enones (Scheme 2.32) showed that the reaction can be carried out in THF, and the catalyst generated *in situ* from CuCl (5 mol%), the imidazolium salt (5 mol%), and NaO^{t}Bu (10 mol%), to form the [Cu(O^{t}Bu)(NHC)] as the catalysis initiating species. In this case, stable boron enolate products are formed which need to be hydrolysed by methanol to the ketone products [114].

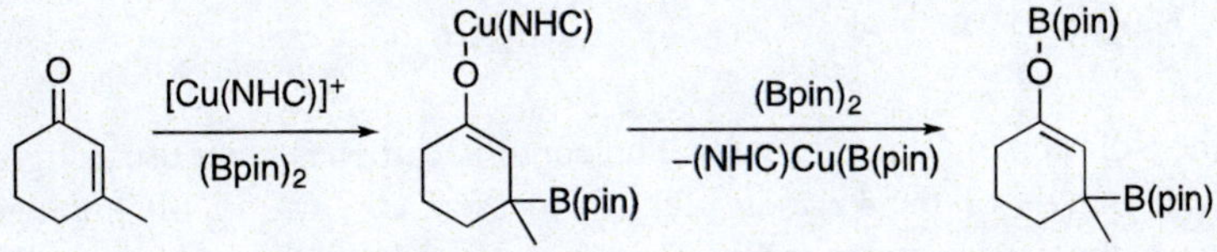

Scheme 2.32 Borylation of enone mediated by Cu-NHC complexes

2.12 Chloroesterification and Chloroacetylation Reactions

The rare chloroesterification of terminal alkynes (*i.e.* the addition of Cl–COOR across the CC triple bond) catalysed by [$RhCl(CO)(PPh_3)_2$] gave *Z*-β-choro-α,β-unsaturated esters (Scheme 2.33) [115].

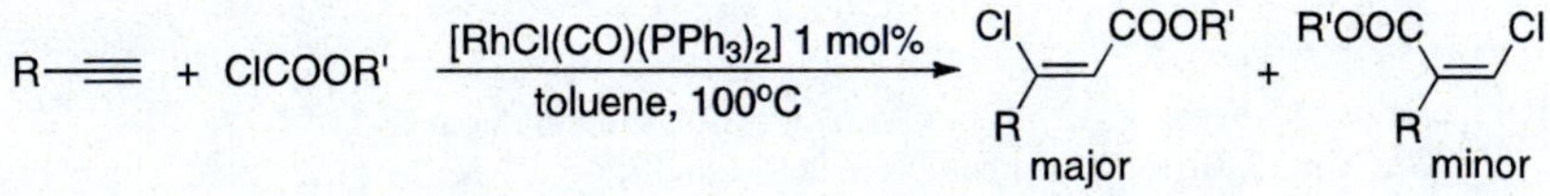

Scheme 2.33 Products in the rhodium catalysed-chroroesterification of alkynes

The chloroesterification of terminal alkynes or conjugated enynes by ClCOOMe was catalysed by complex [RhCl(IPr)(1,5-cod)] (1 mol%) under similar conditions as the [$RhCl(CO)(PPh_3)_2$]. High conversions (70–80%) were obtained with high regio- and stereo-selectivity. Here too, Markovnikov, *Z*-additions products were dominant. Under similar conditions the complex [$RhCl(1,5\text{-cod})(PPh_3)$] gave lower conversions (*ca.* 20%) [116].

Chloroacylation of terminal aryl, alkyl or alkenyl alkynes [*i.e.* the addition of RC(=O)–Cl across the CC triple bond] with aromatic acyl chlorides was catalysed by [IrCl(cod)(IPr)] (5 mol%) in good conversions (70–94%) in toluene (90°C, 20 h). *Z*-addition products were observed only. Internal alkynes were unreactive. Surprisingly, a phosphine/[$Ir(\mu\text{-}Cl)(1,5\text{-cod})]_2$ system under the same conditions provides decarbonylation products (Scheme 2.34) [117].

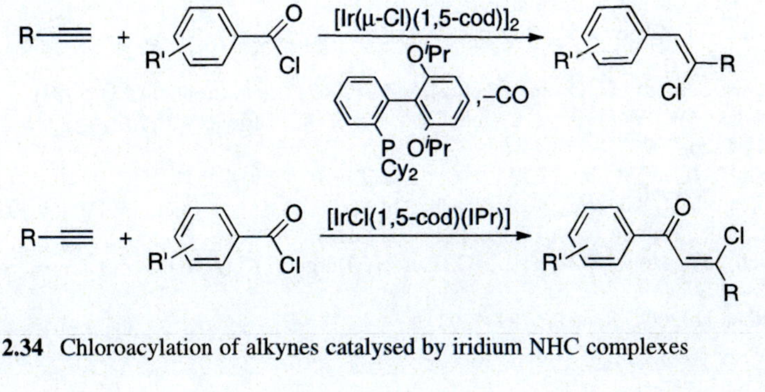

Scheme 2.34 Chloroacylation of alkynes catalysed by iridium NHC complexes

2.13 Conclusion

Since the isolation of relatively stable nucleophilic carbenes, their use as ligands for almost all metals of the Periodic Table and their applications in homogeneous catalysis have risen exponentially. This is attested by the increasing number of researchers from diverse backgrounds that are entering this area of study. The deeper understanding of the steric and electronic characteristics of the carbene ligands, and the way they interact with the metal in a catalytic species promise new developments in the rational design of more active, selective and stable catalysts for numerous applications [118]. We believe that exciting developments in this area will appear in the near future.

References

1. Praetorius JM, Wang R, Crudden CM (2009) Eur J Inorg Chem 13:1746–1751
2. Yu X-Y, Sun H, Brian PO, James BR (2009) Eur J Inorg Chem 13:1752–1758
3. Huang J, Stevens ED, Nolan SP (2000) Organometallics 19:1194–1197
4. Hillier AC, Lee HM, Stevens ED, Nolan SP (2001) Organometallics 20:4246–4252
5. Herrmann WA, Frey GD, Herdtweck E, Steinbeck M (2007) Adv Synth Catal 349:1677–1691
6. Allen DP, Crudden CM, Calhoun LA, Wang R, Decken A (2005) J Organomet Chem 690:5736–5746
7. Baskakov D, Herrmann WA, Herdtweck E, Hoffmann SD (2007) Organometallics 26:626–632
8. Herrmann WA, Baskakov D, Herdtweck E, Hoffmann SD, Bunlaksananusorn T, Rampf F, Rodefeld L (2006) Organometallics 25:2449–2456
9. Lee HM, Smith DC Jr, He Z, Stevens ED, Yi CS, Nolan SP (2001) Organometallics 20:794–797
10. Dinger MB, Mol JC (2003) Eur J Inorg Chem 2827–2833
11. Dharmasena UL, Foucault HM, dos Santos EN, Fogg DE, Nolan SP (2005) Organometallics 24:1056–1058
12. Grasa GA, Moore Z, Martin KL, Stevens ED, Nolan SP, Paquet V, Lebel HJ (2002) J Organomet Chem 658:126–131
13. Jazzar RFR, Macgregor SA, Mahon MF, Richards SP, Whittlesey MK (2002) J Am Chem Soc 124:4944–4945
14. Edwards MG, Jazzar RFR, Paine BM, Shermer DJ, Whittlesey MK, Williams JMJ, Edney DD (2004) Chem Commun 90–91
15. Chantler VL, Chatwin SL, Jazzar RFR, Mahon MF, Saker O, Whittlesey MK (2008) Dalton Trans 2603–2614
16. Vazquez-Serrano LD, Owens BT, Buriak JM (2002) Chem Commun 2518–2519
17. Sprengers JW, Wassenaar J, Clement ND, Cavell KJ, Elsevier CJ (2005) Angew Chem Int Ed 44:2026–2029
18. Jurcik V, Nolan SP, Cazin CSJ (2009) Chem Eur J 15:2509–2511
19. Heckenroth M, Kluser E, Neels A, Albrecht M (2007) Angew Chem Int Ed 46:6293–6296
20. Nanchen S, Pfaltz A (2006) Chem Eur J 12:4550–4558
21. Powell MT, Hou DR, Perry MC, Cui X, Burgess K (2001) J Am Chem Soc 123: 8878–8879
22. Gunther, H (2004) Synlett 1789–1793

23. Perry MC, Cui X, Powell MT, Hou DR, Reibenspies JH, Burgess K (2003) J Am Chem Soc 125:113–123
24. Fan Y, Cui X, Burgess K, Hall B (2004) J Am Chem Soc 126:16688–16689
25. Corma A, Puebla-Gutierrez E, Iglesias M, Monge A, Perez-Ferreras S, Sanchez F (2006) Adv Synth Catal 348:1899–1907
26. Hillier AC, Lee HM, Stevens ED, Nolan SP (2001) Organometallics 20:4246–4252
27. Gnanamgari D, Moores A, Rajaseelan E, Crabtree RH (2007) Organometallics 26: 1226–1230
28. Gnanamgari D, Sauer ELO, Schley ND, Butler C, Incarvito CD, Crabtree RH (2009) Organometallics 28:321–325
29. Danopoulos AA, Winston S, Motherwell WB (2002) Chem Commun 1376–1377
30. Peris E, Mata JA, Crabtree RH, Poyatos M (2003) Organometallics 22:1110–1114
31. Crabtree RH, Faller JW (2002) Organometallics 21:3596–3604
32. Peris E, Mas-Marza E, Sanau M (2005) Inorg Chem 44:9961–9967
33. Stylianides N, Danopoulos AA, Tsoureas N (2005) J Organomet Chem 690:5948–5958
34. Wang CY, Fu CF, Liu YH, Peng SM, Liu ST (2007) Inorg Chem 46:5779–5786
35. Turkmen H, Pape T, Hahn FE, Cetinkaya B (2008) Eur J Inorg Chem 5418–5423
36. Yang L, Kruger A, Neels A, Albrecht M (2008) Organometallics 27:3161–3171
37. Danopoulos AA, Fiddy SG, Evans J, Neisius T, Newton MA, Tsoureas N, Tulloch AAD (2007) Chem Eur J 13:3652–3659
38. Miecznikowski JR, Crabtree RH (2004) Organometallics 23:629–631
39. Kuhl S, Schneider R, Fort Y (2003) Organometallics 22:4184–4186
40. Peris E, da Costa AP, Viciano M, Sanau M, Merino S, Tejeda J, Royo B (2008) Organometallics 27:1305–1309
41. Guillena G, Ramon DJ, Yus M (2007) Angew Chem Int Ed 46:2358–2364
42. Crabtree RH, Faller JW (2002) Organometallics 21:3596–3604
43. Castarlenas R, Esteruelas MA, Onate E (2008) Organometallics 27:3240–3247
44. Berthon-Gelloz G, Buisine O, Briere JF, Michaud G, Sterin S, Mignani G, Tinant B, Declercq JP, Chapon D, Marko IE (2005) J Organomet Chem 690:6156–6168
45. Sprengers JW, Mars MJ, Duin MA, Cavell KJ, Elsevier CJ (2003) J Organomet Chem 679:149–152
46. De Bo G, Berthon-Gelloz G, Tinant B, Marko IE (2006) Organometallics 25:1881–1890
47. Poyatos M, Mas-Marza E, Mata JA, Sanau M, Peris E (2003) Eur J Inorg Chem 1215–1221
48. Mas-Marza E, Sanau M, Peris E (2005) Inorg Chem 44:9961–9967
49. Jimenez MV, Perez-Torrente JJ, Bartolome MI, Gierz V, Lahoz FJ, Oro LA (2008) Organometallics 27:224–234
50. Chaulagain MR, Mahandru GM, Montgomery J (2006) Tetrahedron 62:7560–7566
51. Díez-González S, Nolan SP (2008) Acc Chem Res 41:349–358
52. Díez-González S, Scott NM, Nolan SP (2006) Organometallics 25:2355–2358
53. Kaur H, Zinn FK, Stevens ED, Nolan SP (2004) Organometallics 23:1157–1160
54. Díez-González S, Stevens ED, Scott NM, Petersen JL, Nolan SP (2008) Chem Eur J 14:158–168
55. Faller JW, Fontaine P (2006) Organometallics 25:5887–5893
56. Cesar V, Bellemin-Laponnaz S, Wadepohl H, Gade LH (2005) Chem Eur J 11:2862–2873
57. Schneider N, Kruck M, Bellemin-Laponnaz S, Wadepohl H, Gade LH (2009) Eur J Inorg Chem 493–500
58. Thomas RL, Souza FES, Marder TB (2001) J Chem Soc Dalton Trans 1650–1656
59. Mann G, John KD, Baker RT (2000) Org Lett 2:2105–2108
60. Baker RT, Nguyen P, Marder TB, Westcott SA (1995) Angew Chem Int Ed 34:1336–1338
61. Dai C, Robins EG, Scott AJ, Clegg W, Yufit DS, Howard JAK, Marder TB (1998) Chem Commun 1983–1984
62. Ramirez J, Corberan R, Sanau M, Peris E, Fernandez E (2005) Chem Commun 3056–3058

63. Corberan R, Ramirez J, Poyatos M, Peris E, Fernandez E (2006) Tetrahedron Asymm 17:1759–1762
64. Lillo V, Mata J, Ramirez J, Peris E, Fernandez E (2006) Organometallics 25:5829–5831
65. Lillo V, Mas-Marza E, Segarra AM, Carbo JJ, Bo C, Peris E, Fernandez E (2007) Chem Commun 3380–3382
66. Peris E, Corberan R, Lillo V, Mata J, Fernandez E (2007) Organometallics 26:4350–4353
67. Laitar DS, Tsui EY, Sadighi JP (2006) Organometallics 25:2405–2408
68. Lillo V, Fructos MR, Ramirez JB, Maseras F, Diaz-Requejo MM, Perez PJ, Fernandez E (2007) Chem Eur J 13:2614–2621
69. Lillo V, Mata JA, Segarra AM, Peris E, Fernandez E (2007) Chem Commun 2184–2186
70. Khramov DM, Rosen EL, Vu PD, Lynch VM, Bielawski CW (2008) Tetrahedron 64:6853–6862
71. Lee Y, Hoveyda AH (2009) J Am Chem Soc 131:3160–3161
72. Beller M, Seayad J, Tillack A, Jiao H (2004) Angew Chem Int Ed 43:3368–3398
73. Muller TE, Yus M, Mizuki T (2008) Chem Rev 108:3795–3892
74. Takaki K, Koizumi S, Yamamoto Y, Komeyama K (2007) Tetrahedron Lett 48:4643–4644
75. Field LD, Messerle BA, Vuong KQ, Turner P (2005) Organometallics 24:4241–4250
76. Burling S, Field LD, Li HL, Messerle BA, Turner P (2003) Eur J Inorg Chem 3179–3184
77. Bauer EB, Andavan GTS, Hollis TK, Rubio RJ, Cho J, Kuchenbeiser GR, Helgert TR, Letko CS, Tham FS (2008) Org Lett 10:1175–1178
78. Cho J, Hollis TK, Helgert TR, Valente EJ (2008) Chem Commun 5001–5003
79. Gischig S, Togni A (2005) Eur J Inorg Chem 4745–4754
80. Houghton J, Dyson G, Douthwaite RE, Whitwood AC, Kariuki BM (2007) Dalton Trans 3065–3073
81. Munro-Leighton C, Delp SA, Blue ED, Gunnoe TB (2007) Organometallics 26:1483–1493
82. Munro-Leighton C, Delp SA, Alsop NM, Blue ED, Gunnoe TB (2008) Chem Commun 111–113
83. Kinder RE, Zhang Z, Widenhoefer RA (2008) Org Lett 10:3157–3159
84. Malyshev DA, Scott NM, Marion N, Stevens ED, Ananikov VP, Beletskaya IP, Nolan SP (2006) Organometallics 25:4462–4470
85. Delp SA, Munro-Leighton C, Goj LA, Ramirez MA, Gunnoe TB, Petersen JL, Boyle PD (2007) Inorg Chem 46:2365–2367
86. Zhang Z, Widenhoefer RA (2008) Org Lett 10:2079–2081
87. Marion N, Ramon RS, Nolan SP (2009) J Am Chem Soc 13:448–449
88. De Fremont P, Singh R, Stevens ED, Petersen JL, Nolan SP (2007) Organometallics 26:1376–1385
89. Zhang Z, Du Lee S, Fisher AS, Widenhoefer RA (2009) Tetrahedron 65:1794–1798
90. Viciu MS, Stevens ED, Petersen JL, Nolan SP (2004) Organometallics 23:3752–3755
91. Biffis A, Tubaro C, Buscemi G, Basato M (2008) Adv Synth Catal 350:189–196
92. Marion N, Diez-Gonzalez S, de Fremont P, Noble AR, Nolan SP (2006) Angew Chem Int Ed 45:3647–3650
93. Sato Y, Yoshin T, Mori M (2003) Org Lett 31–33
94. Sato Y, Yoshino T, Mori M (2005) J Organomet Chem 690:5753–5758
95. Visentin F, Togni A (2007) Organometallics 26:3746–3754
96. Rosenblade SJ, Ros A, Monge D, Alcarazo M, Alvarez E, Lassaletta JM, Fernandez R (2007) Organometallics 26:2570–2578
97. Bonnet LG, Douthwaite RE, Kariuki BM (2003) Organometallics 22:4187–4189
98. Toselli N, Martin D, Buono G (2008) Org Lett 10:1453–1456
99. Lee K, Brown MK, Hird AW, Hoveyda AH (2006) J Am Chem Soc 128:7182–7184
100. Van Veldhuizen JJ, Campbell JE, Giudici RE, Hoveyda AH (2005) J Am Chem Soc 127:6877–6882
101. Lee Y, Akiyama K, Gillingham DG, Brown MK, Hoveyda AH (2008) J Am Chem Soc 130:446–447

102. Yasar S, Ozdemir I, Cetinkaya B, Renaud JL, Bruneau C (2008) Eur J Org Chem 2142–2149
103. Plietker B, Dieskau A, Moews K, Jatsch A (2008) Angew Chem Int Ed 47:198–201
104. Barczak NT, Grote RE, Jarvo ER (2007) Organometallics 26:4863–4865
105. Fraser PK, Woodward S (2001) Tetrahedron Lett 42:2747–2749
106. Winn CL, Guillen F, Pytkowicz J, Roland S, Mangeney P, Alexakis A (2005) J Organomet Chem 690:5672–5695
107. Arnold PL, Rodden M, Davis KM, Scarisbrick AG, Blake AJ, Wilson C (2004) Chem Commun 1612–1613
108. Clavier H, Coutable L, Toupet L, Guillemin J-C, Mauduit M (2005) J Organomet Chem 690:5237–5254
109. Lee K, Hoveyda AH (2009) J Org Chem 74:4455–4462
110. Lee K, Brown MK, Hird AW, Hoveyda AH (2006) J Am Chem Soc 128:7183–7184
111. Martin D, Kehrli S, d'Augustin M, Clavier H, Mauduit M, Alexakis A (2006) J Am Chem Soc 128:8416–8417
112. Matsumoto Y, Yamada K, Tomioka K (2008) J Org Chem 73:4578–4581
113. Zhang T, Shi M (2008) Chem Eur J 14:3759–3764
114. Lillo V, Prieto A, Bonet A, Diaz-Requejo MM, Ramirez J, Perez PJ, Fernandez E (2009) Organometallics 28:659–662
115. Hua R, Shimada S, Tanaka M (1998) J Am Chem Soc 120:12365–12366
116. Sim H, Baek JY (2008) Synlett 551–554
117. Iwai T, Fujihara T, Terao J, Tsuji Y (2009) J Am Chem Soc 131:6668–6669
118. For a recent review on Late Transition Metal-NHC complexes and catalysis see: Díez-González S, Marion N, Nolan SP (2009) Chem Rev 109:3612–3676

Chapter 3
N-Heterocyclic Carbene Complexes in Olefin Metathesis

Xinjun Luan, Reto Dorta, Anita Leitgeb, Christian Slugovc, Sascha Tiede, and Siegfried Blechert

Abstract Olefin metathesis is now a synthetic tool found ubiquitously in various fields involving synthesis. Of its many variations, three are prominently used: (1) catalytic ring closing metathesis (RCM) is an extremely powerful method for the construction of carbon-carbon double bonds in organic chemistry; (2) ring opening metathesis polymerisation (ROMP) where polymers are formed by use of the energy released from cyclic strain; and (3) cross metathesis (CM) where non-cyclic partners are coupled through C–C double bond formation. These important transformations and variations on these themes mediated by second generation ruthenium complexes bearing a NHC ligand will be presented in the following sections.

Authors of Section 3.1: Xinjun Luan and Reto Dorta – Section 3.2: Anita Leitgeb and Christian Slugovc – Section 3.3: Sascha Tiede and Siegfried Blechert.

R. Dorta (✉)
Organic Chemistry Institute, University of Zurich, Winterthurerstrasse 190, 8057 Zurich, Switzerland
e-mail: dorta@oci.uzh.ch

C. Slugovc (✉)
Institute for Chemistry and Technology of Materials (ICTM), Graz University of Technology, 8010 Graz, Austria
e-mail: slugovc@tugraz.at

S. Blechert (✉)
Technische Universität Berlin, Institut für Chemie, Sekretariat C3, Straße des 17, Juni 135 10623, Berlin, Germany
e-mail: blechert@chem.tu-berlin.de

C. S. J. Cazin (ed.), *N-Heterocyclic Carbenes in Transition Metal Catalysis and Organocatalysis*, Catalysis by Metal Complexes 32,
DOI 10.1007/978-90-481-2866-2_3,

3.1 Ru–NHC Complexes in Ring Closing Metathesis

3.1.1 Introduction

3.1.1.1 Scope

This section provides a general overview on the influence of *N*-heterocyclic carbenes (NHCs) as ligands for ruthenium catalysts in the field of ring closing metathesis (RCM). Since their introduction, these second-generation ruthenium metathesis catalysts have undergone extensive modifications and the current account puts particular focus on how NHC ligand alterations influence reactivity and stability of the catalysts. Due to the rapid development in this field and the tremendous research efforts of the last decade, this account only highlights specific examples, mostly focusing on NHC systems with better catalytic properties or that can be used in novel applications, and NHCs with unusual ligand frameworks.

3.1.1.2 Ring Closing Metathesis

Ring closing metathesis (RCM) reactions promoted by transition metal catalysts have gained enormous importance in synthetic organic chemistry [1, 2]. RCM reactions are achieved by intramolecular cyclisations of acyclic dienes to form unsaturated cyclic systems. The process is thought to proceed according to the catalytic cycle outlined in Scheme 3.1 (the Chauvin mechanism) [3]. The key intermediate is a metallacyclobutane, which can undergo cycloreversion either towards the products or back to the starting materials. The driving force for this transformation is the release of the highly volatile olefin, commonly ethylene, from the reaction mixture.

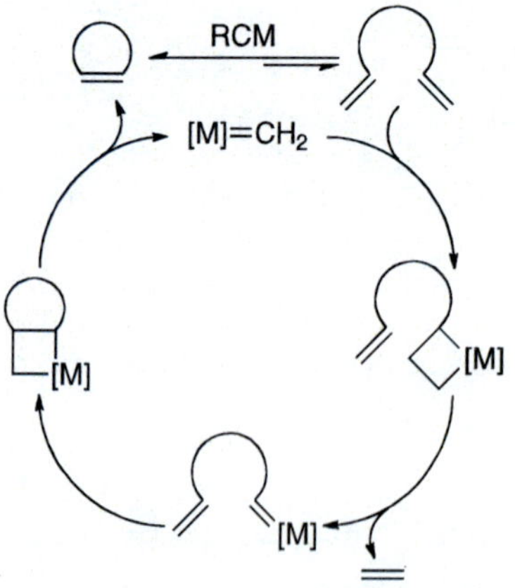

Scheme 3.1 Mechanism for ring closing metathesis

Some selected examples of standard metathesis transformations are shown in Scheme 3.2. These reactions are commonly considered as benchmark reactions when testing the activity of new metathesis catalysts.

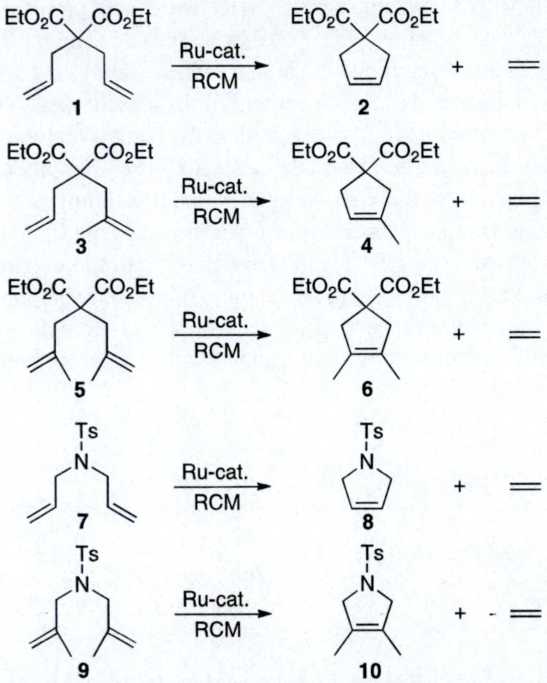

Scheme 3.2 Benchmark transformations in ring closing metathesis

3.1.1.3 Standard Catalysts

Although olefin metathesis had, soon after its discovery, been successfully used in industrial chemistry and polymer chemistry [2], early examples for organic chemistry applications (including RCM) utilised poorly defined and inefficient mixtures of pre-catalyst [4, 5]. The gain in interest of RCM as a prominent transformation in synthetic organic chemistry is mainly due to the introduction of the well-defined Mo-based complex of Schrock (**11**) [6, 7] and the Ru-based complexes of Grubbs (**12** and **13**) [8–10] (Fig. 3.1). These discoveries have triggered extensive follow-up work. Nowadays, RCM using catalysts **11–13** and their derivatives gives access to an impressive range of unsaturated carbo- and hetero-cyclic ring systems.

Complex **13**, which is known as Grubbs I, is generally less reactive than Schrock's molybdenum catalyst **11**, but has greater functional group tolerance and simplified handling characteristics [11, 12]. Therefore, enormous efforts have been made towards developing modified ruthenium-based catalysts that have increased reactivity and/or even better stability. One of the most important breakthroughs in this regard was the replacement of one tricyclohexylphosphine ligand with one *N*-heterocyclic carbene (NHC). This not only resulted in a more stable catalyst, but it also gave a more reactive and active RCM ruthenium catalyst [13]. This second-generation ruthenium catalyst can be easily prepared by replacing one labile phosphine ligand of the Grubbs I pre-catalyst **13** with a non-labile NHC ligand, and prototypical catalysts **14** [13] and **15** [14, 15] are obtained (Fig. 3.2). For reasons that still remain unclear, catalyst **14** containing an *N*-heterocyclic carbene with a saturated backbone (SIMes), outperforms its congener **15** that incorporates its unsaturated NHC counterpart (IMes). Both these complexes show significantly better catalytic behaviour than Grubbs I [13, 14]. It should be noted that catalyst **14** is commonly called Grubbs II. Soon after the discovery of these second-generation ruthenium catalysts, phosphine-free NHC-ruthenium complex **16** (Hoveyda-Grubbs catalyst) [16, 17] and derivatives thereof were reported [18, 19] (Fig. 3.3). Catalyst **16** shows similar metathesis activity to **14** and **15**, but is considered to be more robust at higher temperature. In the following section, the performance of selected ruthenium complexes with modified NHC ligands will be evaluated by using catalysts **14** or **16** as reference systems.

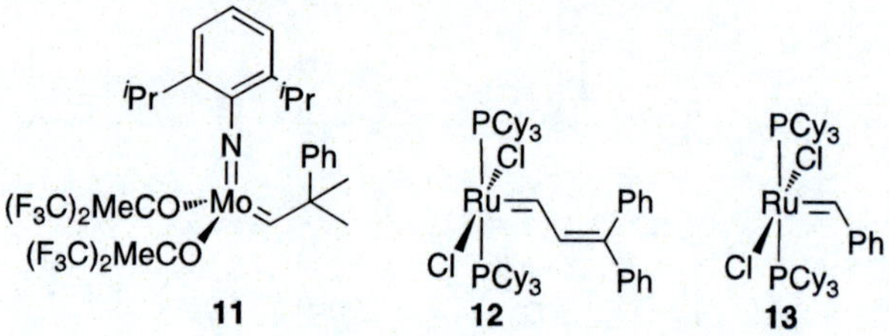

Fig. 3.1 Metathesis catalysts introduced by Schrock (**11**) and Grubbs (**12**, **13**)

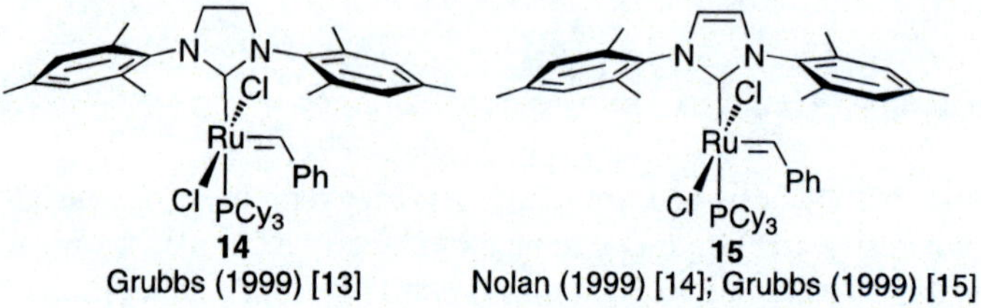

Fig. 3.2 First examples of second-generation NHC–Ru catalysts

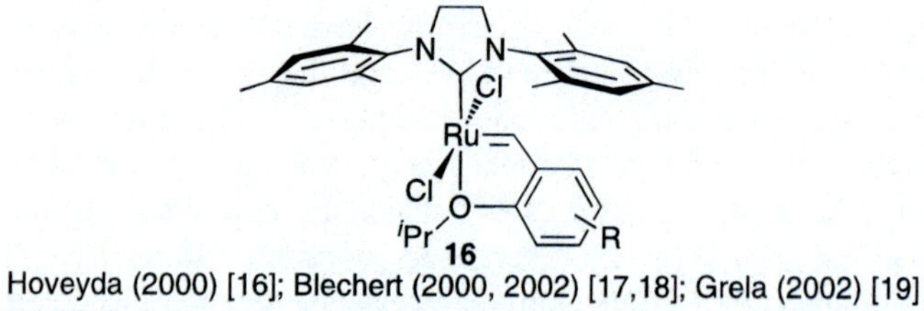

Fig. 3.3 Examples of phosphine-free second-generation NHC–Ru catalysts

3.1.2 NHC Modifications in Second-Generation Ru-Catalysts

3.1.2.1 Five-Membered Central *N*-Heterocycle

The majority of modifications to NHC–Ru metathesis catalysts have pertained to relatively simple changes of the steric and electronic demands of the NHC ligand. Such modifications of the NHC ligands can relatively easily be introduced by changing the side-arm substituents or by modifying the central heterocyclic ring. The changes to the *N*-heterocyclic carbene ligand can render the Ru-catalyst more active, but can just as easily lead to less active Ru-species. That small modifications can indeed lead to rather large reactivity differences had already been observed for the original SIMes- respectively IMes-derived complexes **14** and **15** [13–15].

Modifications of the Side Chains

In the course of studying a large number of examples where the side chains of the imidazol- and imidazolidin-2-ylidene were altered, several research groups found that NHCs bearing exclusively alkyl side chains did not provide catalysts with better characteristics when compared to SIMes- and IMes-derived systems **14** and **15**. While Herrmann and co-workers showed that an unsaturated NHC bearing cyclohexyl wing tips could be incorporated into a second-generation catalyst that was active in metathesis [20–23], more recent studies showed that similar complexes were either very difficult to prepare or were unstable and showed dramatically decreased catalytic properties [24–26] (complexes **17–19**, Fig. 3.4).

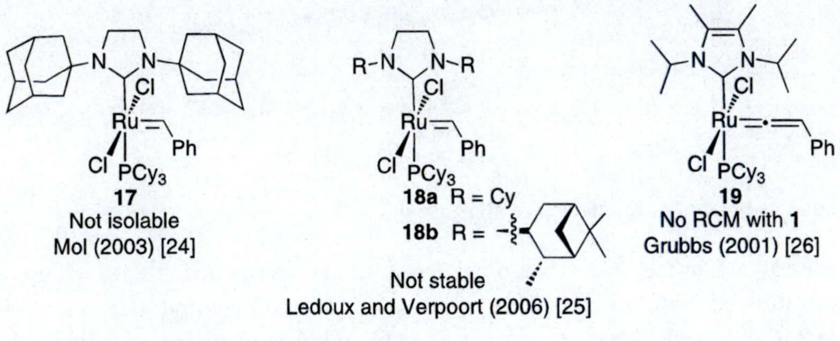

Fig. 3.4 Complexes **17–19** bearing alkyl side chains

However, similar NHC architectures employing aromatic side chains have shown more encouraging results. In 2000, Nolan and co-workers reported the synthesis and characterisation of the NHC–Ru complex **20** bearing a sterically more demanding *N,N'*-bis-[2,6-(di-*iso*-propyl)phenyl]imidazol-2-ylidene (IPr) ligand [27, 28] (Fig. 3.5). Standard RCM substrate **1** was used to test the catalytic performance of **20**. The ring closure was found to be complete after 15 min by using 5 mol% **20** as catalyst at room temperature. Under identical conditions, **15** [which contains the IMes (IMes = *N,N'*-bis-[2,4,6-(trimethyl)phenyl]imidazol-2-ylidene) ligand that has less bulky mesityl sidearm substituents] showed lower conversion (92%). Catalyst **20** also performed better than **15** in the macrocyclisation to give disubstituted olefin **21** (Fig. 3.5) The bulkier nature of the NHC ligand though comes at a price and the authors observed inferior reactivity when trying to ring-close bulky diene substrates **5** and **9** [29].

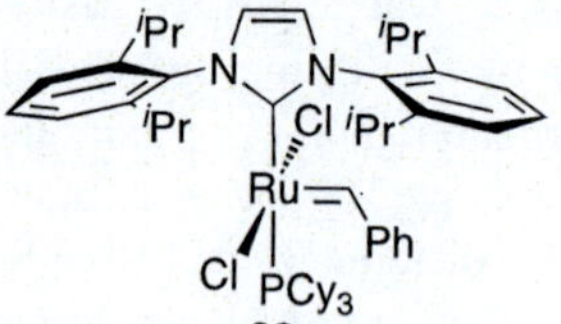

20
Nolan (2000) [28]

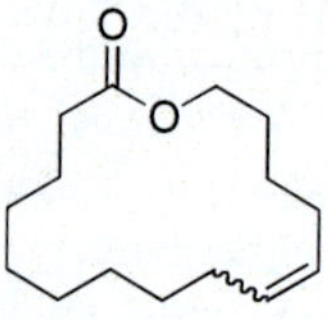

21

Fig. 3.5 Catalyst **20** and the RCM product **21**

In 2001, Fürstner reported the preparation and characterisation of the NHC–Ru complex **22** containing *N,N'*-bis[2,6-(diisopropyl)phenyl]imidazolidin-2-ylidene (SIPr) [29] (Fig. 3.6), which is the congener of complex **20**. Subsequently, Mol and co-workers revealed that complex **22** was a highly active metathesis initiator [30]. More recent comparative studies showed that catalyst **22** could catalyse the RCM of **1** faster than any other NHC–Ru catalyst, while it was not stable enough to obtain complete conversion in the RCM of **3** and was inefficient for the construction of the tetrasubstituted double bond of cyclic olefin **6** [31].

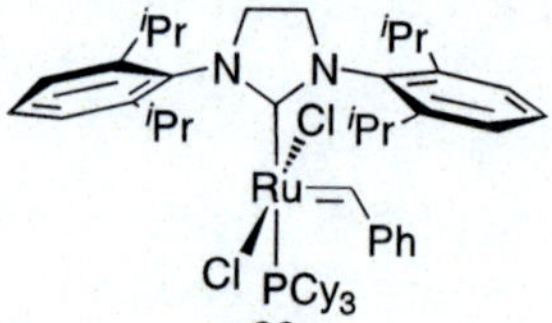

22
Fürstner (2001) [29]

Fig. 3.6 Catalyst **22** bearing the bulky SIPr ligand

Recent work in the Dorta group has led to the synthesis of a family of new and stable, saturated NHC ligands incorporating substituted naphthyl side chains. Precatalysts **23a–g** were prepared as an inseparable mixture of *anti* and *syn* conformers that do not interconvert when the NHC ligand is bound to ruthenium [32–34] (Fig. 3.7). Complexes **23a–g** were then benchmarked against Grubbs II catalyst **14** and two catalysts (**23f** and **23g**) were identified that clearly outperformed Grubbs II (**14**) in the RCM for normal substrates (**1** and **7**). Complex **23b,** which incorporates a sterically less demanding NHC ligand, showed high activity with challenging substrate **5** (97% conversion). In order to understand whether the relative orientation of the naphthyl side chains has an effect on the reactivity of these catalysts, the authors more recently synthesised phosphine-free ruthenium catalysts **24a–b** [35].

Chromatographic workup enabled the separations of *anti* and *syn* isomers and pure ***anti*-24a**, ***syn*-24a** and ***anti*-24b** where obtained. RCM studies with these pre-catalyst and standard substrates **1**, **3**, **7** and **9** indicated that an *anti* arrangement of the naphthyl side chains gave better activity in these RCM reactions.

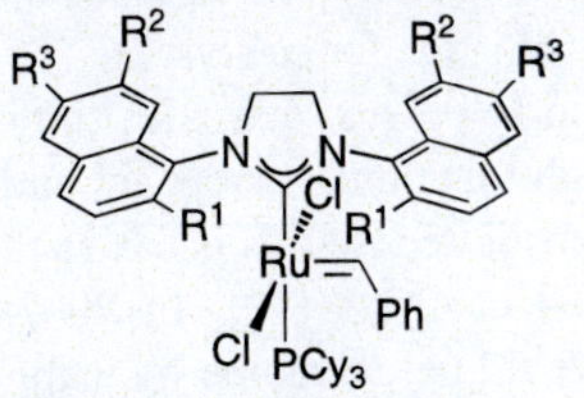

23a, R1 = Me, R2 = R3 = H
23b, R1 = R2 = Me, R3 = H
23c, R1 = iPr, R2 = R3 = H
23d, R1 = R3 = iPr, R2 = H
23e, R1 = iPr, R2 = H, R3 = Ad
23f, R1 = R2 = iPr, R3 = H
23g, R1 = Cy, R2 = R3 = H
Dorta (2008) [32]; (2009) [34]

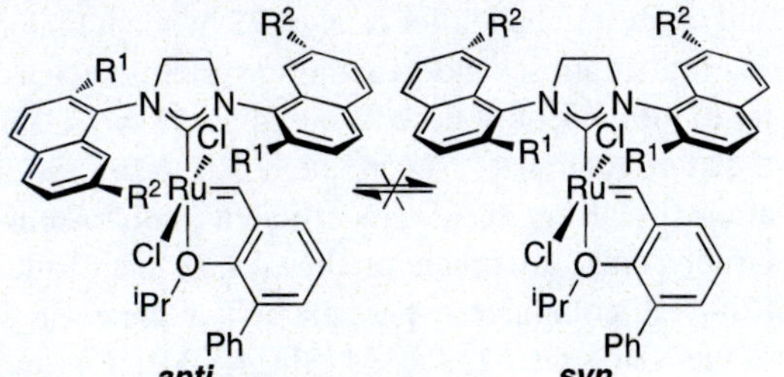

anti ***syn***
24a, R1 = R2 = iPr
24b, R1 = Cy, R2 = H
Dorta (2009) [35]

Fig. 3.7 The NHC–Ru catalysts containing substituted naphthyl side chains

In addition, Blechert and co-workers reported on the highly effective RCM forming tetrasubstituted olefins (such as **6** and **10**) in C_6F_6 by using pre-catalyst **23a** prepared *via* a different procedure [36].

Some attempts have also been made to vary the *para*- or *meta*-substituents on the phenyl ring of the SIMes ligand, however none of the resulting NHC–Ru complexes showed better RCM activity than the original Grubbs II **14** when tested with standard substrates **1** and **7** [37–40].

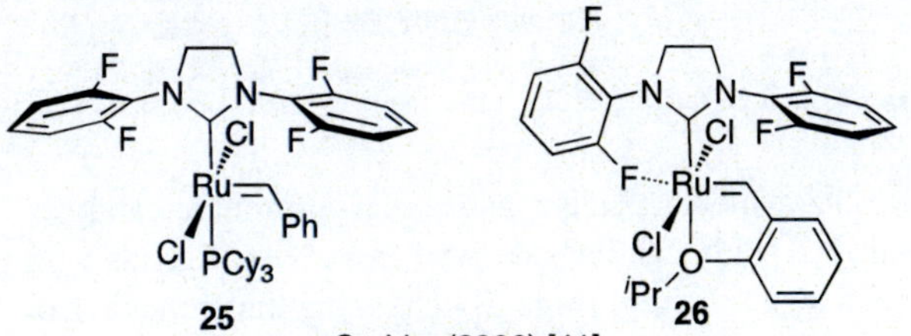

25 **26**
Grubbs (2006) [41]

Fig. 3.8 Catalysts **25** and **26** containing fluorinated NHC

In contrast, substituting the *ortho*-methyl groups of SIMes with ortho-fluoride atoms profoundly alters the catalytic metathesis performance. In 2006, Grubbs and co-workers reported the synthesis of the fluorinated NHC–Ru catalysts **25** and **26** [41] (Fig. 3.8). Catalytic tests in the RCM of **1** to form **2** showed that the phosphine-free catalyst **26** was slower than the standard catalyst **16**, which was consistent with theoretical investigations suggesting the electron-withdrawing fluoride atoms would lead to a decrease in catalyst activity [42]. However, in contrast to the computational prediction, the fluorinated phosphine-containing catalyst **25** demonstrated a significantly increased reaction rate compared to Grubbs II catalyst **14**. X-Ray analysis of **25** and **26** helped the authors understand the discrepancy in activities of **25** and **26**. Indeed, a strong ruthenium-fluoride interaction was observed in complex **26** and as a result, it is believed that the same interaction accelerates the rate-limiting phosphine dissociation and catalyst initiation for catalyst **25**, while it inhibits the initiation for catalyst **26**.

Ruthenium second-generation systems such as **14** and **16** possess activities similar to molybdenum-based complexes **11**, yet display high functional group, air, and moisture tolerance. Nevertheless, RCM to form tetrasubstituted olefins is easier to attain by using the more efficient molybdenum-based catalysts [13]. In 2007, Grubbs and co-workers predicted that this challenge would be overcome by using Ru–NHC complexes where the bulk at the *ortho* positions of the *N*-bound aryl rings of the NHCs were reduced [43] (Fig. 3.9). Following this hypothesis, complexes **27** and **28a–b** were synthesised and tested in the two benchmark transformations to prepare **6** and **10**. All three of the new catalysts performed significantly better than either **14** or **16**. Catalyst **27** converted substrate **5** faster than **28a**, as was expected because saturated NHCs commonly promote RCM with better activity than their unsaturated analogues [44]. Interestingly, catalyst **28b** featuring a modified benzylidene ether moiety (Blechert-type modification) initiated more rapidly and showed 90% conversion for **5** in CD_2Cl_2 at 30°C after just 1.5 h.

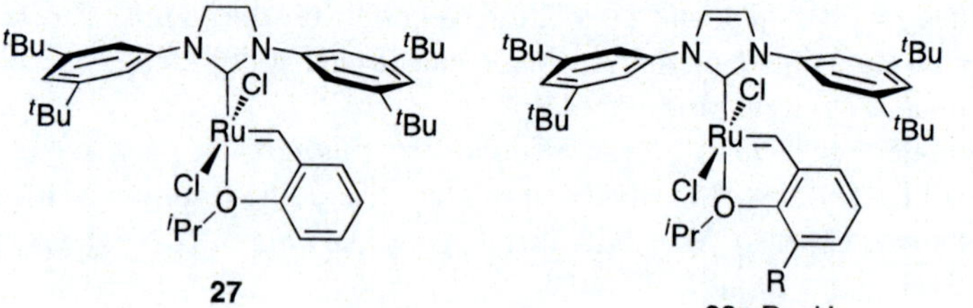

27

Grubbs (2007) [43]

28a R = H
28b R = Ph

Fig. 3.9 Catalysts **27** and **28** lacking ortho-substituents on the *N*-aryl side chains

In order to obtain more efficient and easier-to-prepare catalysts, Schrodi and Grubbs continued to study catalysts derived from NHC ligands with reduced bulk at the *ortho* positions of the aryl rings. By changing the aromatic side chains of the NHC ligand in Grubbs II from mesityl (SIMes) to mono-*ortho*-substituted phenyl groups, complexes **29a–c** and the analogous phosphine-free complexes **30a–c** were prepared [45] (Fig. 3.10). In the RCM study of dimethyl dimethallylmalonate **5**, these modified Grubbs II catalysts (**29a–c**) gave higher conversions than phosphine-free catalysts **30a–c** (CH_2Cl_2, 40°C). With the increase of the *ortho* substituents' size, the observed catalytic activity progressed in the following order: **29a** > **29b** > **29c** and **30a** > **30b** > **30c**. When the RCM reactions of **5** and **9** were conducted at elevated temperature (C_6D_6, 60°C), catalyst **30a** quantitatively afforded tetrasubstituted products **6** and **10** within 30 min. The addition of water or performing the reactions in air did not significantly affect the yields of the RCM.

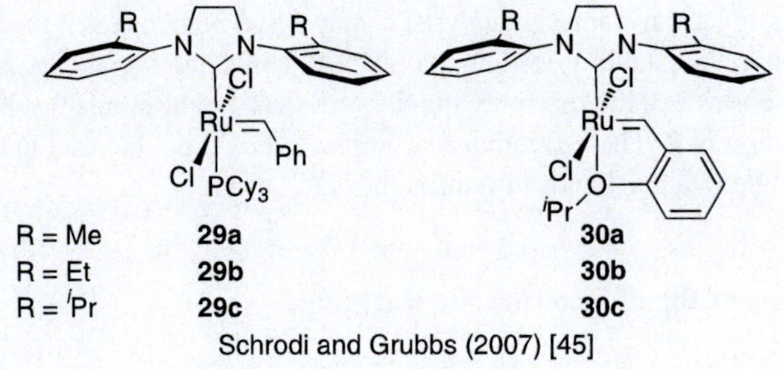

Fig. 3.10 Ruthenium catalysts with mono-ortho-substituted NHCs

In light of the successful application of catalysts **14** and **15** in olefin metathesis, Fürstner [29, 46, 47], Mol [24], Blechert [48] and Verpoort [25, 49, 50] have examined the metathesis activity of catalysts incorporating NHC ligands that combine one alkyl and one mesityl (or 2,6-diisopropylphenyl) side chain. The results demonstrated that the introduction of one aliphatic group into the NHC framework decreased the catalytic activity in RCM by comparison with standard catalyst systems **14** and **15**. The anticipated higher activity through higher electron-donation of the NHC was therefore not observed. However, steric effects played an important role in influencing the metathesis activity and increasing the size of the alkyl side chains led to lower RCM activity.

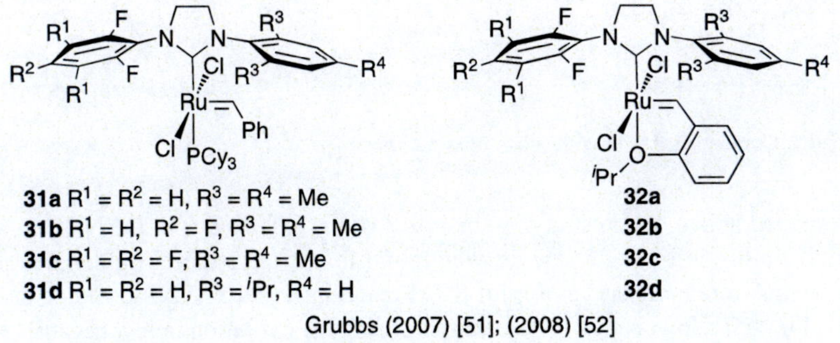

Fig. 3.11 Complexes **31a–d** and **32a–32d** with unsymmetrical NHCs

Encouraged by the significant rate enhancement in the catalytic performance of catalyst **25** bearing *ortho*-fluorinated aryl groups on the NHC ligand, Grubbs and co-workers more recently reported a series of ruthenium-based olefin metathesis catalysts coordinated with unsymmetrical NHC ligands [51, 52] (Fig. 3.11). With complexes **31a–d** and **32a–d** in hand, the authors compared their catalytic activity in the RCM reaction of **1**, **3**, and **5** using their standard reaction protocol [31]. The study revealed that catalyst **31a–c** affected the formation of **2** more efficiently than Grubbs II **14**, with catalyst **31b** being the most efficient of all (>97% conversion in 9 min). On the other hand, complexes **32a–c** were less efficient than parent complex **16**, possibly due to a fluorine–ruthenium interaction [41]. Complexes **31d** and **32d** bearing the more bulky NHC ligand turned out to be very poor olefin metathesis catalysts. Catalyst **31d** demonstrated the highest initial activity, albeit lowest stability, leading to only 59% conversion. In the RCM of the more challenging substrate **3**, the reactivity trends were found to be similar to those observed for the formation of **2**. These fluorinated complexes could not be used in the formation of tetrasubstituted double bonds of **6** through RCM.

Modifications of the *N*-Heterocyclic Backbone

The backbone effect of the NHC ligands was questioned in the development of ruthenium-based olefin metathesis catalysts. The earliest example of such a modification was reported by Grubbs and co-workers in their original paper containing Grubbs II complex **14** [13]. Although complex **33** (Fig. 3.12) showed excellent or even superior reactivity to **14**, no follow-up research on this catalyst has since appeared. Subsequent reports from Fürstner [29] and Köhler [53] were less convincing, and neither complexes **34** nor **35** gave positive enhancement for RCM reactions when compared to the benchmark catalysts **14** and **16**.

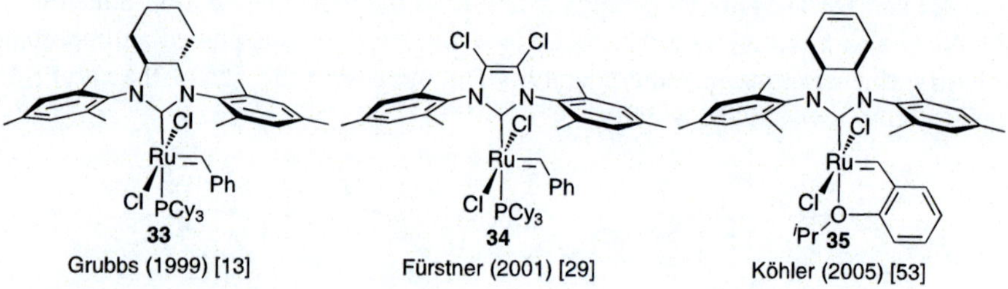

Fig. 3.12 Complexes **33–35** with substituted backbones

As noted above, while studying the effect of the side chains of the NHC ligands on ruthenium olefin metathesis, it was found that catalysts bearing smaller-sized aryl groups were more successful for RCM reactions to form tetrasubstituted double bonds [43, 45]. However, several attempts during the evolution failed, because these complexes were either difficult to synthesise or unstable (Fig. 3.13).

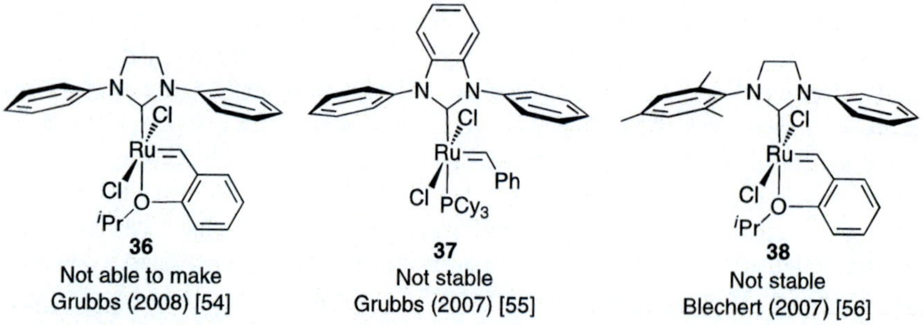

Fig. 3.13 Examples of Ru-complexes bearing NHCs with phenyl side chains

For instance, complex **36** containing only phenyl side chains failed to form even in trace amounts [54], and complexes **37** and **38** were rather unstable and decomposed to metathesis-inactive compounds through the activation of the *ortho* C–H bonds of these groups by the metal [55, 56].

In order to suppress this unwanted decomposition pathway, Grubbs and coworkers reasoned that restricting the rotation of the *N*-phenyl rings by placing bulky substituents on the backbone of the *N*-heterocycle would keep the aryl C–H bonds away from the ruthenium centre. Following this principle, they prepared two very stable complexes **39** and **40** by installing *gem*-dimethyl groups on the *N*-heterocycle, and metathesis activity was tested by using benchmark substrates **1**, **3** and **5** [54] (Fig. 3.14). Both **39** and **40** efficiently catalysed the RCM from **1** to **2**, but at a slower rate than catalyst **16**. For trisubstituted substrate **3**, the RCM with **39** or **40** was clearly slower than the cyclisation with **16**. Remarkably, catalyst **39** was highly effective in the RCM to form the hindered tetrasubstituted olefin **6**.

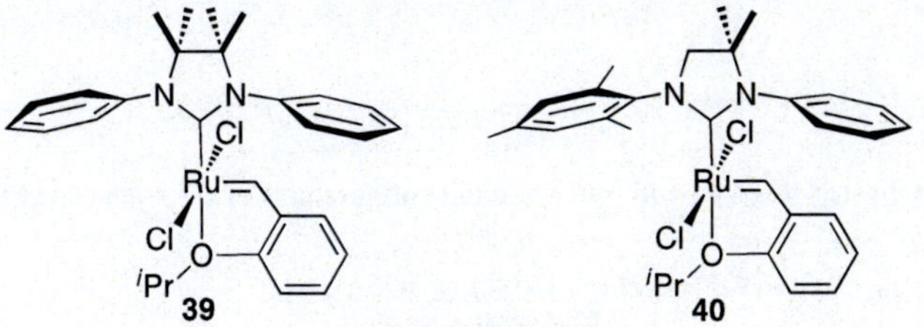

39 **40**

Grubbs (2008) [54]

Fig. 3.14 Stable catalysts **39** and **40** with NHCs containing substituted *N*-heterocyclic backbones

Unfortunately, complexes **39** and **40** are still more prone to decomposition than catalyst **16**. Therefore, Grubbs sought to investigate a series of new ruthenium catalysts bearing NHCs with varying degrees of *N*-heterocyclic backbone and aryl side chain substitution, and catalysts **16** and **30a** were chosen as basic catalyst structures [57]. In 2009, complexes **41a–c** and **42a–c** were prepared to attempt to understand how the degree of substitution on the backbone influences catalyst activity and lifetime (Fig. 3.15).

41a R = R' = H, R" = Me
41b R = Me, R' = R" = H
41c R = Me(H), R' = H(Me), R" = Me

42a R = R' = H, R" = Me
42b R = Me(H), R' = H(Me), R" = H
42c R = R' = R^3 = Me

Grubbs (2009) [57]

Fig. 3.15 Catalysts **41a–c** and **42a–c** with varying degrees of *N*-heterocyclic backbone substitution

Again, standard substrates **1**, **3** and **5** were utilised to monitor the new catalysts' behaviour in RCM catalysis. It was found that increased backbone substitution led to increased catalyst lifetimes but decreased rates of reaction.

Even more recently, Grisi and co-workers reported two catalysts bearing NHC ligands with *syn* (**43**) and *anti* (**44**) methyl groups on the *N*-heterocyclic backbone [58]. These two catalysts represent the possible configurations of the same NHC ligand used as isomeric mixture in catalyst **42b** (Fig. 3.16). The catalytic performance of both **43** and **44** was evaluated in the RCM reactions of **1**, **3**, and **5**, with catalyst **29a** taken as benchmark system. It was found that *syn*-catalyst **43** was more active than the parent pre-catalyst **29a**, while *anti*-catalyst **44** was less active than **29a**.

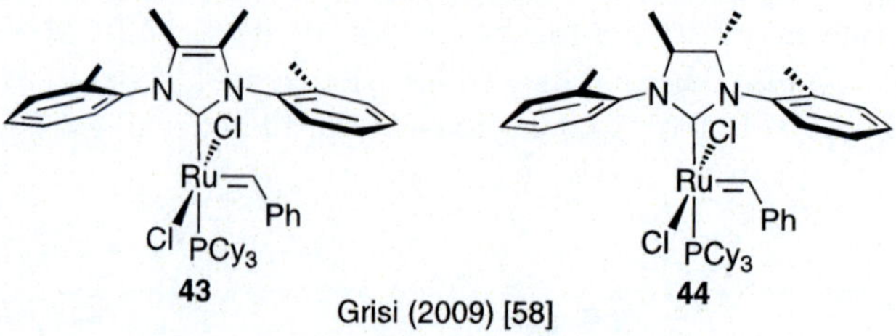

Fig. 3.16 Catalysts **43** and **44** with *syn* and *anti* configurations of the *N*-heterocyclic backbone

Modifications of the *N*-Heterocyclic Ring Elements

In addition to modifications of the side chains and of the backbone of the five-membered *N*-heterocyclic carbenes, one may also vary the classical 1,3-dinitrogen containing heterocycle.

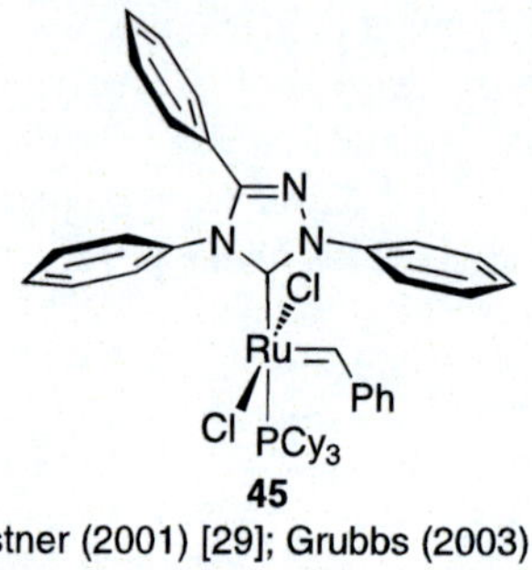

Fig. 3.17 Catalyst **45** with a triazole-based NHC

As early as 2001, Fürstner and co-workers reported the preparation and full characterisation of complex **45** [29] (Fig. 3.17), which incorporates a triazole-based *N*-heterocycle that was previously isolated by Enders [59]. A few years later, complex **45** was also prepared using another pathway by Grubbs [60]. When tested in the standard transformation from **5** to **6**, complex **45** provided a surprisingly high yield of 80% in just 2 h. However, higher conversion upon prolonging the reaction time was not observed. Further studies revealed that complex **45** was unstable in solution. Even at room temperature and under an inert atmosphere, significant amounts of complex **45** underwent decomposition. The short lifetime of complex **45** was therefore found limiting. Whether triazole structures with bulkier side chains (e.g. mesityl) show better stability and catalytic behaviour remains to be seen.

In 2007, Grubbs and co-workers reported new ruthenium olefin metathesis catalysts **46–48**, that make use of cyclic (alkyl)(amino)carbenes (CAAC) [61] (Fig. 3.18). This family of carbenes was designed and prepared by Bertrand [62] by replacing one amino group of a typical NHC with an alkyl group. The greater electron-donating ability and tunable steric environment led Grubbs to explore the utility of these entities as ligands in ruthenium olefin metathesis catalysis. Catalysts **46** and **47** were synthesised in good yields, and evaluated in the RCM reactions to form **2**, **4**, and **6**. They showed lower activity relative to **14** and **16**, and both elevated temperatures and extended reaction times were required to complete the RCM reactions of **1** and **3**. Attempts to decrease the steric bulk of the *N*-aryl ring met with limited synthetic success and the introduction of an *N*-mesityl group was unsuccessful, while complex **48** could only be generated in low yield (18%). Nevertheless, the catalytic results demonstrated that RCM activities for the formation of **2** and **4** with catalyst **48** were comparable to those achieved with standard catalysts **14** and **16**. Complexes **46–48** showed no reactivity in the conversion of **5–6**.

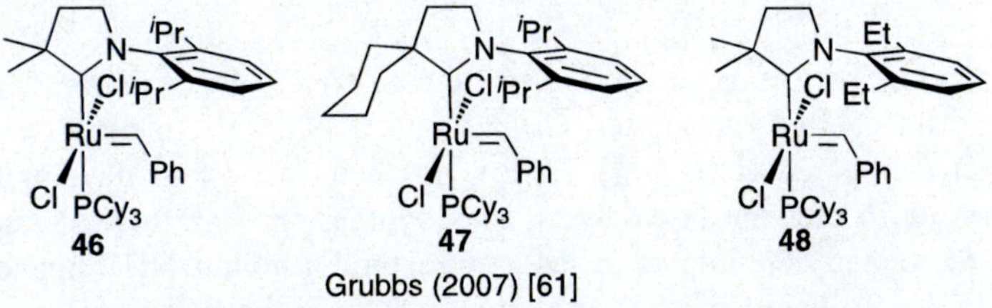

Fig. 3.18 Catalysts **46–48** bearing cyclic (alkyl)(amino)carbenes

Despite the numerous reports concerning NHC–Ru olefin metathesis initiators, a complex incorporating a carbene that has only one exocyclic substituent adjacent to the carbenic centre was not reported until 2008. Studies by Grubbs and co-workers led to the development of ruthenium-based catalysts bearing such carbene ligands, in this case incorporating thiazole-2-ylidenes [63] (Fig. 3.19).

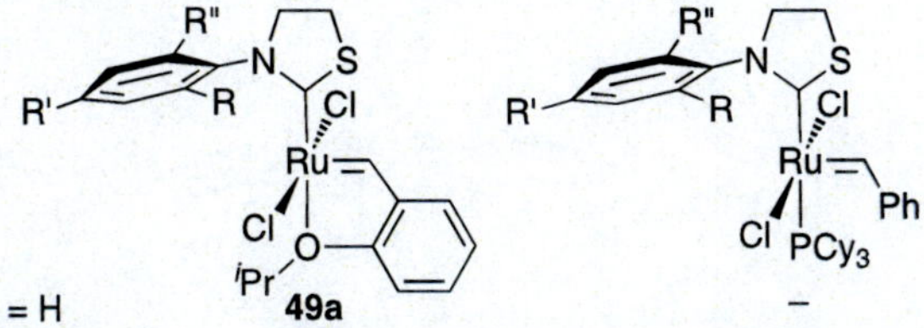

R = R' = R" = H	**49a**	–
R = Me, R' = R" = H	**49b**	–
R = R' = R" = Me	**49c**	**50c**
R = R³ = Et, R' = H	**49d**	**50d**
R = R³ = ^{i}Pr, R' = H	**49e**	–

Grubbs (2008) [63]

Fig. 3.19 Complexes **49a–e** and **50c–d** with sulphur-containing central *N*-heterocycle

The phosphine-free complexes **49a–e** were prepared in a straightforward manner in moderate to high yields by using the corresponding 3-aryl-4,5-dimethylthiazole-2-ylides. In contrast, the phosphine-containing analogues were not as stable as complexes **49a–e**, and only complexes **50c** and **50d** could be isolated. According to ^{1}H and ^{31}P NMR spectroscopic studies, complexes **50a–b** underwent complete decomposition at room temperature in 2 and 18 h respectively, and complex **50e** could not be formed at all. Standard substrates **1**, **3**, and **5** were employed to evaluate the efficiency of the new complexes in RCM. Due to their decreased steric protection, these complexes were not as active as the standard catalysts **14** and **16**. Suffering from a high decomposition rate, complexes **50c–d** enabled the cyclisation of **1** and **3** with <50% conversion. The slower initiation of complexes **49a–e** prevented the ring-closure of **1** and **3** at 30°C, and elevated temperatures (50°C) were required to obtain >97% (for **1**) and >96% (for **3**) conversions. A longer induction period was observed for complex **49e** that only managed 73% conversion for **3** after the same time. Furthermore, complexes **49a–e** and **50c–d** were not able to ring-close substrates leading to tetrasubstituted double bonds (*i.e.* **6**).

3.1.2.2 Four-Membered Central *N*-Heterocycle

The outstanding performances of five-membered NHC ligands in organometallic chemistry and catalysis prompted Grubbs and co-workers to develop a novel stable four-membered NHC [64]. Following their interest in developing new ruthenium olefin metathesis catalysts, they synthesised and fully characterised complex **51** to study the impact of the architecturally unique NHC ligand on the activity of the Ru-based catalyst [65] (Fig. 3.20). In the RCM of **1** at 40°C in CH_2Cl_2 with **51** (5 mol% catalyst), the reaction reached completion within 20 min, whereas less than 10 min are required for standard catalysts **14** and **16**. It should be noted that catalysts **14** and **16** are able to complete the RCM of **1** with only 1 mol% catalyst at 30°C.

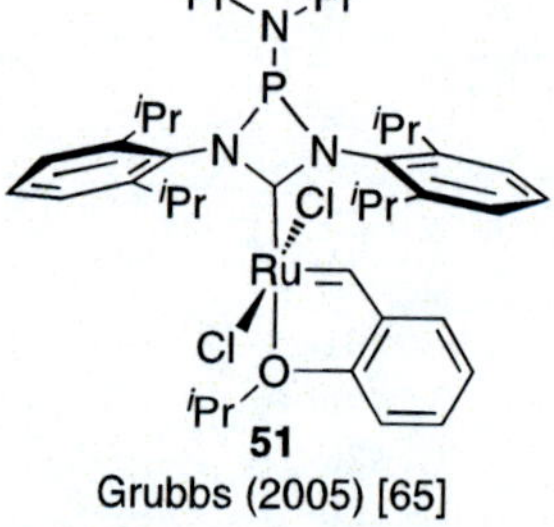

51

Grubbs (2005) [65]

Fig. 3.20 Catalyst **51** with a four-membered NHC

3.1.2.3 Six-Membered Central *N*-Heterocycle

In addition to the example of a four-membered *N*-heterocyclic carbene shown above, variations of the ring size in NHCs were also reported with studies on the six-membered NHC ligand complexes **52** by Grubbs and co-workers [66], and **53** by Buchmeiser [67] which bears the fast initiating benzylidene ether chelate developed by Grela [19] (Fig. 3.21). The reactivity of complexes **52** and **53** was screened in the RCM of benchmark substrate **1** to form **2**. Both catalysts promoted this transformation, albeit showing lower reactivity than catalysts **14** and **16**. For example, the RCM of **1** at 50°C with **52** resulted in 72% conversion after 30 min and 83% after 1 h. In comparison, the reaction proceeded to completion with catalyst **14** after 10 min under identical conditions. From X-ray structural studies on complexes **52** and **53**, it is apparent that the mesityl groups bend more toward the metal centre than in the parent complexes containing SIMes. Grubbs and co-workers therefore assumed that the higher steric pressure exerted by the mesityl side chains when the six-membered central *N*-heterocycle is used might disfavour olefin binding or metallacyclobutane formation during metathesis.

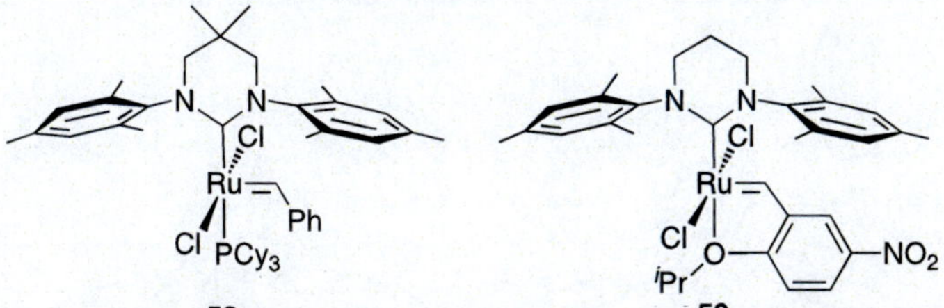

52
Grubbs (2004) [66]

53
Buchmeiser (2004) [67]

Fig. 3.21 Catalysts **52** and **53** with six-membered NHCs

3.1.2.4 Seven-Membered Central *N*-Heterocycle

In 2009, Buchmeiser and co-workers reported the synthesis of a novel ruthenium complex **54** based on a seven-membered NHC ligand [68] (Fig. 3.22). To examine the catalytic activity of complex **54** in the RCM reaction, the authors subjected the complex to a series of typical RCM reactions by using substrates **1**, **3**, and **5**. Pre-catalyst **54** showed only moderate reactivity with **1** and **3** and no reaction occurred with **5**.

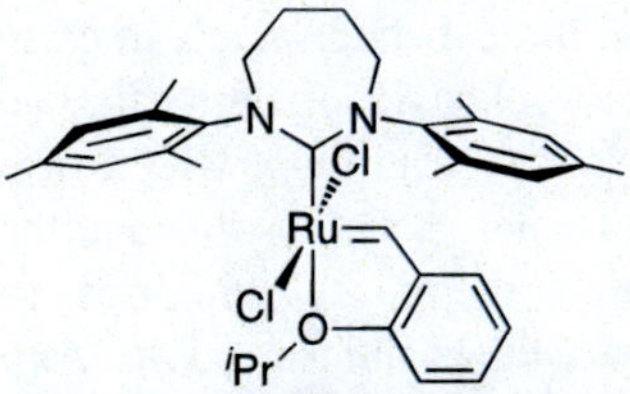

54
Buchmeiser (2009) [68]

Fig. 3.22 Complex **54** with a seven-membered NHC

3.1.3 NHCs for Asymmetric RCM

The development of chiral ruthenium metathesis catalysts is still in its early stages and certainly represents a great challenge. Early on in the development of second-generation ruthenium catalysts, it was recognised that the NHC ligands could be suitable candidates for introducing chirality to these complexes. During the past decade, two main classes of ruthenium catalysts with chiral NHC ligands have been introduced by the groups of Grubbs and Hoveyda, respectively.

3.1.3.1 Chiral Monodentate NHCs

The pioneering work on enantioselective ruthenium olefin metathesis was carried out by Grubbs and co-workers in 2001 [69] (Fig. 3.23). Catalysts **55a–b** and **56a–b** were designed and prepared from C_2-symmetric NHC ligands with a combination of chiral backbone and mono-*ortho*-substituted aryl side chains, a motif that was expected to form a chiral environment around the metal centre.

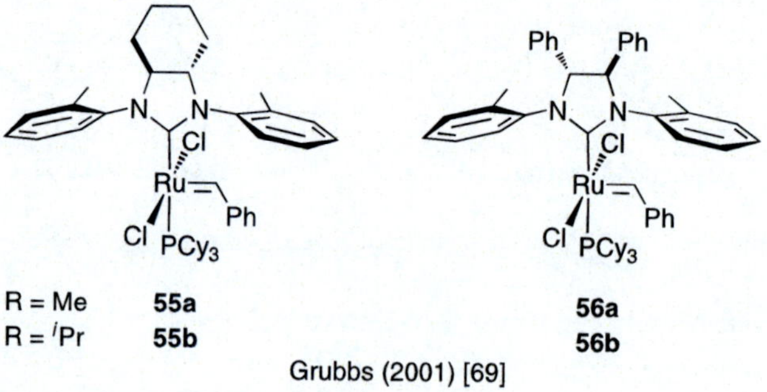

Fig. 3.23 Catalysts **55a–b** and **56a–b** bearing C_2-symmetrical monodentate NHCs

Subsequently, these catalysts were evaluated in the enantioselective desymmetrisation of achiral trienes, and three distinct trends in catalyst selectivity were found. Firstly, catalysts **56a–b** with two phenyl moieties on the backbone of the *N*-heterocycle exhibited higher enantioselectivity than those with a fused cyclohexyl group as the backbone **55a–b**. Secondly, mono-*ortho*-substituted aryl side chains induced greater enantioselectivity than symmetrical mesityl wing tips. Thirdly, changing the halide ligands from Cl^- to I^- increased the enantioselectivity. As a result, catalyst **56b** turned out to be the most effective. For example, **56b** in the presence of NaI was able to promote the desymmetrisation of **57** to give chiral dihydrofuran **58** in up to 82% conversion and 90% *ee* (Scheme 3.3).

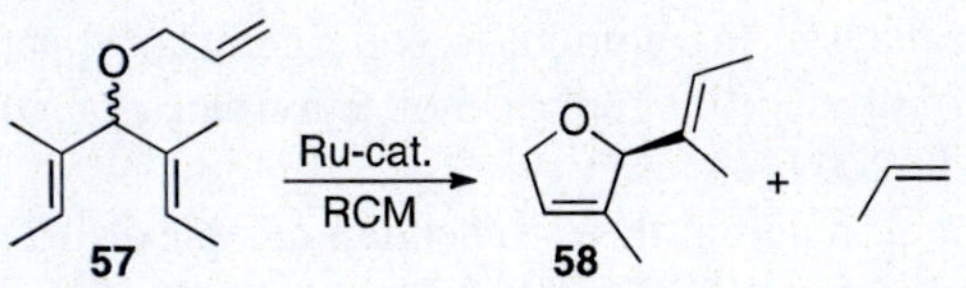

Scheme 3.3 Asymmetric RCM of **57**

Further detailed investigations towards new chiral ruthenium catalysts that could enhance enantioselectivity and expand the substrate scope in asymmetric RCM were reported by Grubbs and co-workers in 2006 [70] (Fig. 3.24). Catalysts **59** and **61**, which are close derivatives of **56** incorporating additional substituents on the aryl ring *para* to the *ortho*-isopropyl group, maintained similar enantioselectivity than **56b**. However, incorporation of an isopropyl group on the side chain *ortho* to the *ortho*-isopropyl group **60** led to an increase in enantioselectivity for a number of substrates.

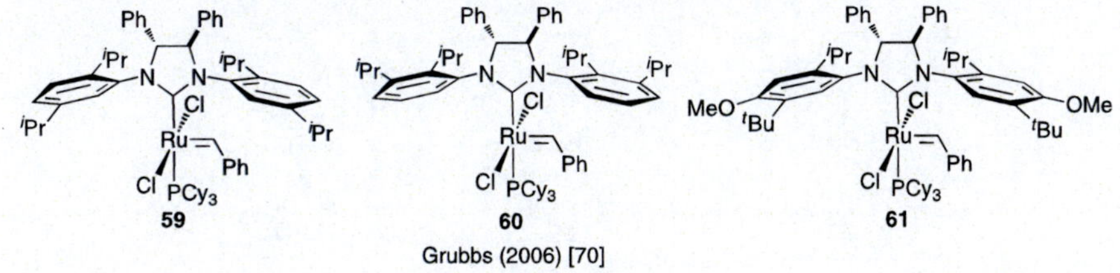

Fig. 3.24 Catalysts **59–61** with different side chain substitutions

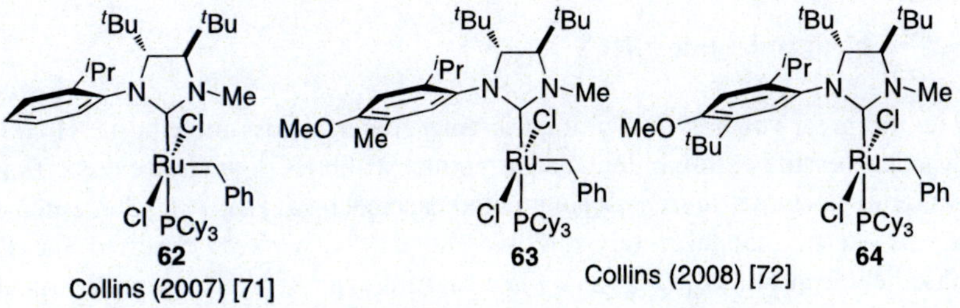

Fig. 3.25 Catalysts **62–64** bearing C_1-symmetrical monodentate NHCs

Encouraged by the elegant strategy shown above, Collins and co-workers recently designed and synthesised a new family of Ru-based olefin metathesis catalysts possessing chiral C_1-symmetric monodentate NHC ligands [71, 72] (Fig. 3.25). The new ligand modifications were operated on the backbone and one side chain of the NHC ligands in Grubbs' chiral catalysts. The 1,2-diphenyl backbone of the NHCs was replaced by a 1,2-di-*tert*-butyl unit, and one aryl side chain was exchanged with a less sterically demanding methyl group. With catalysts **62–64** in hand, the authors subjected these to a series of achiral trienes under similar reaction conditions optimised for catalysts bearing C_2-symmetric NHCs. Catalyst **62** promoted asymmetric RCM to product **58** in 81% yield and 82% *ee*, which is comparable to Grubbs' chiral catalysts. In addition, to enhance catalyst stability, a benzyl or *n*-propyl group was introduced to replace the methyl side chain of the NHC ligands by the same group in 2009 [73].

In 2008, Grisi et al. reported three ruthenium complexes **65–67** bearing chiral, symmetrical monodentate NHC ligands with two *N*-(S)-phenylethyl side chains [74] (Fig. 3.26). Three different types of backbones were incorporated into the *N*-heterocyclic moiety of the ligands. When achiral triene **57** was treated with catalysts **65–67** under identical reaction conditions, a dramatic difference was observed. As expected, the absence of backbone chirality in complex **65** makes it completely inefficient for inducing enantioselectivity in the formation of **58**. Similarly, the mismatched chiral backbone framework of complex **66** was not able to promote asymmetric RCM of **57**. In contrast, appreciable albeit low selectivity (33% *ee*) was observed when the backbone possessed *anti* stereochemistry.

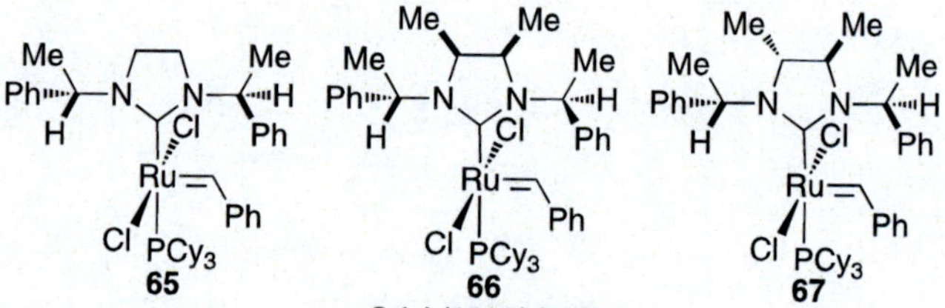

65 **66** **67**

Grisi (2008) [74]

Fig. 3.26 Catalysts **65–67** bearing chiral NHCs with alkyl side chains

3.1.3.2 Chiral Bidentate NHCs

Despite the great success in asymmetric ring opening/cross metathesis, Hoveyda catalysts possessing chiral, bidentate C_1-symmetrical NHC ligands are less efficient in promoting asymmetric ring closing metathesis reactions [75, 76]. After extensive variations on the model catalyst, Hoveyda and co-workers reported the first Ru-based catalyst **68** [77] (Fig. 3.27) bearing this type of NHC ligand that is able to perform asymmetric RCM of **69**. Moderate enantioselectivity was observed when compared to Grubbs' chiral catalyst **60** (Scheme 3.4).

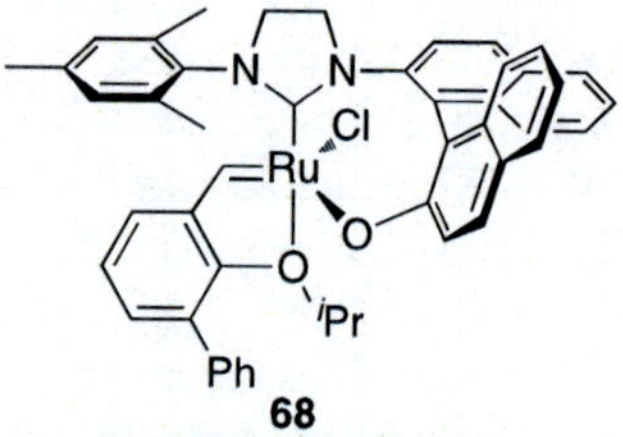

68

Hoveyda (2003) [77]

Fig. 3.27 Catalyst **68** with a C_1-symmetrical bidentate NHC

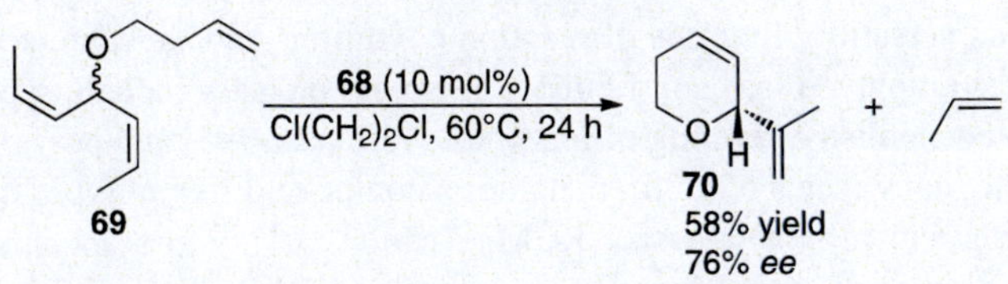

Scheme 3.4 Asymmetric RCM of **69**

3.1.4 *Conclusions and Outlook*

Ten years have passed since the advent of second-generation ruthenium olefin metathesis catalysts that incorporate a NHC ligand. Within these last 10 years, a large number of NHC ligands have been developed and tested in metathesis reactions. While these modified ligands have in most cases not proven superior to the original SIMes and IMes architectures, the intense research that has gone into finding modified NHCs for ruthenium-based catalysts will without doubt help the development of NHC catalysis outside the field of ruthenium metathesis. As a case in point, we might mention chiral variations of NHC ligands that have been first introduced for olefin metathesis and are now making their way into various other catalytic applications [78–87]. Further detailed studies on catalyst design and on mechanistic aspects of the metathesis reaction in general and of RCM in particular may well lead to significant industrial implementations of such NHC-containing catalysts (see Chapter 14).

3.2 NHC-Complexes for Ring Opening Metathesis Polymerisation

3.2.1 *Olefin Metathesis*

Since the first reports on olefin metathesis in the 1960s [88, 89], this elementary C–C bond forming reaction has sparked an enormous activity in organometallic research, culminating in the award of the Nobel Prize to Yves Chauvin, Richard Schrock and Robert Grubbs in 2005.

Olefin metathesis is a rearrangement reaction that includes a transition metal carbene and an olefin. After the catalytic cycle, a new olefin and a new active metal carbene are formed (Scheme 3.5) [90].

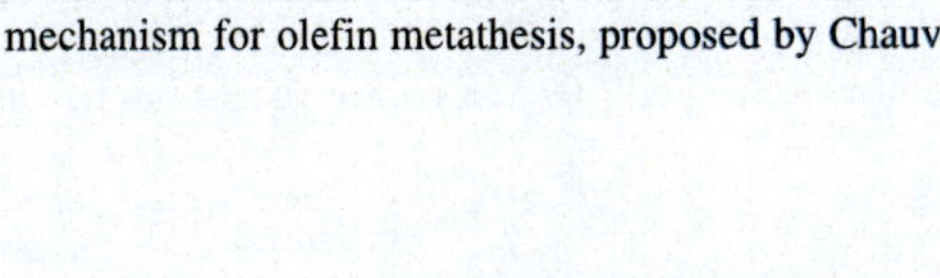

Scheme 3.5 General mechanism for olefin metathesis, proposed by Chauvin in 1971

Due to its versatility in terms of reaction conditions and substrate scope, olefin metathesis is employed in many fields of chemical industry including pharmaceutics [91], oleochemistry, refining of renewable resources [92] and polymer chemistry [93]. The many facets of olefin metathesis comprise different types of reactions such as ring closing metathesis (RCM, Section 3.1), cross metathesis (CM, Section 3.3) or ring rearrangement metathesis (RRM) [94], giving access to sophisticated, elegant synthetic pathways in organic chemistry. In the field of macromolecular chemistry various polymer architectures are facilitated by the use of acyclic diene metathesis polymerisation (ADMET) [95], alternating diene metathesis polycondensation (ALTMET) [96], ring opening metathesis polymerisation (ROMP) or ring opening insertion metathesis polymerisation (ROIMP) [97].

3.2.2 ROMP – Basics

ROMP is without doubt the most important incarnation of olefin metathesis in polymer chemistry [98]. Preconditions enabling this process involve a strained cyclic olefinic monomer and a suitable initiator. The driving force in ROMP is the release of ring strain, rendering the last step in the catalytic cycle irreversible (Scheme 3.6). The synthesis of well-defined polymers of complex architectures such as multi-functionalised block-copolymers is enabled by living polymerisation, one of the main benefits of ROMP [92, 98].

Norbornene derivatives are very popular monomers for ROMP due to a comparably high ring strain and good functionalisability. The latter is needed to append any desired functional unit to the monomer [98, 99].

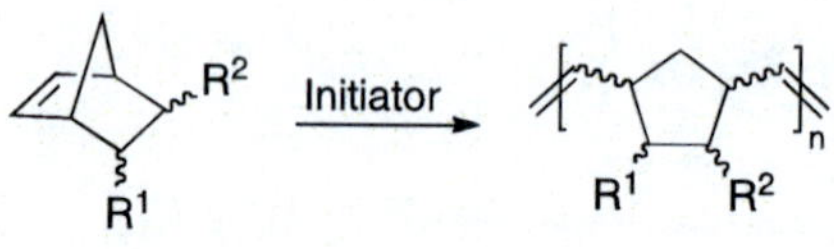

Scheme 3.6 ROMP of a functionalised norbornene derivative

3.2.2.1 ROMP Initiators

In order to perform ROMP successfully, the use of an appropriate initiator is crucial. With the development of the so-called "Schrock-initiators" – tungsten or molybdenum alkylidene based systems with various ligands – the first well-defined initiator performing metathesis at high initiation and propagation rates was born [100]. However, the applications of these systems were restricted by their limited functional group tolerance and high sensitivity towards oxygen and moisture [98]. The ruthenium based "first-generation-Grubbs-catalyst" **71** (Fig. 3.28) featuring two neutral phosphine ligands solved these problems to a significant extent. Living polymerisation was now achieved (also in protic media) under ambient conditions.

Still, a rather low activity compared to the Schrock catalysts eventually led to the development of the "second-generation-catalysts" **72** (Fig. 3.28) where one of the labile PCy_3 ligands is replaced by a bulky and strongly σ-donating *N*-heterocyclic carbene (NHC) ligand [101]. The resulting initiator exhibits a completely different kinetic behaviour within the catalytic cycle [102]. Living polymerisation is no longer provided. With the invention of phosphine-free "third-generation-catalysts" **73** (Fig. 3.28), a pyridine derivative was introduced as the labile ligand. This led to fast and complete initiation and eventually to an increase of activity by six orders of magnitude [103, 104]. As an alternative to the Grubbs-type initiators, another family of ruthenium based metathesis initiators exists, featuring an indenylidene-type carbene ligand instead of the benzylidene (Fig. 3.28, **74a–c**) [105]. Performance of these complexes in ROMP can be described as similar to their benzylidene counterparts [106, 107].

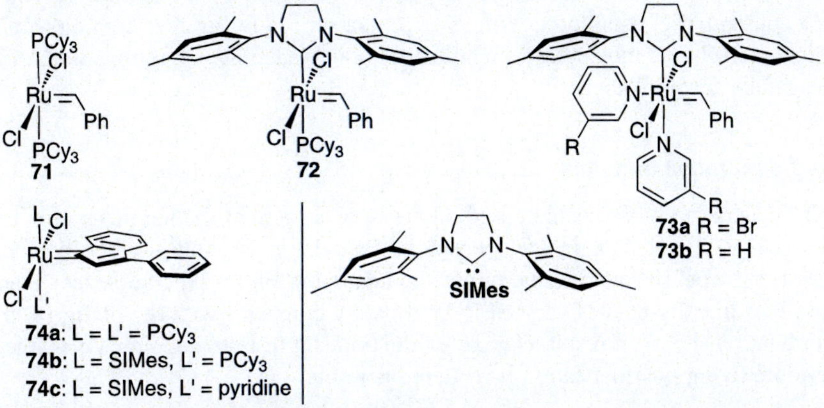

Fig. 3.28 ROMP initiators, first, second and third generation: Grubbs catalysts (complexes **71**, **72**, and **73**, respectively); **74a–c**: 3-phenyl-indenylidene replaces the benzylidene carbene

Considering all given factors, ROM polymerisation can now be identified as the polymerisation method of choice for the synthesis of highly defined, tailor-made specialty polymers.

Catalyst Activity

The progress of a ROM polymerisation depends on two rate constants, namely k_i (initiation) and k_p (propagation). Whereas the initiation rate is determined by the dissociation of a labile ligand to give a free coordination site, propagation depends on whether the approaching olefin or the labile ligand will rather (re)coordinate to the ruthenium centre [102]. This results in different reaction behaviour of each catalyst generation.

First Generation Initiators

Phosphine dissociation is not rate determining, resulting in a high initiation rate. The rate determining step is the coordination of the olefin that is competing with recoordination of the phosphine. This leads to low activity and therefore to a comparatively slow, but well-defined polymerisation progress. Living polymerisation is possible.

Second Generation Initiators

In second generation initiators, the electron donating NHC ligand exerts a strong influence on the initiation step. Experiments showed that the dissociation rate of the phosphine is decreased by almost two orders of magnitude compared to first generation complexes and becomes the rate determining step in the metathesis cycle. In contrast, propagation is enhanced immensely (four orders of magnitude) due to an obvious selectivity in favour of olefin coordination instead of phosphine recoordination [108].

Summarising, second generation initiators exhibit a comparatively low initiation rate. Once the polymerisation has been started, propagation in contrast is very fast. As a consequence, resulting polymers are characterised by a rather high molecular weight and a broad molecular weight distribution. Generally, no living polymerisation can be accomplished.

Third Generation Initiators

Third generation initiators are based on the NHC system of second generation initiators, but do not contain any phosphine ligand. Instead, one or two pyridine ligands are weakly bound to the ruthenium centre (*cf.* Fig. 3.28, complexes **73** and **74c**). Pyridine dissociates very easily and hardly competes with the olefin for the coordination site. As a result, complete initiation and fast propagation are enabled, therefore living polymerisation is rendered possible.

3.2.3 NHC Ligands for ROMP Initiators

Many parallel developments in organometallic chemistry focusing on various transition metals have eventually led to the use of *N*-heterocyclic carbenes in metathesis catalysts. A new episode of a successful story has been started.

3.2.3.1 Why Carbenes as Ancillary Ligands?

One major role of the tricyclohexylphosphine ligands in first generation catalysts (Fig. 3.28) is steric and electronic stabilisation of the reactive catalytic species [109]. Nucleophilic carbene ligands are believed to act as phosphine mimics imposing different steric and electronic influences to the transition metal [110, 111]. Common NHC degradation processes were eventually prohibited by flanking the NHC with sterically demanding groups, turning them into isolable free carbene ligands with numerous application possibilities in transition metal chemistry [112].

The idea to use *N*-heterocyclic carbenes as ancillary ligands also in ruthenium based metathesis catalysts/initiators was followed by several groups in the late 1990s. First, both phosphines were exchanged yielding almost inert complexes with undesirably low metathesis activity [113, 114]. The later synthesised mixed NHC/phosphine systems of type **72** or **74b** (Fig. 3.28) brought a real breakthrough in terms of activity and stability of metathesis catalysts [115]. Initially, these initiators often bore an unsaturated NHC backbone, which turned out to be disadvantageous in terms of polymerisation activity.

Carbene ligands can replace phosphines due to similar electronic properties. The development of NHC design concepts featuring different substituents and backbones eventually culminated in the most prominent derivative, the SIMes (SIMes = *N,N'*-bis[2,4,6-(trimethyl)phenyl]imidazolidin-2-ylidene) ligand that is used in the second and later also third generation catalysts (complexes **72, 73, 74b** and **74c** in Fig. 3.28) [105, 109, 114, 116].

3.2.3.2 Different NHC Design Motifs Used in ROMP Initiators

NHC ligands used for metathesis catalysts generally consist of an imidazole derived carbene that is decorated with miscellaneous functionalities. The backbone of the imidazole moiety represents only one of the tuneable sites. Saturation or functionalisation of the double bond in the NHC backbone affords numerous derivatives (Fig. 3.29). Additionally, the fragments bound to the imidazole nitrogens can be of various shapes. The favourable mesityl group combines the required bulkiness (see previous section) with a certain flatness of the phenyl rings. Therefore **IMes** (*N,N'*-bis-[2,4,6-(trimethyl) phenyl]imidazol-2-ylidene) and its saturated analogue **SIMes** were among the first described NHC ligands used in ruthenium based olefin metathesis catalysts/initiators.

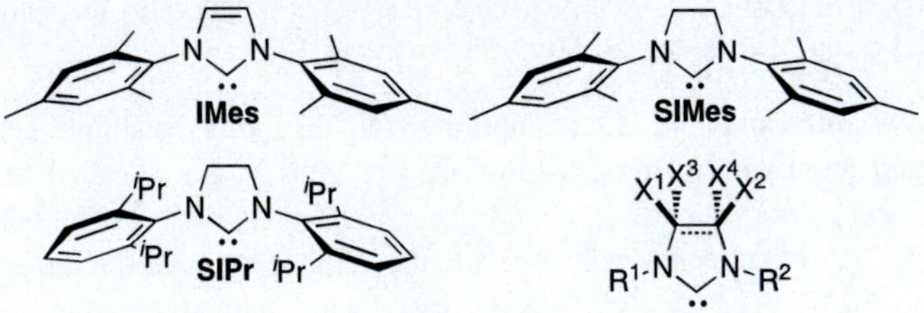

Fig. 3.29 NHC ligands – principal layout and frequently used derivatives

IMes, SIMes and SIPr in ROMP

Comparing the performance in ROMP of 1,5-cyclooctadiene (COD) of second generation catalysts featuring either the SIMes, SIPr or the IMes ligand reveals that the saturated system is highly superior in metathesis activity. The following illustration shows conversion of the monomer to the polymer product as a function of time (Fig. 3.30).

Using SIMes or SIPr as co-ligand, 50% conversion within only 1 min and complete conversion after less than 10 min was accomplished. In contrast, the initiator featuring the unsaturated NHC ligand which is less electron donating, requires 20 min for 50% conversion and more than 1 h for complete polymerisation of 1,4-cyclooctadiene (COD) [117, 118]. This example shows in an impressively clear way, that a minor change (at first glance) can have major effects on the polymerisation. Therefore, design and further development of the NHC ligand in second and third generation metathesis catalysts is still a crucial field of investigation in current metathesis polymerisation research.

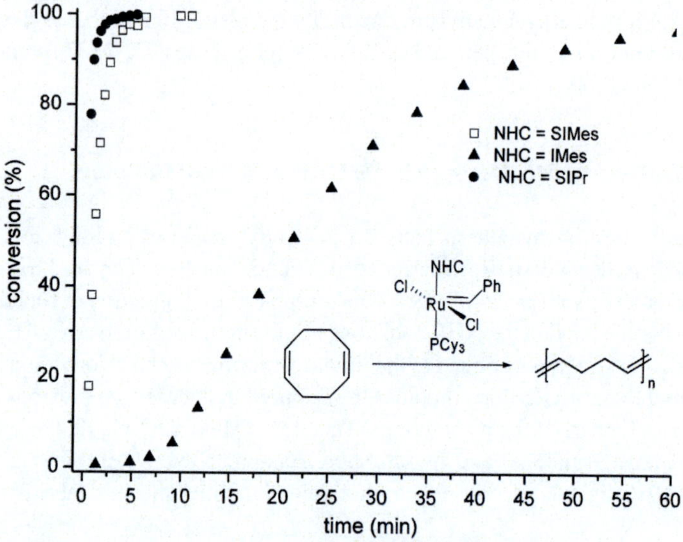

Fig. 3.30 ROMP of COD: conversion *vs.* time plot comparing second generation initiators bearing the NHC ligands IMes, SIMes and SIPr; redrawn from [117]

One successful example of alteration in the NHC ligand is the simple replacement of the mesityl groups by *iso*-propyl substituted phenyls in the so-called SIPr ligand (Fig. 3.30), and its corresponding second generation initiator [119]. The bulkier substituents induce a further increase of activity to the initiator, especially at lower temperatures (up to 30°C). The catalyst bearing this ligand outperforms Grubbs second generation catalyst **72** (Fig. 3.28) due to increased initiation [117, 120]. On the other hand, the sterically demanding *iso*-propyl moieties cause increased sensitivity at higher temperatures as bimolecular degradation processes apparently occur [101, 120].

3.2.4 REMP – Access to Cyclic Polymers

REMP, the acronym for ring expansion metathesis polymerisation is a special case of ROMP, where the growing polymer chain stays attached to the catalyst at both ends until a macrocycle is released. This requires that the active carbene be tethered to a permanent ligand of the complex. Again, judicious design of NHC ligands is the key issue in the development of potent catalysts for this special application.

3.2.4.1 Dissymmetric NHC Ligands for REMP Catalysts

Catalysts with an unsymmetrical NHC ligand featuring a vinylic side chain have the unique ability to metathesise their own ligand to form a metallacycle as shown in Scheme 3.7 [119]. Ring opening metathesis will then incorporate the monomers, e.g. cyclooctene, into that cycle until a cyclic polymer is cleaved by another intramolecular metathesis step. The catalyst is recovered and can restart this "endless route to cyclic polymers" [121].

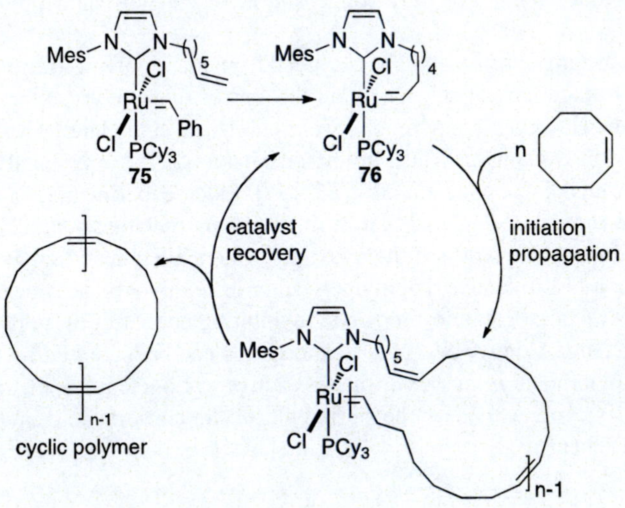

Scheme 3.7 REMP: Intramolecular metathesis of pre-catalyst **75** to form catalyst **76**; incorporation of monomers, release of a cyclic polymer and catalyst recovery

Stability and metathesis activity of REMP catalysts are strongly influenced by the size of the metallacycle or rather the length of the side chain of the NHC ligand. To address stability matters (storage and during polymerisation) short tethers that will form seven- or eight-membered metallacycles are advantageous. Yet, these catalysts exhibit clearly lower activities than complexes with longer tethers. A veritable reactivity boost can be achieved by using the saturated NHC system. Therefore, a pleasant compromise is reached with a saturated version of **76**, providing high activity and satisfying stability, as well as synthetic accessibility. The tether comprises six carbon atoms between the ruthenium centre and the NHC nitrogen. In a representative reaction, a 50% conversion in the polymerisation of cyclooctene was reached within 12 min, compared to 120 min for its unsaturated analogue **76** [122].

3.2.5 *Alternating ROM Copolymers – NHC Makes the Difference*

Controlling the exact architecture of polymers has always attracted attention in macromolecular chemistry. Successful synthesis of alternating copolymers using ring opening metathesis polymerisation is of great interest also from a mechanistic perspective. NHC ligands were found to be ideal to tune the selectivity of the metathesis initiators.

3.2.5.1 Alternating Copolymers Caused by Steric and Electronic Selectivity

Synthesis of strictly alternating copolymers can be achieved *via* various polymerisation techniques including poly-condensation or Ziegler-Natta polymerisations [123, 124].

ROMP was long considered to be excluded from these polymerisation techniques due to an assumed lack of selectivity on the part of the catalyst to the monomeric double bonds. However, some reports were published in the late 1990s, stating the general possibility using molybdenum based catalysts [125] or co-catalytic systems benefiting from solvent cage effects [126, 127]. Later, experiments employing sensibly chosen monomers showed that in ring opening metathesis copolymerisation, steric demand and ring strain of the reaction partners are crucial factors for achieving alternation [128]. Hence, copolymerisation of highly strained but rather bulky norbornene derivatives with cyclooctene (exhibiting neither of these properties), is a promising setup. Control over rate of alternation can be increased by keeping the ratio of the monomers in favour of the less active cyclooctene, so that it will enter the metathesis cycle whenever the steric bulk of the norbornene derivative is less favourable (Scheme 3.8).

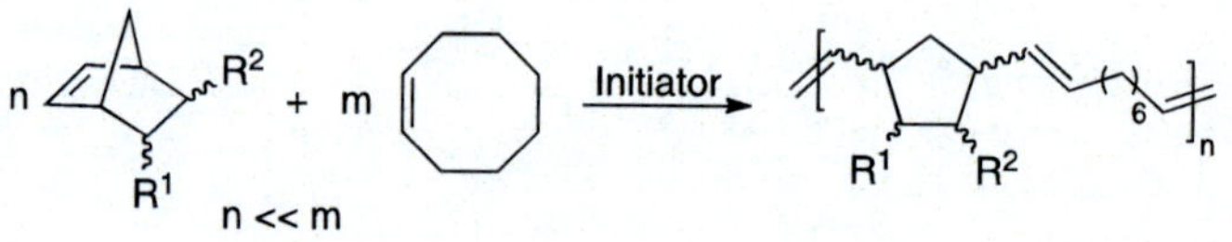

Scheme 3.8 Alternating ROMP: monomers exhibiting different steric demand and ring strain

One can assume that the strained norbornene will always be consumed preferentially, unless the steric hindrance is maximised, and enough cyclooctene is present in the immediate vicinity of the active centre of the macroinitiator. In that special case, the relatively inactive monomer will insert and undergo the ring opening process. The resulting linear polymer fragment is much more flexible than the norbornene derived fragment. Hence, steric stress is reduced and the next monomer to be consumed will be a norbornene, followed by a cyclooctene and so forth, resulting in a perfectly alternating copolymer chain.

NHC Ligands Promoting Alternation

In second generation metathesis initiators the NHC ligand is rotating more or less freely, depending on the nature of the active carbene, configuration of ancillary ligands and the steric bulk of the NHC substituents.

Increasing steric bulk at one of the NHC side groups will cause interference with the active site of the complex and generate an amplifying effect on alternation control. A methyl substituted one-carbon spacer like in 1-mesityl-3-((1'R)-1-phenylethyl)-4,5-dihydroimidazol-2-ylidene is enough to dramatically improve the alternation rate [129]. In Fig. 3.31 the steric impact on monomer coordination for different situations is rationalised.

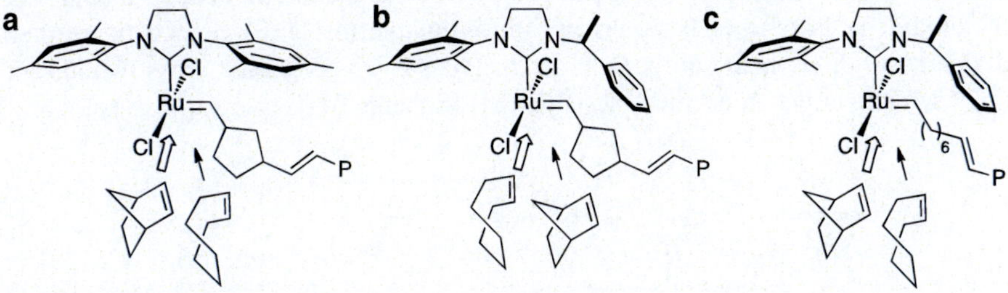

Fig. 3.31 Steric control in alternating ROMP: Tendencies of norbornene and cyclooctene to give productive olefin metathesis upon coordination are illustrated by a *thick arrow* (preferred monomer) or a *thin arrow* (less favoured monomer); (**a**) only minor steric hindrance: SIMes greatly favours the polymerisation of the strained norbornene; (**b**) the rotating phenylethyl-group induces a sterically more congested active site, leading to preferred incorporation of the smaller cyclooctene; (**c**) the flexible and small cyclooctene derived polymer fragment permits the incorporation of the bulky norbornene

Alternating ROMP is not restricted to a "norbornene–cyclooctene-couple". Cyclopentene can also be used as the "flexible" monomer [129]. This is particularly interesting, as cyclopentene is known to be problematic in ring opening metathesis homopolymerisation, because of its low ring strain. Moreover, the short distance between the polymeric double bonds encourages chain degradation by backbiting which leads to undesired broad molecular weight distributions.

3.2.6 *Further Applications*

NHC-bearing ruthenium carbene complexes are used to design latent and switchable initiators. In this research field emphasis is given to the development of initiators, which are inactive at room temperature in the presence of the corresponding monomer and can be activated upon a proper stimulus such as heat [130] or light [131]. Once activated, a high polymerisation activity is desired which is provided by NHC co-ligands [132].

In general and summarising, the use of NHC ligands in ruthenium based initiators dramatically enhanced the scope of olefin metathesis polymerisation methods in the last decade. The high degree of tuneability of steric and electronic factors provided in NHC ligands will certainly lead to further exciting developments in initiator design, tackling challenges such as stereoregular polymerisation.

3.3 Cross Metathesis

3.3.1 Introduction

The olefin cross metathesis (CM) can be described as the intermolecular metathesis of alkylidene fragments between two different olefins [133]. It can be further divided into three main subtypes: cross metathesis, ring opening cross metathesis (ROCM) and enyne cross metathesis (ECM) (Scheme 3.9).

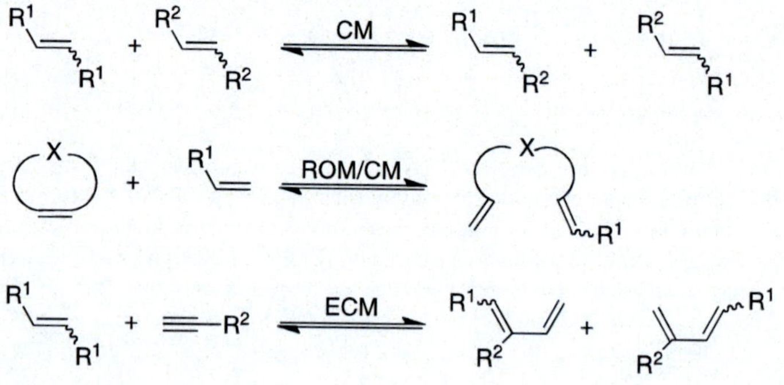

Scheme 3.9 Types of cross metathesis

The catalysts that will be discussed in this section are shown in Fig. 3.32.

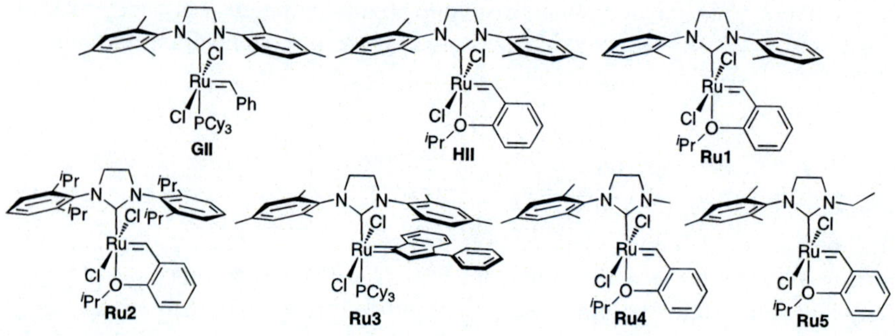

Fig. 3.32 Ru-metathesis catalysts used in this chapter

3.3.2 Olefin Cross Metathesis

During recent years, cross metathesis has found a wide range of applications in total synthesis. CM has been the key step in the syntheses of (-)-lasubine II [134], (+)-7a-*epi*-7-deoxycasuarine [135], and melithiazole C [136] to name just a few examples. It has been used for the modification of tetrapyrrolic macrocycles [137] as well as erythromycin derivatives [138], the dimerisation of steroids [139] and the synthesis of prostaglandin analogues [140].

Efficiency and selectivity of the CM reaction and the activity of the catalyst are three key issues concerning any kind of CM. It is vital to suppress side reactions like the competing dimerisation in order to achieve high yields of cross-product. One way to suppress dimerisation is to add one of the CM partners as a dimer and in excess to the second partner [141]. In 2003, Grubbs and co-workers published a general model for the reactivity of CM-partners [142]. They ranked the olefins according to their ability to homodimerise in four different categories with type I being the most reactive and type IV the least reactive.

Type I olefins are sterically unhindered and electron-rich, type II–IV include increasingly electron-deficient and/or sterically hindered olefins. For practical purposes it is vital to use cross-partners with different levels of reactivity to achieve high selectivities. In general it can be said that the higher the difference in reactivity, the lower the necessary excess of the cross-partner (Scheme 3.10).

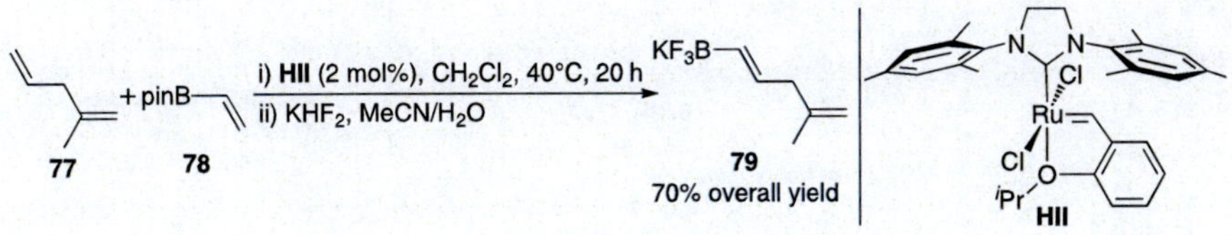

Scheme 3.10 CM to synthesise an organotrifluoroborate (pin = pinacol)

With this rule in mind the outcome of CM-reactions can often be predicted. In the synthesis of organotrifluoroborate **79** [143] the terminal double bond is a type I substrate, while the 1,1-disubstituted olefin can be considered type III. The reaction of 2-methyl-1,4-pentadiene **77** with type II cross-partner **78** furnishes **79** efficiently (only 2 mol% catalyst used) in good yields after two steps.

Another way for regioselective CM is to protect one of the double bonds. In the synthesis of Amphidinol 3 (AM3) one of the double bonds was masked as iodoolefin **80** (Scheme 3.11) [144]. Reductive removal of the iodine was followed by another CM leading to **85**. The synthesis of AM3 also yields an example of degradation by CM, which led to a structural revision for the stereocentre at C2 (Scheme 3.11, (2)).

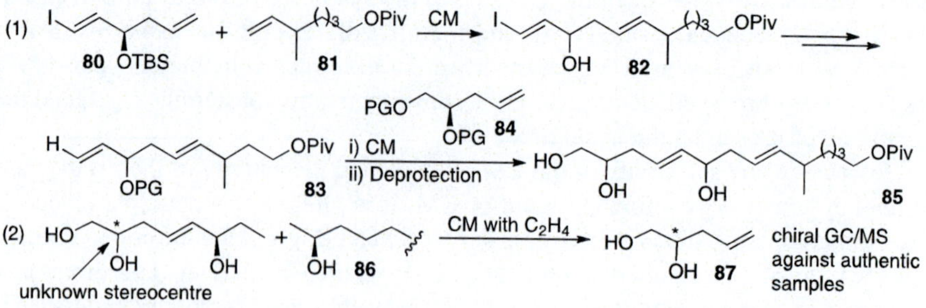

Scheme 3.11 Combinatorial use of CM and structural revision of AM3 by CM-degradation (PG = protecting group, Piv = pivaloyl, TBS = *tert*-butyl silyl)

Studies by the Grubbs group using sterically hindered olefins led to surprising results [145]. Use of the sterically less demanding catalyst **Ru1** for the CM between **88** and **89** shows activities superior to the **HII** catalyst. On the contrary, cross-metatheses leading to trisubstituted olefins like **92** show higher yields with sterically more hindered catalyst **Ru2** (Scheme 3.12).

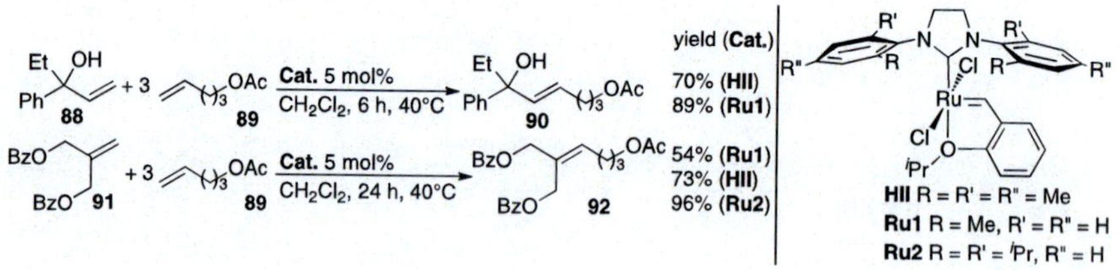

Scheme 3.12 CM of sterically demanding substrates

The authors believe this is due to competition of productive and unproductive CM-pathways. While the less encumbered **Ru1** allows larger residues for the generation of disubstituted olefins, it also favours unproductive pathways with 1,2-disubstituted substrates, diminishing the number of effective turnover events. As the size of the ligand decreases, so does the number of unproductive pathways relative to the productive.

The first cross metathesis to form a tetra-substituted olefin was achieved recently [146]. Howell and co-workers used lactams as substrates for CM with mono- and di-substituted olefins. The authors suggest that the limitations of the method are primarily due to steric reasons. Varying the electron density of the lactam showed no great influence on the reactivity while steric influences like α-branched allylic cross-partners or a methyl-group in the C4-position of the lactam both led to no reaction (Scheme 3.13).

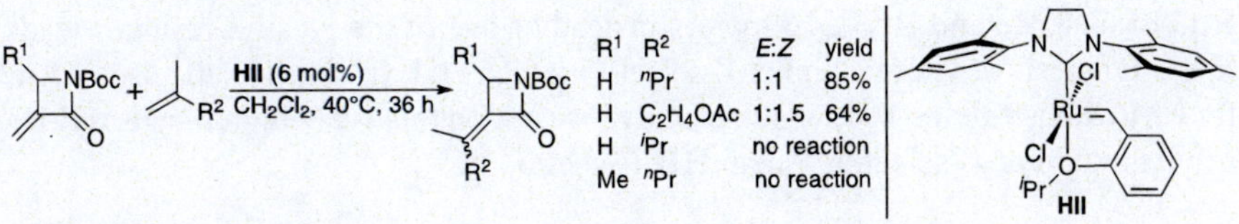

Scheme 3.13 CM to form tetra-substituted olefins (Boc = *tert*-butyloxycarbonyl)

The choice of the desired CM-partner directly influences the choice of the catalyst [147]. Comparing the **GII** and the **HII** catalysts shows that the latter has access to a much broader spectrum of cross-partners [148]. It is possible to use electron deficient cross-partners like acroleine, perfluorinated olefins, acrylonitrile, or other α,β-unsaturated carbonyls, whereas **GII** leads to no reaction or very low conversions due to side-reactions in these cases.

There are two factors which are important for the efficiency: the initiating ability of the catalyst to form the reactive species, which may need increased temperatures, and the amount of catalyst-consuming side reactions, which would increase with a rise in temperature.

Another possibility to achieve more efficient CM is the use of microwave irradiation [149]. Its use sometimes leads to drastically shortened reaction times and higher yields compared to thermal heating while the *E*/*Z*-ratio remains unaffected.

The choice of solvent may have a critical impact on efficiency too. In metathesis, dichloromethane, 1,2-dichloroethane and toluene are the solvents most commonly used. There are examples that show much higher yields in ring closing metathesis (RCM) when using fluorinated solvents [150]. An impressive effect of hexafluorobenzene as a solvent for CM is the modification of the steroid **93**: the use of 1,2-dichloroethane leads to a very low yield and significant amounts of dimerisation while the same reaction proceeds in 90% yield in C_6F_6 (Scheme 3.14) [151].

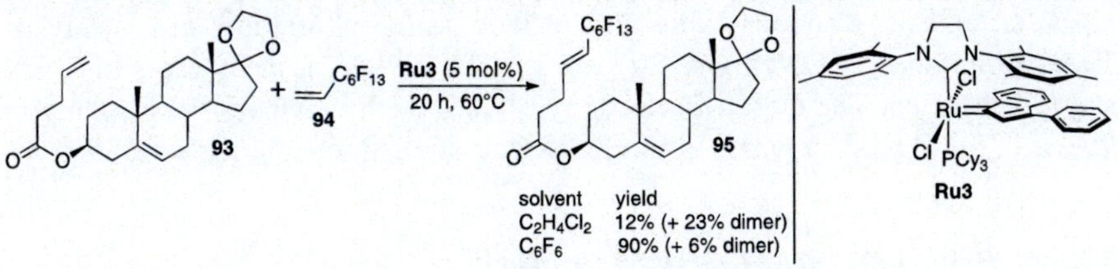

Scheme 3.14 Successful application of hexafluorobenzene as a solvent in CM

CM has, in most cases, a good to excellent *E*-selectivity. This is primarily due to steric reasons in the metallacycle intermediate of the metathesis. The high *E*-selectivity of the CM makes it an ideal method for the stereocontrolled synthesis of stilbenes [152], while there is still no highly *Z*-selective Ru-based catalyst known [153].

The use of unsymmetrically substituted Ru-complexes can lead to some interesting selectivities in CM [154]. The modified **HII**-catalysts **Ru4** and **Ru5**, where

Ru4 bears a Me and **Ru5** a Et moiety instead of one of the mesityl-residues leads, in the synthesis of **98,** to a higher *Z*-selectivity than **HII**. In contrast, the use of **Ru5** for CM with acroleine **100**, which usually reacts with relatively high *Z*-selectivities, results in higher *E*-selectivity than **HII** (Scheme 3.15).

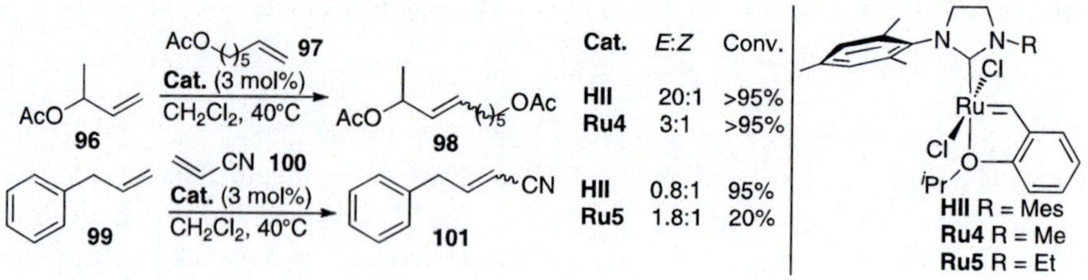

Scheme 3.15 Use of unsymmetrically substituted Ru-catalysts in CM

While allylhalides are commonly used as cross-partners, leading to good yields and showing in general good to excellent *E*-selectivities [155], the use of vinylhalides poses a much greater challenge. While most other vinyl compounds like vinylsulfones [156], vinyl boronates [157], vinyl phosphine oxides [158], vinyl oxazoles [159], vinyl silanes [133] or acrylonitrile [133] can be used without greater difficulties, vinyl halides tend to deactivate the catalyst. This also applies to vinylesters and vinyl carbonates [160] as well as to vinyl sulfoxides [156].

CM products from vinylhalides are highly desirable especially because of the possible use in metal catalysed coupling reactions. Johnson and co-workers. performed detailed studies of the possible deactivation pathways [161]. The Fischer-carbene complexes of the vinyl halides have an increased stability compared to their alkylidene counterparts and the Fischer carbenes may be deactivated either by migration of the phosphine or by elimination of HX leading to a carbide.

Despite those challenges, both Johnson [161] and Grela [162] performed several cross metathesis reactions with vinylhalides using phosphine free catalysts. Turnover numbers (TON) above 20 were very few, while in many cases the TON stayed below ten. The diastereoselectivity of CMs with vinylhalides is slightly in favour of the *Z* product which is similar to their acrolein-counterparts.

3.3.3 Ring Opening Cross Metathesis

One way to toggle problems due to low reactivity is to increase the reactivity of the cross partner by using a strained ring-system. This so-called ROCM competes with the ring opening polymerisation ROMP, so it should be noted that the cross partner needs to be more reactive than the ring-opened substrate. Norbornene type substrates are among the most common used due to their strain energies which resemble cyclobutenes (27.2 kcal/mol compared to 29.8 kcal/mol) [163]. The total synthesis of (+)-asteriscanolide includes an impressive example of ROCM which is followed as a cascade by a Cope rearrangement (Scheme 3.16) [164].

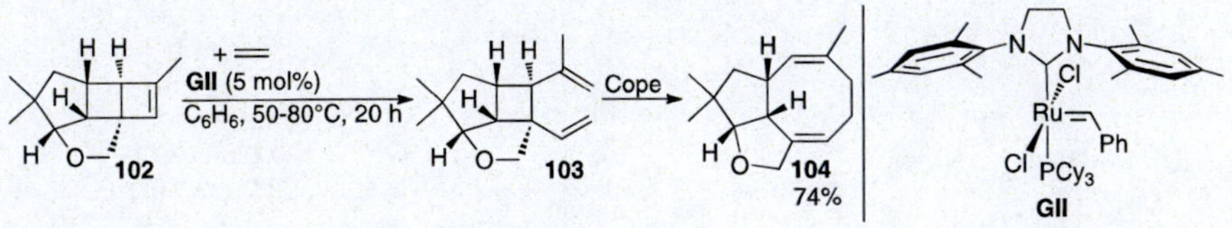

Scheme 3.16 ROCM/Cope-rearrangement cascade

ROCM also offers the possibility of a ring expansion. Tadano and co-workers used this approach successfully as a key step in the total synthesis of (+)-mycoepoxydiene, though only in moderate yield [165] (Scheme 3.17).

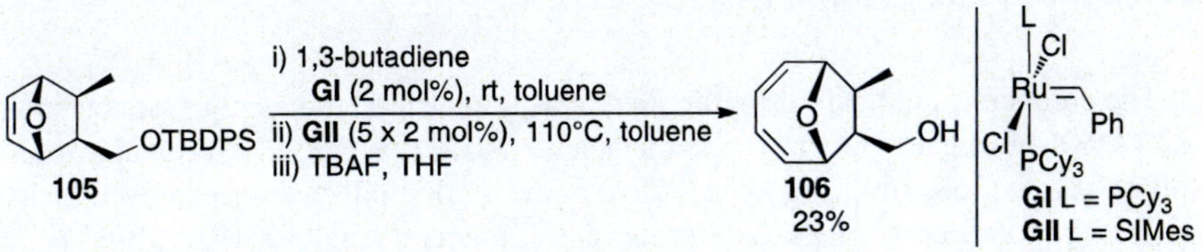

Scheme 3.17 Ring-expansion using ROCM [TBDPS = *tert*-butyldiphenylsilyl]

One drawback compared to CM is the rise of a new regioselectivity issue because two different regioisomers can be formed during the ROCM process. An interesting example for overcoming those problems has been published recently [166]. The formation of either isomer **109a–c** or **110a–c** can be controlled by the choice of the cross-partner: the electron deficient cross-partner methylacrylate **108a** yielded exclusively **109a**, while electron rich phenyl vinyl sulphide **108c** formed the opposite regioisomer **110c** (Table 3.1).

Table 3.1 Regioselectivity in ROCM reactions

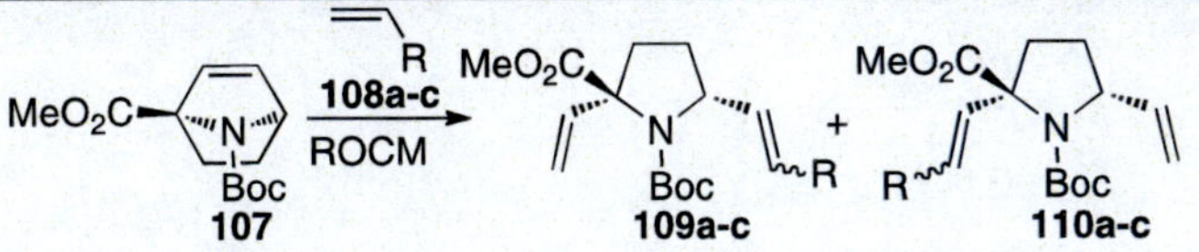

Entries	R	Olefin	Catalyst	Solvent	T (°C)	Yield (%)	Ratio **109:110**
1	CO_2Me	**108a**	**HII**	PhMe	80	98	100:0
2	Ph	**108b**	**GII**	$CHCl_3$	55	71	36:64
3	SPh	**108c**	**GII**	$CHCl_3$	55	89	0:100

An interesting way to control the stereoselectivity of metathesis-reactions is by intramolecular H-bonding between the chlorine ligands at the Ru-centre and an OH-moiety in the substrate [167]. With this concept and enantiomerically enriched allylic alcohols as substrates, the use of an achiral Ru–NHC complex can result in high diastereoselectivities like in the ROCM of **111–112** (Scheme 3.18). If non-H-bonding substrates are used, the selectivity not only decreases but proceeds in the opposite sense (product **113** and **114**).

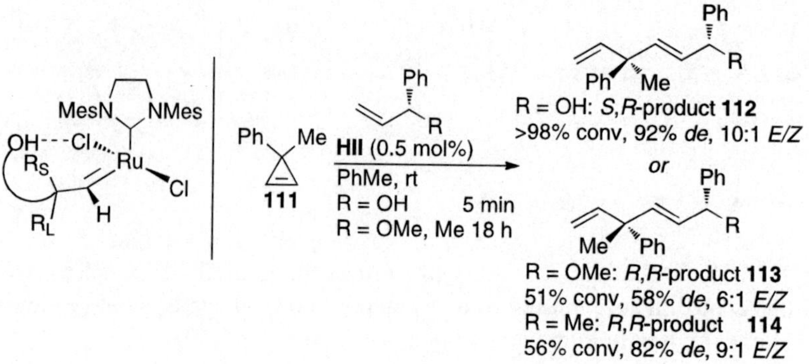

Scheme 3.18 Stereoinduction by H-bonding in ROCM

The number of publications in the field of asymmetric ring opening cross metathesis (AROCM) and asymmetric cross metathesis (ACM) using Ru–NHC complexes is still quite low [168, 169] (Fig. 3.33). Most of the work published concerns AROCM by Hoveyda and co-workers. The group uses bidentate catalysts which bind both with the NHC carbon atom as well as with a phenolate-moiety to the Ru-metal centre. Grubbs and co-workers have also designed chiral catalysts, even before Hoveyda, but up to 2006 they had not used them in ACM or AROCM. Contrary to **116–118**, **115** resembles **GII** instead of **HII**, bearing no styrene ether but a phosphine. The transfer of stereoinformation in monodentate **115** is realised through interaction between the chiral backbone and the bulky substituted aryl rings.

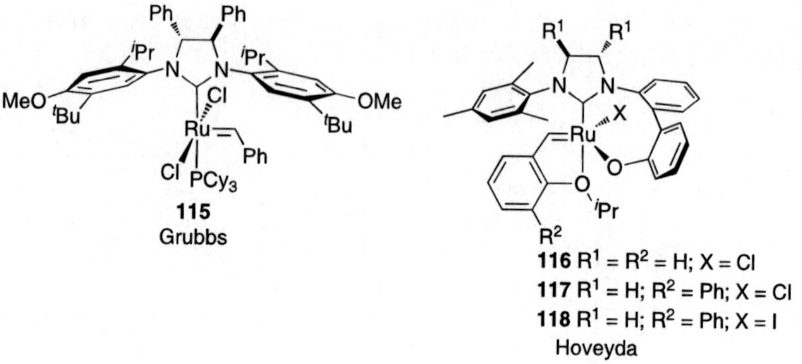

Fig. 3.33 Examples of chiral Ru-catalysts used for AROCM

The Hoveyda system gives in most cases good or very good enantioselectivities for AROCM. During the years several modifications were made (Scheme 3.19). The introduction of bulky substituents in the NHC-backbone (Entry 1) and substitution of Cl using I (Entry 2) led to increased stereoinduction while substituents *ortho* to the ether of the styrene increased the initiating abilities, making lower reaction temperatures possible. **115** was only scarcely used in AROCM, leading generally to lower *ee*'s than its counterparts **116–118**. Still, Grubbs and co-workers were the only group who published a successful ACM. However, the CM lacked reactivity and achieved at most a mediocre enantioselectivity (Entry 3).

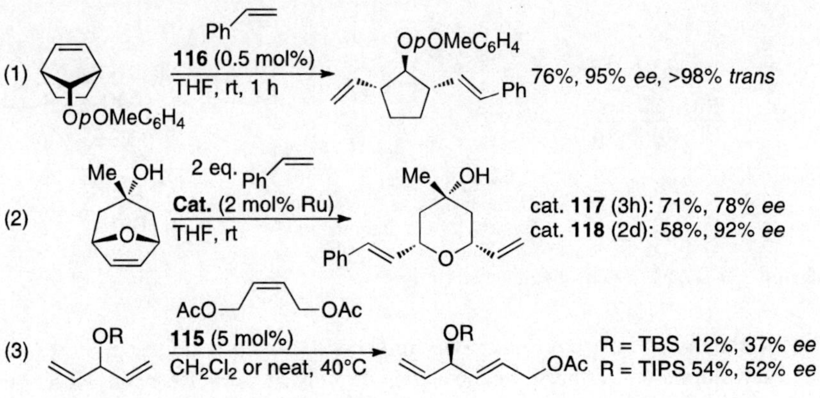

Scheme 3.19 Examples of AROCM and ACM

3.3.4 *Enyne Cross Metathesis*

The enyne cross metathesis was first developed in 1997 [170, 171]. Compared to CM it benefits from its inherent cross-selectivity and in theory it is atom economical, though in reality the alkene cross-partner is usually added in excess. The inability to control product stereochemistry of ECM reactions is the main weakness of the method. ECM reactions are often directly combined with other transformations like cyclopropanation [172], Diels-Alder reactions [173], cyclisations [174] or ring closing metathesis [175].

Use of an ethylene atmosphere can be beneficial to increase catalyst selectivity and reactivity [176] but it can also lead to the unwanted production of the corresponding 1,3-butadiene due to competition of ethylene with the desired cross-partner [177]. On the other hand ECM with ethylene as the cross-partner can be used as a very convenient method for the production of 2,3-disubstituted butadiene systems. The successful application of this methodology has for example been achieved by Mori *et al.* in the total synthesis of anolignan A (Scheme 3.20) [178].

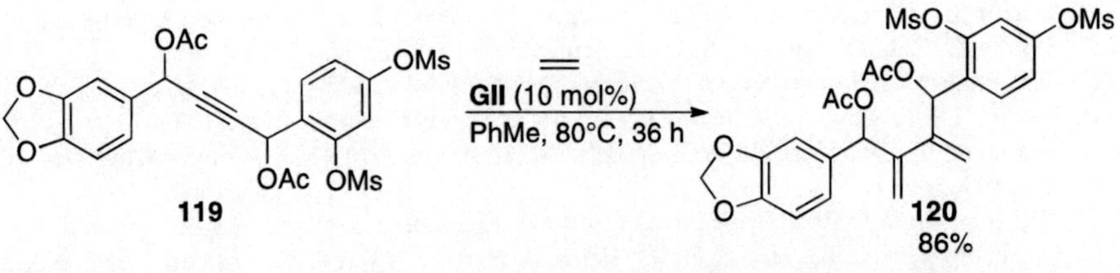

Scheme 3.20 ECM with ethylene to synthesise anolignan A precursor **120**

With the right choice of catalyst even very sterically demanding transformations are possible. Catalyst **Ru1**'s low steric congestion proved it superior to others in the enyne cross metathesis to yield highly substituted 1,3-dienes [179]. It catalyses the reaction of methylenecyclobutane **121** with a range of different cross-partners at 0°C with excellent yields (Scheme 3.21).

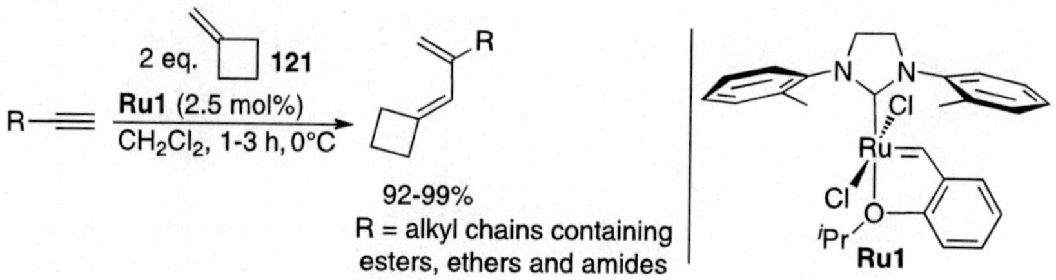

Scheme 3.21 ECM of sterically hindered substrates

Still, the control of stereoselectivity remains the main issue to be addressed in ECM-development. Furthermore, no asymmetric ECM using Ru-catalysts has been published until now.

References

1. Trnka TM, Grubbs RH (2001) Acc Chem Res 34:18–29
2. Grubbs RH (ed) (2003) Handbook of metathesis. Wiley, Weinheim
3. Hérisson JL, Chauvin Y (1971) Macromol Chem 141:161–176
4. Villemin D (1980) Tetrahedron Lett 21:1715–1718
5. Tsuji J, Hashiguchi S (1980) Tetrahedron Lett 21:2955–2958
6. Schrock RR, Murdzek JS, Bazan GC, Robbins J, DiMare M, O'Regan M (1990) J Am Chem Soc 112:3875–3876
7. Bazan GC, Schrok RR, Cho HN, Gibson VC (1991) Macromolecules 24:4495–4502
8. Fu GC, Grubbs RH (1992) J Am Chem Soc 114:7324–7325
9. Fu GC, Grubbs RH (1993) J Am Chem Soc 115:3800–3801
10. Wu Z, Nguyen ST, Grubbs RH, Ziller JW (1995) J Am Chem Soc 117:5503–5511
11. Schwab P, France MB, Ziller JW, Grubbs RH (1995) Angew Chem Int Ed Engl 34:2039–2041
12. Schwab P, Grubbs RH, Ziller JW (1996) J Am Chem Soc 118:100–110
13. Scholl M, Ding S, Lee CW, Grubbs RH (1999) Org Lett 1:953–956
14. Huang J, Stevens ED, Nolan SP, Petersen JL (1999) J Am Chem Soc 121:2674–2678
15. Scholl M, Trnka TM, Morgan JP, Grubbs RH (1999) Tetrahedron Lett 40:2247–2250
16. Garber SB, Kingsbury JS, Gray BL, Hoveyda AH (2000) J Am Chem Soc 122:8168–8179
17. Gessler S, Randl S, Blechert S (2000) Tetrahedron Lett 41:9973–9976
18. Wakamatsu H, Blechert S (2002) Angew Chem Int Ed 41:2403–2405
19. Grela K, Harutyunyan S, Michrowska A (2002) Angew Chem Int Ed 41:4038–4040
20. Weskamp T, Schattenmann WC, Spiegler M, Herrmann WA (1998) Angew Chem Int Ed 37:2490–2493
21. Weskamp T, Kohl FJ, Herrmann WA (1999) J Organomet Chem 582:362–365
22. Weskamp T, Kohl FJ, Hieringer W, Gleich D, Herrmann WA (1999) Angew Chem Int Ed 38:2416–2419
23. Ackermann L, Fürstner A, Weskamp T, Kohl FJ, Herrmann WA (1999) Tetrahedron Lett 40:4787–4790
24. Dinger MB, Nieczypor P, Mol JC (2003) Organometallics 22:5291–5296

25. Ledoux N, Allaert B, Pattyn S, Mierde HV, Vercaemst C, Verpoort F (2006) Chem Eur J 12:4654–4661
26. Louie J, Grubbs RH (2001) Angew Chem Int Ed 40:247–249
27. Jafarpour L, Schanz HJ, Stevens ED, Nolan SP (1999) Organometallics 18:5416–5419
28. Jafarpour L, Stevens ED, Nolan SP (2000) J Organomet Chem 606:49–54
29. Fürstner A, Ackermann L, Gabor B, Goddard R, Lehmann CW, Mynott R, Stelzer F, Thiel OR (2001) Chem Eur J 7:3236–3253
30. Dinger MB, Mol JC (2002) Adv Synth Catal 344:671–677
31. Ritter T, Hejl A, Wenzel AG, Funk TW, Grubbs RH (2006) Organometallics 25:5740–5745
32. Luan X, Mariz R, Gatti M, Costabile C, Poater A, Cavallo L, Linden A, Dorta R (2008) J Am Chem Soc 130:6848–6858
33. Vieille-Petit L, Luan X, Mariz R, Blumentritt S, Linden A, Dorta R (2009) Eur J Inorg Chem 1861–1870
34. Vieille-Petit L, Luan X, Gatti M, Blumentritt S, Linden A, Clavier H, Nolan SP, Dorta R (2009) Chem Commun 3783–3785
35. Gatti M, Vieille-Petit L, Luan X, Mariz R, Drinkel E, Linden A, Dorta R (2009) J Am Chem Soc 131:9498–9499
36. Rost D, Porta M, Gessler S, Blechert S (2008) Tetrahedron Lett 49:5968–5971
37. Süßner M, Plenio H (2005) Chem Commun 5417–5419
38. Leuthauser S, Schmidts V, Thiele CM, Plenio H (2008) Chem Eur J 14:5465–5481
39. Balof SL, P'Pool S, Berger NJ, Valente EJ, Shiller AM, Schanz HJ (2008) Dalton Trans 5791–5799
40. Schoeps D, Buhr K, Dijkstra M, Ebert K, Plenio H (2009) Chem Eur J 15:2960–2965
41. Ritter T, Day MW, Grubbs RH (2006) J Am Chem Soc 128:11768–11769
42. Occhipinti G, Bjørsvik HR, Jensen VR (2006) J Am Chem Soc 128:6952–6964
43. Berlin JM, Campbell K, Ritter T, Funk TW, Chlenov A, Grubbs RH (2007) Org Lett 9:1339–1342
44. Sanford MS, Love JA, Grubbs RH (2001) J Am Chem Soc 123:6543–6544
45. Stewart IC, Ung T, Pletnev AA, Berlin JM, Grubbs RH, Schrodi Y (2007) Org Lett 9:1589–1592
46. Fürstner A, Krause H, Ackermann, L, Lehmann CW (2001) Chem Commun 2240–2241
47. Prühs S, Lehmann CW, Fürstner A (2004) Organometallics 23:280–287
48. Vehlow K, Maechling S, Blechert S (2006) Organometallics 25:25–28
49. Ledoux N, Allaert B, Linden A, Voort PVD, Verpoot F (2007) Organometallics 26:1052–1056
50. Ledoux N, Linden A, Allaert B, Mierde HV, Verpoort F (2007) Adv Synth Catal 349:1692–1700
51. Vougioukalakis GC, Grubbs RH (2007) Organometallics 26:2469–2472
52. Vougioukalakis GC, Grubbs RH (2008) Chem Eur J 14:7545–7556
53. Weigl K, Köhler K, Dechert S, Meyer F (2005) Organometallics 24:4049–4056
54. Chung C, Grubbs RH (2008) Org Lett 13:2693–2696
55. Hong SH, Chlenov A, Day MW, Grubbs RH (2007) Angew Chem Int Ed 46:5148–5151
56. Vehlow K, Gessler S, Blechert S (2007) Angew Chem Int Ed 46:8082–8085
57. Kuhn KM, Bourg JB, Chung CK, Virgil SC, Grubbs RH (2009) J Am Chem Soc 131:5313–5320
58. Grisi F, Mariconda A, Costabile C, Bertolasi V, Longo P (2009) Organometallics 28:4988–4995
59. Enders D, Breuer K, Raade G, Runsink J, Teles JH, Melder JP, Ebel K, Brode S (1995) Angew Chem Int Ed Engl 34:1021–1023
60. Trnka TM, Morgan JP, Sanford MS, Wilhelm TE, Scholl M, Choi TL, Ding S, Day MW, Grubbs RH (2003) J Am Chem Soc 125:2546–2558
61. Anderson DR, Lavallo V, O'Leary DJ, Bertrand G, Grubbs RH (2007) Angew Chem Int Ed 46:7262–7265

62. Lavallo V, Canac Y, Prasang C, Donnadieu B, Bertrand G (2005) Angew Chem Int Ed 44:5705–5709
63. Vougioukalakis GC, Grubbs RH (2008) J Am Chem Soc 130:2234–2245
64. Despagnet-Ayoub E, Grubbs RH (2004) J Am Chem Soc 126:10198–10199
65. Despagnet-Ayoub E, Grubbs RH (2005) Organometallics 24:338–340
66. Yun J, Marinez ER, Grubbs RH (2004) Organometallics 23:4172–4173
67. Yang L, Mayr M, Wurst K, Buchmeiser MR (2004) Chem Eur J 10:5761–5770
68. Kumar PS, Wurst K, Buchmeiser MR (2009) Organometallics 28:1785–1790
69. Seiders TJ, Ward DW, Grubbs RH (2001) Org Lett 3:3225–3228
70. Funk TW, Berlin JM, Grubbs RH (2006) J Am Chem Soc 128:1840–1846
71. Fournier PA, Collins SK (2007) Organometallics 26:2945–2949
72. Fournier PA, Savoie J, Stenne B, Bédard M, Grandbois A, Collins SK (2008) Chem Eur J 14:8690–8695
73. Savoie J, Stenne B, Collins SK (2009) Adv Synth Catal 351:1826–1832
74. Grisi F, Costabile C, Gallo E, Mariconda A, Tedesco C, Longo P (2008) Organometallics 27:4649–4656
75. Van Veldhuizen JJ, Garber SB, Kingsbury JS, Hoveyda AH (2002) J Am Chem Soc 124:4954–4955
76. Gillingham DG, Kataoka O, Garber SB, Hoveyda AH (2004) J Am Chem Soc 126:12288–12290
77. Van Veldhuizen JJ, Gillingham DG, Garber SB, Kataoka O, Hoveyda AH (2003) J Am Chem Soc 125:12502–12508
78. Luan X, Mariz R, Robert C, Gatti M, Blumentritt S, Linden A, Dorta R (2008) Org Lett 10:5569–5572
79. Chaulagain MR, Sormunen GJ, Montgomery J (2007) J Am Chem Soc 129:9568–9569
80. Martin D, Kehrli S, d'Augustin M, Clavier H, Mauduit M, Alexakis A (2006) J Am Chem Soc 128:8416–8417
81. Xu L, Shi Y (2008) J Org Chem 73:749–751
82. Matsumoto Y, Yamada K, Tomioka K (2008) J Org Chem 73:4578–4581
83. Lee KS, Hoveyda AH (2009) J Org Chem 74:4455–4462
84. Lee Y, Li B, Hoveyda AH (2009) J Am Chem Soc 131:11625–11633
85. Lee Y, Hoveyda AH (2009) J Am Chem Soc 131:3160–3161
86. May TL, Brown MK, Hoveyda AH (2008) Angew Chem Int Ed 47:7358–7362
87. Brown MK, May TL, Baxter CA, Hoveyda AH (2007) Angew Chem Int Ed 46:1097–1100
88. Truett WL, Johnson DR, Robinson IM, Montague BA (1960) J Am Chem Soc 82:2337–2340
89. Calderon N, Chen HY, Scott KW (1967) Tetrahedron Lett 34:3327–3329
90. Hérisson JL, Chauvin Y (1971) Makromol Chem 141:161–176
91. Kenneth CD, Fogg DE (2007) NATO Sci Ser II: Math Phys Chem 243:285–303
92. Meier MAR, Fokou PA (2009) J Am Chem Soc 131:1664–1665
93. Bielawski CW, Grubbs RH (2007) Prog Polym Sci 32:1–29
94. Boehrsch V, Neidhoefer J, Blechert S (2006) Angew Chem Int Ed 45:1302–1305
95. Buchmeiser M (2000) Chem Rev 100:1565–1604
96. Demel S, Slugovc C, Stelzer F, Fodor-Csorba K, Galli G (2003) Macromol Rapid Commun 24:636–641
97. Choi T, Rutenberg IM, Grubbs RH (2002) Angew Chem 114:3995–3997
98. Slugovc C (2004) Macromol Rapid Commun 25:1283–1297
99. Slugovc C, Demel S, Riegler S, Hobisch J, Stelzer F (2004) Macromol Rapid Commun 25:475–480
100. Schrock RR, Hoveyda AH (2003) Angew Chem 115:4740–4782
101. Courchay FC, Sworen JC, Wagener KB (2003) Macromolecules 36:8231–8239
102. Sanford MS, Love JA, Grubbs RH (2001) J Am Chem Soc 123:6543–6554
103. Love JA, Morgan JP, Trnka TM, Grubbs RH (2002) Angew Chem 114:4207–4209
104. Slugovc C, Demel S, Stelzer F (2002) Chem Commun 2572–2573
105. de Frément P, Clavier H, Montembault V, Fontaine L, Nolan SP (2008) J Mol Catal A 283:108–113

106. Leitgeb A, Burtscher D, Bauer T, Slugovc C (2009) Chimica Oggi 27:30–32
107. Burtscher D, Lexer C, Mereiter K, Winde R, Karch R, Slugovc C (2008) J Polym Sci Part A: Polym Chem 46:4630–4635
108. Sanford MS, Ulman M, Grubbs RH (2001) J Am Chem Soc 123:749–750
109. Huang J, Stevens ED, Nolan SP, Petersen JL (1999) J Am Chem Soc 121:2374–2678
110. Lappert MF (1988) J Organomet Chem 358:185–213
111. Herrmann WA, Goosen LJ, Spiegler MJ (1997) J Organomet Chem 547:357–366
112. Ardeuengo AJ III, Gamper SF, Calabrese JC, Davidson FJ (1994) J Am Chem Soc 116:4391–4393
113. Westkamp T, Schattenmann WC, Herrmann WA (1998) Angew Chem 110:2631–2633
114. Scholl M, Trnka TM, Morgan JP, Grubbs RH (1999) Tetrahedron Lett 40:2247–2250
115. Frenzel U, Nuyken O (2002) J Polym Sci Part A: Polym Chem 40:2895–2916
116. Scholl M, Ding S, Lee CW, Grubbs RH (1999) Org Lett 6:953–956
117. Ritter T, Hejl A, Wenzel AG, Funk TW, Grubbs RH (2006) Organometallics 25:5740–5745
118. Demel S, Schoefberger W, Slugovc C, Stelzer F (2003) J Mol Catal A 200:11–19
119. Fürstner A, Ackermann L, Gabor B, Goddard R, Lehmann CW, Mynott R, Stelzer F, Thiel OR (2001) Chem Eur J 7:3236–3653
120. Dinger MB, Mol JC (2002) Adv Synth Catal 344:671–677
121. Bielawski CW, Benitez D, Grubbs RH (2002) Science 297:2041–2044
122. Boydston AJ, Xia Y, Kornfield JA, Gorodetskaya IA, Grubbs H (2008) J Am Chem Soc 130:12775–12782
123. Brookhard M, Rix F-C, DeSimone JM (1992) J Am Chem Soc 114:5894–5895
124. Pellecchia C, Pappalardo D, D'Arco M, Zambelli A (1996) Macromolecules 19:1158–1162
125. Wu Z, Grubbs RH (1995) Macromolecules 28:3502–3508
126. Al Samak B, Carvill AJ, Hamilton JG, Rooney JJ, Thompson JM (1997) Chem Commun 2057–2058
127. Amir-Ebrahimi V, Rooney JJ (2004) J Mol Catal A 208:115–121
128. Ilker MF, Coughlin EB (2002) Macromolecules 35:54–58
129. Vehlow K, Wang D, Buchmeiser MR, Blechert S (2008) Angew Chem Int Ed 47:2615–2618
130. Gstrein X, Burtscher D, Szadkowska A, Barbasiewicz M, Stelzer F, Grela K, Slugovc C (2007) J Polym Sci Part A: Polym Chem 45:3494–3500
131. Wang D, Wurst K, Knolle W, Decker U, Prager L, Naumov S, Buchmeiser MR (2008) Angew Chem 120:3311–3314
132. Szadkowska A, Grela K (2008) Curr Org Chem 12:1631–1647
133. For reviews on cross metathesis, see: (a) Connon SJ, Blechert S (2003) Angew Chem Int Ed 42:1900–1923; (b) Grubbs RH (2004) Tetrahedron 60:7117–7140; (c) Grubbs RH, Handbook of Metathesis 1st edition, Wiley-VCH, Weinheim 2003; (d) Nicolaou KC, Bulger PG, Sarlah D (2005) Angew Chem Int Ed 44:4490–4527
134. Zaja M, Blechert S (2004) Tetrahedron 60:9629–9634
135. Koch D, Maechling S, Blechert S (2007) Tetrahedron 63:7112–7119
136. Gebauer J, Arseniyadis S, Cossy J (2007) Org Lett 9:3425–3427
137. Liu X, Sternberg E, Dolphin D (2008) J Org Chem 73:6542–6550
138. Hsu MC, Junia AJ, Haight AR, Zhang W (2004) J Org Chem 69:3907–3911
139. Edelsztein VC, DI Chenna PH, Burton G (2009) Tetrahedron 65:3615–3623
140. Sheddan NA, Mulzer J (2006) Org Lett 8:3101–3104
141. O'Leary DJ, Blackwell HE, Washenfelder RA, Grubbs RH (1998) Tetrahedron Lett 39: 7427–7430
142. Chatterjee AK, Choi TL, Sanders DP, Grubbs RH (2003) J Am Chem Soc 125: 11360–11370
143. Esteban J, Costa AM, Vilarrasa J (2008) Org Lett 10:4843–4846
144. Oishi T, Kanemoto M, Swasono R, Matsumori M, Murata M (2008) Org Lett 10: 5203–5206
145. Stewart IC, Douglas CJ, Grubbs RH (2008) Org Lett 10:441–444
146. Liang Y, Raju R, Le T, Taylor CD, Howell AR (2009) Tetrahedron Lett 50:1020–1022

147. For a thorough review of Ru-NHC-catalysts for metathesis, see: Samojłowicz C, Bieniek M, Grela K (2009) Chem Rev 109:3708–3742; for ruthenium indenylidene-complexes in cross metathesis, see: Boeda F, Bantreil X, Clavier H, Nolan SP (2008) Adv Synth Catal 350:2959–2966; For HII-types systems, see: Schrodi Y, Pederson RL (2007) Aldrichimica Acta 40:45–52
148. Hoveyda AH, Gillingham DG, Van Veldhuizen JJ, Kataoka O, Garber SB, Kingsbury JS, Harrity JPA (2004) Org Biomol Chem 2:8–23
149. Coquerel Y, Rodriguez J (2008) Eur J Org Chem 1125–1132
150. (a) Rost D, Porta M, Gessler S, Blechert S (2008) Tetrahedron Lett 49:5968–5971; (b) Grandbois A, Collins SK (2008) Chem Eur J 14:9323–9329
151. Samojłowicz C, Bieniek M, Zarecki A, Kadyrov R, Grela K (2008) Chem Commun 6282–6284
152. (a) Cheikh L, Cariou R, Fischmeister C, Dixneuf PH (2007) Adv Synth Catal 349:546–550; (b) Ferré-Filmon K, Delaude L, Demonceau A, Noels AF (2005) Eur J Org Chem 3319–3325; (c) Velder J, Ritter S, Lex J, Schmalz HG (2006) Synthesis 273–278
153. For a Z-selective Mo-Catalyst, see: Flook MM, Liang AJ, Schrock RR, Müller P, Hoveyda AH (2009) J Am Chem Soc 131:7962–7963
154. Vehlow K, Maechling S, Blechert S (2006) Organometallics 25:25–28
155. Selected examples: (a) Bingcan L, Das SK, Roy R (2002) Org Lett 4:2723–2726; (b) Bandini M, Cozzi PG, Licciulli S, Umani-Ronchi A (2004) Synthesis 409–414; (c) Thibaudeau S, Fuller R, Gouverneur V (2004) Org Biomol Chem 2:1110–1112
156. (a) Michrowska A, Bieniek M, Kim M, Klajn R, Grela K (2003) Tetrahedron 59:4525–4531; (b) Bieniek M, Kołoda D, Grela K (2006) Org Lett 8:5689–5692
157. Morrill C, Funk TW, Grubbs RH (2004) Tetrahedron Lett 45:7733–7736
158. (a) Vinokurov N, Michrowska A, Szmigielska A, Drzazga Z, Wójciuk, Demchuk OM, Grela K, Pietrusiewicz KM, Butenschön H (2006) Adv Synth Catal 348:931–938; (b) Bisaro F, Gouverneur V (2005) Tetrahedron 61:2395–2400
159. Hoffman TJ, Rigby JH, Arseniyadis S, Cossy J (2008) J Org Chem 73:2400–2403
160. Caskey SR, Stewart MH, Kivela JE, Sootsman JR, Johnson MJA, Kampf JW (2005) J Am Chem Soc 127:16750–16751
161. (a) Macnaughtan ML, Johnson MJA, Kampf JW (2007) J Am Chem Soc 129:7708–7709; (b) Macnaughtan ML, Gary JB, Gerlach DL, Johnson MJA, Kampf JW (2009) Organometallics 28:2880–2887
162. Sashuk V, Samojłowicz C, Szadkowska A, Grela K (2008) Chem Commun 2468–2470
163. Lowry TH, Richardson KS (1987) Mechanism and theory in organic chemistry, 3rd edn. Harper Collins, New York, Chapter 2
164. Limanto J, Snapper ML (2000) J Am Chem Soc 122:8071–8072
165. Takao K, Yasui H, Yamamoto S, Sasaki D, Kawasaki S, Watanabe G, Tadano K (2004) J Org Chem 69:8789–8795
166. Carreras J, Avenzoa A, Busto JH, Peregrina JM (2009) J Org Chem 74:1736–1739
167. Hoveyda AH, Lombardi PJ, O'Brien RV, Zhugralin AR (2009) J Am Chem Soc 131:8378–8379
168. (a) Van Veldhuizen JJ, Garber SB, Kingsbury JS, Hoveyda AH, et al (2002) J Am Chem Soc 124:4954–4955; (b) Van Veldhuizen JJ, Gillingham DG, Garber SB, Kataoka O, Hoveyda AH (2003) J Am Chem Soc 125:12505–12508; (c) Gillingham DG, Kataoka O, Garber SB, Hoveyda HA (2004) J Am Chem Soc 126:12288–12290; (d) Van Veldhuizen JJ, Campbell JE, Giudici RE, Hoveyda AH (2005) J Am Chem Soc 127:6877–6882; (e) Giudici RE, Hoveyda AH (2007) J Am Chem Soc 129:3824–3825; (f) Gillingham DG, Hoveyda AH (2007) Angew Chem Int Ed 46:3860–3864
169. Berlin JM, Goldberg SD, Grubbs RH (2006) Angew Chem Int Ed 45:7591–7595
170. (a) Stragies R, Schuster M, Blechert S (1997) Angew Chem Int Ed 36:2518–2520; (b) Kinoshita A, Sakakibara N, Mori M (1997) J Am Chem Soc 119:12388–12389
171. For reviews on enyne metathesis, see: (a) Mori M (2007) Adv Synth Catal 349:121–135; (b) Diver ST, Giessert AJ (2004) Chem Rev 104:1317–1382; (c) Poulsen CS, Madsen R (2003) Synthesis 1–18

172. Murelli RP, Catalán S, Gannon MP, Snapper ML (2008) Tetrahedron Lett 49:5714–5717
173. (a) Kotha S, Mandal K, Banerjee S, Mobin SM (2007) Eur J Org Chem 1244–1255; (b) Arimitsu S, Fernández B, del Pozo C, Fustero S, Hammond GB (2008) J Org Chem 73:2656–2661
174. Kalbarczyk KP, Diver ST (2009) J Org Chem 74:2193–2196
175. (a) Kotha S, Singh K (2007) Eur J Org Chem 5909–5916; (b) Park S, Kim M, Lee D (2005) J Am Chem Soc 127:9410–9415
176. Giessert AJ, Diver ST (2003) Synthesis 466–471
177. Stragies R, Voigtmann U, Blechert S (2000) Tetrahedron Lett 41:5465–5468
178. Mori M, Tonogaki K, Nishiguchi N (2002) J Org Chem 67:224–226
179. Clark DA, Basile BS, Karnofel WS, Diver ST (2008) Org Lett 10:4927–4929

Chapter 4
N-Heterocyclic Carbene Complexes in Polymerisation, Oligomerisation and Telomerisation Reactions

David S. McGuinness and Kingsley J. Cavell

Abstract Over the past decade significant advances have been made in the fields of polymerisation, oligomerisation and telomerisation with metal-NHC catalysts. Complexes from across the transition series, as well as lanthanide examples, have been employed as catalysts for these reactions. Recent developments in the use of metal-NHC complexes in α-olefin polymerisation and oligomerisation, CO/olefin copolymerisation, atom-transfer radical polymerisation (ATRP) and diene telomerisation are discussed in subsequent sections.

4.1 Polymerisation and Oligomerisation of α-Olefins

It was several years after NHCs started appearing in catalysis that they were first employed in olefin polymerisation. An international patent with a priority date of 1998 appears to be the first time NHCs were tested as ligands in this regard, in this instance on Cr (Section 4.1.1) [1]. Given the interest in non-metallocene catalysed olefin polymerisation [2, 3], it is perhaps surprising that metal-NHC complexes have not featured more largely in this area. Perhaps one of the reasons these complexes have not found wider application in alkene polymerisation is the propensity of alkyl-metal NHC complexes to decompose *via* alkyl-imidazolium elimination (Scheme 4.1) [4, 5]. Alkyl-metal or metal-hydride intermediates are involved in each step of the polymerisation cycle. Accordingly, this reaction represents a route to decomposition of carbene-based catalysts [4]. Methods do exist to limit this reaction, however, including introduction of steric bulk to the carbene and protection

D. S. McGuinness (✉)
School of Chemistry, University of Tasmania, Private Bag 75, Hobart, Tasmania 7000, Australia
e-mail: David.McGuinness@utas.edu.au

K. J. Cavell (✉)
School of Chemistry, Cardiff University, Main Building, Park Place, CF10 3AT Cardiff, UK
e-mail: Cavellkj@cf.ac.uk

C. S. J. Cazin (ed.), *N-Heterocyclic Carbenes in Transition Metal Catalysis and Organocatalysis*, Catalysis by Metal Complexes 32,
DOI 10.1007/978-90-481-2866-2_4, © Springer Science+Business Media B.V. 2011

via donor-functionalisation and chelation [5]. In addition, the metal employed in catalysis has a bearing, as carbene-alkyl elimination is formally a reductive process (Scheme 4.1). While it can be facile for late transition metals, early transition metals may be less susceptible, as they tend to favour higher oxidation states. As a result of these considerations, active catalysts from right across the transition series have now been reported, as discussed below.

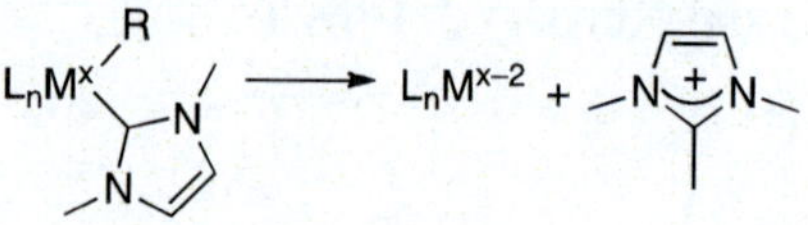

Scheme 4.1 Carbene-alkyl reductive elimination

Oligomerisation and polymerisation with NHC complexes has been carried out with a range of monomers, the most common being ethylene, propene, styrene and norbornene. The copolymerisation of CO/olefin is discussed in Section 4.2, while polymerisation of polar monomers *via* the ATRP mechanism is covered in Section 4.3. In the current section, activities for ethylene polymerisation have been calculated in terms of $kg_{(product)} \cdot mol_{(cat)}^{-1} \cdot bar_{(ethylene)}^{-1} \cdot h^{-1}$ ($kg \cdot mol^{-1} \cdot bar^{-1} \cdot h^{-1}$). In keeping with past reviews of ethylene polymerisation [2, 3], the following descriptors of activity have been used, bearing in mind the limitations of comparing different systems under different conditions: very high (>1 000); high (1 000–100); moderate (100–10); low (10–1); very low (<1). With all other monomers, activities are reported as turnover frequencies (TOF, h^{-1}), and are given with respect to the number of turnovers of the monomer.

4.1.1 *Early Transition Metals and Lanthanides*

Simple monodentate NHCs are somewhat susceptible to dissociation when coordinated to early transition metals [6], so in most cases multidentate chelating ligands are employed in which the carbene is tethered to a strongly coordinating anchoring group. This is not universally the case however, and simple monodentate NHC complexes of Zr **1** (Fig. 4.1) have been studied [7]. The complexes were activated with MAO and tested for ethylene polymerisation, leading to moderate activities between 7 and 75 $kg \cdot mol^{-1} \cdot bar^{-1} \cdot h^{-1}$ for linear polyethylene.

In all other cases, multidentate ligands have been employed with the early transition metals. Alkyl Zr and Hf complexes of a diamido-NHC ligand **2** (Fig. 4.1) were investigated [6] and treatment of these complexes with $[Ph_3C][B(C_6F_5)_4]$ generated monomethyl cations which were slightly active for 1-hexene polymerisation. The Zr cation was also moderately active in ethylene polymerisation (125 $kg \cdot mol^{-1} \cdot bar^{-1} \cdot h^{-1}$). The active species was thought to be short lived however, rapidly decomposing to an inactive complex. Ethylene polymerisation with the Hf cation was not reported.

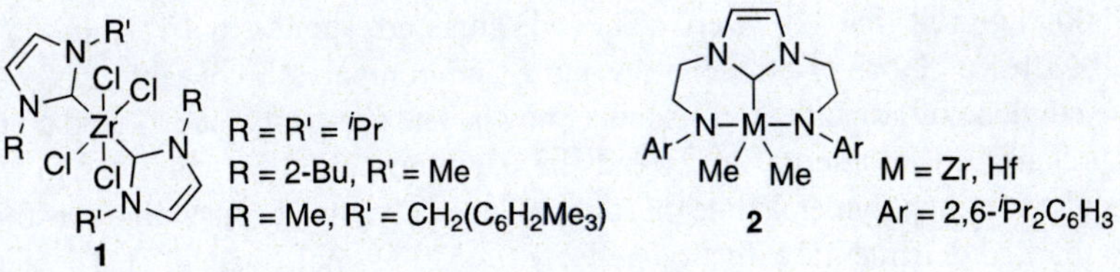

Fig. 4.1

A number of highly active ethylene polymerisation catalysts have resulted from the combination of functionalised NHC ligands with Ti, the first of these was the bis(phenolate)carbene ligated complex **3** [8]. Upon activation with modified MAO (MMAO), this species gave an activity of 290 kg·mol^{-1}·bar^{-1}·h^{-1} in the one test reported, making it one of the most active carbene-based olefin polymerisation catalysts known. In later work the same complex was evaluated with straight MAO activation, and activities of up to *ca.* 100 kg·mol^{-1}·bar^{-1}·h^{-1} were reported for linear polyethylene production [9].

The Ti(III) complex **4** (Fig. 4.2) was tested as part of a broader survey of ethylene polymerisation with complexes of the bis(carbene)pyridine ligands [10]. Along with MAO the catalyst generated was highly active, with productivities up to 790 kg·mol^{-1}·bar^{-1}·h^{-1} being obtainable. The V analogue **5** (Fig. 4.2) was also prepared and was somewhat more active than Ti, yielding very high activities (>1 000 kg·mol^{-1}·bar^{-1}·h^{-1}) again after MAO activation. The highest activity achieved with V was 1 450 kg·mol^{-1}·bar^{-1}·h^{-1}, which corresponds to an ethylene TOF of 51 000 h^{-1}. The polyethylene produced by both systems was of very high molecular weight (>2 M), while the melting temperatures of the polymers (Ti, 141°C; V, 144°C) were consistent with high density polyethylene with no significant branching.

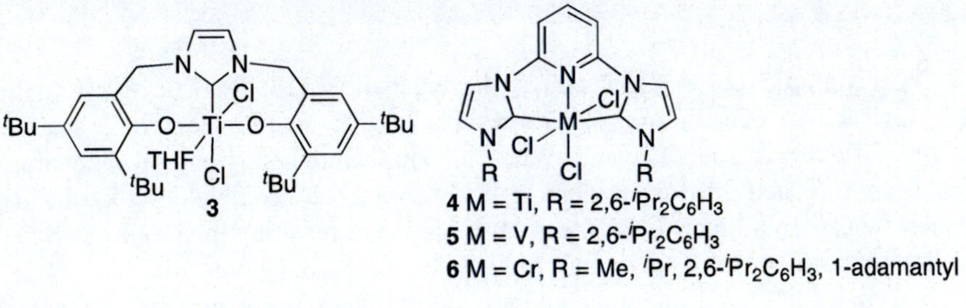

Fig. 4.2

The Cr analogues **6** (Fig. 4.2) of the bis(carbene)pyridine systems were found to be exceptionally active for the oligomerisation of ethylene [10, 11]. Activation with MAO led to optimal results, and complexes with Me, ^{i}Pr and 2,6-$^{i}Pr_2C_6H_3$ substitution gave rise to a modified Schulz-Flory distribution of α-olefins along with formation of some low molecular weight polymer. The distribution of oligomers was dependent upon ligand substitution, indicating that the ligand remains coordinated in the active species. Early screening revealed that the 2,6-$^{i}Pr_2C_6H_3$ substituted complex was the most active, and with limited optimisation exceptionally high activities up to 40 000 kg·mol^{-1}·bar^{-1}·h^{-1} were achieved. This corresponds to a TOF in ethylene of 1.4 M h^{-1} at 1 bar of ethylene pressure, and compares favourably to the most active ethylene oligomerisation systems known. The adamantyl substituted complex was prepared in an attempt to extend the usefulness of these catalysts to higher molecular weight polymer. While this did produce polyethylene, the large substituents had a marked detrimental effect on activity (*ca.* 70 kg·mol^{-1}·bar^{-1}·h^{-1}). The activity of the catalysts decreases over a 30 min run, and this deactivation was dependent on the temperature; above 50°C a very rapid deactivation was observed.

In follow up mechanistic studies [12], it was shown that catalysis occurs *via* an extended metallacycle mechanism as opposed to conventional Cossee-Arlman linear chain growth. This mechanism gives rise to a distorted Schulz-Flory distribution of oligomers due to less favourable product release at the early stages of metallacycle growth (Cr–C_4, Cr–C_6). Significant amounts of branched olefins (vinylidenes) and linear internal olefins were produced at high catalyst loadings, and these result from secondary incorporation of α-olefins into the growing metallacycles. This latter observation led to a study of α-olefin homo-oligomerisation with these catalysts [13]. Olefins from propene to 1-octene were oligomerised, predominately to dimers, *via* a metallacycle which for the most part produced vinylidene oligomers (Scheme 4.2). The results of this work were compared to similar results with the Cr/SiO_2 Phillips catalyst, and provide some degree of support for a metallacycle mechanism with this commercial polymerisation catalyst.

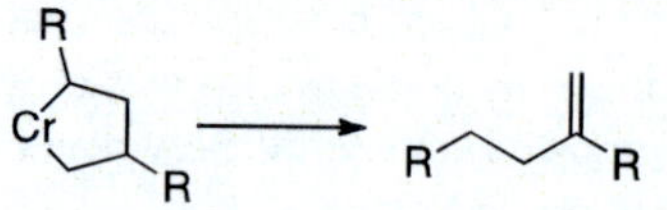

Scheme 4.2 Metallacycle route to vinylidene dimers

Other examples of Cr–NHC catalysts have generally been less effective, and the very high activity of complexes **6**, *via* a metallacycle mechanism, seems to be specific to the pyridylbis(carbene) ligand. When a central thiophene donor was incorporated, as in **7**, activity for ethylene polymerisation was completely absent [12]. Bidentate carbene-donor ligands were also prepared, resulting in complexes **8** and **9**. The activity in each case was low-moderate (**8**: 8 kg·mol^{-1}·bar^{-1}·h^{-1}; **9**: 11 kg·mol^{-1}·bar^{-1}·h^{-1}), and oligomerisation occurred *via* a linear growth mechanism (Cossee-Arlman) (Fig. 4.3).

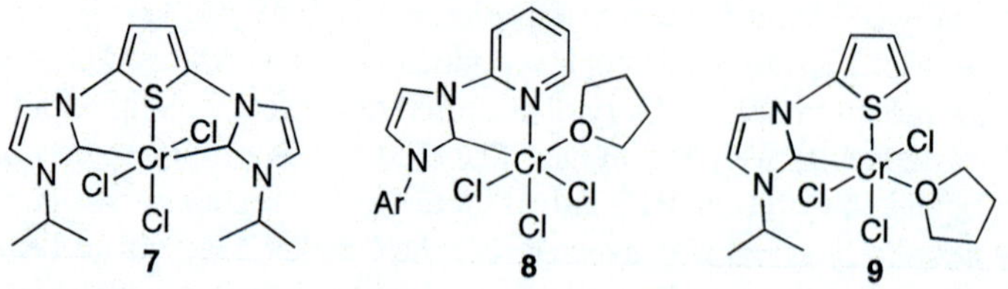

Fig. 4.3

A patent published in 2000 from Borealis [1] seems to be the first time that NHC complexes were evaluated for olefin polymerisation. In this disclosure the Cr(II) complex **10** (Fig. 4.4) was activated with MAO and tested both as a discrete complex and supported on SiO_2, $MgCl_2/SiO_2$ and polystyrene:divinylbenzene. The highest activity reported for ethylene polymerisation was 67 kg·mol^{-1}·bar^{-1}·h^{-1}. Interestingly, run times up to around 5 h are reported, indicating that the active catalyst has a high stability. It is not possible to judge if this is a result of complexation with the carbene ligand, however, all of the longer runs are conducted with supported catalysts, and it is well known that silica supported alkyl chromium complexes are active for ethylene polymerisation in the absence of additional ligands [14]. The discrete complex gave an activity of 29 kg·mol^{-1}·bar^{-1}·h^{-1} over a 40 min run. Around the same time as this patent appeared, Jolly and co-workers independently prepared a similar Cr(III) complex, **11** [15] (Fig. 4.4). Although not quantified, **11** was reported to give a low activity for ethylene polymerisation after treatment with MAO. In another patent from Borealis, Cr(III) complexes of structure **12** (Fig. 4.4) were reported to polymerise ethylene after treatment with MAO [16]. The activities reported were very low below 1 kg·mol^{-1}·bar^{-1}·h^{-1} in each case. Theopold and co-workers also looked at bidentate ligands, in this case the chelating bis(carbene) complexes of Cr(II) and Cr(III), **13** [17] (Fig. 4.4). The Cr(III) complex displayed a low activity (1.8 kg·mol^{-1}·bar^{-1}·h^{-1}) when activated with MAO, while the Cr(II) analogue was inactive. In the absence of additional ligation, Cr salts in combination with MAO display high activity for ethylene oligomerisation [18]. Thus, it seems in some cases NHC ligands can act as very effective poisons toward catalysis.

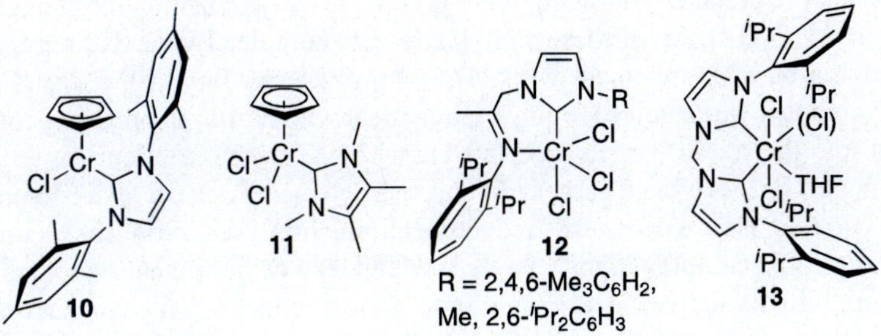

Fig. 4.4

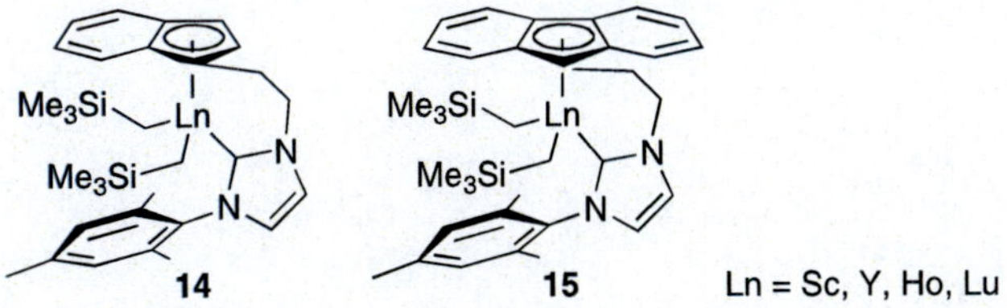

Fig. 4.5

There has also been some interest in NHC-lanthanide complexes as polymerisation catalysts. Indenyl and fluorenyl functionalised NHC complexes of structures **14** and **15** (Fig. 4.5) were evaluated for isoprene polymerisation following activation with $[Ph_3C][B(C_6F_5)_4]/Al^iBu_3$ [19, 20]. The Sc complexes of both ligands were inactive. The Y, Lu and Ho complexes of the fluorenyl substituted ligand **15** were catalysts for the living polymerisation of isoprene with high selectivity for 3,4 polymerisation. The Ho complex was the most active, showing a TOF of 167 h^{-1}, followed by Y (125 h^{-1}) and Lu (83 h^{-1}). The indenyl functionalised complexes **14** displayed lower activities, but did follow the same trend, Ho>Y>Lu. The indenyl functionalised complexes also led to lower 3,4 selectivity in the polymer.

4.1.2 *Mid-Transition Metals*

Given the success of the Grubbs-type NHC–Ru catalysts in metathesis polymerisation (Chapter 3), it is somewhat surprising that more research has not been done on mid-transition metal carbene complexes for coordination-insertion polymerisation. At this stage however, there are only a few reported attempts with the metals Co, Fe and Ir.

The results with Co and Fe are thus far not very promising. Complexes of the form **16** (Fig. 4.6) were disclosed in a patent of 2001, and were tested for ethylene polymerisation following MAO activation [16]. The Co complex was reported to give a low activity (2 $kg \cdot mol^{-1} \cdot bar^{-1} \cdot h^{-1}$), while the Fe complex was inactive. Bis(carbene)pyridine complexes of Fe(II), Fe(III) and Co(II) of structure **17** (Fig. 4.6) were prepared following success with the early transition metal analogues (Section 4.1.1) [10]. All of these complexes were completely inactive for ethylene polymerisation after treatment with MAO, however, the fate of the ligand in the Fe(III) complex was studied (following treatment with MAO) in order to gain some insight into the stability of the alkylated complex. The imidazolium cation corresponding to the methylated carbene ligand was detected, suggesting decomposition of the complex *via* carbene-alkyl reductive elimination. As such, it seems unlikely the alkylated Fe complex is sufficiently stable to lead to significant turnovers, even if it were initially active. At the same time, Danopoulos [21] also prepared Co(II) and Co(I) complexes of the same ligand, and likewise reported that these are not active in ethylene polymerisation.

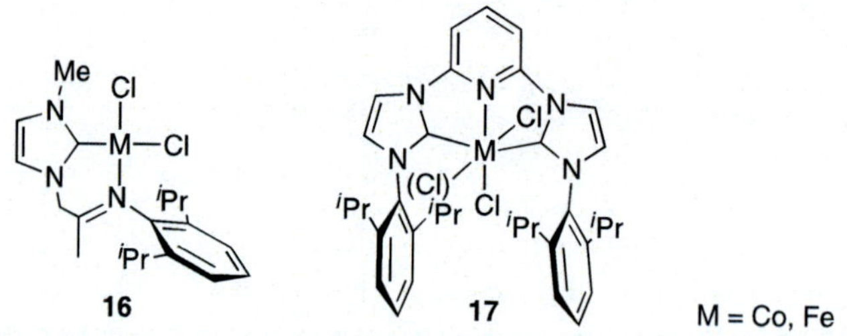

Fig. 4.6

More success has been had with Ir complexes incorporating permethylcyclopentadiene and NHC ligands. Complexes **18–20** (Fig. 4.7) were evaluated for norbornene polymerisation following activation with MAO [22]. Complex **19** was the most active, giving a TOF of 12 220 h^{-1} over 10 min, followed by **18** (TOF = 3 220 h^{-1}), while **20** was inactive, indicating that a hemilabile pendant group seems essential. Analysis (NMR) of the polymers formed with **18** and **19** shows that polymerisation proceeds *via* an addition (coordination-insertion) mechanism.

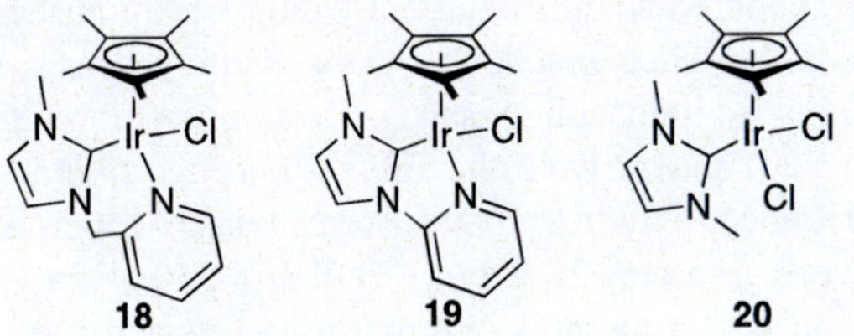

Fig. 4.7

4.1.3 Late Transition Metals

Olefin dimerisation with Ni–NHC complexes became a topic of interest following reports of Ni(II) phosphine complexes being employed in imidazolium-based ionic liquid solvents [23, 24]. It had previously been established that alkyl-Ni(II) complexes containing NHC ligands can rapidly decompose *via* imidazolium formation (Scheme 4.1) [5], and it was thus of interest to explore the effect that an excess of the imidazolium cation would have on this reaction.

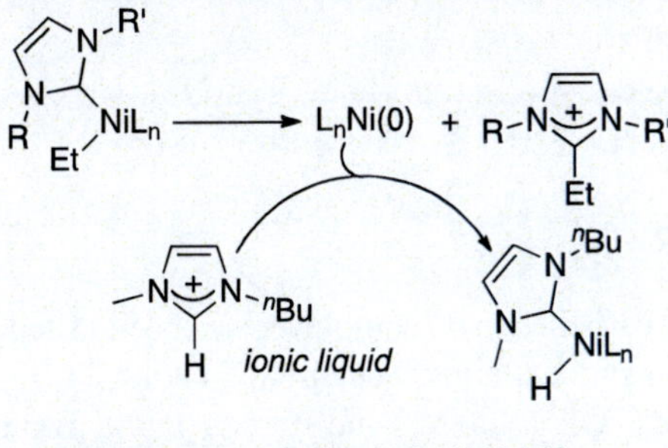

Scheme 4.3 Regeneration of Ni–H active species in ionic liquid

When dicarbene complexes of the form **21** were tested for 1-butene or propene dimerisation, upon activation with $AlEt_2Cl$ or MAO in toluene, rapid deactivation took place yielding Ni(0) [25]. It was shown that this decomposition did indeed involve carbene-hydride and carbene-alkyl reductive elimination. Some dimerisation was evident at −15°C (TON = 50), however decomposition of the intermediate Ni species seemed too rapid for effective catalysis. In contrast, when the complexes were tested in an imidazolium based ionic liquid, rapid dimerisation of propene and 1-butene was observed ($TOF_{propene}$ = 75 000 h^{-1}; $TOF_{1\text{-}butene}$ < 7 000 h^{-1}). This effect was ascribed to the ability of the imidazolium-based ionic liquid to oxidatively add to the Ni(0) decomposition product, effectively regenerating a Ni(II) hydride active catalyst (Scheme 4.3). Product distributions resulting from the different catalysts were very similar, suggesting a common active species was resulting from incorporation of the ionic liquid. This however does not represent a long-term solution to the problem, as eventually all of the imidazolium cation would become alkyl substituted at the two-position, whereupon its stabilising effect would be diminished. It was, in fact, subsequently shown that such reactions can be employed as a synthetic tool to catalytically functionalise imidazolium salts at the two-position [26].

Interestingly, a 2001 patent from the IFP [27] seems to demonstrate that high activities can be obtained in hydrocarbon solvents when $AlEtCl_2$ is employed as the activator. In this work, complex **22** led to a TOF of 8 370 h^{-1} in 1-butene dimerisation. A later study in which the bulky complex **23** was activated with MAO concluded that some ethylene dimerisation was occurring, but this was not quantified [28]. It would seem that the catalyst gave a very low activity in this case.

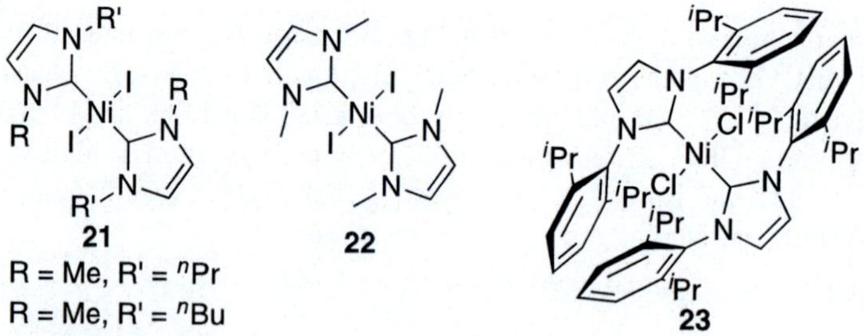

21

R = Me, R' = ^{n}Pr
R = Me, R' = ^{n}Bu
R = Me, R' = ^{i}Pr
R = ^{i}Pr, R' = ^{i}Pr

22

23

Fig. 4.8

A number of mono-carbene Ni(II) complexes has been tested for oligomerisation and polymerisation of various olefins. The allyl-Ni cation **24** (Fig. 4.9) was found to be active for butadiene (TOF = 400 h^{-1}) and styrene polymerisation (TOF = 323 h^{-1}) [29]. When a less bulky *N,N'*-dimethyl carbene analogue was employed in styrene conversion, a higher activity resulted and styrene oligomers were formed instead (85% head-to-tail dimers). In the case of butadiene polymerisation, it was speculated that catalysis might be due to a carbene-free allyl-Ni complex. Neutral Ni(II)-benzyl complexes of the form **25** were tested with ethylene and norbornene, the best results being obtained with the weakly coordinating triflate ligand, perhaps suggesting cation formation is necessary prior to olefin insertion [30]. Ethylene was very slowly (TOF = 0.5 h^{-1}) converted to butenes, while norbornene underwent additional polymerisation with TOFs up to *ca.* 3 500 h^{-1}.

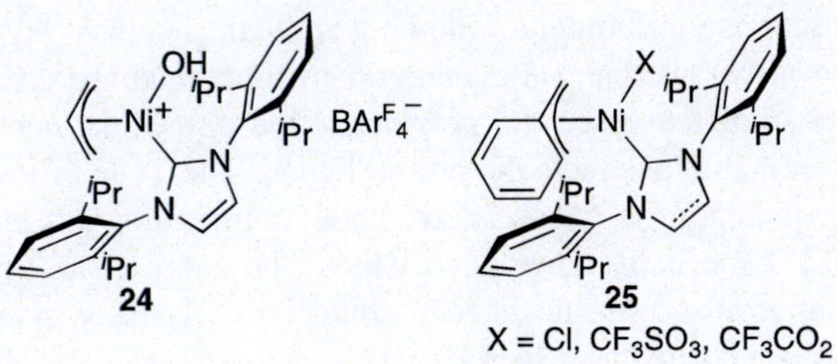

Fig. 4.9

NHC–Ni(II) complexes incorporating cyclopentadienyl or indenyl ligands have been evaluated for polymerisation in a number of cases. Upon treatment with MAO, complex **26** (Fig. 4.10) was moderately active (15 kg·mol^{-1}·bar^{-1}·h^{-1}) for the dimerisation of ethylene to 1-butene [31]. Pietrzykowski and co-workers tested closely related cyclopentadienyl complexes such as **27** (Fig. 4.10) for styrene polymerisation following activation with MAO, resulting in atactic polystyrene with TOFs up to 1 480 h^{-1} [32]. In this work the polymers were studied by MALDI-TOF MS which showed there to be no methyl end groups present. This result suggests that the mechanism is not one of coordination-insertion, and the authors suggested a cationic mechanism initiated by [(NHC)(Cp)Ni]$^+$[MAO]$^-$. A complex in which the indenyl and carbene ligands are tethered (**28**) (Fig. 4.10) has also been prepared and tested for styrene polymerisation [33]. After treatment with $NaBPh_4$ this system produced polystyrene, albeit with a very low activity (225 turnovers after 24 h). In light of the aforementioned results of Pietrzykowski, it seems likely that catalysis proceeds *via* a cationic mechanism, again initiated by a Ni cation formed by bromide abstraction.

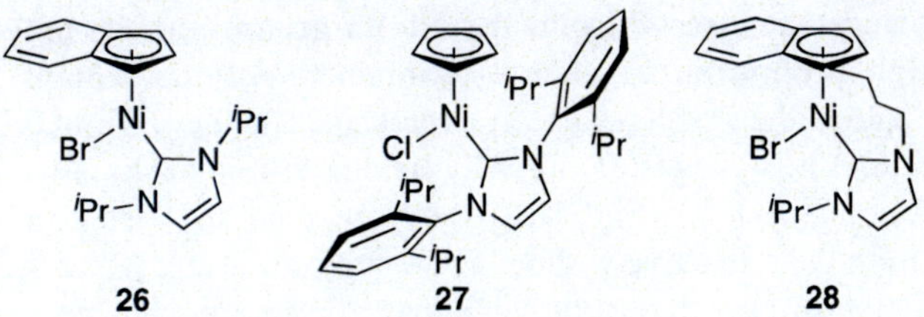

Fig. 4.10

The remainder of the work on Ni(II) complexes involves the use of chelating ligands in which the carbene is functionalised with pendant heteroatom donor(s). The picolyl-functionalised NHC dicationic complex **29** (Fig. 4.11) was tested for ethylene polymerisation after treatment with MAO [34]. This complex was found to be highly active in a preliminary test (330 kg·mol^{-1}·bar^{-1}·h^{-1}), giving predominantly linear polyethylene. Unfortunately this work does not seem to have been followed up. The same system was active for norbornene polymerisation (TOF = 24 400 h^{-1} over 1 h). Maximum activity was achieved at 80°C whereafter thermal deactivation became significant, although the nature of this deactivation was not studied. The phenoxide-functionalised carbene complex **30** (Fig. 4.11) was also reported to polymerise norbornene following activation with MAO [35], although no details were provided. The salicylaldiminato-functionalised NHC complex **31** (Fig. 4.11) was studied for styrene polymerisation upon treatment with $NaBPh_4$ [36], which presumably abstracts the bromo ligand. The TON of styrene was only 500 after 12 h, indicating a poorly active catalyst (or a low concentration of the active species). The mechanism was not discussed, however it is difficult to envisage how a coordination-insertion mechanism could be initiated as such, and again a cationic mechanism seems most likely.

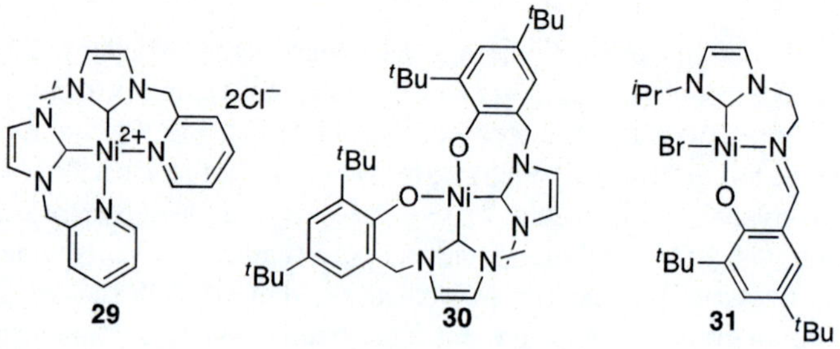

Fig. 4.11

Waymouth has investigated chelating NHC-enolates as ligands for Ni(II) catalysed ethylene polymerisation in a number of studies [37, 38]. Phenyl-Ni complexes of the form **32** (Fig. 4.12) catalyse polyethylene formation with moderate activity up to 27 $kg{\cdot}mol^{-1}{\cdot}bar^{-1}{\cdot}h^{-1}$ in the absence of any co-catalyst [37]. Linear polyethylene of fairly low molecular weight (M_n = 1 000 – 7 100 $g{\cdot}mol^{-1}$) is produced. The authors showed that insertion into the Ni–Ph group initiates polymerisation, giving a high proportion of phenyl terminated polymer chains. Following chain termination the resultant species (presumed to be a Ni-hydride) rapidly decomposes, and it is suggested this may be due to imidazolium formation. The related allyl complexes **33** (Fig. 4.12) were prepared and tested in a follow up study [38]. Both were inactive without a co-catalyst present, which was put down to poor initiation of polymerisation *via* ethylene insertion into the Ni-allyl bond. With diethylzinc present, the nitro-substituted complex displayed a low activity (1.2 $kg{\cdot}mol^{-1}{\cdot}bar^{-1}{\cdot}h^{-1}$) for ethylene polymerisation, again giving fairly low molecular weight polymers. Interestingly the catalyst does not suffer from the same rapid deactivation as catalyst **32** does. The authors speculate that $ZnEt_2$ may stabilise the Ni complex against decomposition, although no mechanism for this stabilisation is suggested.

Allyl-Pd(II) complexes featuring a single NHC ligand have been explored for norbornene polymerisation. The first such disclosure appears in a patent of 2006 [39] in which complexes of the form **34** (Fig. 4.12) are activated with $AgSbF_6$ and $LiB(C_6F_5)_4$. The resulting cationic complexes catalyse the polymerisation of norbornene and polar-substituted norbornenes with very high activity. The best result reported, for instance, involved activation with $LiB(C_6F_5)_4$, and a TOF of 750 000 h^{-1} in norbornene polymerisation was obtained in a 15 min run. The catalysts were not only tolerant to polar monomer functionality, but catalysis could also be carried out in water. The polymerisation of norbornene and acrylates was also claimed, although no polymer characterisation was provided, such that the level of acrylate incorporation is not clear.

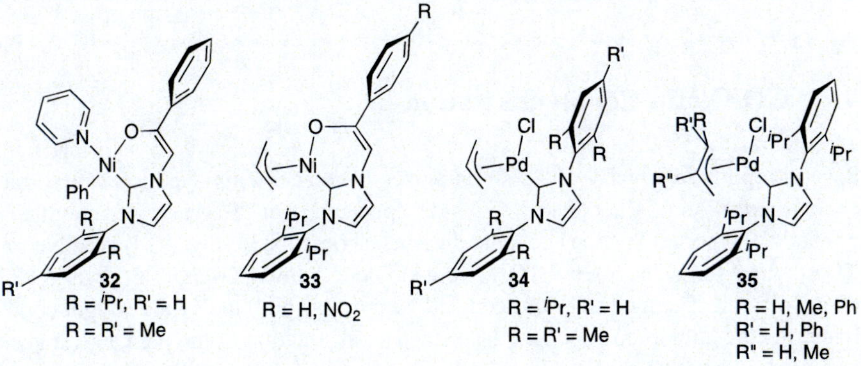

Fig. 4.12

Chung and co-workers investigated analogues of **34** with a broader range of substituents on the aryl group [40]. Again polymerisation of norbornene and polar-functionalised norbornenes was tested, with $AgSbF_6$ and $AgBF_4$ activation. The activity was dependent upon carbene ligand substitution (although no clear trend was evident) and the counterion, with SbF_6^- being better than BF_4^-. With norbornene a TOF of 2 060 h^{-1} could be achieved over a 12 h run. Ester–functionalised monomers were polymerised much more slowly (TOF < 15 h^{-1}). In a follow up study the same authors explored the effect of substituents on the allyl ligand, as in complexes of the form **35** (Fig. 4.12) [41]. It was found that increased steric bulk on the allyl ligand leads to improved activities, although an explanation for this effect was not offered. One possibility may be that steric bulk on the allyl group favours formation of the η^1 form, which is likely required before norbornene insertion can occur. This work also showed that the carbene ligand is essential for activity to be observed (allyl-Pd chlorides are inactive), and that these catalysts are air stable. The picolyl-functionalised carbene complex **36** (Fig. 4.13) has also been tested for norbornene polymerisation (following activation with MAO) [42], and similar high activities to those reported for **34** were observed. Related carbene-imine complexes of palladium, such as **37** (Fig. 4.13), were reported in the patent literature for ethylene polymerisation (MAO activation), but the catalysts are at best poorly active and the results seem unreliable [16]. Complex **37** produces polyethylene in one run (4 kg·mol^{-1}·bar^{-1}·h^{-1}), but shows no activity in a number of other very similar tests.

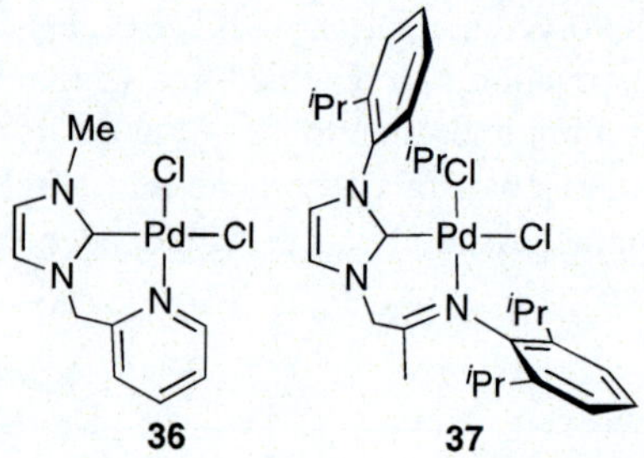

Fig. 4.13

4.2 CO/Olefin Copolymerisation

Several reports in which NHC–Pd complexes have been employed to catalyse the copolymerisation of alkenes with CO have appeared over the years. Herrmann and co-workers reported that the chelating dicarbene complex **38** (Fig. 4.14) is active for CO/ethylene [43]. The highest TON [(mol ethylene + mol CO)·mol Pd^{-1}] was 3 075 after a 4 h run. The modest TONs coupled with a very high molecular weight copolymer led the authors to conclude that only a small fraction of the pre-catalyst goes on to form an active species. Low molecular weight (M_n = 3 790) CO/norbornene copolymer resulted when complex **39** (Fig. 4.14) was tested by Chen and Lin [44]. The catalyst displayed only a very low activity, yielding 330 turnovers after 3 days.

Finally, the phosphine-functionalised carbene complex **40** (Fig. 4.14) has been tested following activation with $[H(Et_2O)][BAr_4^F]$ [45], but again a very low activity was achieved (TON ≈ 129 after 2 h). The poor performance of these catalysts may also reflect their susceptibility to reductive elimination (of 2-acylimidazolium salt) and generation of Pd(0) [4, 5]. Interest in CO/alkene seems to have dwindled recently, and as such no further reports on carbene complexes catalysing this reaction have appeared since 2003.

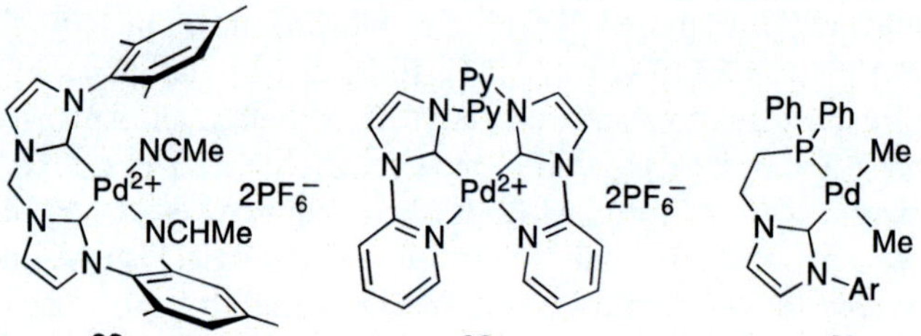

Fig. 4.14

4.3 Atom Transfer Radical Polymerisation

There are a number of reports of NHC complexes of mid-late transition metals being used as catalysts for atom transfer radical polymerisation (ATRP) of acrylates and styrene. Grubbs reported Fe(II) complexes of a simple monodentate carbene,

41 (Fig. 4.15), are active for ATRP of both styrene and methylmethacrylate (MMA) [46]. Polymerisation was well controlled with polydispersities ranging from 1.05 to 1.47. The rates of polymerisation ($k_{obs} \approx 1–4 \times 10^{-5}\ s^{-1}$) showed the complexes to be more active than phosphine and amine ligated Fe complexes, and were said to rival Cu-based ATRP systems. It was quite recent that Cu(I) complexes of NHCs were tested as ATRP catalysts [47]. In this work, tetrahydropyrimidine-based carbenes were employed to yield mono-carbene and di-carbene complexes **42** and **43** (Fig. 4.15), which were tested for MMA polymerisation. The mono-carbene complex **42** gave relatively high polydispersities (1.4–1.8) and a low initiation efficiency (0.5), both indicative of poor catalyst control. The di-carbene complex **43** led to uncontrolled radical polymerisation, which was ascribed to the insolubility of the complex.

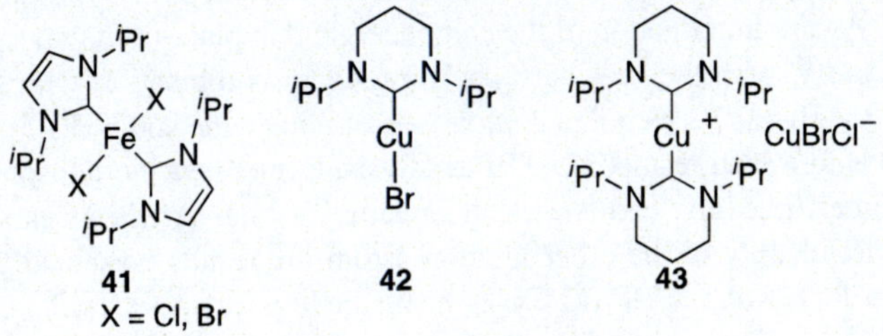

Fig. 4.15

A range of Ru–NHC complexes have been studied in some detail, and their ability to catalyse ATRP has been compared to phosphine-ligated analogues. Most of these complexes are based on catalysts which have been employed in olefin metathesis (Chapter 3). Catalyst **44** (Fig. 4.16) and the di-carbene analogue **45** (Fig. 4.16) were found to polymerise MMA in a fairly controlled manner, with rate constants, k_{obs}, of $3–12 \times 10^{-5}\ s^{-1}$ [48]. There was, however, no real improvement over the first generation Grubbs catalyst $[RuCl_2(CHPh)(PCy_3)_2]$, and in fact the bis-phosphine complex was more active ($k_{obs} = 21 \times 10^{-5}\ s^{-1}$). Styrene was also polymerised with these complexes, giving polydispersities that were broader than with the Grubbs catalyst, while polymerisation of *n*-butylacrylate was *via* an uncontrolled redox-initiated free radical process.

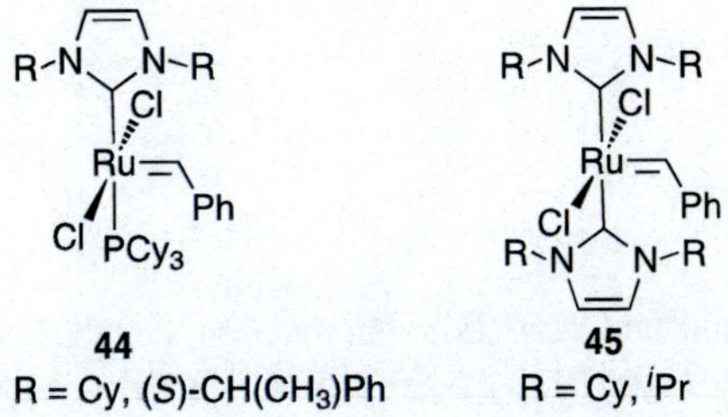

Fig. 4.16

Ruthenium-NHC complexes incorporating the *para*-cymene ligand, **46** (Fig. 4.17), are active for both MMA and styrene polymerisation (k_{obs} = 3–10 × 10^{-6} s^{-1}) [49]. For MMA polymerisation, a combination of R = Mes and R′ = H or Cl was most efficient, while R = Cy and R′ = H worked best for styrene polymerisation. For styrene, other R, R′ combinations led to uncontrolled free radical polymerisation. With R = Mes, some degree of styrene metathesis was also observed. The authors claim that these catalysts are much more versatile than [$RuCl_2(CHPh)(PR_3)_2$], although this does not appear to be borne out in the results. Other researchers investigated the cationic complex **47** (Fig. 4.17), and compared this to the tricyclohexylphosphine analogue [50]. Styrene, MMA and alkyl acrylates were polymerised, with the highest activities achieved with **47** (k_{obs} styrene = 1.8 × 10^{-5} s^{-1}, cf. 5.1 × 10^{-6} s^{-1} for phosphine complex). The same researchers tested O^N chelate Ru–NHC complexes **48** and **49** (Fig. 4.17), and their analogues with tricyclohexylphosphine in place of the carbene [51]. Both catalysts were tested as neutral complexes and as cations after chloride abstraction with $AgBF_4$. Again, introduction of the carbene ligand in place of the phosphine led to increased activity, with the best catalyst being **49** in combination with $AgBF_4$ (k_{obs} MMA = 15.7 × 10^{-5} s^{-1}). Overall, a comparison of phosphine and NHC ligands on Ru catalysts leads to mixed results in ATRP catalysis. Sometimes the phosphine appears best while other times the carbenes lead to superior catalysts. It appears to depend upon the exact identity of the other ligation. From the results presented, however, it would appear that [$RuCl_2(CHPh)(PCy_3)_2$] gives the highest activities.

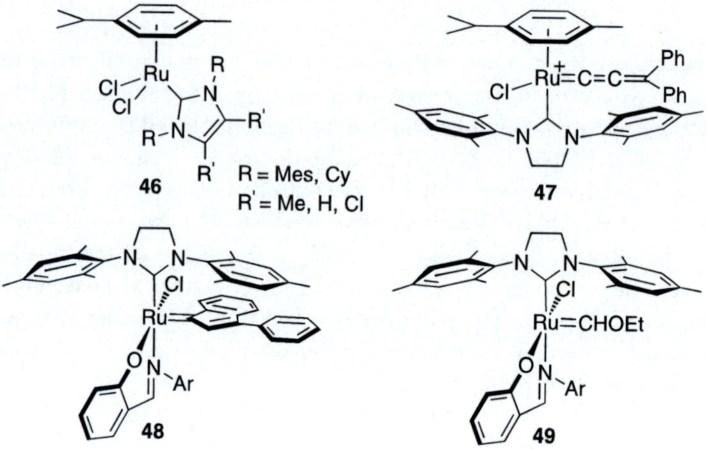

Fig. 4.17

4.4 Telomerisation

Telomerisation is an important, atom efficient reaction, which generates functionalised dienes from 1,3-diene feedstocks. The reaction, which typically employs a palladium based catalyst, comprises coupling two molecules of a conjugated 1,3-diene with addition of one molecule of a nucleophile, HNu, yielding mixtures of functionalised *cis/trans* dienes (telomers) predominantly consisting of 'linear' (*n*) and 'branched' (*iso*) isomers (Scheme 4.4); by-products include trienes (from linear 1,3-diene dimerisation, without addition of the nucleophile) and the product from Diels-Alder addition of two 1,3-diene molecules (4-vinylcyclohexene with 1,3-butadiene) [52–56].

1,3-Butadiene is the most frequently used 1,3-diene for telomerisation. Non-symmetrical 1,3-dienes (isoprene, 2-methylbuta-1,3-diene) increase the number of possible products due to the possibility of head-to-head, tail-to-tail, or mixed head/tail couplings making selectivity a difficult problem. A wide range of nucleophiles have been investigated, including water, alcohols, amines, carboxylic acids and compounds with active methylene groups [52, 57]. Products from telomerisation with water (hydrodimerisation) or ammonia give rise to long chain alcohols or amines (used in detergent manufacture), other telomerisation products are useful as plasticisers, diesel fuel additives, and intermediates in fine chemical and natural product syntheses [58–60]. Telomerisation may also be carried out enantioselectively [60].

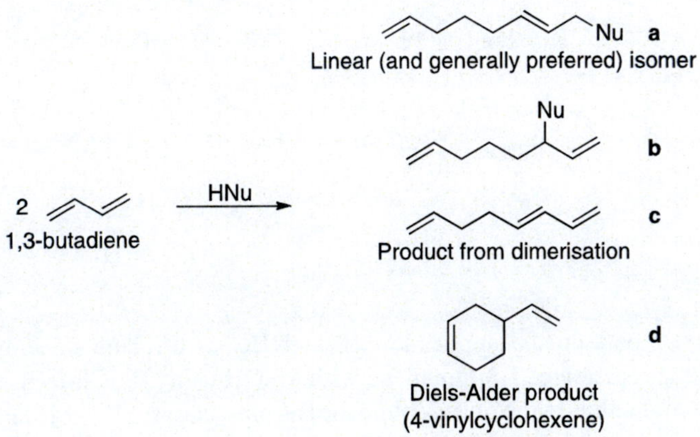

Scheme 4.4 Products from the telomerisation of 1,3-butadiene with a nucleophile

Since telomerisation was first reported in 1967 [61, 62] there have been many examples described using a range of Pd-phosphine/phosphite catalyst systems [58, 63]. The first significant industrial application of telomerisation, by the Kuraray company in Japan, was the hydrodimerisation of 1,3-butadiene with water in aqueous sulfolane using a Pd/phosphonium salt catalyst system [54]. The resulting octa-2,7-dienol was hydrogenated to 1-octanol for use as feedstock for plasticiser production. Studies by Beller and co-workers showed that by controlling reaction conditions Pd-phosphine systems could be tuned to produce reasonable performance [58]. However, high catalyst loadings, modest conversions, poor selectivity and significant quantities of by-product have mitigated against commercialisation of telomerisation [55, 59, 64].

4.4.1 Pd–NHC Complexes as Telomerisation Catalysts

N-Heterocyclic carbenes have provided valuable alternatives to phosphines as ligands in catalysis. NHCs are powerful donor ligands forming strong bonds with transition metals, and therefore, less prone to dissociate from the metal centre [65, 66]. Additionally, the wing-like distribution of the N-substituents on NHCs, for example in **50** and **51** (Fig. 4.18), affords a steric environment distinct from that of cone shaped phosphines [66, 67]. However, NHCs are known to undergo reductive coupling reactions, with loss of the NHC as imidazolium salt [4, 5]; the reaction represents a potential decomposition pathway. To date, application of Pd–NHC catalyst systems in telomerisation has focused on the reaction of 1,3-butadiene with alcohols (and phenols) [64, 68–72] and amines [73]. Activities and selectivities have been impressive [68], and Pd–NHC catalysts have recently been applied on pilot plant scale for production of >15 000 kg of 1-octene from 1,3-butadiene and MeOH, used as co-monomers in polyethylene formation [71, 72, 74].

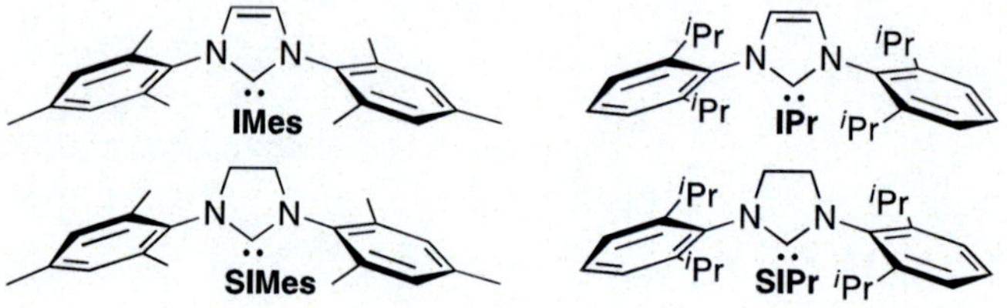

50 **51**

Fig. 4.18

The first report on the application of an NHC as a ligand in telomerisation described the reaction of 1,3-butadiene with MeOH using [PdClMe(tmiy)$_2$] (tmiy = 1,3,4,5-tetramethylimidazolidin-2-ylidene) as pre-catalyst [75, 76]. The resulting system showed low activity (TOF *ca.* 50 h^{-1}) but high selectivity for linear telomer. Since that time various Pd–NHC complexes have been employed as pre-catalysts in telomerisation; for example, complexes **52–62** (Fig. 4.19) [64, 68–71, 73, 77].

The complexes [Pd(NHC)(dvds)] (dvds = 1,3-divinyltetramethyldisiloxane), **52–56**, which are readily prepared by the reaction of commercially available Pd(0)-dvds solution with free NHC [68, 69, 78], are direct analogues of several phosphine systems. The "benchmark" catalyst, generated from complex **52**, is highly effective for telomerisation of 1,3-butadiene with methanol (0.001 mol% Pd, temperature 70–90°C; TONs of 96 000, with almost complete conversion and high selectivity for linear telomer, regioselectivity) [68, 69]. Low loadings of **52** (0.00005 mol%) in the presence of 80-fold excess of IMes·HCl resulted in TONs of 1.54 × 10^6; selectivity remained unaffected [68] [Excess imidazolium salt allows regeneration of active catalyst, *via* oxidative addition, should reductive elimination of the carbene ligand occur during catalysis] [4, 5]. At lower temperature (50°C) regioselectivity is even more marked; linear product is formed with <1% impurities, although conversion is reduced (57%, TON = 57 000, TOF = 3 560 h^{-1}) [69]. Variations in the NHC ligand (**52–56**) allow a useful comparison of steric and electronic influences [68]. Changes in backbone substituents (H, Me, Cl) in complexes **52–54** provide variation in ligand basicity [79]; steric influences were probed using complexes **55** and **56** [68]. Selected data for complexes **52–59** are presented in Table 4.1.

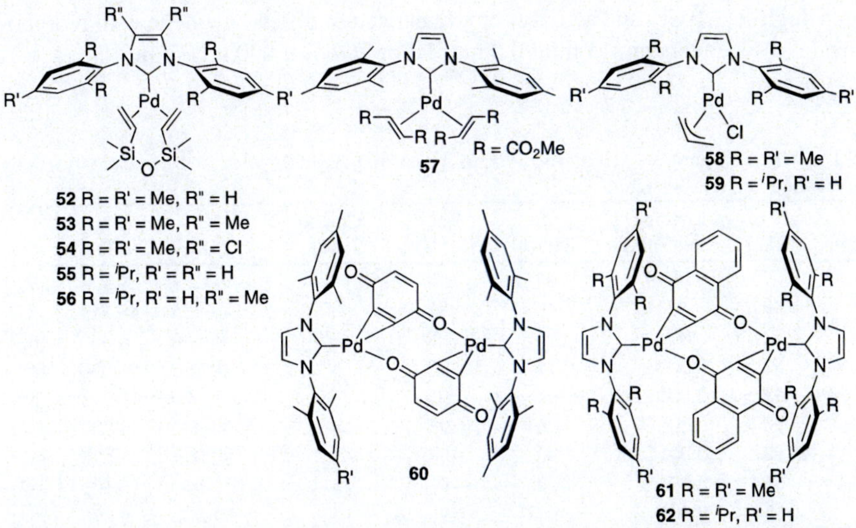

Fig. 4.19

Comparing the most basic Me_2IMes with least basic Cl_2IMes, indicates variation in performance due to electronic effects is minor (Table 4.1, entries 5 and 6) [68]. However, steric hindrance was found to have a significant impact on both activity and selectivity. Comparing Pd(IPr/Me_2IPr)(dvds) complexes **55** and **56** with the benchmark, **52**, shows reduced conversion, and lower regio- and chemo-selectivity (proportion of telomers **a** and **b** to by-products **c** and **d**, Scheme 4.4); the greater effect is apparent with Me_2IPr as ligand, where only 2% conversion is observed (Table 4.1, entries 7 and 8) [68]. Furthermore, the combination of bulky ligand with bulky nucleophile (e.g. secondary alcohols) is sufficient to push the product distribution from telomers to the dimerisation product, **c** [78]. Catalysts derived from [Pd(IMes)(dimethylfumarate)$_2$] **57**, and [Pd(SIMes)(*p*-benzoquinone)]$_2$ **60**, have also been investigated (Table 4.1, entries 1 and 2) [77]. Complexes **52–56** were studied in the telomerisation of 1,3-butadiene with a variety of alcohols and phenols. Effective telomerisation occurs with primary alcohols, however, conversion and chemoselectivity is reduced for secondary alcohols; with phenols, conversions were also lower although chemoselectivity was high [68]. For sterically hindered phenols (e.g. 2,4,6-trimethylphenol) regioselectivity was poor. [Pd(η^3-allyl)Cl(NHC)] complexes **58** and **59** showed activities comparable to [Pd(NHC)(dvds)] complexes **52–56** for telomerisation of 1,3-butadiene with MeOH (Table 4.1, entries 3–11) [80, 81]. However, **59** was inactive for telomerisation of 1,3-butadiene with amines [73]. Removal of the chloride ion from **59** with $AgPF_6$ or $AgBF_4$ leads to cationic complexes of the type **63** (Scheme 4.5), that gave highly active catalysts. Cationic complexes generated *in situ* from **59** and sodium salts of non-coordinating anions showed significantly higher activities than preformed complexes for telomerisation of 1,3-butadiene with amines; best results were achieved using a PF_6 salt of **59**, which gave high selectivity and activity for telomerisation of 1,3-butadiene with morpholine (100% conversion, 15 min; 0.2 mol% Pd; TOF = 2 000 h^{-1}) [73].

Table 4.1 Comparative telomerisation reactions of butadiene with methanol catalysed by preformed Pd-NHC systems

Entry	Cat.	Loading mol% Pd	Yield (%)	Regio. (%)	Chemo. (%)	TON	TOF	Ref
1	**57**	0.0049	84	98	–	8 574	8 574	[77]
2	**60**	0.0034	89	97	–	13 215	13 215	[77]
3	**52**	0.001	96	98	>99	96 000	6 000	[68]
4	**52**	0.00005	77	98	99	1 540 000	96 250	[68]
5	**53**	0.001	93	98	99	93 000	5 813	[68]
6	**54**	0.001	96	98	>99	96 000	6 000	[68]
7	**55**	0.001	90	92	97	90 000	5 625	[68]
8	**56**	0.001	2	91	–	2 000	125	[68]
9	**58**	0.001	94	98	99	94 000	5 875	[68]
10	**58**	0.0001	89	98	98	890 000	55 625	[68]
11	**59**	0.001	46	92	96	46 000	2 875	[68]

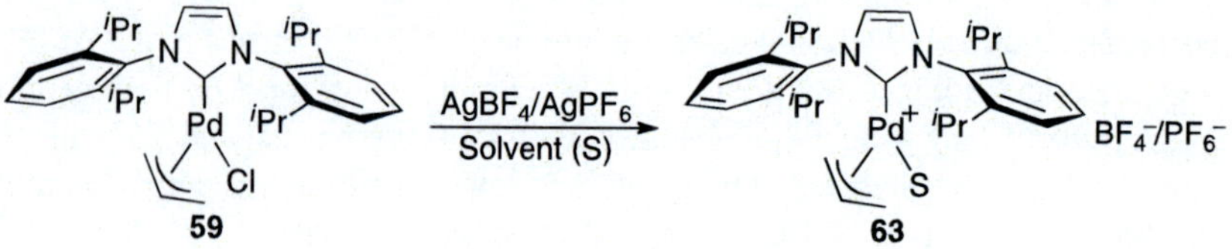

Scheme 4.5 Activation of **59** by chloride-abstraction

Preformed complexes of type **52**, **53** and **59**, and *in situ* catalyst systems based on IMesX (X = CO_2, $MeSO_3H$, HCl, HBr, HI) and IPrHCl, have been tested for telomerisation of butadiene with primary and secondary amines. Under optimised conditions, and low catalyst loadings, excellent activities and selectivities were observed [82].

Table 4.2 Telomerisation reactions of butadiene with methanol catalysed by *in situ* generated Pd–NHC systems

Entry	Ligand	Yield (%)	Regio.(%)	Chemio.(%)	TON	TOF	Ref
1	 **64**	0	–	–	–	–	[70]
2	 **65**	73	98	96	73 000	4 560	[70]
3	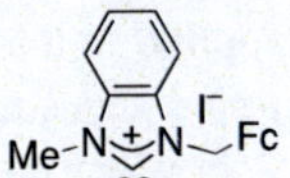 **66**	8	99	75	8 000	500	[70]
4	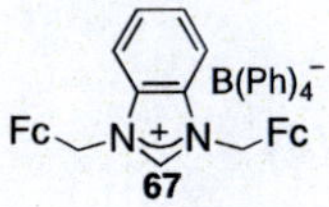 **67**	88	97	97	88 000	5 500	[70]
5	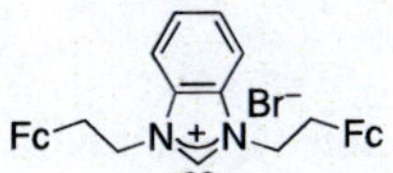 **68**	82	97	97	82 000	5 125	[70]
6	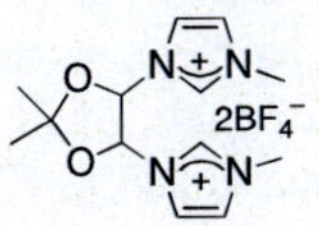 **69**	0	–	–	–	–	[69]
7	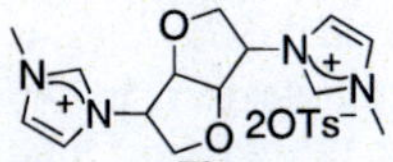 **70**	0	–	–	–	–	[69]
8	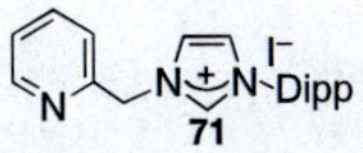 **71**	0	–	–	–	–	[78]
9	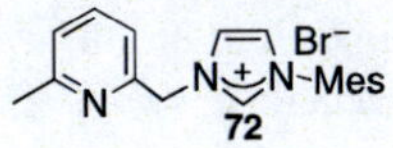 **72**	0	–	–	–	–	[78]
10	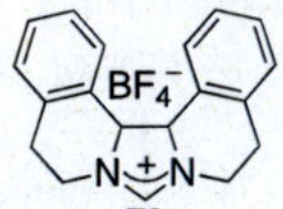 **73**	5	>98	71	7 000		[68]

In situ derived systems, in general, performed similarly to preformed complexes, in telomerisation of butadiene with MeOH, Tables 4.1 and 4.2 [68, 70, 71, 77, 78]. *In situ* systems may be generated from free NHC or from imidazolium salt in combination with an appropriate Pd(0) or Pd(II) source. Typically, 2–4 equivalents of imidazolium salt relative to Pd have been used [68, 70, 77]. *In situ* catalysts derived from mono- and bis-Fc-substituted (Fc = ferrocenyl) imidazolium and benzimidazolium salts (**64–68**) (Table 4.2) showed interesting telomerisation activities ascribed to the steric bulk of the Fc substituents [70]. Unsymmetrical salts **65** and **66** bearing *N*-Fc and *N*-Me substituents showed variable conversions but very high regioselectivities (Table 4.2, entries 2 and 3); the latter decreased somewhat when methylene or ethylene groups separated the Fc and benzimidazolium centres, salts **67** and **68**. No activity was observed using **64**, most likely due to extreme steric crowding. *In situ* testing of potentially chelating ligands derived from bis-imidazolium salts **69** and **70** [69], and picolyl-functionalised imidazolium salts **71** and **72** [77], showed no telomerisation activity. Chiral monoimidazolium salt **73** showed good regioselectivity (>98%), but poor overall yield and chemoselectivity (5% and 71%, respectively; TON = 7 000 [68]). An *in situ* catalyst system [Pd(acac)$_2$/IMes·HCl] has recently been applied in telomerisation of isoprene with glycerol and polyethylene glycol [83]. Activity was modest and chemoselectivity was poor, but selectivity to linear monotelomer was excellent.

4.4.2 NHCs and Imidazolium-Based Ionic Liquids

The relationship between imidazolium salts (ionic liquid solvents, ILs) and NHC ligands provides a unique situation specific to catalyst systems employed in imidazolium-based ILs [4, 5, 25, 86]. Accordingly, unforeseen reactions may occur when catalysts are used in imidazolium ILs, particularly in the presence of bases capable of deprotonating the C^2–H (or the C^{4,5}-H) position of the imidazolium ring. For example, catalytically active [PdBr$_2$(bmiy)$_2$] complexes (bmiy = 1-*n*-butyl-3-methylimidazolidin-2-ylidene) have been isolated from [bmim]Br under Heck reaction conditions using Pd(OAc)$_2$ as pre-catalyst [84]. Furthermore, oxidative addition of imidazolium C^2–H bonds to M(0) complexes yielding [M^{II}(hydrido)(NHC)] (M = Ni, Pd, Pt) complexes is a facile process, and may occur for reactions undertaken in IL solvents [25, 85, 86]. Oxidative addition of imidazolium C^{4,5}-H protons to M(0) is also known [87, 88]. A report on hydrodimerisation of 1,3-butadiene using [PdCl$_4$]$^-$ in [bmim]Cl provides a further example of non-innocent behaviour of imidazolium ILs [89]. A [PdCl$_2$(*N*-methylimidazole)$_2$] complex, formed *via* elimination of 1-butene from bmim, was isolated from the reaction mixture and is thought to be the active catalyst.

The non-innocent behaviour of ILs may not always be beneficial to catalysis [5]. For example, in the telomerisation of 1,3-butadiene with methanol, using Pd-phosphine catalyst systems, in [bmim]$^+$ and [emim]$^+$ (emim = 1-ethyl-3-methylimidazolium) salts of non-coordinating anions as solvents [90, 91], very low activities were observed, and it was separately found that addition of even small amounts of 1,3-dialkylimidazolium salts was sufficient to poison the Pd-phosphine catalyst; probably forming catalytically inactive L$_2$Pd-(bmiy/emiy) complexes, in which coordination sites required for catalysis were blocked [91]. Running the reactions in 1-*n*-butyl-2,3-dimethylimidazolium ILs, where the imidazolium C$_2$ position was blocked with a methyl group, resulted in activity comparable to the Pd-phosphine reference systems run in MeOH, with improved chemoselectivity and regioselectivity [91]. Telomerisation of 1,3-butadiene with methanol [91, 92], and with HNEt$_2$ [93] in ILs using Pd-phosphine systems yields results which although satisfactory, are markedly inferior to Pd–NHC catalysed systems in molecular solvents [73].

4.4.3 Telomerisation Mechanism

A monoligated species L-Pd(0) is believed to be the active intermediate in telomerisation, with the metal centre undergoing oxidation and reduction during each catalytic cycle. A ligand, L, that binds strongly to Pd, and is appropriately bulky to prevent agglomeration of low-valent Pd(0) species formed during the catalytic cycle, is therefore desirable. NHCs should therefore be excellent ligands for telomerisation. Whilst mechanistic investigations on Pd–NHC catalysed telomerisation are rare [68], the mechanism and intermediates formed probably closely parallel that for Pd-phosphine catalysed telomerisation [66, 67]. The mechanism for telomerisation of 1,3-butadiene with methanol catalysed by Pd-phosphine systems has been elucidated by Jolly and co-workers [57], with recent complementary studies by Beller and co-workers to explain the observed regioselectivity of the reaction (Scheme 4.6) [58]. Alternative mechanisms, invoking bimetallic intermediates bridged by supporting ligands and/or η^3,η^3-octadiendiyl proto-telomers (for which intermediates have also been isolated and characterised), have been proposed [94, 95]. However, bimetallic species are considered unlikely in the presence of bulky NHC ligands and hence such mechanisms are probably less relevant here. The most likely and generally accepted mechanism is that described in Scheme 4.6. A more

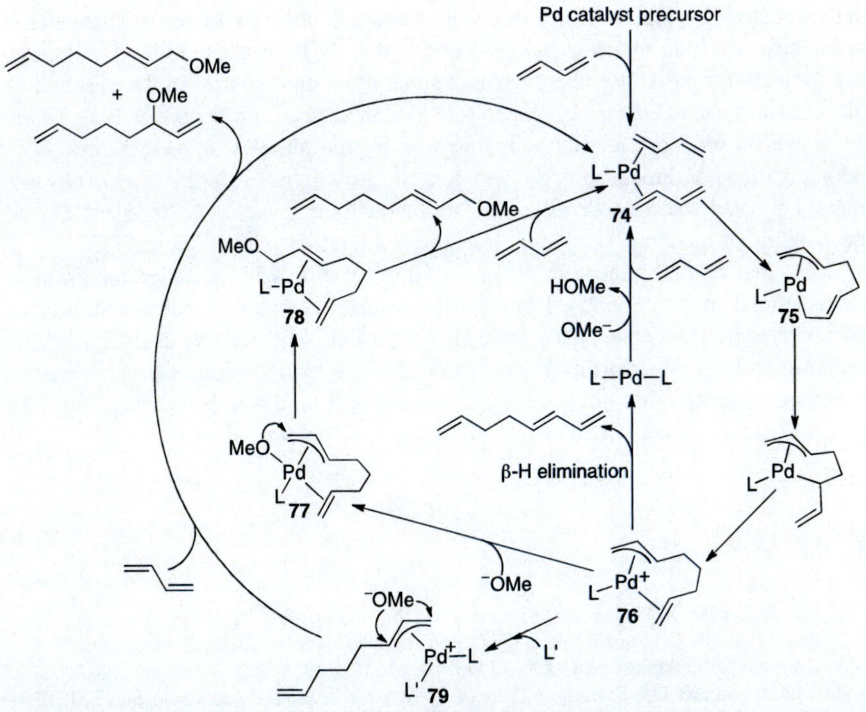

Scheme 4.6 Proposed mechanism for the telomerisation of 1,3-butadiene with methanol

comprehensive discussion of the telomerisation mechanism has recently appeared and only a brief summary is provided here [56].

In high concentrations of 1,3-butadiene, any weakly bound ligand(s) on the Pd(0) pre-catalyst will be displaced by two molecules of 1,3-butadiene; telomerisation then proceeds by oxidative coupling of the 1,3-butadiene molecules to form a [$Pd^{II}(\eta^3,\eta^1$-octadiendiyl)L] complex, **75** [57, 96–101]. Electron rich ligands promote this step. Coordination of methoxide gives a square planar intermediate, **77**; alternatively, external (intermolecular) attack may occur [102]. Importantly, for regioselectivity, the π-accepting η^2-alkene moiety of the η^3,η^2-octadienyl ligand (**75** or **76**) will direct nucleophilic attack of methoxide to the ligand *trans* to itself, and thus to C_1 rather than C_3 of the η^3-allyl moiety [102]. A rigid conformation engendered by the η^3,η^2-octadienyl ligand and a sterically demanding ligand, L should also direct attack to C_1, **77** [58]. Displacement of the η^2-alkene moiety of the η^3,η^2-octadienyl chelate, in **75** or **76** gives an η^3-octadienyl ligand, **78**. Methoxide attack on the η^3-allyl moiety of **78** is now not directed to C_1 and attack at C_3 produces the branched telomer, which is liberated in the presence of excess 1,3-butadiene to regenerate **74**, and poor regioselectivity is observed. The linear dimerisation product 1,3,7-octatriene is generated by ß-hydrogen elimination of the C_4–H of the η^3-octadienyl ligand.

Complex **76** is considered to be the key telomerisation/dimerisation "switch", and is critical in selectivity for linear telomers, *i.e.* the intermediate allows both chemo- and regioselective control to be achieved [57, 58]. Having a π-acceptor alkene ligand *trans* to C_1, in combination with an NHC ligand (**76**, L = NHC), will result in electronically directed nucleophilic attack at C_1 and formation of linear telomers. These effects account for the high regioselectivity observed when NHC ligands are used. Although regioselectivity is likely to be primarily under electronic control in the presence of NHC ligands, steric effects also play a role in both chemo- and regioselectivity. Steric influence on chemoselectivity is highlighted by the almost complete switch from telomers to linear dimerisation product 1,3,7-octatriene (as a mixture of *cis*/*trans* isomers) observed when a combination of the sterically demanding NHC ligand **51** and nucleophile of moderate steric bulk (secondary alkoxide) were employed [78].

It is clear that telomerisation is an exciting reaction with considerable promise for synthesis of functionalised organic molecules, as a stand-alone reaction or as part of a reaction scheme. NHCs have proved particularly effective as ligands in this reaction and many opportunities now exist for the exploitation of this chemistry. It can be expected that novel catalyst systems will be developed as new reaction schemes are designed aimed at specific target molecules.

References

1. Jens K-J, Tilset M, Voges MH, Blom R, Froseth M (2000) WO 00/01739
2. Britovsek GJP, Gibson VC, Wass DF (1999) Angew Chem Int Ed 38:428–447
3. Gibson VC, Spitzmesser SK (2003) Chem Rev 103:283–315
4. (a) McGuinness DS, Saendig N, Yates BF, Cavell KJ (2001) J Am Chem Soc 123:4029–4040; (b) McGuinness DS, Cavell KJ (2000) Organometallics 19:4918–4920

5. Cavell KJ, McGuinness DS (2004) Coord Chem Rev 248:671–681
6. Spencer LP, Fryzuk MD (2005) J Organomet Chem 690:5788–5803
7. Niehues M, Kehr G, Erker G, Wibbeling B, Fröhlich R, Blacque O, Berke H (2002) J Organomet Chem 663:192–203
8. Aihara H, Matsuo T, Kawaguchi H (2003) Chem Commun 2204–2205
9. Liu N, Zhang D (2009) Organometallics 28:499–505
10. McGuinness DS, Gibson VC, Steed JW (2004) Organometallics 23:6288–6292
11. McGuinness DS, Gibson VC, Wass DF, Steed JW (2003) J Am Chem Soc 125:12716
12. McGuinness DS, Suttil JA, Gardiner MG, Davies NW (2008) Organometallics 27:4238–4247
13. McGuinness DS (2009) Organometallics 28:244–248
14. McDaniel MP (1985) Adv Catal 33:47–98
15. Döhring A, Göhre J, Jolly PW, Kryger B, Rust J, Verhovnik GPJ (2000) Organometallics 19:388–402
16. Tilset M, Andell O, Dhindsa A, Froseth M (2002) WO 02/49758
17. Kreisel KA, Yap GPA, Theopold KH (2006) Organometallics 25:4670–4679
18. McGuinness DS, Overett M, Tooze RP, Blann K, Dixon JT, Slawin AMZ (2007) Organometallics 26:1108–1111
19. Wang B, Wang D, Cui D, Gao W, Tang T, Chen X, Jing X (2007) Organometallics 26:3167–3171
20. Wang B, Cui D, Lv K (2008) Macromolecules 41:1983–1988
21. Danopoulos AA, Wright JA, Motherwell WB, Ellwood S (2004) Organometallics 23:4807–4810
22. Xiao X-Q, Jin G-X (2008) J Organomet Chem 693:3363–3368
23. Chauvin Y, Einloft S, Olivier H (1995) Ind Eng Chem Res 34:1149–1155
24. Ellis B, Keim W, Wasserscheid P (1999) Chem Commun 337–338
25. McGuinness DS, Mueller W, Wasserscheid P, Cavell KJ, Skelton BW, White AH, Englert U (2002) Organometallics 21:175–181
26. Cavell KJ (2008) Dalton Trans 47:6676–6685
27. Olivier-Bourbigou H, Commereuc D, Harry S (2001) US 01/0047121
28. MacKinnon AL, Baird MC (2003) J Organomet Chem 683:114–119
29. Cámpora J, de la Tabla LO, Palma P, Álvarez E, Lahoz F, Mereiter K (2006) Organometallics 25:3314–3316
30. Sujith S, Noh EK, Lee BY, Han JW (2008) J Organomet Chem 693:2171–2176
31. Sun HM, Shao Q, Hu DM, Li WF, Shen Q, Zhang Y (2005) Organometallics 24:331–334
32. Buchowicz W, Kozioł A, Jerzykiewicz LB, Lis T, Pasynkiewicz S, Pecherzewska A, Pietrzykowski A (2006) J Mol Catal A 257:118–123
33. Sun HM, Hu DM, Wang YS, Shen Q, Zhang Y (2007) J Organomet Chem 692:903–907
34. Wang X, Liu S, Jin G-X (2004) Organometallics 23:6002–6007
35. Zhang D, Kawaguchi H (2006) Organometallics 25:5506–5509
36. Li W, Sun H, Chen M, Wang Z, Hu D, Shen Q, Zhang Y (2005) Organometallics 24:5925–5928
37. Ketz BE, Ottenwaelder XG, Waymouth RM (2005) Chem Commun 5693–5695
38. Benson S, Payne B, Waymouth RM (2007) J Polym Sci A 45:3637–3647
39. Goodall BL, McIntosh LH (2006) US 7041758
40. Jung IG, Seo J, Chung YK, Shin DM, Chun S-H, Son SU (2007) J Polym Sci A 45:3042–3052
41. Jung IG, Lee YT, Choi SY, Choi DS, Kang YK, Chung YK (2009) J Organomet Chem 694:297–303
42. Wang X, Liu S, Weng L-H, Jin G-X (2006) Organometallics 25:3565–3569
43. Gardiner MG, Herrmann WA, Reisinger C-P, Schwarz J, Spiegler M (1999) J Organomet Chem 572:239–247
44. Chen JCC, Lin IJB (2000) Organometallics 19:5113–5121
45. Tsoureas N, Danopoulos AA, Tulloch AAD, Light ME (2003) Organometallics 22:4750–4758
46. Louie J, Grubbs RH (2000) Chem Commun 1479-1480
47. Bantu B, Wang D, Wurst K, Buchmeiser MR (2005) Tetrahedron 61:12145–12152
48. Simal F, Delfosse S, Demonceau A, Noels AF, Denk K, Kohl FJ, Weskamp T, Herrmann WA (2002) Chem Eur J 8:3047–3052

49. Delaude L, Delfosse S, Richel A, Demonceau A, Noels AF (2003) Chem Commun 1526–1527
50. Opstal T, Verpoort F (2003) Polym Bull 50:17–23
51. Opstal T, Verpoort F (2003) Angew Chem Int Ed 42:2876–2879
52. Keim W, Behr A, Röper M (1982) Alkene and alkyne oligomerisation, cooligomerisation and telomerization reactions. In: Comprehensive organometallic chemistry, vol 8. Pergamon Press, Oxford
53. Behr A (1984) In: Ugo R (ed) Aspects of homogeneous catalysis, vol 5. Dordrecht, Reidel
54. Takacs JM (1995) Transition metal allyl complexes: Telomerization of dienes. In: Hegedus LS (ed) Comprehensive organometallic chemistry II, vol 12. Pergamon Press, Oxford
55. Yoshimura N (2000) Telomerization (hydrodimerization) of olefins. In: Cornils B, Herrmann WA (eds) Applied homogeneous catalysis with organometallic compounds: A comprehensive handbook, Special Workbench ed. Wiley-VCH, Weinheim.
56. Nielsen DJ, Cavell KJ (2006) Pd–NHC complexes as catalysts in telomerization and aryl amination reactions. In: Nolan SP (ed) N-Heterocyclic carbenes in synthesis. Wiley-VCH, Weinheim, pp 73–102
57. Jolly PW (1985) Angew Chem Int Ed Engl 24:283–295
58. Vollmuller F, Krause J, Klein S, Mägerlein W, Beller M (2000) Eur J Inorg Chem 1825–1832
59. Hénin F, Bessmertnykh A, Serra-Muns A, Muzart J, Baillia H (2004) Eur J Org Chem 511–520
60. Keim W, Koehnes A, Roethel T (1990) J Organomet Chem 382:295–301
61. Takahashi S, Shibano T, Hagihara N (1967) Tetrahedron Lett 26:2451–2453
62. Smutny EJ (1967) J Am Chem Soc 89:6793–6794
63. Vollmüller F, Mägerlein W, Klein S, Krause J, Beller M (2001) Adv Synth Catal 343:29–33
64. Rottger D, Beller M, Jackstell R, Klein H, Wiese K-D (2005) US2005038273, WO2100803
65. Herrmann WA (2002) Angew Chem Int Ed 41:1290–1309
66. Scott NM, Nolan SP (2005) Eur J Inorg Chem 10:1815–1828
67. Hillier AC, Sommer WJ, Yong BS, Petersen JL, Cavallo L, Nolan SP (2003) Organometallics 22:4322–4326
68. Jackstell R, Harkal S, Jiao H, Spannenberg A, Borgmann C, Röttger D, Nierlich F, Elliot M, Niven S, Cavell KJ, Navarro O, Viciu MS, Nolan SP, Beller M (2004) Chem Eur J 10:3891–3900
69. Jackstell R, Andreu MG, Frisch A, Selvarkumar K, Zapf A, Klein H, Spannenberg A, Röttger D, Briel O, Karch R, Beller M (2002) Angew Chem Int Ed 41:986–989
70. Jackstell R, Frisch A, Beller M, Röttger D, Malaun M, Bildstein B (2002) J Mol Catal A: Chem 185:105–112
71. Beller M, Jackstell R, Klein H, Rottger D, Weise K-D, Maschmeyer D, Tuchlenski A, Kaizik A, Fernandez SS (2004) US2004242947, WO03031379
72. Beller M, Jackstell R, Klein H, Rottger D, Maschmeyer D, Fernandez SS (2005) US2005065387, WO03031378
73. Viciu MS, Zinn FK, Stevens ED, Nolan SP (2003) Organometallics 22:3175–3177
74. Beller M (2005) Personal communication
75. McGuinness DS (1997) B.Sc. Honours Thesis, University of Tasmania, Hobart
76. McGuinness DS, Cavell KJ, Skelton BW, White AH (1999) Organometallics 18:1596–1605
77. Clément NDS (2005) Ph.D. thesis, Cardiff University, Cardiff
78. Harkal S, Jackstell R, Nierlich F, Ortmann D, Beller M (2005) Org Lett 7:541–544
79. Magill AM, Cavell KJ, Yates BF (2004) J Am Chem Soc 126:8717–8724
80. Viciu MS, Navarro O, Germaneau RF, Kelly RA III, Sommer WJ, Marion N, Stevens ED, Cavallo L, Nolan SP (2004) Organometallics 23:1629–1635
81. Viciu MS, Germaneau RF, Navarro-Fernandez O, Stevens ED, Nolan SP (2002) Organometallics 21:5470–5472
82. Grotevendt A, Bartolome M, Nielsen DJ, Spannenberg A, Jackstell R (2007) Cavell K J, Oro LA, Beller M. Tetrahedron Lett 48:9203–9207
83. Gordillo A, Pachón LD, de Jesus E, Rothenburg G (2009) Adv Synth Catal 351:325–330
84. Xu L, Chen W, Xiao J (2000) Organometallics 19:1123–1127

85. Clément NDS, Cavell KJ, Jones C, Elsevier CJ (2004) Angew Chem Int Ed 43:1277–1279
86. McGuinness DS, Cavell KJ, Yates BF, Skelton BW, White AH (2001) J Am Chem Soc 123:8317–8328
87. Bacciu D, Cavell KJ, Fallis IA, Ooi L (2005) Angew Chem Int Ed 44:5282–5284
88. Bacciu D (2007) Ph.D. thesis, Cardiff University, Cardiff
89. Dullius JEL, Suarez PAZ, Einloft S, de Souza RF, Dupont J, Fischer J, De Cian A (1998) Organometallics 17:815–819
90. Dupont J, Spencer J (2004) Angew Chem Int Ed 43:5296–5297
91. Magna L, Chauvin Y, Niccolai GP, Basset JM (2003) Organometallics 22:4418–4425
92. Chauvin Y, Magna L, Niccolai GP, Basset JM (2002) EP 1201634 A1
93. Fonseca GS, de Souza RF, Dupont J (2002) Catal Commun 3:377–380
94. Behr A, Ilsemann Gv, Keim W, Krüger C, Tsay Y-H (1986) Organometallics 5:514–518
95. Keim W, Röper M (1981) J Org Chem 46:3702; Keim W, Kurtz KR, Röper M (1983) J Mol Catal A: Chem 20:129–138
96. Büch HM, Binger P, Benn R, Krüger C, Rufinska A (1983) Angew Chem Int Ed Engl 22:774–775
97. Döhring A, Jolly PW, Mynott R, Schick K-P, Wilke G (1981) Z Natursforsch B 36:1198–1199
98. Storzer U, Walter O, Zevaco T, Dinjus E (2005) Organometallics 24:514–520
99. Döhring A, Goddard R, Hopp G, Jolly PW, Kokel N, Krüger C (1994) Inorg Chim Acta 222:179–192
100. Benn R, Bussemeier B, Holle S, Jolly PW, Mynott R, Tkatchenko I, Wilke G (1985) J Organomet Chem 279:63–86
101. Szabó KJ (2000) Chem Eur J 6:4413–4421
102. Prinz T, Driessen-Hölscher B (1999) Chem Eur J 5:2069–2076

Chapter 5
N-Heterocyclic Carbene Complexes in Cyclisation Reactions

Janis Louie

Abstract Cycloaddition reactions hold a prominent place in the arsenal of synthetic methods currently available to organic chemists. Although some cycloadditions can be promoted simply by heat, light, Lewis acids, high pressure or sonication, many cycloadditions require the use of a transition metal to catalyse the reaction. Transition metal catalysed cycloaddition is an attractive approach as it provides an opportunity to obtain regio- and/or enantio-pure cyclic products. The use of *N*-heterocyclic carbenes (NHCs) as ligands is a rapidly expanding field. As will be discussed in this chapter, these ligands have been used in conjunction with several metals in the synthesis of a wide range of carbocycles and heterocycles through cycloaddition reactions.

5.1 Introduction

A central focus in modern organic synthesis has been the development of highly efficient catalytic processes for the syntheses of natural and unnatural compounds of medicinal interest or intermediates useful for functional materials. A particularly attractive approach is to apply transition metal catalysed cyclisation reactions for the transformation of simple starting materials into monocyclic, bicyclic and polycyclic scaffolds that can be further elaborated into specific targets.

In the last 10 years, *N*-heterocyclic carbenes (NHCs) have emerged as a group of promising ligands in the design of new catalyst systems for several transition metal catalysed reactions, *i.e.* cross-coupling, metathesis and cyclisation reactions. NHC ligands show strong σ-electron-donating and a slight π-backbonding tendency, the former property imparts high stability to their respective ligated complexes [1]. As will be described in this chapter, these ligands when compared with phosphines or amines ligands enhance the reactivity of transition metal catalysts.

J. Louie (✉)
Department of Chemistry, University of Utah, Salt Lake City, Utah 84112-0850, USA
e-mail: louie@chem.utah.edu

C. S. J. Cazin (ed.), *N-Heterocyclic Carbenes in Transition Metal Catalysis and Organocatalysis*, Catalysis by Metal Complexes 32,
DOI 10.1007/978-90-481-2866-2_5,

5.2 Cyclopropanation

Organic chemists have always been fascinated by the cyclopropane subunit [2], which has played and continues to play a prominent role in organic chemistry. Its strained structure, interesting bonding characteristics, and value as an internal mechanistic probe have attracted the attention of the physical organic community. While the cyclopropane ring is a highly strained entity, it is nonetheless found in a wide variety of naturally occurring compounds including terpenes, pheromones, fatty acid metabolites and unusual amino acids. The prevalence of cyclopropane-containing compounds with biological activity, whether isolated from natural sources or rationally designed pharmaceutical agents, has inspired chemists to find novel and diverse approaches to their synthesis.

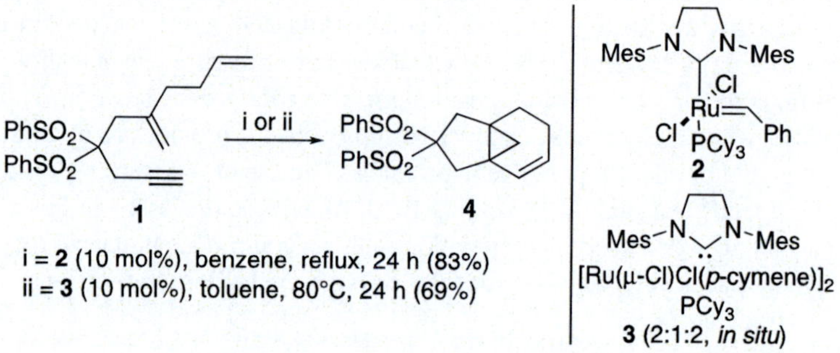

Scheme 5.1 Tandem cyclopropanation/ring closing metathesis of dienyne using Grubbs'catalyst

Among these approaches, an interesting tandem cyclopropanation/ring closing metathesis of dienyne **1**, which uses Grubbs'catalyst **2** or an *in situ* non-carbenic ruthenium complex **3**, has been reported (Scheme 5.1) [3]. Using 10 mol% catalyst loading in toluene or benzene, the product **4** was obtained in a good yield.

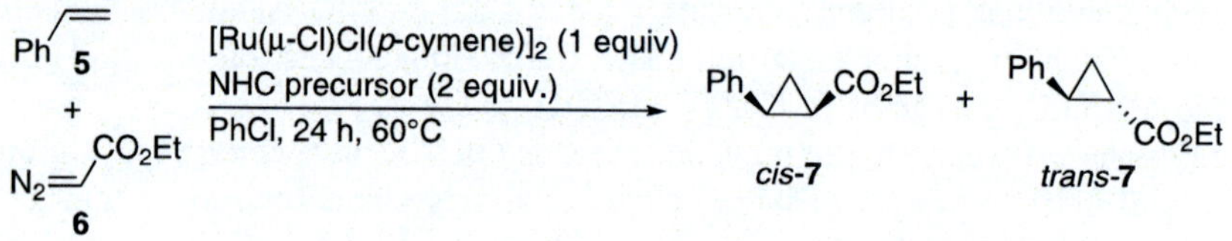

Scheme 5.2 Cyclopropanation of styrene and ethyl diazoacetate using ruthenium-NHC complexes

The use of stoichiometric ruthenium-NHC complexes generated *in situ* from [Ru(μ-Cl)Cl(*p*-cymene)]$_2$, an imidazolium salt [4] or an imidizol(idin)ium-2-carboxylate [4] has been applied in the cyclopropanation of styrene **5** with ethyl diazoacetate (EDA) **6** (Scheme 5.2). No base was necessary when imidazolium-2 carboxylate were employed. The diastereoselectivity was low and the *cis*/*trans* ratio was around 50/50 (Table 5.1). Although the diastereoselectivity was moderate, the reaction was highly chemoselectivity as possible side reactions (homologation, dimerisation and metathesis) were totally or partially suppressed.

Table 5.1 Preparation of cyclopropane compounds as described in Scheme 5.2

NHC Precursor	Base	*cis/trans* ratio	Product Yield (%)
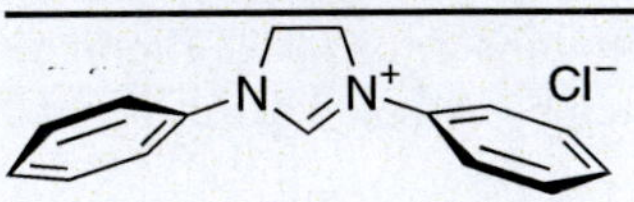	KO^{t}Bu	0.52	83
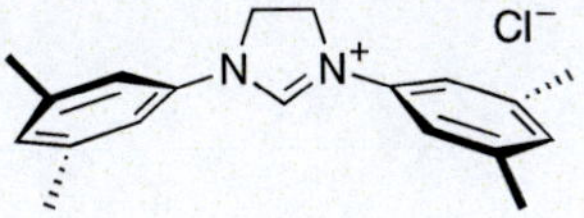	KO^{t}Bu	0.50	80
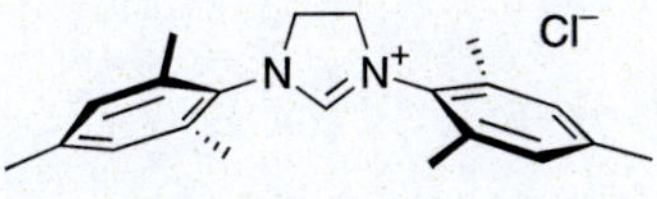	KO^{t}Bu	0.56	81
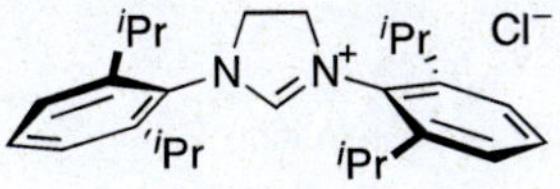	KO^{t}Bu	0.56	80
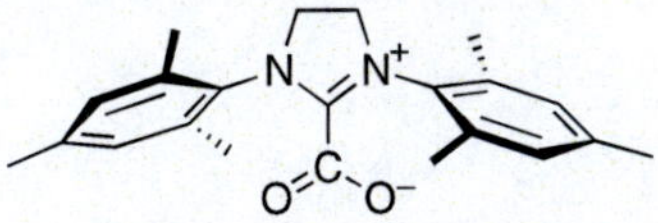	–	0.59	83
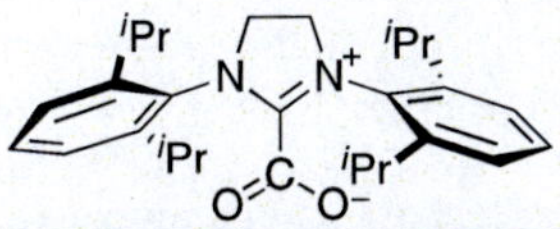	–	0.63	75

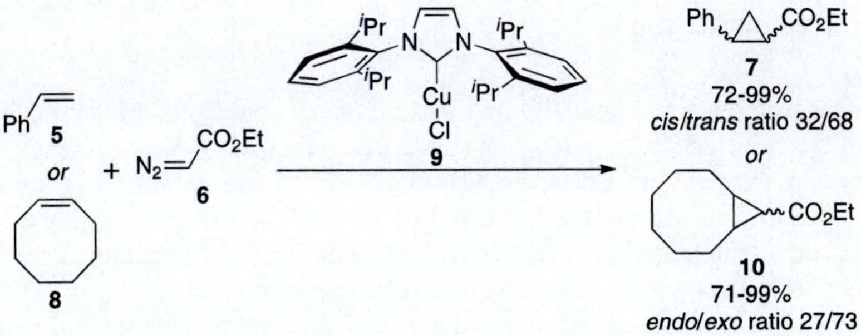

Scheme 5.3 Cyclopropanation of alkenes and ethyl diazoacetate using Cu–NHC complexes

A NHC–Cu complex **9** has also been used in the cyclopropanation of **5** and cyclooctene **8** using EDA **6** (Scheme 5.3) [5]. Complex **9** was isolated prior to use and, as in the case of NHC–Ru complex, the cyclopropanation reaction did not display high diastereoselectivity. However, products **7** and **10** were obtained in good to excellent yields depending on the ratio between the alkenes and EDA. Improved yields were obtained when alkenes were used in six- or ten-fold excess.

Cyclopropanes **13** have been prepared from a NHC–rhodium catalysed decarbonylation of cyclobutanones **11** (Scheme 5.4) [6]. The isolated complex **12** proved to be a chemoselective decarbonylation catalyst as it specifically activated the C–C bond of cyclobutanone even in the presence of a aldehydic C–H bond. A plausible mechanism for this reaction involves initial insertion of the rhodium between the carbonyl carbon and the α-carbon followed by carbonyl extrusion and reductive elimination (Scheme 5.4).

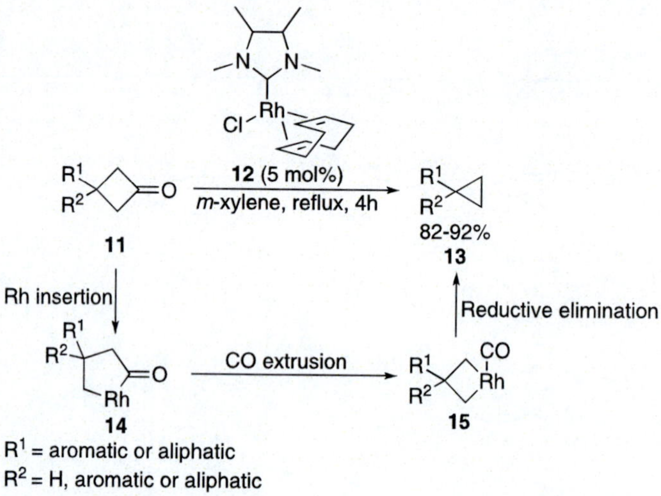

Scheme 5.4 Chemoselective decarbonylation catalysed by Rh–NHC complex

5.3 Cycloaddition

Cycloaddition reactions catalysed by transition metal complexes are an important tool in the construction of a wide range of carbo- and hetero-cyclic systems, such as benzene, pyridines, triazoles, etc. [7]. In general, these reactions are extremely atom-efficient and involve the formation of several C–C bonds in a single step. Among the innumerable possible catalytic systems for the cycloaddition reaction the NHC-metal complexes have received special attention [7c].

The NHCs have been used as ligands of different metal catalysts (*i.e.* copper, nickel, gold, cobalt, palladium, rhodium) in a wide range of cycloaddition reactions such as [4+2] (see Section 5.6), [3+2], [2+2+2] and others. These NHC-metal catalysts have allowed reactions to occur at lower temperature and pressure. Furthermore, some NHC–TM catalysts even promote previously unknown reactions. One of the most popular reactions to generate 1,2,3-triazoles is the 1,3-dipolar Huisgen cycloaddition (reaction between azides and alkynes) [8]. Lately, this [3+2] cycloaddition reaction has been aided by different $[Cu(NHC)_n]X$ complexes [9]. The reactions between electron-rich, electron-poor and/or hindered alkynes **16** and azides **17** in the presence of low NHC-copper **18–20** loadings (in some cases even ppm amounts were used) afforded the 1,2,3-triazoles **21** regioselectively (Scheme 5.5; Table 5.2).

Catalyst **18** has also been used in the reaction between alkynes **16** and azides **17** generated *in situ* from the corresponding halides **22** (Scheme 5.5).

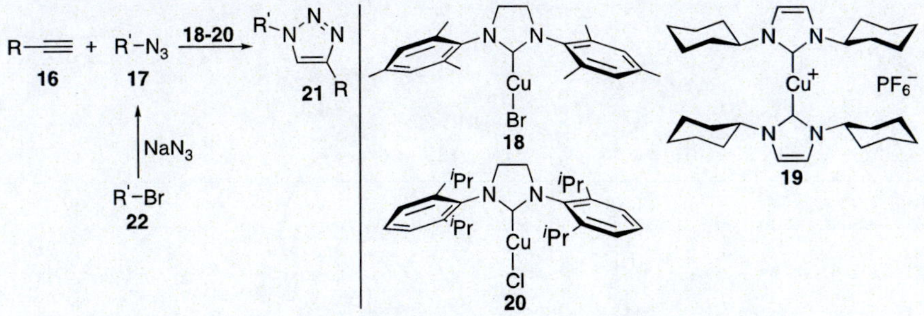

Scheme 5.5 1,3-Dipolar Huisgen cycloaddition catalysed by copper-NHC catalyst

Table 5.2 Preparation of 1,2,3-triazoles as described in Scheme 5.5

R	R'	Cu cat.	Yield (%)
Ph	$PhCH_2$	**18**	98 (94)[a]
		19	99
		20	98
Ph	4-CN–$PhCH_2$	**18**	93 (97)[a]
		20	93
4-MeO-Ph	C_7H_{15}	**18**	93
		19	95
		20	96
1-Cyclohexene	4-NO_2–$PhCH_2$	**18**	93
		19	92
		20	98
$(CH_3)_2CHOH$	$PhCH_2CH_2$	**18**	94
		20	97
$SiMe_3$	$PhCH_2$	**18**	98
CO_2Et	$(CH_2)_3CN$	**19**	96
2-Pyridinyl	$PhCH_2$	**19**	97

[a]Reaction carried out with the *in situ* generated azide

Normally, copper-catalysed Huisgen cycloadditions work with terminal alkynes only. The formation of a Cu-acetylide complex is considered to be the starting point of the catalyst cycle. However, the NHC–Cu complex **18** was able to catalyse the [3+2] cycloaddition of azides **17** and 3-hexyne **23** (Scheme 5.6).

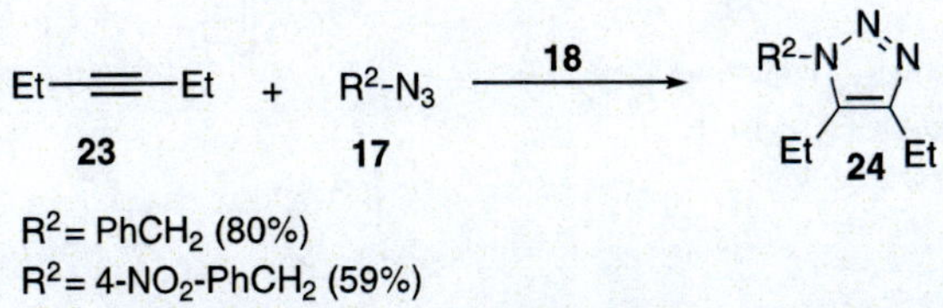

Scheme 5.6 [3+2] Cycloaddition reaction of internal alkynes

The strong σ-donor property of NHC ligands enhances the catalytic activity in [3+2] cycloaddition by promoting the activation of internal alkynes (*i.e.* **26**), which proceeds by the formation of a π-alkyne complex **25** (Scheme 5.7).

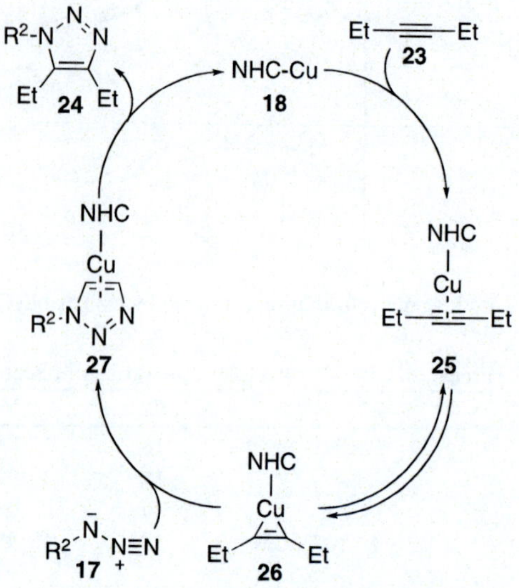

Scheme 5.7 Proposed mechanism of [3+2] cycloaddition reaction of internal alkynes

[3+2] Cycloadditions are among the most powerful strategies for the synthesis of five-membered carbocyclic and heterocyclic ring systems. The [3+2] cycloaddition reaction between diphenylcyclopropenone **28** and internal alkynes **29** can be catalysed by a quinoline-tethered NHC complex **30** of rhodium (Scheme 5.8) [10]. The cyclopentadienone **31** adducts were obtained in moderate to good yields using only 2 mol% of **30**. The reaction also showed to be highly regioselective; only one regioisomer (the aryl group present in the internal alkynes at two-position) was obtained in all cases.

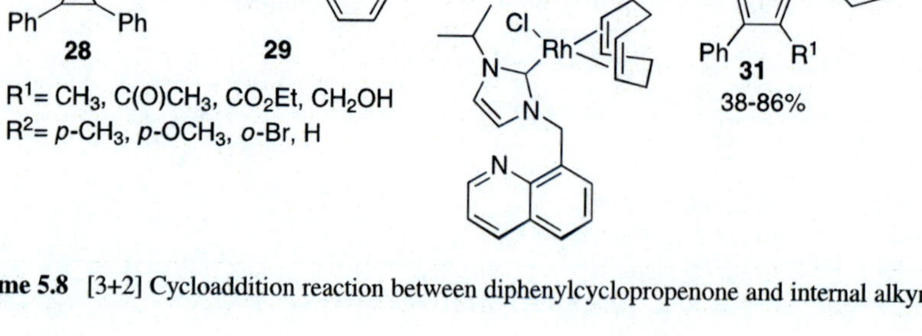

Scheme 5.8 [3+2] Cycloaddition reaction between diphenylcyclopropenone and internal alkynes

Cyclopropyl ketones **32** and cyclopropyl imines **33** can also undergo [3+2] cycloaddition reactions with enones **34** in presence of NHC–Ni complexes to afford the corresponding cyclopentane compounds **35** (Scheme 5.9) [11]. The catalytic system is prepared *in situ* from the use of [Ni(COD)$_2$], SIPr·HCl salt and KO^{t}Bu, the reaction also required the use of Ti(O^{t}Bu)$_4$ as an additive to improve yields and increase reactions rates. In most of the cases, the products **35** were obtained in good to excellent diastereoselectivities.

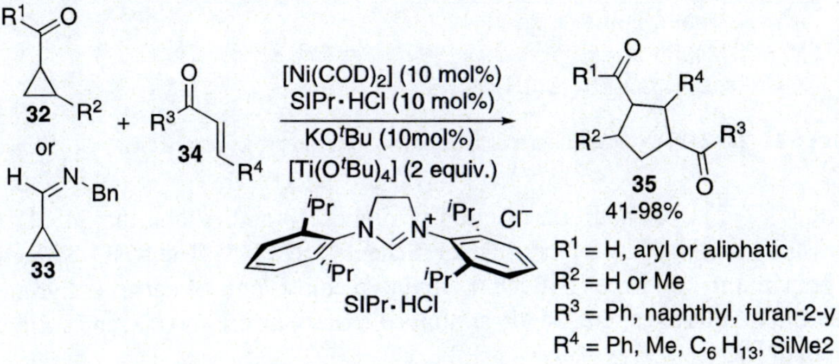

Scheme 5.9 NHC–Ni catalysed [3+2] cycloaddition reaction of cyclopropyl ketones or cyclopropyl imines with enones

[2+2+2] Cycloaddition of alkynes catalysed by transition metals is one of the most efficient and valuable ways to prepare benzene and pyridine systems [12]. Among the possible catalytic systems able to catalyse this reaction, cobalt and iron complexes containing NHCs as ligands have shown high catalytic activity in the intramolecular cyclotrimerisation of triynes **36** (Scheme 5.10) [13]. The reaction was catalysed with low loading of a combination of zinc powder and $CoCl_2$ or $FeCl_3$ with two or three equivalents of IPr carbene, respectively.

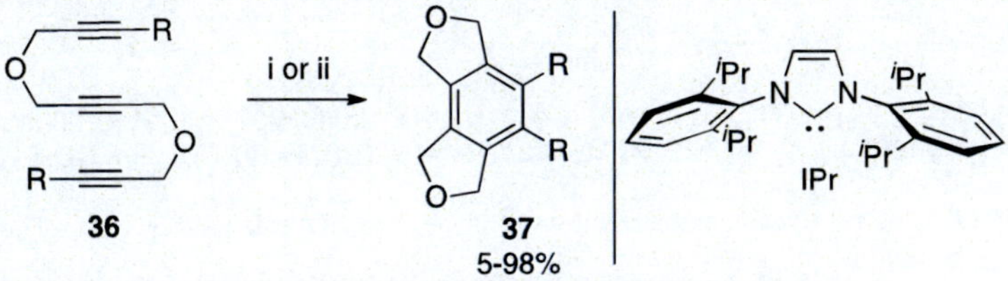

i- $CoCl_2$ (2-3 mol%), IPr (4-6 mol%), Zn (10 mol%), THF, 50°C.
ii- $FeCl_3$ (2 mol%), IPr (2 mol%), Zn (10 mol%), THF, 50°C.

Scheme 5.10 [2+2+2] Cycloaddition of tethered alkynes catalysed by NHC–Co or NHC–Fe

Enediynes **38** undergo [2+2+2] cycloaddition reactions to afford polycyclic cyclohexadienes **39** in presence of a cobalt catalyst (Scheme 5.11) [14]. In this system, the presence of a NHC ligand improved the catalytic power of cobalt when compared with phosphine ligands. In addition to increased yields, lower ligand loadings were required when IPr was used in lieu of PPh_3 (*i.e.* 6 mol% *versus* 200 mol%, respectively).

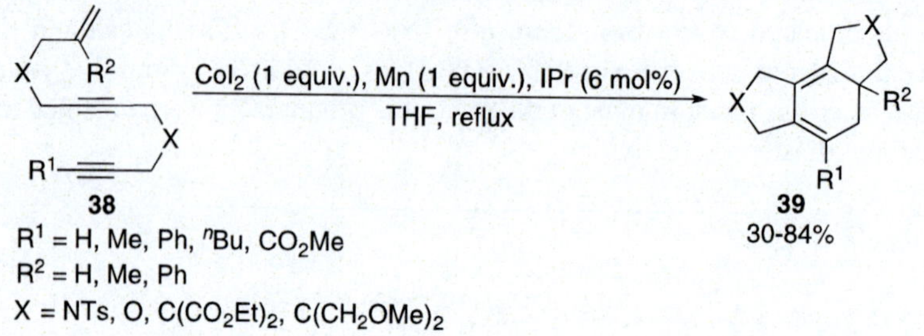

Scheme 5.11 [2+2+2] Cycloaddition reactions of enediynes in presence of a NHC–Co complex

The [2+2+2] cycloaddition reaction of diynes **40** and carbon dioxide **41** were successfully catalysed by a NHC–nickel (Scheme 5.12) [15]. The NHC–Ni complex was prepared *in situ* from $[Ni(COD)_2]$ and two equivalents of carbene. Pyrones **42** were obtained in excellent yields at atmospheric pressure of CO_2 and mild reaction conditions.

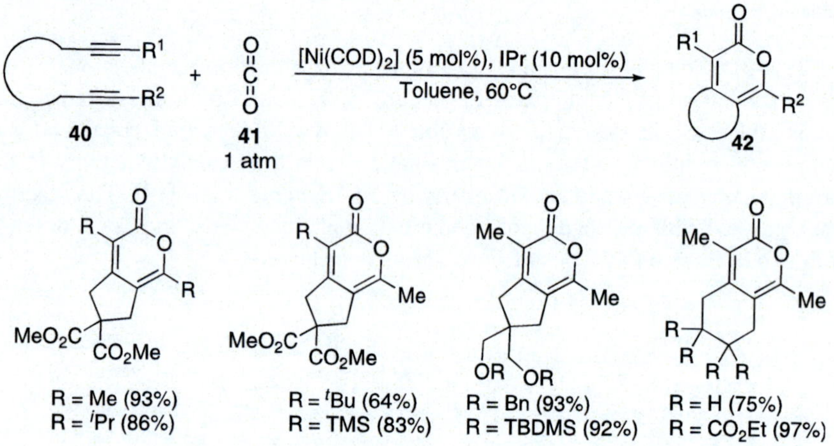

Scheme 5.12 [2+2+2] Cycloaddition reactions of diynes and CO_2 catalysed by NHC–Ni complex

Pyridine compounds **45** can also be produced by the NHC–Ni catalysed cycloaddition between nitriles **43** and diynes **44** (Scheme 5.13) [16]. The SIPr carbene was found to be the best ligand for the nickel complex in this reaction. The reaction required mild reaction conditions and low catalyst loadings, as in the case of cycloaddition of carbon dioxide. In addition to tethered alkynes (*i.e.* diynes), pyridines were prepared from a 3-component coupling reaction with **43** and 3-hexyne **23** (Scheme 5.13). The reaction of diynes **44** and nitriles **43** was also catalysed by a combination of $[Ni(COD)_2]$, NHC salts and ^{n}BuLi, which generates the NHC–Ni catalyst *in situ*. The pyridines **45** were obtained with comparable yields to those obtained when isolated carbenes were employed (Scheme 5.13) [17], although higher catalyst loading was necessary. This *in situ* protocol was also expanded to the cycloaddition between CO_2 and diyne, however the pyrone was obtained only in moderate yield.

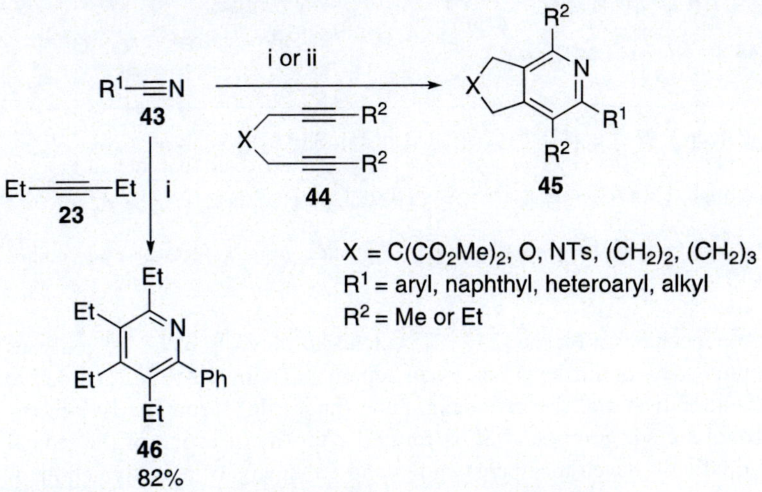

i) [Ni(COD)$_2$] (3 mol%), SIPr (6 mol%), toluene, rt (46-97%).
ii) [Ni(COD)$_2$] (5 mol%), IPr·HCl (10 mol%), ^{n}BuLi (25 mol%), benzene, rt (49-92%)

Scheme 5.13 Pyridine compounds formed by a [2+2+2] cycloadditon reaction catalysed by NHC–Ni complexes

The NHC–nickel catalytic system is also useful in the synthesis of pyridones **48**. The [2+2+2] cycloaddition of diynes **44** and isocyanates **47** affords a wide range of pyridones **48** in excellent yields in presence of [Ni(COD)$_2$]/SIPr catalytic system (Scheme 5.14) [18].

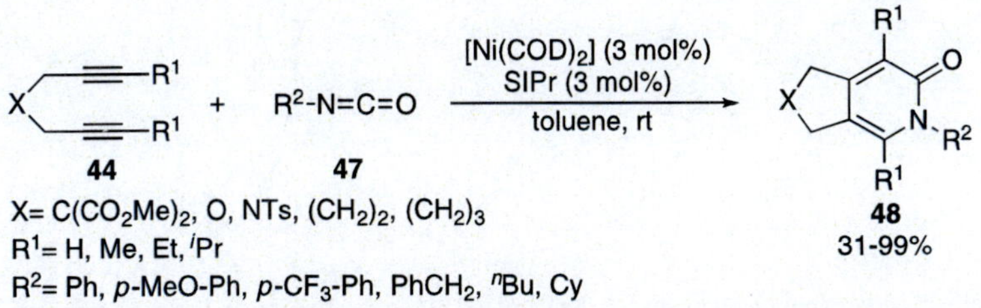

Scheme 5.14 [2+2+2] Cycloaddition of diynes and isocyanates in presence of [Ni(COD)$_2$]/SIPr catalytic system

The reaction described above can also be performed in a fully intermolecular fashion. When monoyne **49** is used in presence of isocyanate **47** and catalytic amount of [Ni(COD)$_2$]/SIPr, pyridone **50** is formed in 90% yield (Scheme 5.15) [18]. However, when an excess of isocyanate **47** is used in the same reaction conditions, the formation of pyrimidinediones **51** is observed rather than the expected pyridines **50** (Scheme 5.15) [19].

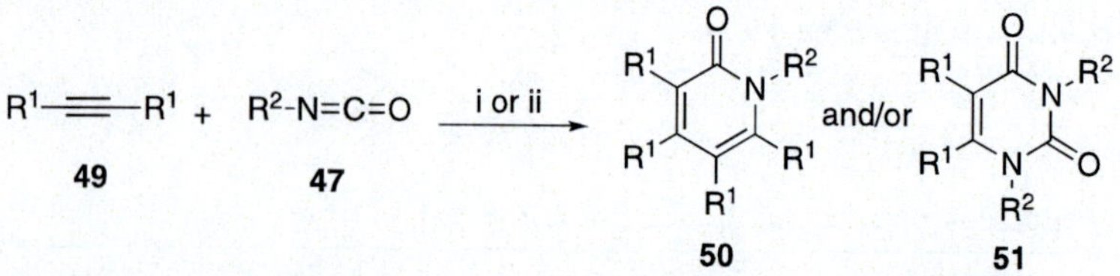

i) **49** (2 equiv.), **47** (1 equiv.), [Ni(COD)$_2$] (3 mol%), SIPr (3mol%), toluene, rt.
only **50** is formed

ii) **49** (1 equiv.), **47** (excess), [Ni(COD)$_2$] (5 mol%), IPr (10 mol%), toluene, rt.

Scheme 5.15 Pyridone and/or pyrimidinediones formation from a [2+2+2] cycloaddition NHC–Ni catalysed

A plausible mechanism for the [2+2+2] cycloaddition reactions between diynes and heterocumulenes (or nitriles) is shown in Scheme 5.16. Initially [2+2] oxidative addition of one alkyne and the heterocumulene (or nitrile) forms the five-membered intermediate **54**; compound **55** is formed after the insertion of the second alkyne and finally the seven-membered compound **55** undergoes reductive elimination to afford the product **56** and regenerate the Ni(0) catalyst.

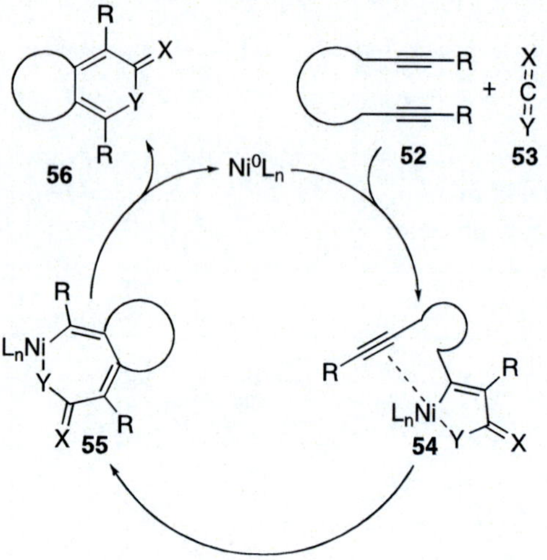

Scheme 5.16 Plausible mechanism for the [2+2+2] cycloaddition reactions between diynes and heterocumulenes

The NHC/Ni-catalysed cycloaddition reactions are highly regioselective. When asymmetric diynes containing one hindered substituent (such as TMS or ^{t}Bu) are used, only one regioisomer is obtained [15, 16]. As shown in Scheme 5.17, the larger substituent is located at the three-position of pyrones **58** or pyridines **59**. This regioselectivity is explained by the initial oxidative coupling of the larger alkynyl unit and CO_2 **41** (or nitrile **43**) followed by insertion of methyl-terminated alkyne. Although binding and oxidative coupling of the least sterically hindered alkynyl unit is favoured, subsequent insertion should be inhibited since it requires the bulky alkynyl unit to be placed adjacent to the imidazolylidene ligand.

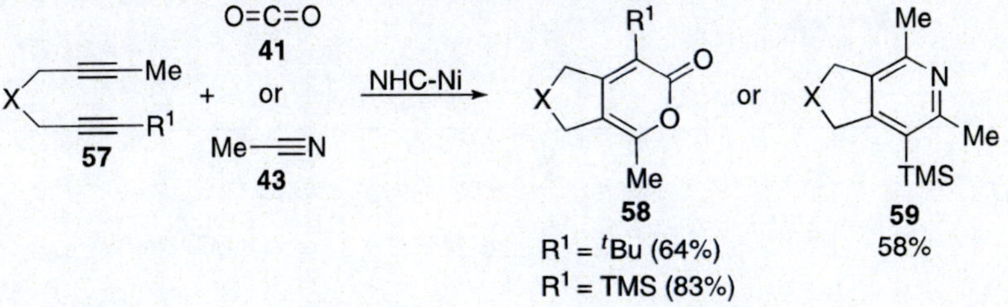

Scheme 5.17 Effect of substituent in the regioselectivity of [2+2+2] cycloadditon reaction

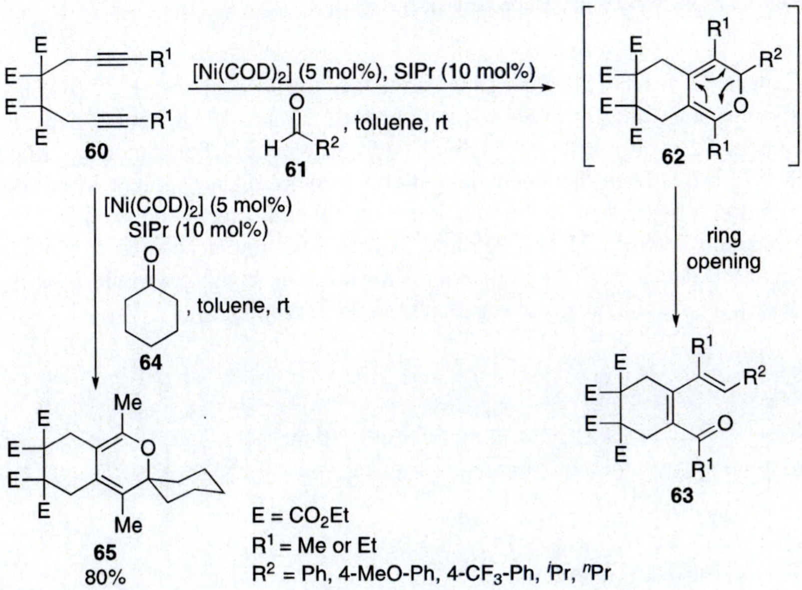

Scheme 5.18 [2+2+2] Cycloadditon reaction between diynes and aldehydes/ketones **64** in the presence of [Ni(COD)$_2$]/SIPr

Diynes **60** react with aldehydes **61** in the presence of [Ni(COD)$_2$]/SIPr catalytic system to afford dienones **63** upon electrolytic ring opening of the expected pyrans **62** (Scheme 5.18) [20]. When cyclohexanone **64** is employed the pyran ring **65** is obtained in good yield.

Diyne **66**, which possesses a three carbon linkage between the alkynes, also reacts with aldehyde **67** in presence of [Ni(COD)$_2$]/SIPr catalytic system (Scheme 5.19). However, the connectivity of the dienone **69** obtained from this diyne was different from those obtained from diyne **60**. In the former case, adduct **69** was obtained from a β-hydride elimination of the nickelacycle **68** instead of a reductive elimination.

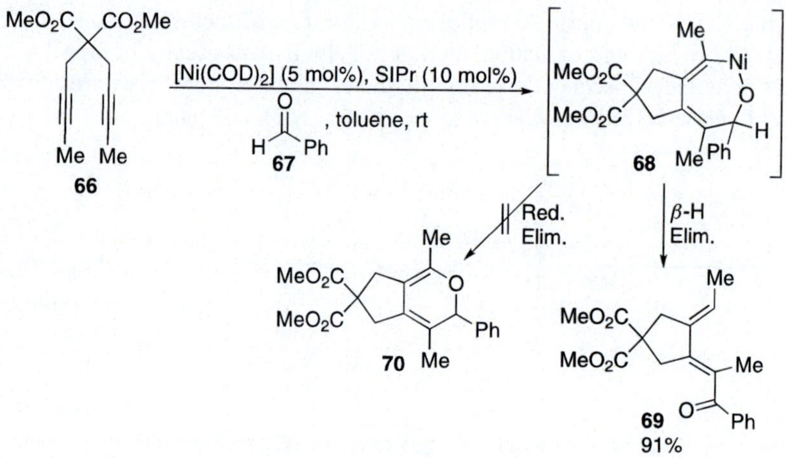

Scheme 5.19 *β*-Hydride elimination of the nickelacycle **68**

Enynes **71** react with aldehydes **61** in the presence of the [Ni(COD)$_2$]/SIPr catalytic system to afford two distinct products **72** and **73** (Scheme 5.20) [20b]. The enone **72** is derived from aldehyde addition with the alkyne moiety while the adduct **73** arises from the aldehyde addition with the alkene moiety. The product distribution is dependent on the substituent on either the alkyne or alkene moieties. The reaction between **71** and ketones **74** led to the unprecedented formation of pyrans **75** (Scheme 5.20). The reaction showed to be highly regioselective; in all the cases, the carbonyl carbon was bound to the olefin.

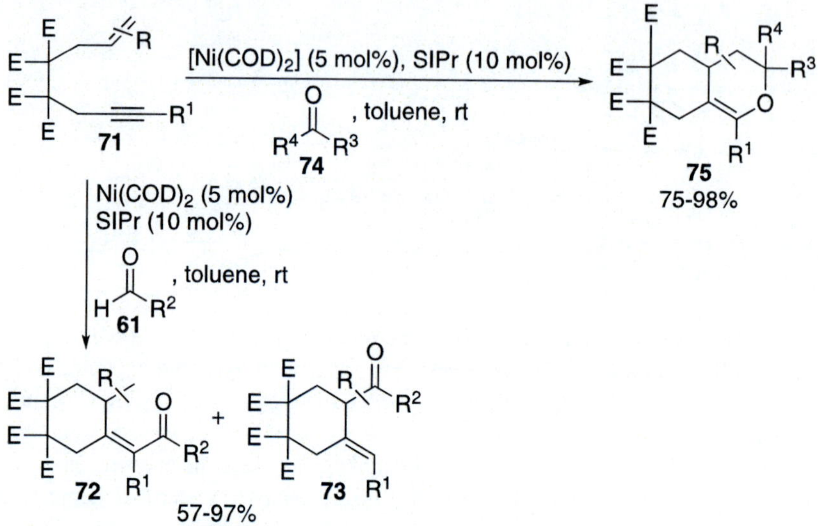

Scheme 5.20 [2+2+2] Cycloaddition reaction of enynes and aldehydes

Macrolactones **77** and/or **78** can be prepared from the reductive cyclisation of ynals **76** in the presence of NHC–nickel complexes (Scheme 5.21) [21]. This macrolactonisation occurs with different selectivity depending on the ligands attached to the nickel. If carbenes such as IMes or IPr are used, the exocyclic olefin **77** is preferentially obtained, however when phosphine ligands are used, the endocyclic adducts **78** are preferentially obtained.

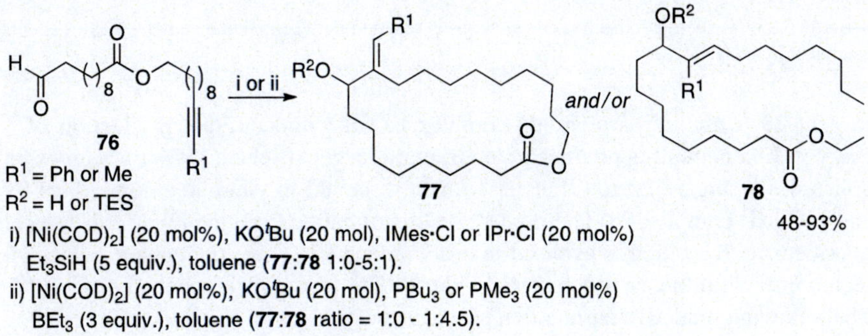

Scheme 5.21 Macrolactones prepared from the reductive cyclisation of ynals in the presence of NHC–nickel complexes

Diynes **44** and enynes **79** undergo a reductive cyclisation in presence of NHC–Pt complex **80** (Scheme 5.22) [22]. The products **81** are obtained in good yields in the presence of 5 mol% of **80**, 25 mol% of $SnCl_2$ and 5 atm of H_2.

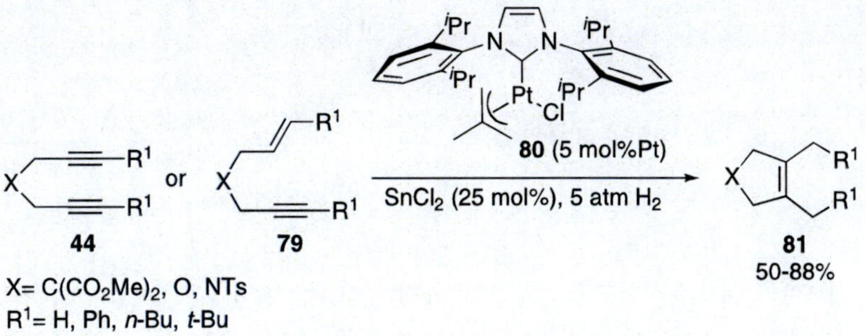

Scheme 5.22 Reductive cyclisation of diynes or enynes catalysed by NHC–Pt complex

A tandem cyclisation/cross-coupling reaction between 6-halo-1-hexene **82** and Grignard reagents **83** is successfully catalysed by a NHC–Co catalytic system (Scheme 5.23) [23].

A possible reaction mechanism involves the initial generation of the carbene **86** *via* the deprotonation of SIEt•HCl **85** by the Grignard reagent. The carbene **86** binds with $CoCl_2$ to generate the complex **87** which is also reduced by the Grignard

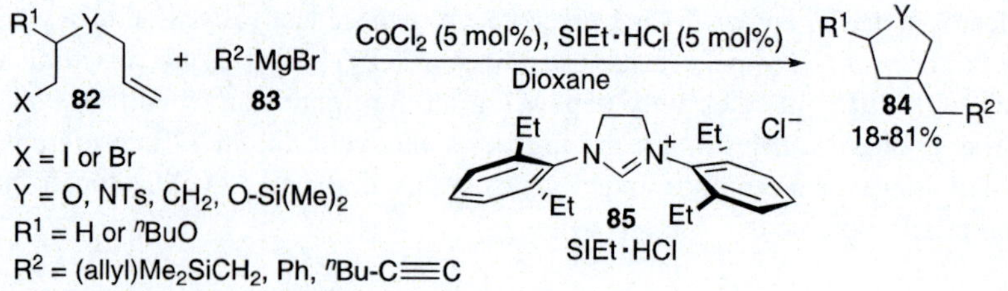

Scheme 5.23 Tandem cyclisation/cross-coupling reaction between alkene and Grignard reagents catalysed by a NHC–Co

reagent **83** to the zero-valent-ate complex **88** with concomitant production of 1 equiv. of homocoupling product **89** of Grignard reagent (Scheme 5.24). Complex **88** undergoes a single electron transfer with substrate **82** to yield an anion radical **89** and cobalt(I) complex **90**. The immediate loss of halide from the **89** affords radical intermediate **91**, which is cyclised into the radical **92**. Then, the cobalt species **90** recombines with the carbon centered radical **92** to form divalent cobalt species **93**. The following reductive elimination provides the product **84** and the Co(0) complex **94**, which is reconverted into **88** by the action of the remaining Grignard reagent.

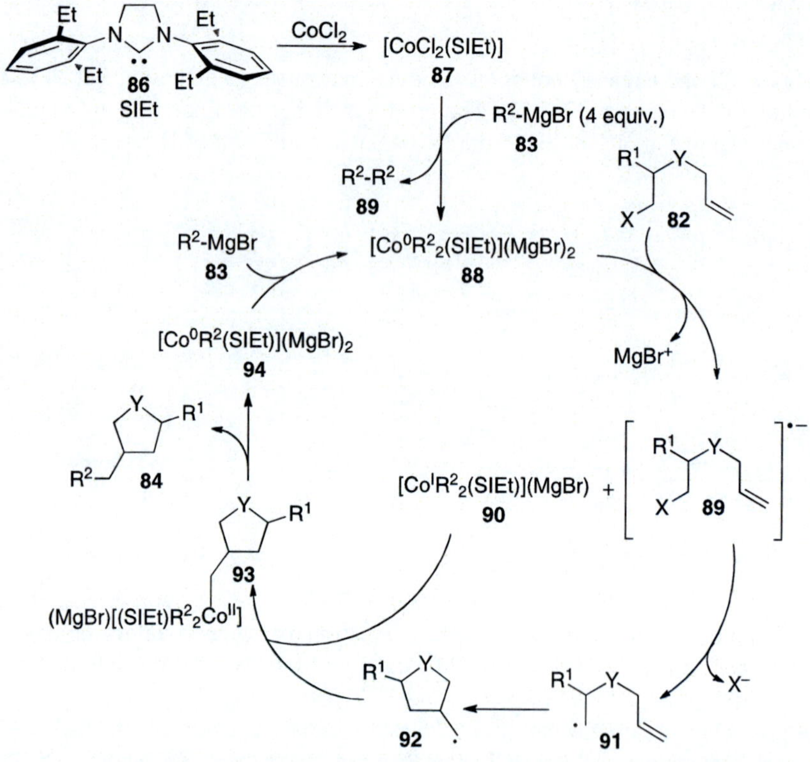

Scheme 5.24 Plausible mechanism for the tandem cyclisation/cross-coupling reaction

Gold catalysis provides an excellent method to construct complex chemical architectures in a mild manner that would be difficult to achieve using other reaction paradigms. NHC–Au complexes have been largely used in cycloisomerisation, rearrangement or isomerisation of allylic acetate, among other reactions [7, 24]. However, more recently, NHC–Au has also been used to catalyse cycloaddition reactions. The [4+2] cycloaddition NHC–Au reaction between cyclopropyl ketone **95** and indoles **96** or carbonyl compounds **98** affords the furan-fused polycyclic compounds **97** and **99**, respectively, in good yields (Scheme 5.25) [25]. This [4+2] cycloaddition proved to be highly regioselective and efficient.

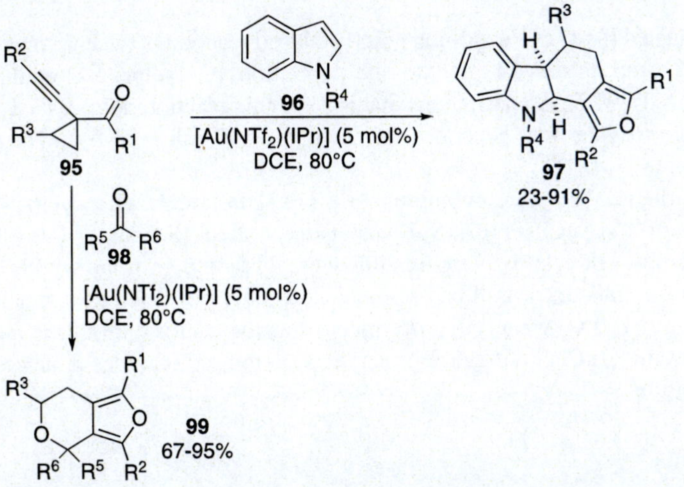

Scheme 5.25 [4+2] cycloaddition reaction catalysed by golden-NHC complex

Cyclopropyl ketones **95** also react with enol ether **100** in presence of 5 mol% of $[Au(NTf_2)(IPr)]$ in a [4+2] cycloaddition reaction to afford the bicycle[3.2.0] heptane skeleton **101** (Scheme 5.26) [26].

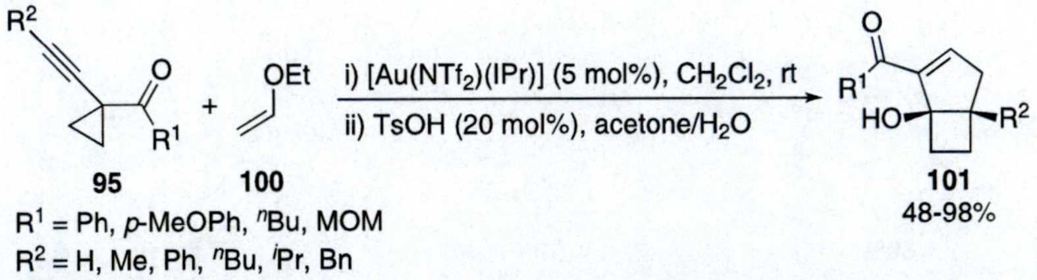

Scheme 5.26 [4+2] cycloaddition reaction catalysed by golden-NHC complex between cyclopropyl ketones and enol ether

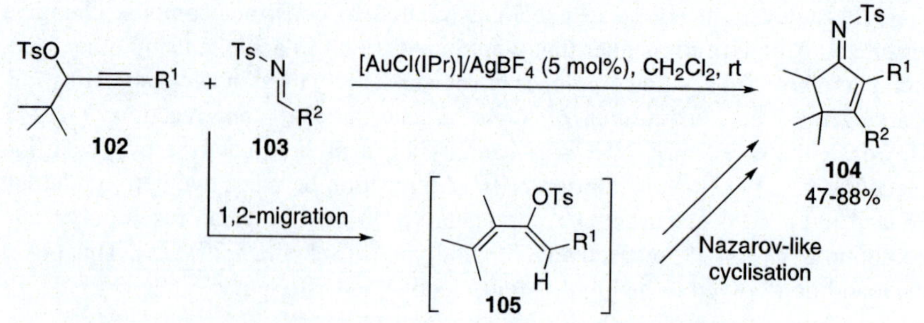

Scheme 5.27 [4+1] Cycloaddition gold-catalysed reaction between propargyl tosylates and imines

An unusual [4+1] cycloaddition gold-catalysed reaction between propargyl tosylates **102** and imines **103** led to the formation of cyclopent-2-enimines **104** (Scheme 5.27) [27]. A possible mechanism for this reaction involves a 1,2-migration of the tosylate that generates the 1,3-diene **105** followed by a Nazarov-like cyclisation.

Allenedienes **106** were submitted to a [4+3] intramolecular cycloaddition in presence of a [AuCl(IPr)]/AgSbF$_4$ catalytic system (Scheme 5.28) [28]. The cycloaddition adducts **107** and/or **108** were obtained in good yields at room temperature. In contrast, this cycloaddition reaction requires a much higher temperature (110°C) when PtCl$_2$ is employed as the catalyst [29]. This fact shows that the use of [AuCl(IPr)]/AgSbF$_4$ catalytic system is critical for the success of this cycloaddition.

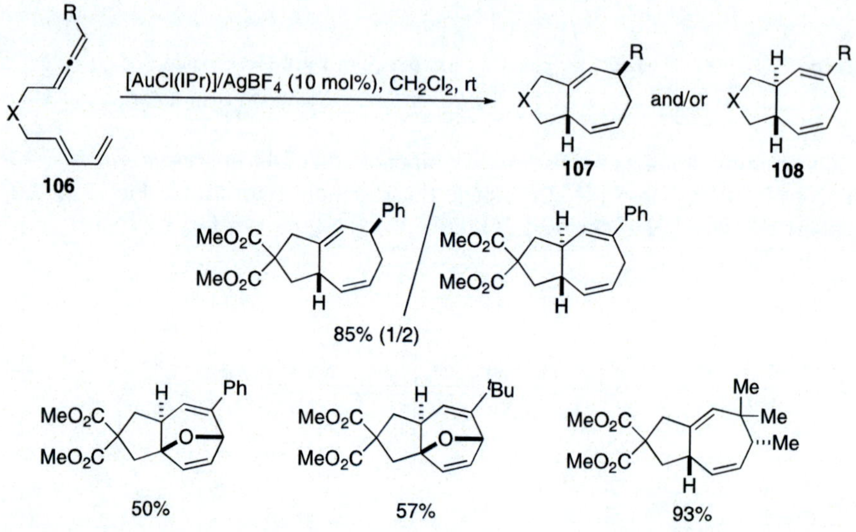

Scheme 5.28 [4+3] Intramolecular cycloaddition catalysed by [AuCl(IPr)]/AgSbF4 catalytic system

5.4 Cycloisomerisation

Transition metal catalysed cycloisomerisation of unsaturated systems is a powerful synthetic tool for the access of a wide range of heterocyclic and carbocyclic motifs [30]. The use of NHCs as ligands for transition metal cycloisomerisation catalysts has been extensively studied [7].

For the last 2 decades ruthenium carbene complexes (Grubbs'catalyst first generation **109** or second generation **110,** Fig. 5.1) have been largely employed and studied in metathesis type reactions (see Chapter 3) [31]. However, in recent years, the benefits of NHC–Ru complexes as catalysts (or pre-catalysts) have expanded to the area of non-metathetical transformations such as cycloisomerisation.

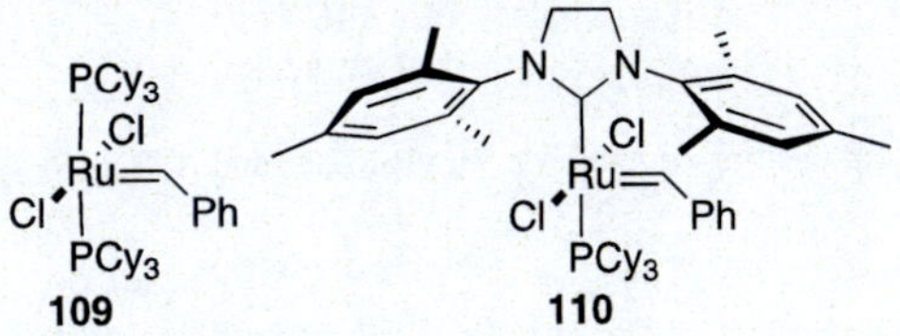

Fig. 5.1 Grubbs'catalysts first (**109**) and second (**110**) generation

1,6-Allenes **111** undergo cycloisomerisation in presence of 20 mol% of catalyst **110** (Scheme 5.29) [32]. Interestingly, when the Grubbs' first generation catalyst **109** is used in this reaction the ring-closing metathesis is observed [33].

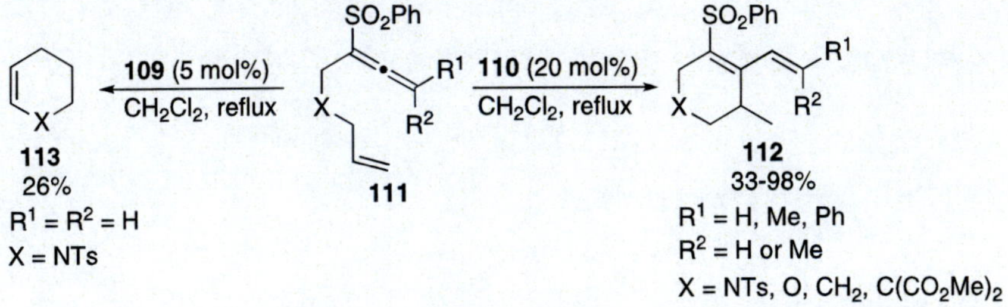

Scheme 5.29 Cycloisomerisation of 1,6-allenes catalysed by Grubbs' second generation catalyst

A mixture of catalyst **110** and vinyl trimethylsilyl enolether **115** has been used in cycloisomerisation of *N*-allyl-*o*-vinylanilines **114** and *N,N*-diallyl-*p*-toluenesulfonamide **115** to afford the corresponding products **118** and **119**, respectively (Scheme 5.30) [34]. It is believed that the active catalyst species is the ruthenium hydride NHC complex **117**.

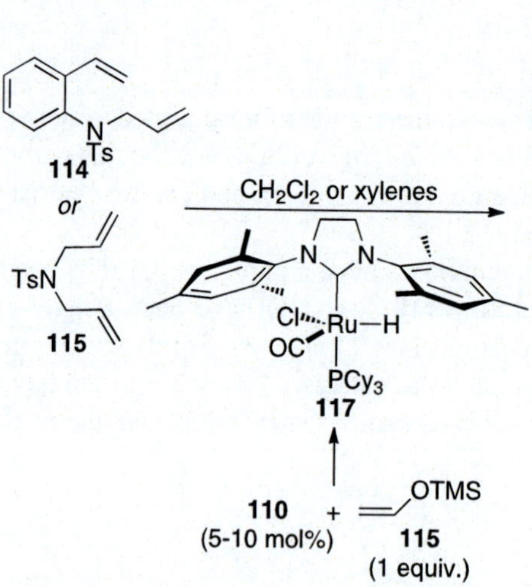

Scheme 5.30 Cycloisomerisation of *N*-allyl-*o*-vinylanilines and *N*,*N*-diallyl-*p*-toluenesulfonamide

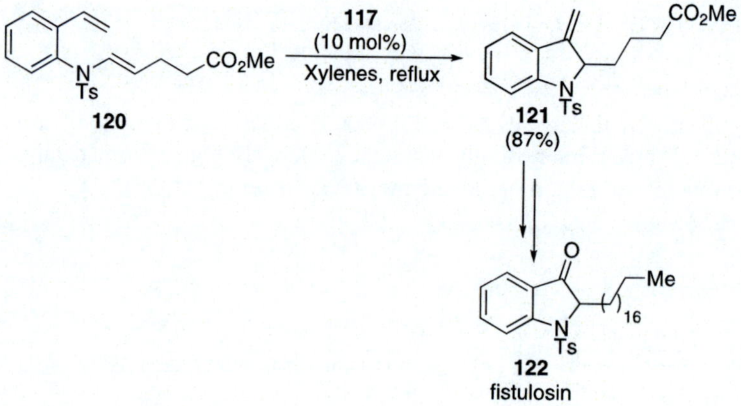

Scheme 5.31 Total synthesis of fistulosin

The same catalytic system mentioned above was also used in the synthesis of compound **121** (Scheme 5.31), which is a key intermediate in the total synthesis of fistulosin **122** [35]. The catalyst **117** promoted the cycloisomerisation of the *N*-vinylaniline **120** to indole ring **121** in 87% yield.

Enynes **79** can also undergo cycloisomerisation reactions in presence of NHC/transition metal complexes (Scheme 5.32). The cycloadduct **124** can be prepared either in presence of complex **110** [36] or in presence of a NHC–Ni complex (prepared *in situ* from a mixture of [Ni(COD)$_2$] and IDTB **123** [37]). In the latter case, the active catalytic species is believed to be a Ni–H species that is generated *via* C–H activation of the carbene ligand.

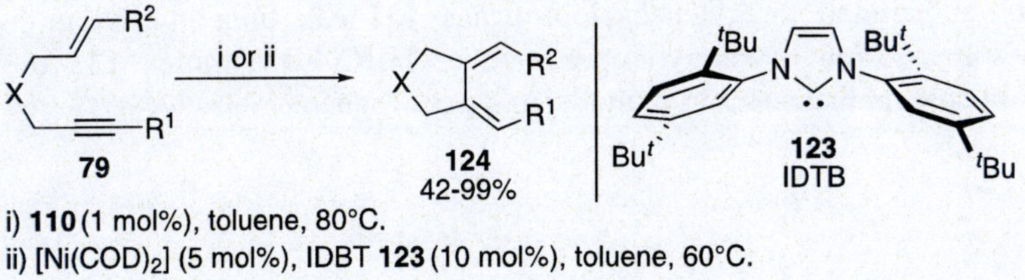

Scheme 5.32 Cycloisomerisations of enynes in presence of NHC–Ru or NHC–Ni complexes

The use of the NHC–Ni catalytic system has also been used to promote the cycloisomerisation of vinyl cyclopropanes **125** to afford the cyclopentene rings **126** in excellent yields (Scheme 5.33) [38]. The reaction required only 1 mol% of $[Ni(COD)_2]$ and 2 mol% of IPr carbene.

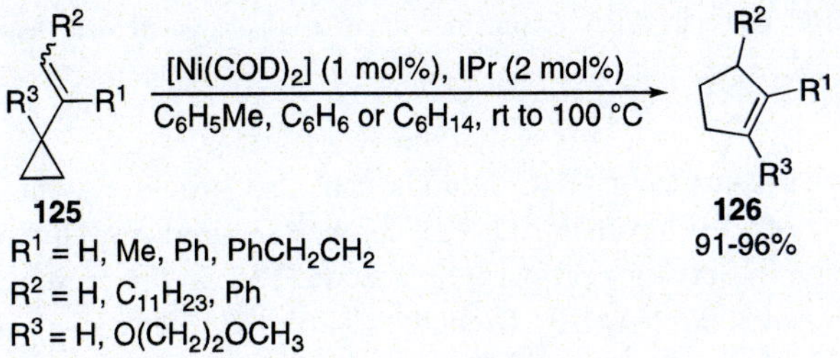

Scheme 5.33 NHC–Ni catalysed cycloisomerisation of vinyl cyclopropanes

Vinyl cyclopropanes tethered to an alkyne chain **127** were also subjected to the cycloisomerisation reaction in presence of the NHC–Ni catalyst system (Scheme 5.34) [39]. The product formation depends on the substrate used and the NHC ligand. When SIPr carbene is used, three different products were obtained depending on the size of the R group attached to the alkyne moiety. If R is small (like a methyl) product **128** is obtained exclusively. If R is Et or ^{i}Pr a mixture of **128** and **129** is obtained in 3:2 to 1:2 ratio, respectively. However, when R is large groups such as ^{t}Bu or TMS only product **130** is obtained. When I*t*Bu carbene **131** is used as the ligand, cycloisomerisation of **127** afforded product **128** exclusively, regardless of substituent size (Scheme 5.34) [39].

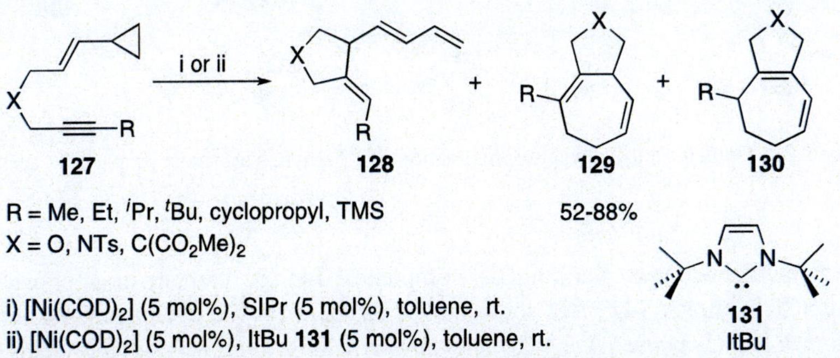

Scheme 5.34 Cycloisomerisation reaction of vinyl cyclopropanes tethered to an alkyne in presence of the NHC–Ni

Aryl-substituted methylenecyclopropanes **132** can undergo intramolecular cycloisomerisation in catalytic presence of NHC–Pd complex **133** to form 1,2-dihydronaphthalenes **134** in moderate yields (Scheme 5.35) [40].

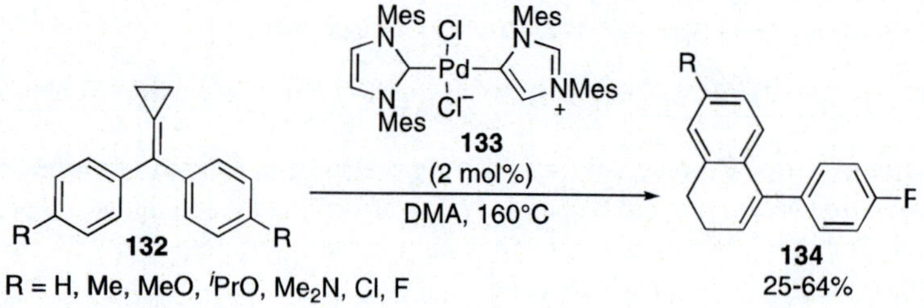

Scheme 5.35 NHC–Pd catalysed cycloisomerisation of aryl-substituted methylenecyclopropanes

Gold catalysts containing NHC ligands can also promote cycloisomerisation reactions. Bicyclo[3.1.0]hexanes **137–139** can be prepared from the cycloisomerisation of 1,5-enynes bearing a propargylic acetate (**135**) in the presence of catalytic amounts of [AuCl(IPr)]/$AgBF_4$ (Scheme 5.36) [41]. The cycloisomerisation reaction of **135** occurs by a 1,3-OAc shift/allene-ene cyclisation/1,2-OAc shift sequence. Experimental results with allenyl acetate **136** support this hypothesis as **139** is obtained in higher ratios than **137** and **138** [41b].

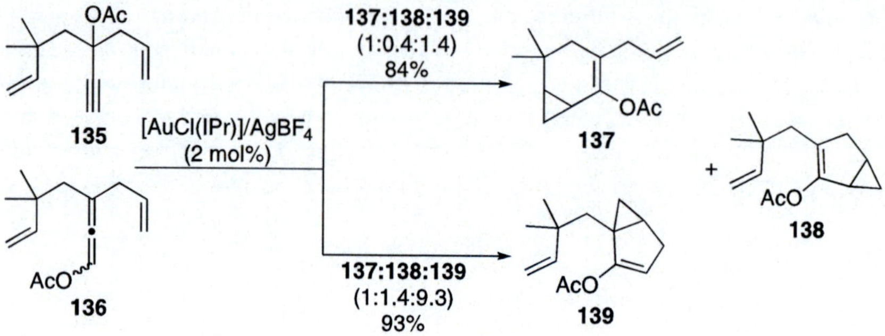

Scheme 5.36 Golden catalysed cycloisomerisation of 1,5-enynes bearing a propargylic acetate

Carbene/diphosphine platinum(II) complexes **141** or carbene/monophosphine platinum(II) complex **142** were shown to be efficient for the cycloisomerisation of 1,6-enynes **140** (Scheme 5.37) [42]. The product **143** is obtained in higher yields and higher enantioselectivity when catalyst **142** is employed rather than catalyst **141**.

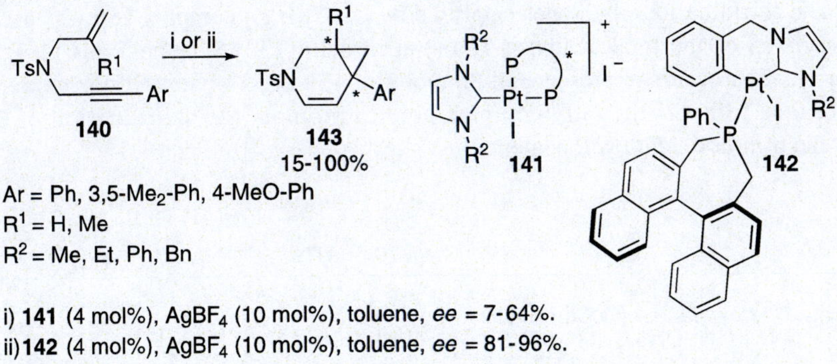

Scheme 5.37 Enantioselective cycloisomerisation of 1,6-enynes catalysed by NHC–Pt complex

5.5 Aziridination

Aziridines are important compounds due to their versatility as synthetic intermediates. In addition, aziridine rings are present in innumerable natural products and biologically active compounds. Nitrene addition to alkenes is one of the most well established methods for the synthesis of aziridines. Photolysis or thermolysis of azides are good ways to generate nitrenes. Nitrenes can also be prepared *in situ* from iodosobenzene diacetate and sulfonamides or the ethoxycarbonylnitrene from the *N*-sulfonyloxy precursor.

Nitrene addition to alkenes can be aided by the use of a transition metal, such as copper, rhodium, ruthenium, iron, cobalt, etc. NHC–Cu catalysts have been used in nitrene addition. For example [Cu(DBM)(IPr)] **147** (DBM = dibenzoylmethane) was successfully employed in the aziridination of aliphatic alkenes **144** in presence of trichloroethylsulfamate ester **145** and iodosobenzene **146** (Scheme 5.38) [43].

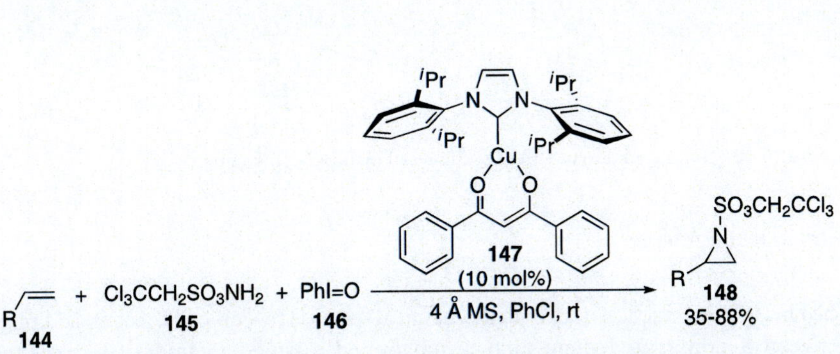

Scheme 5.38 Aziridination of aliphatic alkenes catalysed by a NHC–Cu complex

The aziridination of alkenes catalysed by [CuCl(IPr)] complex **150** was used in a key step of the total synthesis of (+)-agelastatin **152** (Scheme 5.39) [44]. The aziridination occurs in presence of 50 mol% of **150** in 52% yield. It is important to note that **150** was the only complex able to promote the aziridination of **149**, an electron-deficient cyclopentene.

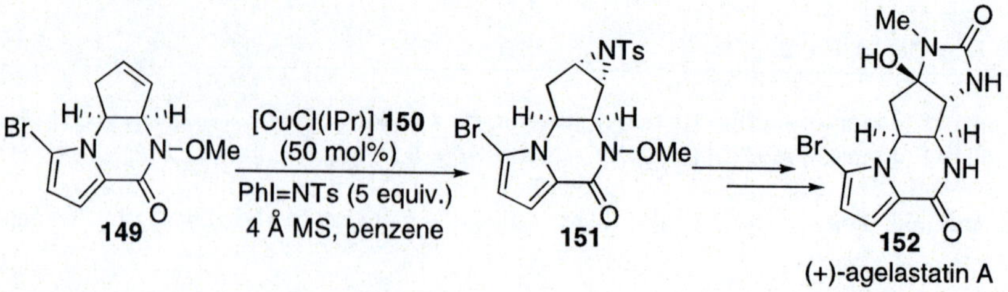

Scheme 5.39 Total synthesis of (+)-agelastatin

Copper(I) complexes containing NHC-phenoxyimine **153** or NHC-phenoxyamine **154** were shown to be good catalyst systems for nitrene addition to alkenes **144** (Scheme 5.40) [45]. The catalyst systems showed to be highly efficient as only 1 mol% catalyst loading was required to afford aziridines **155** in moderate to good yields.

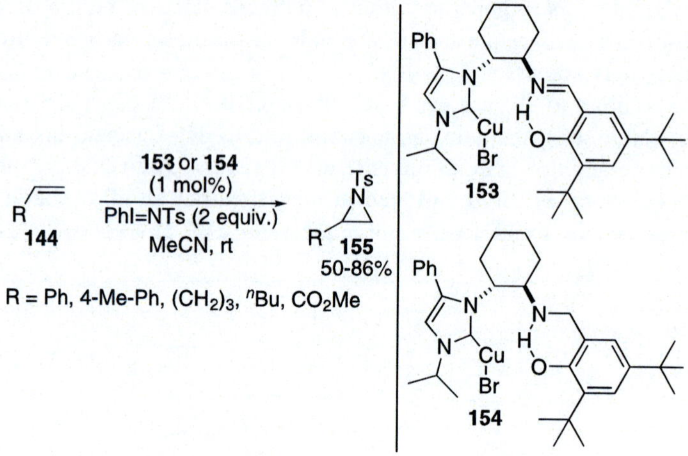

Scheme 5.40 Nitrene addition to alkenes catalysed by NHC–Cu complex

5.6 Diels-Alder

The Diels-Alder reaction is an organic chemical reaction (specifically, a cycloaddition) between a conjugated diene and a substituted alkene, commonly termed the dienophile, to form a substituted cyclohexene system. The reaction can also proceed if the alkene is replaced by an alkyne moiety or even if some of the atoms in the newly-formed ring are not carbon. Due to the high degree of regio- and stereoselectivity (due to the concerted mechanism), the Diels-Alder reaction is a very powerful reaction and is widely used in synthetic organic chemistry. Traditionally, Diels-Alder reactions required thermal conditions; however transition metals have been widely used as Lewis acid to improve reaction rates [46].

A chiral NHC–Ru complex **158** was used in the Diels-Alder reaction between methacrolein **156** and cyclopentadiene **157** (Scheme 5.41) [47]. The adduct **159** was obtained in an excellent yield under mild conditions, albeit with low enantioselectivity.

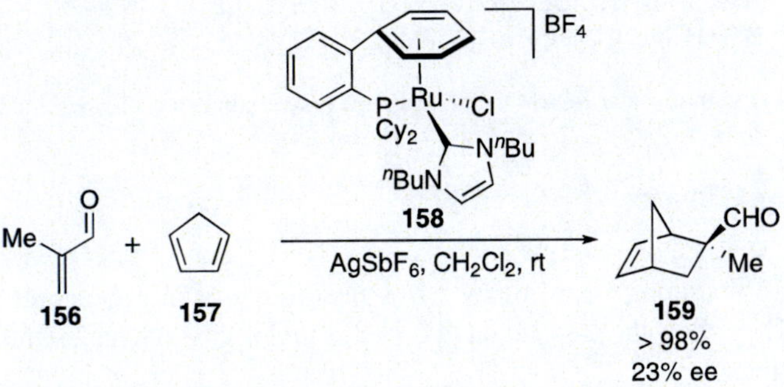

Scheme 5.41 NHC–Ru catalysed Diels-Alder reaction

Diene-ynes **160** undergo an intramolecular [4+2] Diels-Alder reaction in the presence of two different NHC–Rh catalysts (Scheme 5.42) [48]. Catalysts **162** [48a] and **163** [48b] were used in small amounts (1 mol% and 2 mol%, respectively) and the products **161** were obtained in excellent yield under mild conditions. Shorter reaction rates were observed when compared with reaction catalysed by phosphine-based rhodium catalysts.

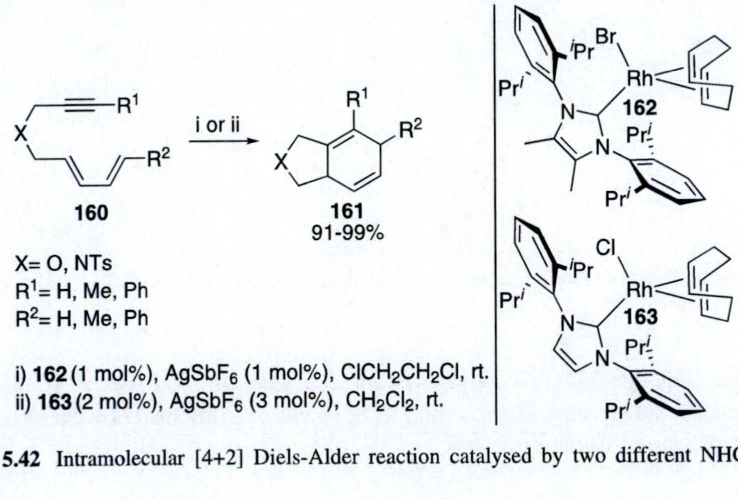

Scheme 5.42 Intramolecular [4+2] Diels-Alder reaction catalysed by two different NHC–Rh catalysts

Catalyst **163** also catalyses the intermolecular [4+2] Diels-Alder-type reaction between alkynes **164** and dienes **165** (Scheme 5.43) [49b]. Products **166** were obtained in an excellent regioselectivity and moderate to excellent yields.

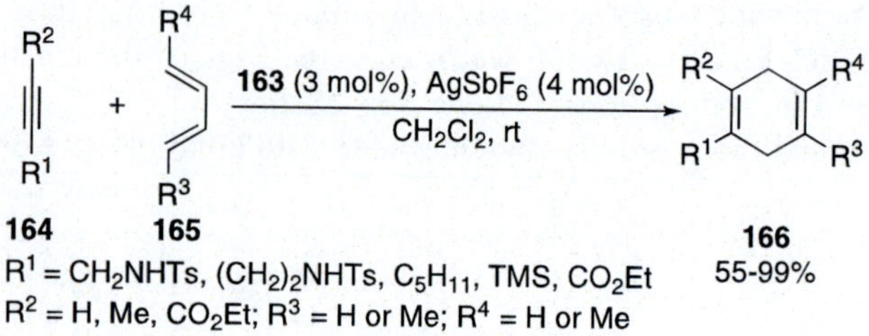

Scheme 5.43 Intermolecular [4+2] Diels-Alder-type reaction between alkynes and dienes

5.7 Ring Closure

Ring closure reactions are in many cases useful synthetic procedures in organic chemistry. They allow the preparation of complex molecules with high stereoselectivity and good yields.

Indole system **168** has been prepared from a ring closure reaction of dihydroisoquinoline derivative **167** catalysed by a NHC–Pd system (Scheme 5.44) [49]. The product **168** was prepared from an intramolecular Buchwald-Hartwig coupling and was used in the total synthesis of *rac*-mangochinine **169**.

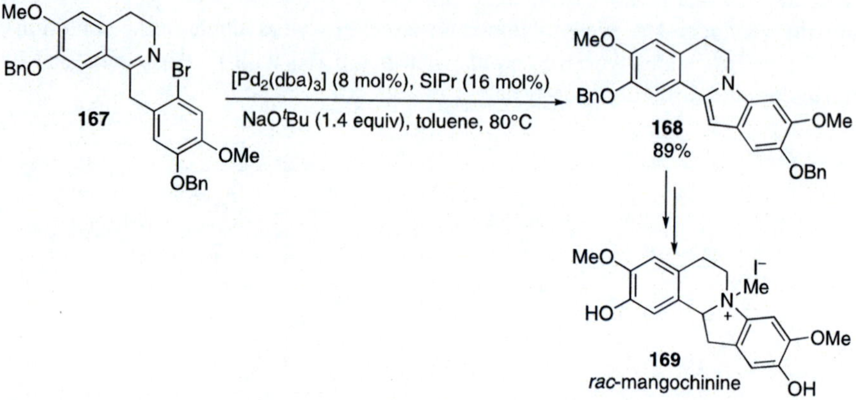

Scheme 5.44 Ring closure reaction of dihydroisoquinoline derivative catalysed by a NHC–Pd system

The NHC–palladium catalyst system can also promote the ring closure of *o*-bromobenzyl ketones **170** (Scheme 5.45) [50]. Benzo[b]furans **171** were successfully prepared using this methodology.

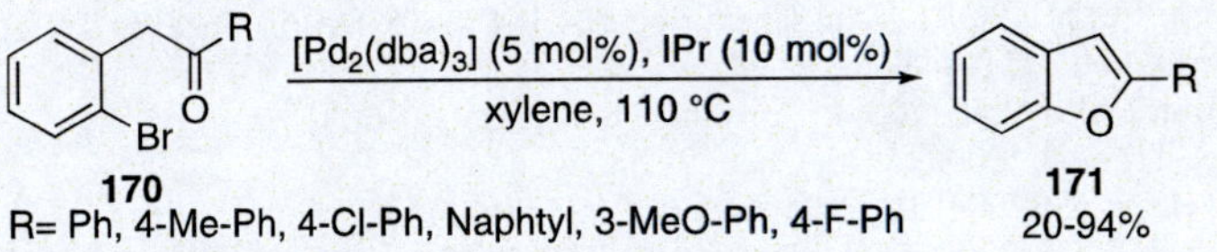

Scheme 5.45 NHC–Pd catalysed ring closure of *o*-bromobenzyl ketones

Bicyclic imidazolidin-2-ones **173** have been prepared in good yields and high diastereoselectivity (Scheme 5.46) [51]. The ring closure of *N,N′*-disubstituted ureas **172** was catalysed by [AuCl(IPr)] complex in presence of a silver salt.

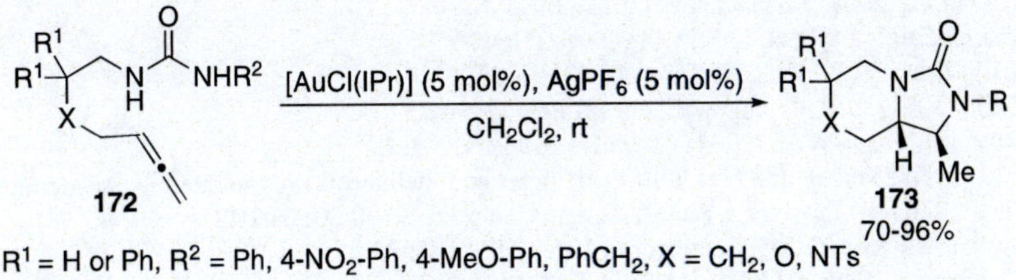

Scheme 5.46 Ring closure of *N,N′*-disubstituted ureas catalysed by [AuCl(IPr)] complex

References

1. (a) Díez-González S, Nolan SP (2007) Coord Chem Rev 251:874–883; (b) Cavallo L et al. (2005) J Organomet Chem 690:5407–5413; (c) Jacobsen H et al. (2009) Coord Chem Rev 253:687–703
2. (a) Patai S, Rappoport Z (1987) The chemistry of the cyclopropyl group. Wiley, New York; (b) Rubin M et al. (2007) Chem Rev 107:3117–3179; (c) Lebel H et al. (2003) Chem Rev 103:977–1050; (d) Donaldson WA (2001) Tetrahedron 57:8589–8627; (e) Wessjohann LA et al. (2003) Chem Rev 103:1625–1648
3. Peppers BP, Diver ST (2004) J Am Chem Soc 126:9524–9525
4. (a) Delaude L et al. (2006) Curr Org Chem 10:203–215; (b) Tudose A et al. (2006) J Organomet Chem 691:5356–5365
5. Fructos MR et al. (2004) J Am Chem Soc 126:10846–10847
6. Matsuda T et al. (2006) Chem Lett 35:288–289
7. (a) Chopade PR, Louie J (2006) Adv Synth Catal 348:2307–2327; (b) Varela JA, Saá C (2003) Chem Rev 103:3787–3802; (c) Díez-González S et al. (2009) Chem Rev 109:3612–3676
8. Huigsen R (1989) Kinetics and reaction mechanisms: Selected examples from the experience of forty years. Pure Appl Chem 61:613–628
9. (a) Díez-González S et al. (2006) Chem Eur J 12:7558–7564; (b) Díez-González S, Nolan SP (2008) Angew Chem Int Ed 47:8881–8884; (c) Díez-González S et al. (2008) Chem Commun 4747–4749
10. Peng HM et al. (2008) Organometallics 27:4484–4493
11. (a) Liu L, Montgomery J (2006) J Am Chem Soc 128:5348–5349; (b) Liu L, Montgomery J (2007) Org Lett 9:3885–3887
12. Saito S, Yamamoto Y (2000) Chem Rev 100:2901–2916
13. Saino N et al. (2005) Org Lett 7:3065–3067
14. Geny A et al. (2009) Adv Synth Catal 351:271–275

15. (a) Louie J et al. (2002) J Am Chem Soc 124:15188–15189; (b) Tekavec TN et al. (2004) Tetrahedron 60:7431–7437
16. McCormick MM et al. (2005) J Am Chem Soc 127:5030–5031
17. Tekavec TN et al. (2006) J Org Chem 71:5834–5836
18. Duong HA et al. (2004) J Am Chem Soc 126:11438–11439
19. Duong HA, Louie J (2006) Tetrahedron 62:7552–7559
20. (a) Tekavec TN, Louie J (2005) Org Lett 7:4037–4039; (b) Tekavec TN, Louie J (2008) J Org Chem 73:2641–2648
21. Knapp-Reed B et al. (2005) J Am Chem Soc 127:13156–13157
22. Jung IG et al. (2006) Organometallics 25:4240–4242
23. (a) Someya H et al. (2007) Org Lett 9:1565–1567; (b) Someya H et al. (2007) Tetrahedron 63:8609–8618
24. Marion N, Nolan SP (2008) Chem Soc Rev 37:1776–1782
25. Zhang G et al. (2008) J Am Chem Soc 130:1814–1815
26. Li G et al. (2008) J Am Chem Soc 130:6945–6946
27. Suárez-Pantiga S et al. (2009) Org Lett 11:13–16
28. Trillo B et al. (2009) Chem Eur J 15:3336–3339
29. Trillo B et al. (2008) Angew Chem Int Ed 47:951–954
30. Diver ST, Geissert AJ (2004) Chem Rev 104:1317–1382
31. (a) Ivin KJ, Mol JC (1997) Olefin metathesis and metathesis polymerization. Academic, San Diego, CA; (b) Fürstner A (2000) Angew Chem Int Ed 39:3012–3043
32. Mukai C, Itoh R (2006) Tetrahedron Lett 47:3971–3974
33. Kinderman SS et al. (2001) Org Lett 3:2045–2048
34. Arisawa M et al. (2006) J Org Chem 71:4255–4261
35. Terada Y et al. (2006) J Org Chem 71:1269–1272
36. Fürstner A et al. (2001) Chem Eur J 7:3236–3253
37. Tekavec TN, Louie J (2008) Tetrahedron 64:6870–6875
38. Zuo G, Louie J (2004) Angew Chem Int Ed 43:2277–2279
39. Zuo G, Louie J (2005) J Am Chem Soc 127:5798–5799
40. Yang Y, Huang X (2008) Synlett 1366–1370
41. (a) Marion N et al. (2006) Chem Commun 2048–2050; (b) Marion N et al. (2009) Chem Eur J 15:3243–3260
42. (a) Brissy D et al. (2007) Organometallics 26:5782–5785; (b) Brissy D et al. (2009) Organometallics 28:140–151; (c) Brissy D et al. (2009) Org Lett 11:2137–2139
43. Xu Q, Appella DH (2008) Org Lett 10:1497–1500
44. Trost BM, Dong G (2006) J Am Chem Soc 128:6054–6055
45. Simonovic S et al. (2009) Eur J Inorg Chem 1786–1795
46. (a) Reymond S, Cossy J (2008) Chem Rev 108:5359–5406; (b) Fringuelli F, Taticchi A (2002) The Diels-Alder reaction: selected practical methods. Wiley, Chichester
47. Faller JW, Fontaine PP (2006) J Organomet Chem 691:5798–5803
48. (a) Gómez FJ et al. (2007) Organometallics 26:4541–4545; (b) Lee SI et al. (2006) J Org Chem 71:91–96
49. Vincze Z et al. (2006) Synthesis 1375–1385
50. Faragó J, Kotschy A (2009) Synthesis 85–90
51. Li H, Widenhoefer RA (2009) Org Lett 11:2671–2674

Chapter 6
N-Heterocyclic Carbene Complexes in Cross-Coupling Reactions

Eloísa Jiménez-Núñez and Manuel Alcarazo

Abstract *N*-heterocyclic carbenes (NHCs) are excellent σ-donor and poor π-acceptor ligands. These properties make them not only behave as good phosphine mimics, but also result in more robust metal complexes. Recognising these advantages, the use of NHCs as spectator ligands in catalysis has become widespread. In this chapter we will focus on the advances in cross-coupling reactions using different metals and NHCs as ancillary ligands, a class of transformation that is a fundamental tool for synthetic chemists.

6.1 Introduction

Transition metal catalysed cross-coupling reactions of organometallic reagents containing Zn, Si, Mg, Sn or B with organic electrophiles in the presence of group 8–10 metals, notably Ni and Pd, are routinely the method of choice, both in academia and industry, for the preparation of C–O, C–S, C–H, C–N and C–C bonds [1].

These processes have flourished, mainly due to their selectivity and versatility, to the point where cross-coupling chemistry is often the initial thinking of organic chemists in synthetic and retro-synthetic approaches [2]. In fact, nowadays it is difficult to find a contribution in fine chemical or natural product synthesis where these molecular assembly tools are not employed. This is often due to the simple preparation and handling of the reaction partners as well as their relative compatibility with several functional groups.

M. Alcarazo (✉)
Max Planck Institut für Kohlenforschung, Kaiser Wilhelm Platz 1,
45470 Mülheim an der Ruhr, Germany
e-mail: alcarazo@mpi-muelheim.mpg.de

C. S. J. Cazin (ed.), *N-Heterocyclic Carbenes in Transition Metal Catalysis and Organocatalysis*, Catalysis by Metal Complexes 32,
DOI 10.1007/978-90-481-2866-2_6,

A very simplified but general scheme for the mechanism of all these transformations is shown below (Scheme 6.1). The first step of the catalytic cycle is the oxidative addition of the organo-halide or -triflate **B** to produce the species **C**. Transmetallation of the appropriate organometallic reagent **D** forms **E** which, upon reductive elimination, provides the desired product and regenerates the catalyst **A**.

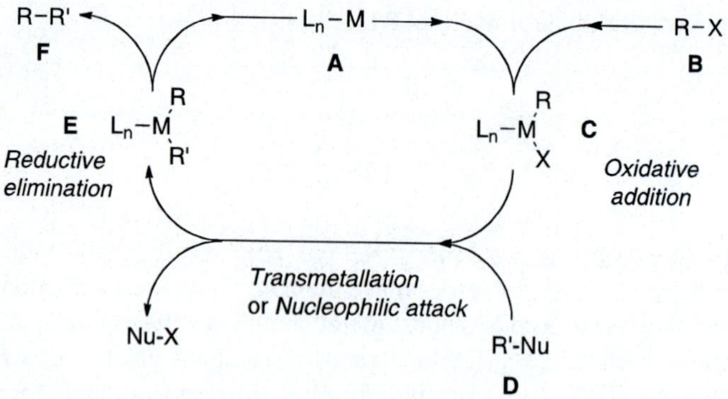

Scheme 6.1 General catalytic cycle for cross-coupling reactions

Although *ligand-free* systems for these transformations are known, the use of strongly σ-donating and sterically demanding ligands such as tertiary phosphines is often preferred to enhance the catalytic efficiency. This can be attributed to the favourable formation of coordinatively unsaturated [M–L_n] species that can effectively undergo oxidative addition reactions. *N*-heterocyclic carbenes, a family of compounds based on low valent C(II), have gained a prominent role in the areas of organometallic chemistry and catalysis due to their powerful donor capacity and the broad steric modulation that can be achieved by changing the substituents on the nitrogen atoms. These properties have seen them emerge as the only family to date that can clearly compete with phosphines as ligands for these processes.

The origin and nature of the electronic and steric properties of NHCs, that render them phosphine surrogates, have already been discussed in depth in the general introduction of this book. Hence, we will focus our attention on the peculiarities that these properties impart to the metal complexes, and therefore the catalytic species.

The current chapter covers literature highlights through July 2009 and addresses issues germane to the synthetic and organic community in general. The main focus of this work will be the study of the fundamental contribution of NHC ligands to this area. The chapter is divided into three sections, the first and more extensive part reviews processes dealing with the formation of carbon–carbon (C–C) bonds. The second provides an overview of reactions in which carbon–heteroatom bonds are formed, specifically C–N, C–B and C–S bonds, while the third one covers miscellaneous couplings that do not fit into the aforementioned sections.

Some general considerations applicable for all reactions that will be described must be presented at this point. It is relatively frequent to use Pd(II) species instead of Pd(0) as pre-catalysts due to their higher stability and easier preparation. The use of Pd(II) is not limited to dihalides but has also been described using related Pd(II) compounds such as alkyl-Pd(II) complexes that, after an induction period, undergo reductive elimination to produce the active catalysts [3]. In addition, it is important to highlight that reductive elimination in all cases occurs from the *cis* complex, therefore, a prior *trans* to *cis* isomerisation must occur if the relative disposition of the pre-catalyst ligands is *trans*. That will consequently generate a longer induction period [4] that now must include both the time required for initial isomerisation and the necessary reduction step from Pd(II) to Pd(0). Where [Pd(0)L(NHC)] complexes have been prepared, no induction time has been observed [5], thus establishing that these species are directly entering the catalytic cycle. It is unclear if the oxidative addition step takes place through an associative mechanism or through a dissociative pathway *via* a corresponding monocoordinated species Pd(0)NHC [6], but it has been shown that the presence of additional carbene ligand acts as a poison quenching the activity on the dicoordinated species [7] (Scheme 6.2).

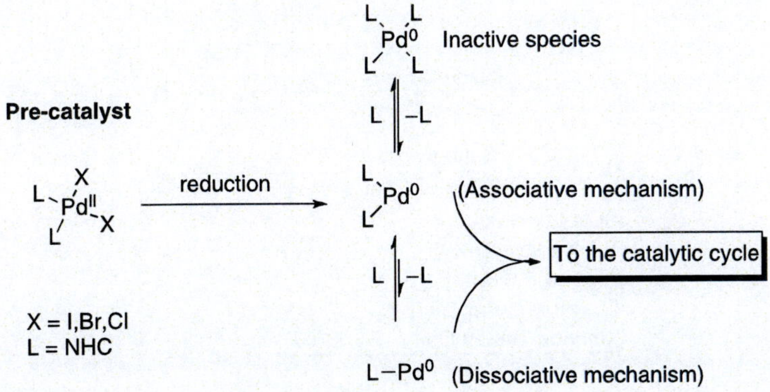

Scheme 6.2 Formation of the catalytic species

When examining pre-catalysts, the coordination of the carbene ligands to the metal centre is a major concern since, in addition to the normal C^2 coordination, C^5-coordinated metal carbenes can also occur [8]. Competing C^5-coordination is influenced by the steric requirements of the metal fragment, the base employed and the presence of a proton at the C^2-position of the imidazole ring [9]. These abnormal carbenes are more strongly donating than the C^2-analogues as only one N flanks the carbene centre. This C^5 binding mode results in many cases in an increase in catalytic activity of the resulting metal complexes [9,10]. It also must be considered that the steric properties of a NHC ligand are modified depending on the coordination position. These coordination modes should be carefully considered when the pre-catalysts are formed *in situ* from the metal salt and the ligand since a detailed knowledge of the nature of the catalytic species is crucial to the optimisation and development of any reaction (Scheme 6.3).

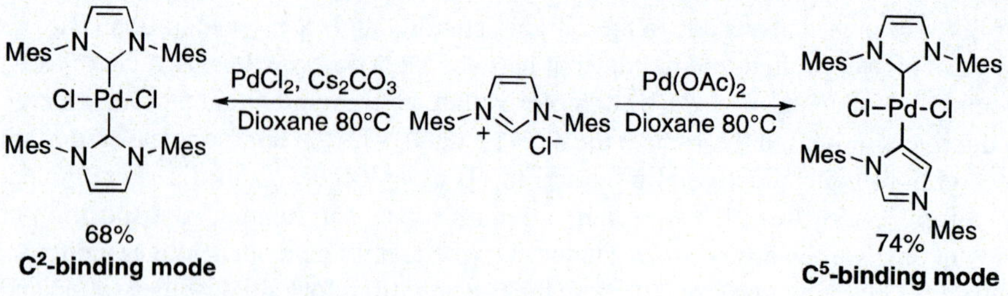

Scheme 6.3 Possible coordination modes of NHCs

6.2 Carbon–Carbon Bond Formation Reactions

Cross-coupling reactions resulting in C–C bond formation are normally classified according to the nature of the organometallic reagent and the organic electrophile. Scheme 6.4 summarises, broadly, the names and reaction partners of the most common of these transformations.

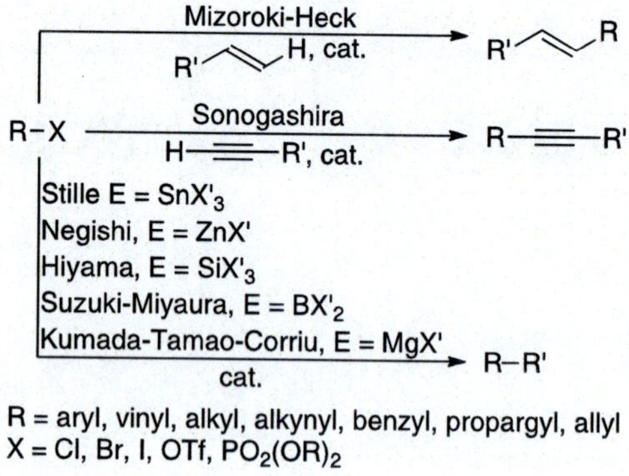

Scheme 6.4 Names and reaction partners for some cross-coupling reactions

6.2.1 Mizoroki-Heck Reaction

The Mizoroki-Heck reaction is a metal catalysed transformation that involves the reaction of a non-functionalised olefin with an aryl or alkenyl group to yield a more substituted alkene [11,12]. The reaction mechanism is described as a sequence of oxidative addition of the catalytic active species to an aryl halide, coordination of the alkene and migratory insertion, β-hydride elimination, and final reductive elimination of the hydride, facilitated by a base, to regenerate the active species and complete the catalytic cycle (Scheme 6.5).

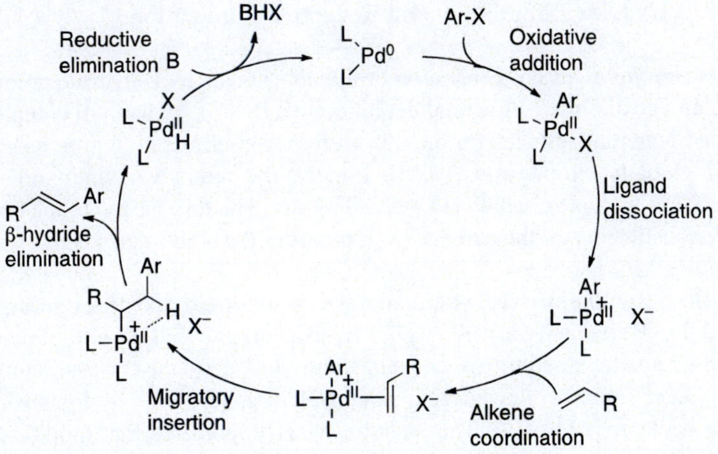

Scheme 6.5 Catalytic cycle for the Mizoroki-Heck reaction

6.2.1.1 Oxidative Addition

The aforementioned strong σ-donor ability of NHCs facilitates the oxidative addition [13] that, in the case of phosphine-based catalysts, is normally the limiting step. This step is easier for the weaker C–I bond in aryl iodides and more difficult for bromides and chlorides. In fact, it has been reported that the Mizoroki-Heck reaction of electron deficient aryl iodides and bromides (activated substrates) is even possible with simple palladium salts, in the absence of any stabilising ligand [14]. Using NHC ligands, it has been shown that even poorly reactive aryl chlorides can undergo oxidative addition at room temperature [5,6]. However, the generalisation of this reactivity to all aryl chlorides still represents a considerable challenge.

6.2.1.2 Alkene Coordination and Migratory Insertion

After oxidative addition, dissociation of a ligand is proposed to occur prior to coordination of the olefin. In Mizoroki-Heck reactions catalysed by phosphine-Pd catalysts, a phosphine ligand is substituted by the alkene, but in the case of NHC–Pd complexes where the binding energy calculated for a NHC is higher than that of the halide anion, the situation could be otherwise [11,15]. Thus, a cationic intermediate is proposed and although coordination of the alkene is disfavoured due to entropy loss it is probably compensated by solvation effects of the cationic species in the highly polar solvents employed in the reaction. This point is in dire need of investigation to clarify the exact mechanism at play in the Pd–NHC mediated Mizoroki-Heck reaction.

6.2.1.3 β-Hydride Elimination and Reductive Elimination

The next step involves the generation of the new alkene by β-hydride elimination, through an agostic interaction, and evolution to a hydride-palladium complex. The calculated potential surfaces for the overall insertion-elimination process are quite flat and globally exothermic [11,15]. Finally, the reductive elimination of the hydride-Pd(II) complex, which is favoured by steric factors related to the bulkiness of the *N*-substituents on the carbene [13], provides the active species that can enter into a new catalytic cycle.

The first example of NHC–Pd complexes in the Mizoroki-Heck reaction was reported by Herrmann in 1995 [16]. *cis*-Biscarbene Pd(II) complexes were synthesised and their activities evaluated after reduction to the corresponding Pd(0) species. Other studies followed, also using NHC ligands, but with ligands bearing different backbones [17–19]. These studies clearly indicated that modification of the NHC electronic properties has an effect on catalytic activity [4,20,21]. From all variations studied, the constitutional change of the ring containing the carbene atom produced the most significant differences [22]. These basic structural changes range from the ring size, such as in the case of tetrahydropyrimidinylidene carbenes **5**, to the nature of the heteroatoms flanking the carbene carbon, as in the case of oxazolinylidene ligands **3** (Scheme 6.6). The effect of the substituents on the nitrogen atoms is thought to be only steric in nature.

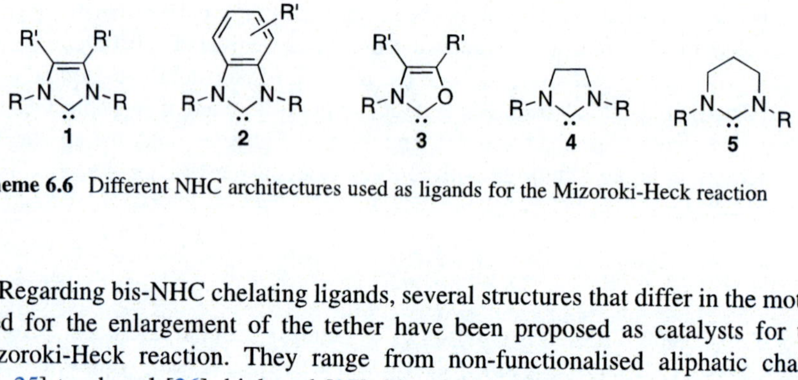

Scheme 6.6 Different NHC architectures used as ligands for the Mizoroki-Heck reaction

Regarding bis-NHC chelating ligands, several structures that differ in the motifs used for the enlargement of the tether have been proposed as catalysts for the Mizoroki-Heck reaction. They range from non-functionalised aliphatic chains [23–25] to phenyl [26], biphenyl [27], binaphthyls [28] and to chains containing additional coordination positions like ethers [29], amines [30], and pyridines in an evolution towards pincer complexes [31–35]. In most cases, the activity of aryl bromides in Mizoroki-Heck transformations was demonstrated to be from moderate to high, while the activation of chlorides was non-existent or poor (Scheme 6.7).

Other classes of complexes that have been studied in depth in the Mizoroki-Heck reaction are those having a bidentate ligand containing both a NHC and a phosphine. The development of these structures was encouraged by early theoretical work from Rösch, who calculated that such ligands should be promising catalysts for this specific transformation [15]. The basic idea was to combine the stabilising effect of the carbene with a more labile Pd–P bond that could coordinate in a reversible way. Thus Herrmann synthesised mixed NHC/phosphine complexes and observed the importance of having bulky substituents on the nitrogen atoms in order to increase their stability to the conditions required for Mizoroki-Heck transformations. Nevertheless, the main drawback of these catalysts was the necessary tuning of the phosphine for each particular reaction [36,37]. Subsequently, Fürstner showed that the activities of neutral and cationic NHC-phosphine complexes containing non-chelating ligands were similar [38], which could be indicative of a similar active species after decoordination of one or two phosphines. Pincer complexes of the type PCP with two phosphino groups and a NHC have also been tested [39] (Scheme 6.8).

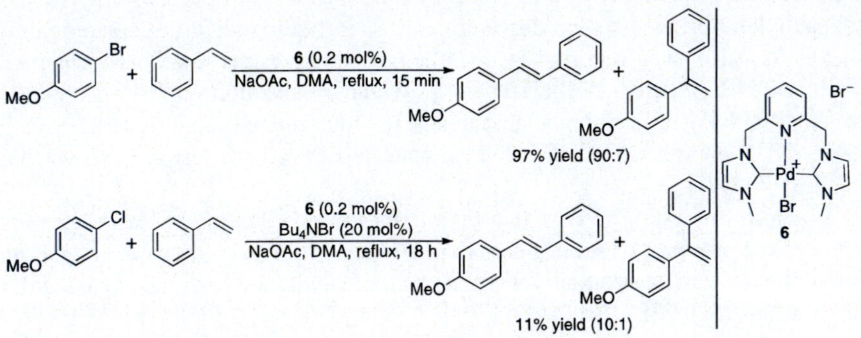

Scheme 6.7 Application of bis-NHC chelating ligands in the Mizoroki-Heck reaction

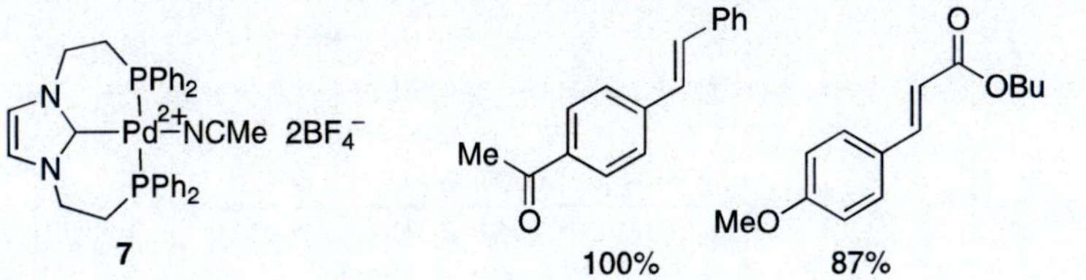

Scheme 6.8 Reactivity of the mixed NHC/phosphine complex **7**

Along with these well-defined complexes, other protocols have been developed to directly involve imidazolium salts with Pd sources and form the active catalysts *in situ*. One of the most popular consists of the use of carbene precursors such as IMes·HCl or IPr·HCl with $Pd(OAc)_2$ or $Pd(dba)_2$ and a base [40]. A mixture of SIPr·HCl and $Pd(OAc)_2$ in a 1:1 ratio was used for the synthesis of resveratrol analogues (MOM protected; MOM = methoxymethylether) through decarbonylative Mizoroki-Heck coupling [41] (Scheme 6.9).

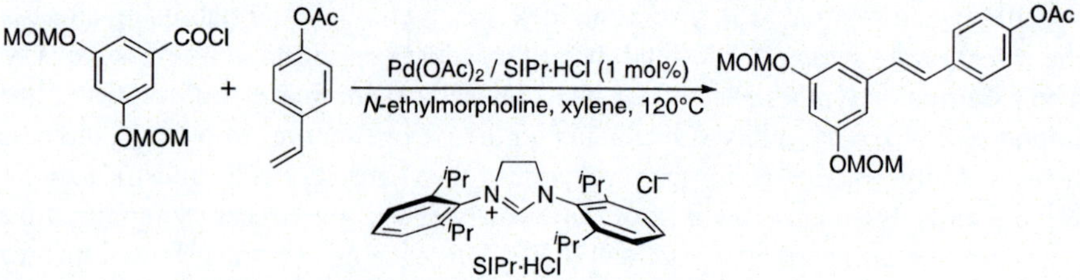

Scheme 6.9 An *in situ* protocol applied to the synthesis of resveratrol analogues

Among these *in situ* protocols are those using ionic liquids as the solvent, or as both the solvent and the ligand. It was shown that the use of $Pd(OAc)_2$ in imidazolium-based ionic liquids forms *in situ* NHC–Pd(II) species [42]. The use of methylene-bridged bis-imidazolium salt ionic liquids to form chelated complexes has also been reported [43], although better results have been obtained when Bu_4NBr is used as the solvent [44] and imidazolium salts were added together with $PdCl_2$ in catalytic amounts [45]. Other related catalytic species such as bis-NHC complexes of silica-hybrid materials have been tested as recyclable catalysts [46,47].

It is also interesting to note that the *in situ* use of some bis-imidazolium salts allowed the presence of several common oxidants in the reaction medium and the possibility to work in open-air vessels, a result of significant relevance to industrial applications (Scheme 6.10) [48].

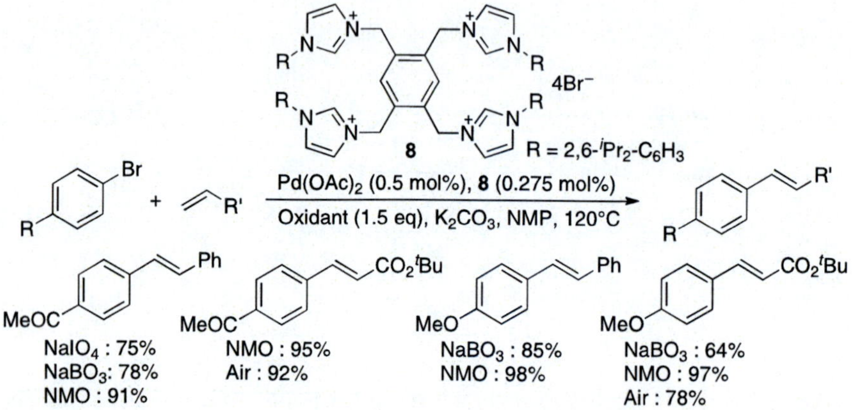

Scheme 6.10 The Mizoroki-Heck transformation in the presence of oxidants

Other species tested *in situ* are chelating bis-imidazolium structures with phenylene spacers [48], and bidentate C–P ligands [49] that achieved, in the case of triarylphosphine-imidazolium salts, very good yields for a wide variety of substituted aryl bromides (Scheme 6.11) [50].

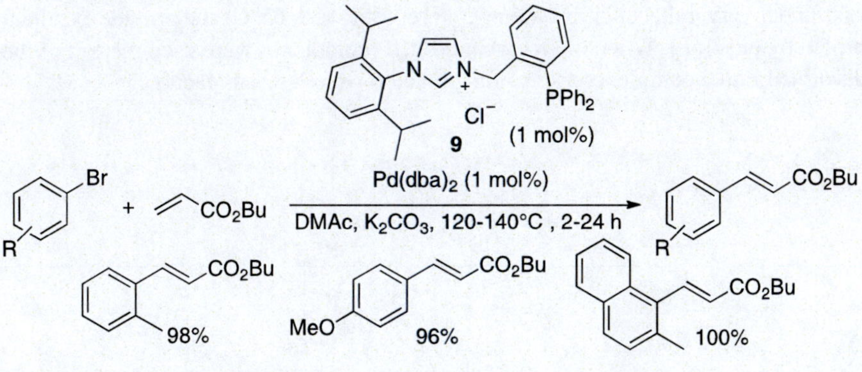

Scheme 6.11 Chelating C–P ligands for the Mizoroki-Heck coupling

As previously stated, CN bidentate complexes with a nitrogen atom incorporated into different heterocycle scaffolds have also been studied. These include NHC-pyridine [51–56], NHC-pyrimidine [57,58], NHC-oxazoline [59], NHC-pyrazole [60], NHC-benzimidazole [27] and NHC-acetylamide complexes [61]. Although they have shown moderate activity in Mizoroki-Heck couplings, the use of nitrogen as an additional donor group has been more successfully applied to nitrogen-containing palladacycles. Oxime- [62], benzylamine- [62–64] and phospha-palladacycles [65] have been prepared and for the case of the palladacycle formally derived from *N,N*-dimethylbenzylamine and IMes·HCl the results of the Mizoroki-Heck reaction with highly hindered aryl bromides ranged from moderate to very good yields (Scheme 6.12) [63,64].

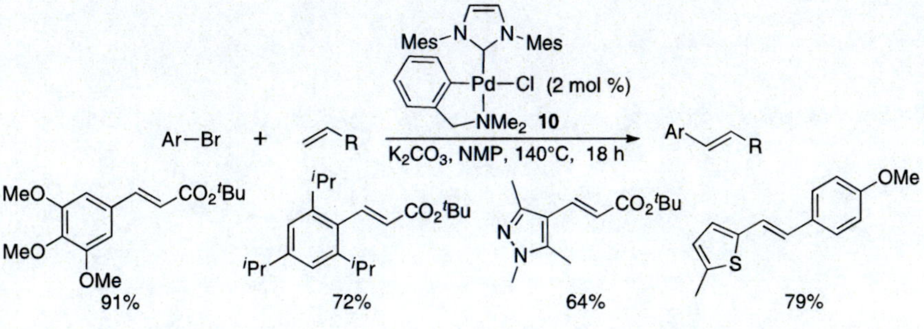

Scheme 6.12 Scope of catalyst **10** containing a palladacycle derived from *N,N*-dimethylbenzylamine

As mentioned in the discussion of the reaction mechanism for this transformation, the active species is a dicoordinate Pd(0) complex, and it is unclear whether an associative or a dissociative process is operative for oxidative addition. In this context, different NHC complexes containing only one carbene ligand have been tested in the Mizoroki-Heck reaction. The most successful are those prepared by Beller, which were able to perform the Mizoroki-Heck reaction of non-activated aryl chlorides with moderate to good yields in ionic liquids (Scheme 6.13). The same compounds have also been applied to the Mizoroki-Heck reaction of aryldiazonium salts under very mild conditions (temperatures of just 50–65°C) and provide excellent product yields [66]. When monocarbene-Pd(II) complexes were used instead of the discussed Pd(0) complexes the results obtained were less satisfactory [67,68].

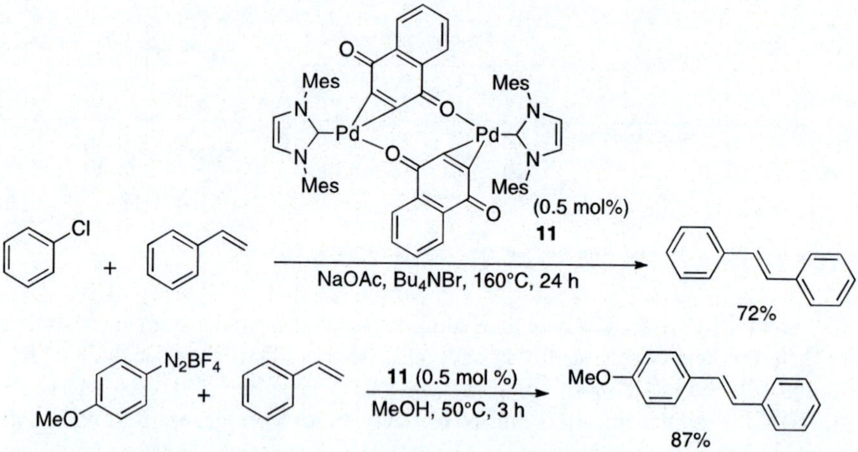

Scheme 6.13 Mizoroki-Heck reaction of non-activated aryl chlorides and diazo compounds using Beller's catalytic systems

Related reactions, that have been catalysed by NHC/Pd systems, are the intramolecular Mizoroki-Heck using catalysts formed *in situ* from imidazolium salts and a Pd(0) source [69], and the arylation of allylic alcohols by a benzothiazole-Pd complex [70,71] (Scheme 6.14).

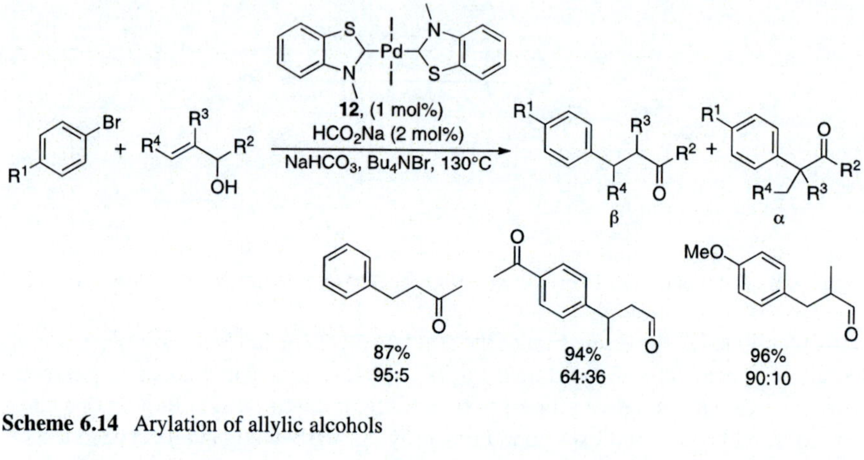

Scheme 6.14 Arylation of allylic alcohols

The catalytic arylation of SEM-protected azoles (SEM = [2-(trimethylsilyl) ethoxy]methyl) with monocarbene-Pd(II) complexes [72], and the reductive Heck or hydroarylation of norbornenes with NHC-phosphine chelating ligands

(Scheme 6.15) are other related processes that can be mentioned alongside the Heck reaction [73]. In addition, Ni-based catalysts have also been reported [74].

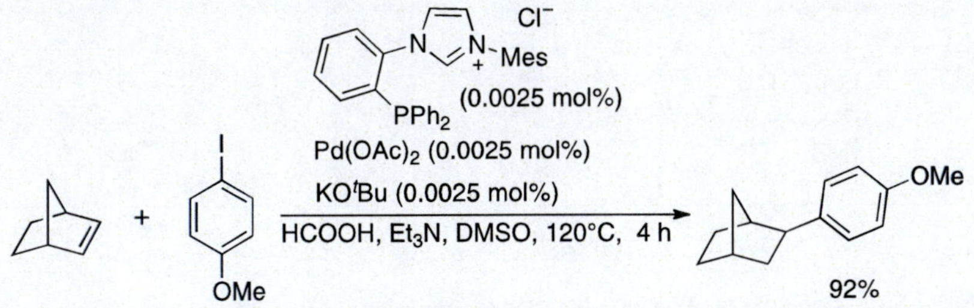

Scheme 6.15 Pd-catalysed hydroarylation or norbornene using chelating NHC/P-ligands

6.2.2 *Reaction with Grignard Substrates*

The main advantage in the use of Grignard reagents for coupling reactions is the facile and direct preparation of these compounds from Mg and the corresponding alkyl or aryl halides. Remarkably, although organomagnesium reagents are often the starting material for the synthesis of organoboron, organozinc or other derivatives frequently employed as reaction partners, their direct application is less developed. This is due to two main reasons; the tendency of Mg derivatives to lead to undesirable homo-coupling products, and the limited functional group tolerance of Grignard reagents. However, when possible, the low cost and toxicity of Mg makes this reaction an excellent process.

Nolan and colleagues published in 1999 the first coupling between aryl chlorides, bromides or iodides and aryl Grignards, in which an *in situ* generated NHC–Pd complex served as catalyst. The reaction proceeded in most cases in excellent yields, however, very sterically hindered products formally derived from aryl halides and aryl Grignards both possessing *ortho*-substituents could not be obtained [75] (Scheme 6.16).

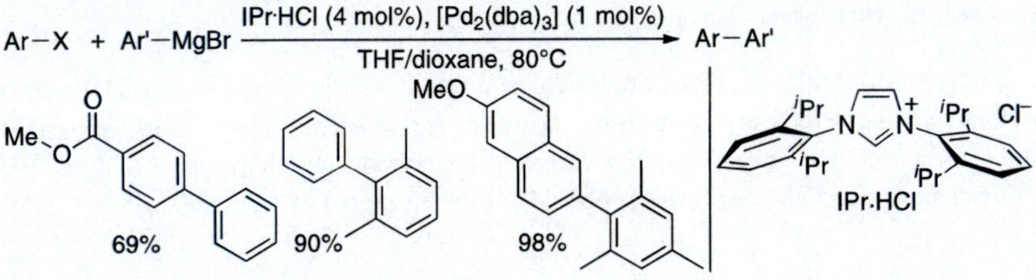

Scheme 6.16 Nolan conditions for the coupling between aryl halides and aryl Grignards

A very elegant extension of this process to alkyl chlorides, even bearing some functionalities, was developed by Beller using the already mentioned well-defined Pd–NHC naphthoquinone complex **11** (Scheme 6.17). The yields are, in general, excellent except when α-substituted alkyl chlorides are employed [76].

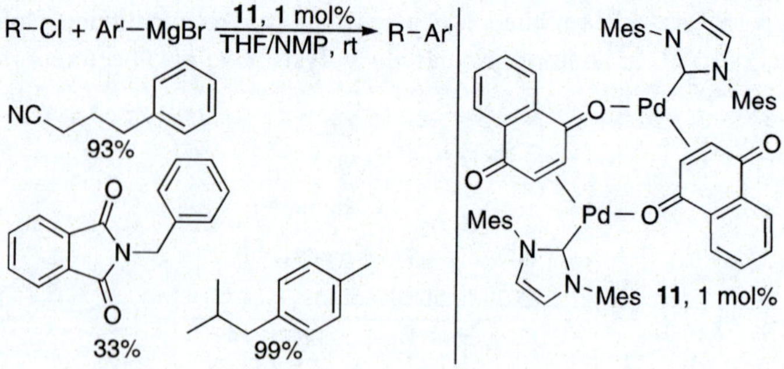

Scheme 6.17 Coupling of alkyl chlorides and Grignards using Beller's catalyst

In 2008, Chen developed a Ni-catalysed version of this reaction using catalysts **13** and **14** containing multidentate-NHC ligands (Scheme 6.18). This interesting method not only allowed for the preparation of some biaryls in good to excellent yields but the catalytic system also tolerates some *N*-based functionalities such as pyridines or pyrimidines [77,78]. A pre-catalyst with similar functional group tolerance has also been described by Organ and co-workers [79] although it is based on Pd instead of the more convenient and inexpensive Ni.

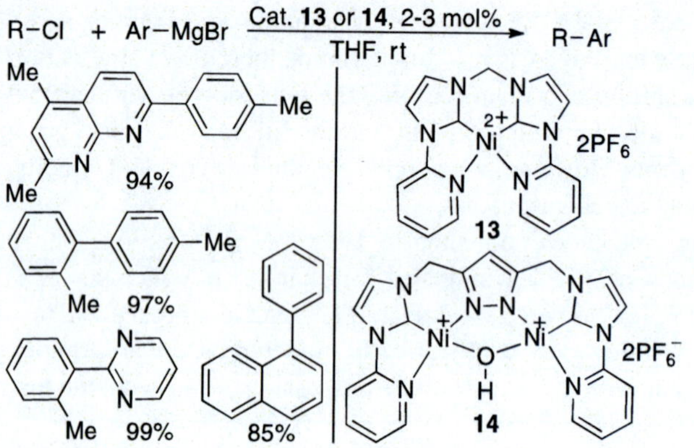

Scheme 6.18 Nickel-catalysed processes using chelating ligands

The cobalt-catalysed reaction between aryl bromides and Grignard reagents assisted by IMes·HCl is also known, however the substrate scope is quite narrow and good yields are only obtained when non-branched long chain alkyl magnesium chlorides are used as coupling reagents [80] (Scheme 6.19).

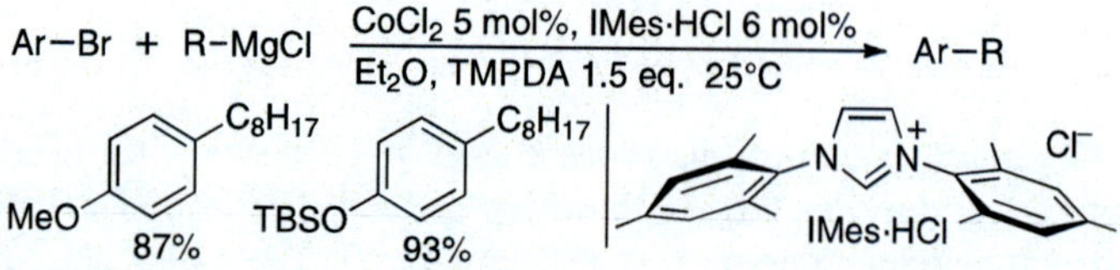

Scheme 6.19 Cobalt-catalysed coupling between aryl bromides and alkyl Grignards

It was not until Cazin and co-workers introduced the well-defined dimer complex $[Pd(\mu\text{-}Cl)(Cl)(SIPr)]_2$ that the Kumada-Tamao-Corriu cross-coupling between severely sterically hindered partners was pursued, allowing the synthesis of tri- and tetra-*ortho* substituted biaryls [81] (Scheme 6.20).

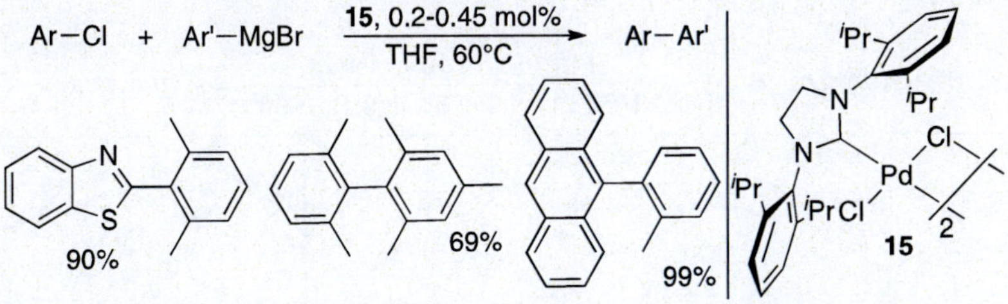

Scheme 6.20 Coupling between very sterically hindered partners using Cazin conditions

In conclusion, further work to increase the scope of this reaction, specifically to obtain higher functional group tolerance, is desirable. In addition, the development of a chiral catalyst that enables the production of enantiopure tetra-*ortho* substituted biaryls would be of significant interest.

6.2.3 Negishi Coupling

The Negishi reaction involves the coupling of organozinc derivatives with organic electrophiles. This process is an extremely useful tool for C–C bond formation due to the high reactivity and availability of the nucleophilic partners, combined with their improved functional group tolerance in comparison to analogous Grignard reagents [1]. However, organozinc reagents are still incompatible with several common functional groups, and their sensitivity to oxygen and especially water is another drawback.

The first really successful Pd–NHC mediated Negishi coupling was developed by Organ and co-workers using IPr·HCl/$Pd_2(dba)_3$ as catalytic mixture at 70°C. Under these conditions the coupling of functionalised alkyl bromides and alkylzinc reagents proceeded smoothly and in high yields, even when branched organozinc compounds were used [82] (Scheme 6.21).

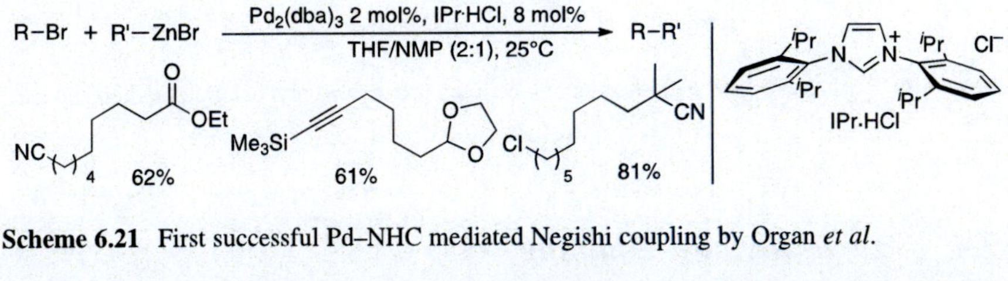

Scheme 6.21 First successful Pd–NHC mediated Negishi coupling by Organ *et al.*

These results were significantly improved by the same authors using the well-defined pre-catalyst **16** (PEPPSI-IPr), that is able to promote the coupling of alkyl halides, aryl halides or alkyl sulphonates with alkylzinc chlorides or bromides [83,84].

Examples of the products obtained, some of them highly sterically hindered, can be seen in (Scheme 6.22). The diversity of products prepared clearly highlights the efficiency of pre-catalyst **16** for the Negishi reaction. Additionally, acetylenic ketones can be obtained in good yields through a carbonylative Negishi cross-coupling of aryl iodides and alkynylzinc reagents using **16** [85].

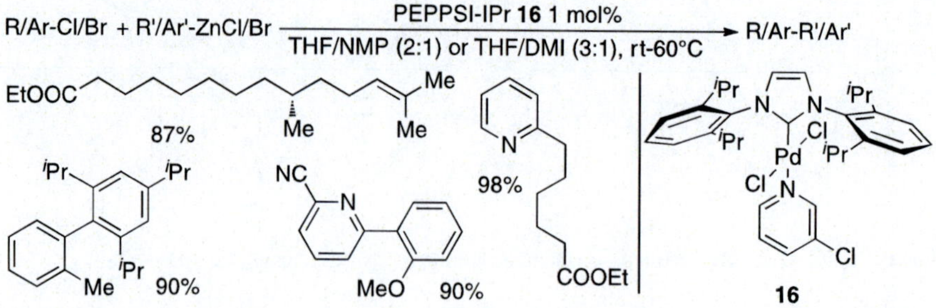

Scheme 6.22 Reaction scope of pre–catalyst **16**. Selected examples

Very recently Chen and co-workers have applied the previously mentioned Ni-based dimetallic pre-catalyst **14** in the Negishi reaction. Remarkable results were obtained even when unactivated aryl chlorides were chosen as reaction partners providing an alternative to the more expensive Pd-based catalysts. The fact that dinuclear pre-catalyst **14** is more active than its mononuclear analogue **13** indicates a possible cooperative effect between the two metal centres [86] (Scheme 6.23).

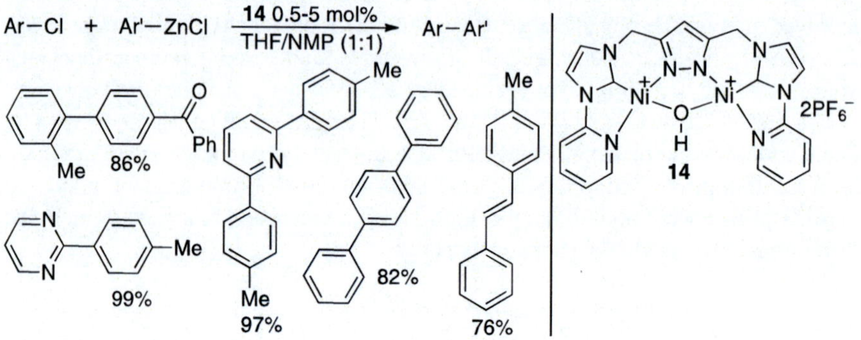

Scheme 6.23 Nickel-catalysed Negishi coupling between unactivated aryl chlorides and organozinc reagents

6.2.4 Suzuki-Miyaura Coupling

The Suzuki-Miyaura reaction, first reported by Suzuki and co-workers in 1979 [87], is the metal-mediated (often palladium) coupling of organic electrophiles such as alkenyl halides or triflates with organoboron reagents in the presence of a base. This coupling is amongst the most widely used protocols for the formation of carbon–carbon bonds because of the availability of a vast number of well defined, non-toxic, air- and moisture-tolerant boronic acids. The reaction is also tolerant to many functional groups allowing the use of elaborated coupling partners. This property makes the process especially suitable for the total synthesis of complex natural products. In addition, it can be carried out using non-dried solvents, including alcohols and water. Again NHCs are a reasonable choice as ancillary ligands for this transformation due to their excellent donor ability that enhances the electron density on the metal, thus facilitating the oxidative addition step even with unfavoured substrates like aryl chlorides or sulphonates. It is also well known that the catalytic activity of a given family of ligands is increased by the introduction of bulkier substituents due to the enhanced propensity of reductive elimination. This further highlights the important beneficial features of the NHCs as ancillary ligands as their steric properties can be widely modulated by changing the substituents on the nitrogen atoms surrounding the carbene carbon.

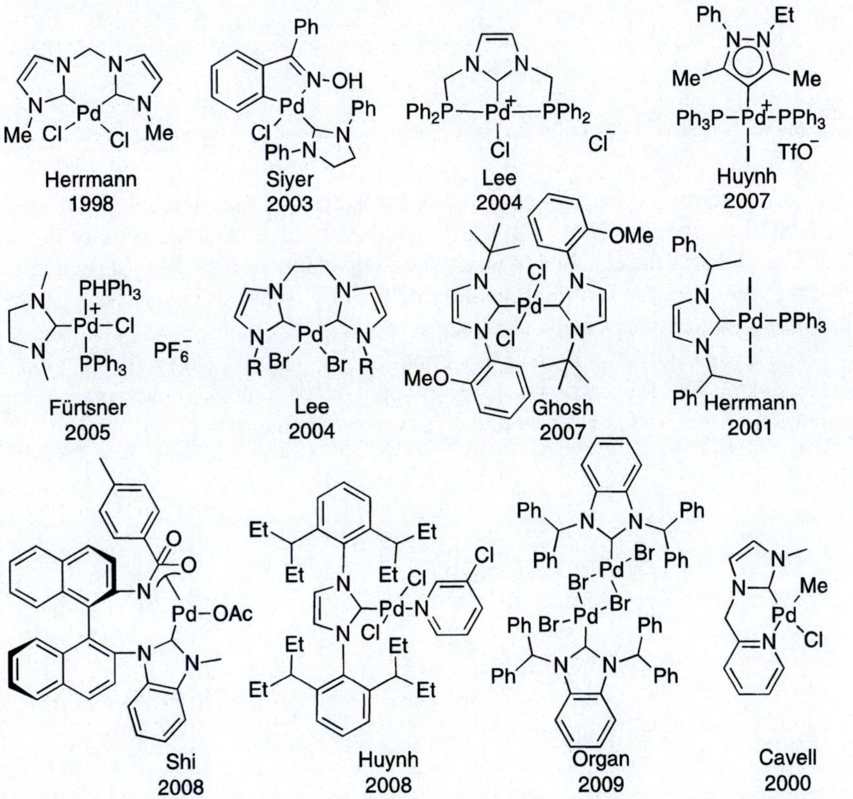

Scheme 6.24 Selected pre-catalysts used for the Suzuki coupling between aryl bromides and iodides with boronic acids

A large number of pre-catalysts have been used for the reaction between aryl bromides or iodides with boronic acids; a selection is depicted in Scheme 6.24

[88–99]. However, the low cost and availability of aryl chlorides make them the most attractive substrates for this transformation and, thus, will be the focus of this section. Special effort will also be made to highlight examples of couplings in which *ortho*-substituted reacting partners are used.

One of the first highly efficient Pd–NHC based catalysts for the coupling between aryl chlorides and boronic acids was developed by Nolan and co-workers [100]. The catalyst was prepared *in situ* from a mixture of IPr·HCl/Pd(dba)$_2$ or IMes·HCl/Pd(OAc)$_2$ and used Cs_2CO_3 as base. A selection of results obtained using this protocol is shown in Scheme 6.25. These, or very similar, reaction conditions have been applied by other authors and give excellent results for more sterically hindered substrates [101–104].

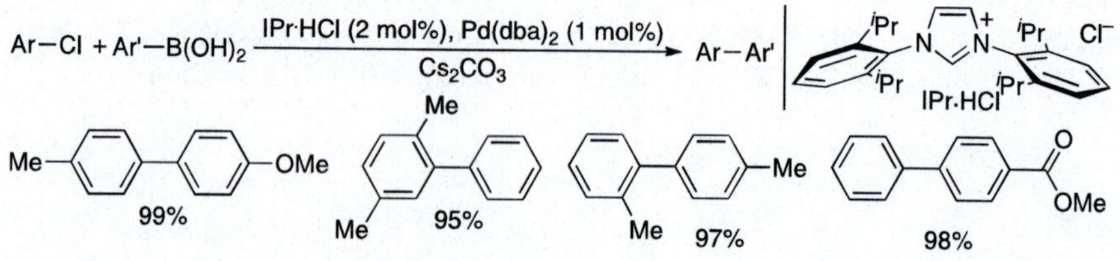

Scheme 6.25 Nolan conditions for the Suzuki coupling between aryl chlorides and boronic acids

More recently Nolan has also reported a series of well-defined [Pd(η^3-allyl)Cl(NHC)] complexes (NHC = IPr, SIPr) with very high catalytic activity for this reaction, allowing the coupling of unactivated aryl chlorides in minutes. The presence of substituents in the terminal position of the allyl scaffold is necessary as they are proposed to favour catalyst activation.

With these catalysts, di- and even some tri-*ortho*-substituted products could be formed using very low catalyst loading (50 ppm – 0.05 mol%). Some representative examples of the products prepared are shown in Scheme 6.26 [105].

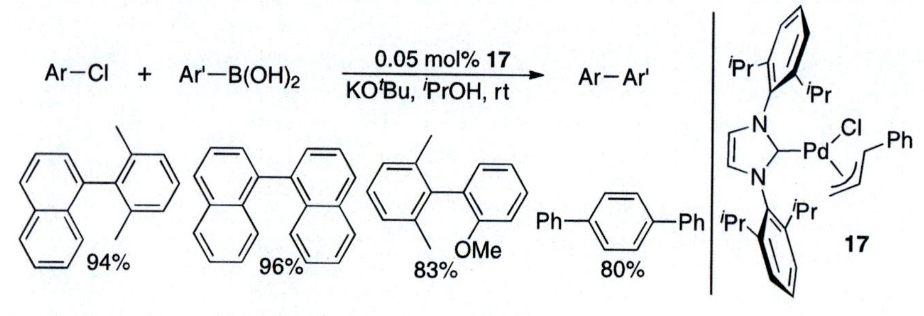

Scheme 6.26 Scope of the well-defined pre-catalyst **17**

The aforementioned PEPPSI-IPr pre-catalyst **16** has also been used in the Suzuki-Miyaura reaction. This pre-catalyst allowed the easy preparation of hindered biaryls and drug-like heteroaromatic compounds in good to excellent yields (Scheme 6.27).

The authors attribute the stability and facile preparation of this pre-catalyst to the stabilising effect of the pyridine ligand that easily dissociates during the activation process once the Pd(II) atom of the pre-catalyst is reduced to Pd(0) [106].

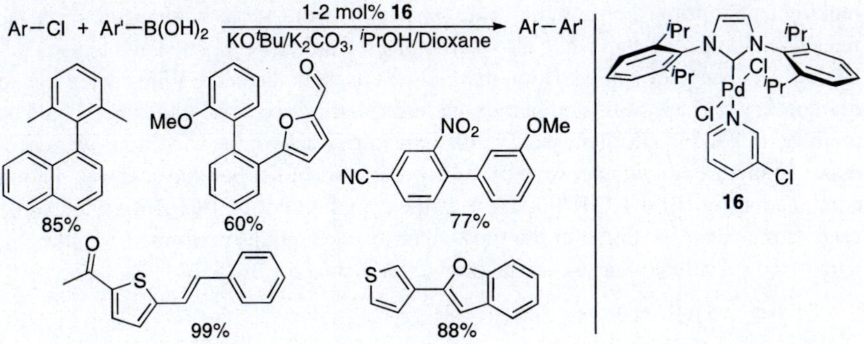

Scheme 6.27 Scope of PEPPSI-IPr pre-catalyst **16** in the Suzuki transformation

Very recently another highly active and well-defined Pd–NHC based pre-catalyst containing a cyclopentadienyl (Cp) ligand **18** has been successfully applied in this transformation. Cp was chosen as stabilising ligand due to its well-known tendency to reductively be removed from Cp–Pd complexes that may help in the transformation of the pre-catalyst into the desired catalytic active species (NHC)Pd(0) [107]. Di- and tri-*ortho* substituted biaryls were obtained in good to excellent yields; however, when the formation of tetra-*ortho* substituted compounds was attempted very poor yields were obtained, even using aryl bromide or iodide substrates (Scheme 6.28).

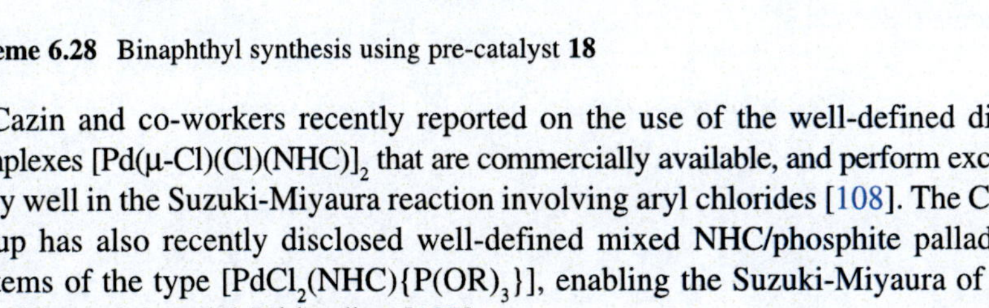

Scheme 6.28 Binaphthyl synthesis using pre-catalyst **18**

Cazin and co-workers recently reported on the use of the well-defined dimer complexes $[Pd(\mu\text{-}Cl)(Cl)(NHC)]_2$ that are commercially available, and perform exceedingly well in the Suzuki-Miyaura reaction involving aryl chlorides [108]. The Cazin group has also recently disclosed well-defined mixed NHC/phosphite palladium systems of the type $[PdCl_2(NHC)\{P(OR)_3\}]$, enabling the Suzuki-Miyaura of aryl chlorides at 0.1 mol% Pd loading [109].

Probably one of the most efficient NHC-based catalytic systems for the Suzuki-Miyaura coupling is the IBiox-Pd complexes introduced by Glorius and colleagues. IBiox ligands are bisoxazoline derived NHCs consisting of a tricyclic rigid backbone bearing two cycloalkyl substituents. The different size and conformations that can be adopted by the cycloalkyl rings make possible the synthesis of ligands with flexible steric bulk (Scheme 6.29a). This flexibility seems to be important in a number of challenging cross coupling reactions such as those involved in the synthesis of very sterically demanding substituted biaryls. It appears that the ligand adapts its steric requirements during the catalytic cycle, providing more steric hindrance during the reductive elimination step while also facilitating the oxidative addition and transmetallation by a change of conformation that reduces its steric demand. With such a ligand design, very sterically hindered biaryls, including tetra-*ortho* substituted ones, could be prepared in good to excellent yields, although strictly anhydrous conditions were necessary in order to avoid undesired proto-deboronation of the boronic acid. In addition combinations of IBiox·HOTf/Pd(OAc)$_2$ or the complex [Pd(μ-Cl)Cl(IBiox)]$_2$ could be used as pre-catalysts. Some of the most significant couplings performed, resulting in tetra-*ortho* substituted biaryls, are highlighted in Scheme 6.29 [110–112].

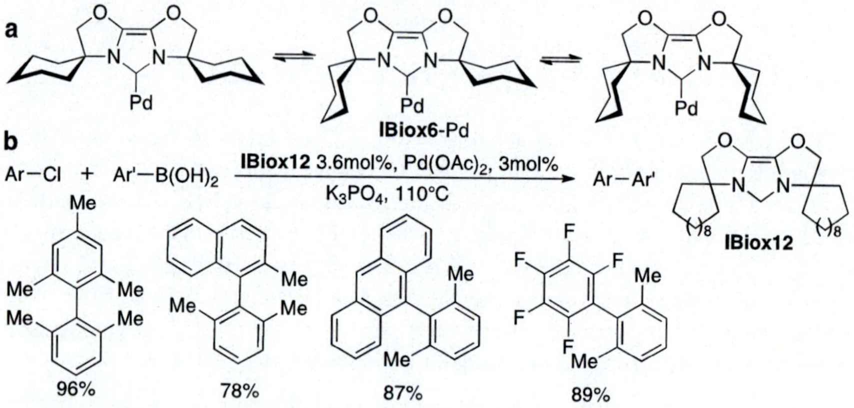

Scheme 6.29 Glorius sterically adaptable IBiox ligands

So far only Pd-based systems have been highlighted in this section; however, the use of other metals such as Ni has clear economic advantages. In this regard, Chiu and co-workers have used a bis-carbene tetradentate ligand to catalyse the coupling of aryl bromides and chlorides with both electron rich and electron poor aromatic rings; however, the reaction with electron poor aryl bromides lead to superior yields (Scheme 6.30) [113].

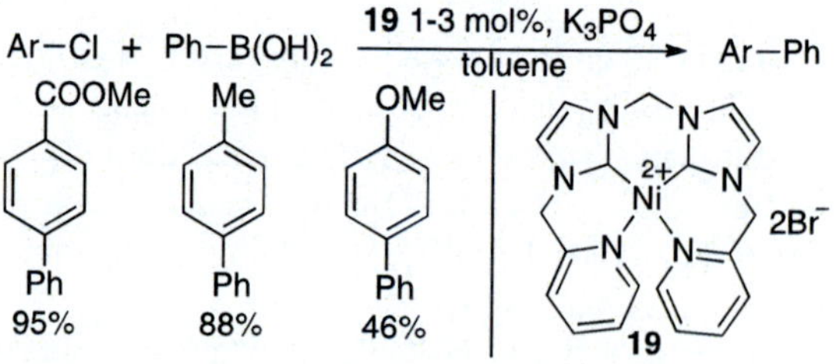

Scheme 6.30 Nickel-catalysed Suzuki transformation

Inamoto and Doi have also worked on pincer-type biscarbene complexes of nickel(II). Pre-catalyst **20** was successfully applied to the Suzuki-Miyaura reaction using aryl bromides and, in some cases, aryl chlorides as coupling partners [114]. Nevertheless, further optimisation of these inexpensive catalyst systems is still necessary in order to increase their activity, thus allowing reactions at lower temperatures and the coupling of more sterically demanding substrates (Scheme 6.31).

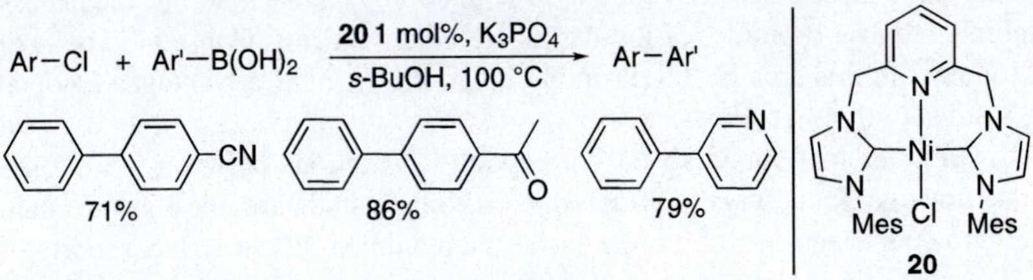

Scheme 6.31 Ni-NHC pincer **20** in Suzuki-Miyaura reaction

Thus far only homogeneous catalytic systems have been discussed. However, heterogeneous catalysts have an intrinsic advantage of being efficient at separating product from catalyst and possibly permitting catalyst recycling. Very recently, Varma and co-workers have reported the preparation of a Pd–NHC complex in the form of organic silica and its application in coupling reactions [115]. The use of this heterogeneous catalyst **21** allowed the reaction of aryl iodides and bromides with excellent yields (Scheme 6.32).

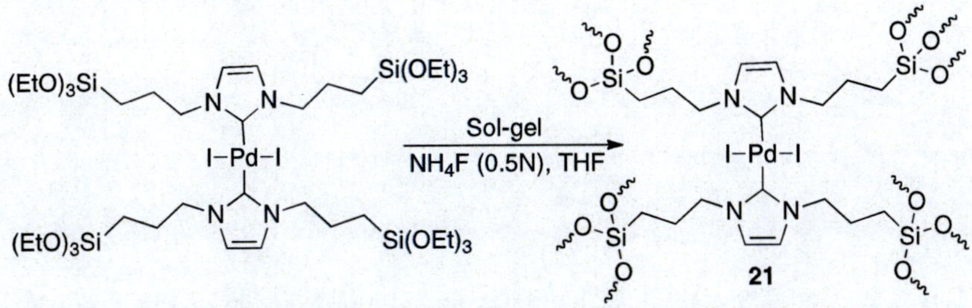

Scheme 6.32 Silica-supported pre-catalyst

Another successful approach to catalyst immobilisation involves attachment of the carbene precursor to a peptide on solid support. Treatment with base generates the corresponding carbenes that undergo *in situ* complexation to Pd(II) centres (Scheme 6.33). Again, the main drawback of this approach was the low reactivity of the catalytic system that only allowed the coupling of aryl iodides and bromides [116]. The reasons for this outcome are in need of further studies.

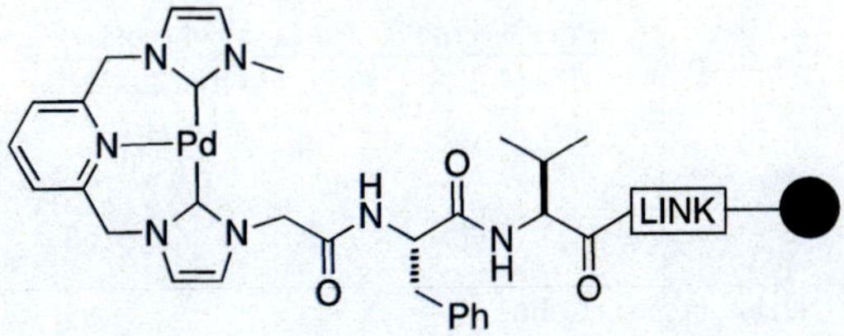

Scheme 6.33 Pre-catalyst immobilisation by attachment to a peptide on solid support

A Au(I)–NHC system has also been used as pre-catalyst for this transformation. However, despite the interest of using Au instead of Pd, the high catalytic charge (20 mol%) necessary to achieve acceptable yields renders this system uncompetitive with those already described [117]. In general, all known heterogeneous catalysts share the same drawback when compared with homogeneous systems; the necessity of unhindered aryl bromides or iodides as coupling reagents. Consequently, further development in this area is still desirable given the intrinsic advantages associated with solid supported catalysts.

All examples mentioned so far correspond to reactions between two aromatic groups, however, couplings in which one or both partners are alkyl groups can be achieved using electron-rich boron-based nucleophiles. Fürstner has reported the use of *B*-alkyl or *B*-allyl methoxy-9-BBN anions for the efficient coupling with some aryl chlorides using an *in situ* prepared IPr·HCl/Pd(OAc)$_2$ system [118]. Some of the results obtained with these easy-to-handle borate-based nucleophiles are shown below (Scheme 6.34).

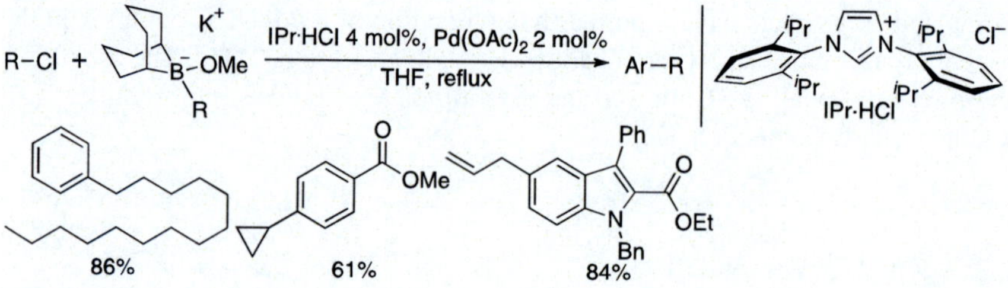

Scheme 6.34 Fürstner protocol for the coupling between aryl chlorides and *B*-alkyl methoxy-9-BBN anions

Recently, Caddick and Cloke have developed an extension of this procedure that allows the use of alkyl bromides as coupling partners. The basic changes consist of a stoichiometric amount of the bulkier KO^{t}Bu instead of KOMe to activate the borane, and the addition of AgOTf to the reaction mixture [119]. These results, although poor in terms of yield, clearly confirm that sp^3–sp^3 Suzuki-Miyaura cross-couplings are possible and should be further developed (Scheme 6.35).

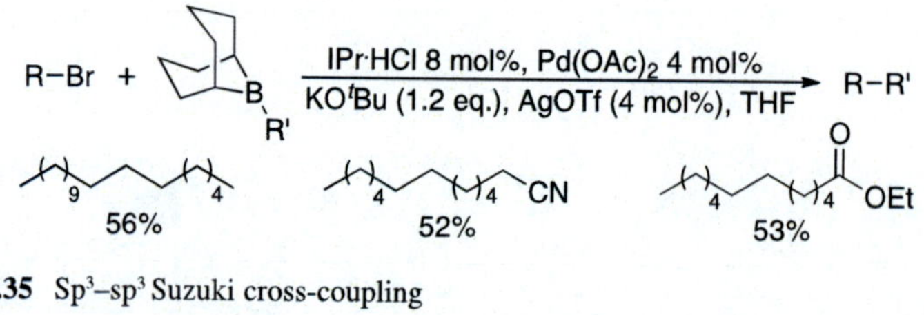

Scheme 6.35 Sp^3–sp^3 Suzuki cross-coupling

6.2.5 Stille Coupling

The Stille reaction has developed as a popular protocol for the formation of C–C bonds due to the air- and moisture-stability as well as functional group compatibility of organotin compounds. Together with the Suzuki-Miyaura coupling it is one of the most powerful methods for the synthesis of molecules containing unsymmetrical biaryl moieties. However, despite its efficiency, this versatile reaction has slowly been displaced by other procedures that avoid the use of highly toxic organostannanes.

The first example of a NHC–Pd catalysed Stille reaction between aryl bromides and aryl stannanes was reported by Herrmann in 1999 [120]. Summarised in Scheme 6.36 are the best results obtained when the well-defined pre-catalyst **22** was employed. Unfortunately, the coupling of aryl chlorides was not possible.

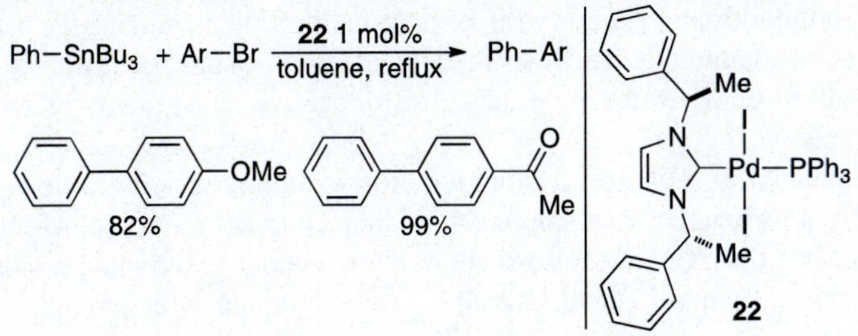

Scheme 6.36 First example of a NHC–Pd catalysed Stille coupling

Subsequently, an *in situ* $Pd(OAc)_2$-imidazolium salt mixture, developed by Nolan and Grasa, has demonstrated its efficiency for the coupling of aryl bromides and even chlorides with aryl and vinylstannanes. This improved reactivity is due to the TBAF additive, whereby F^- anions coordinate to Sn forming hypervalent fluorostannate anions that are more reactive towards transmetallation to Pd [121] (Scheme 6.37).

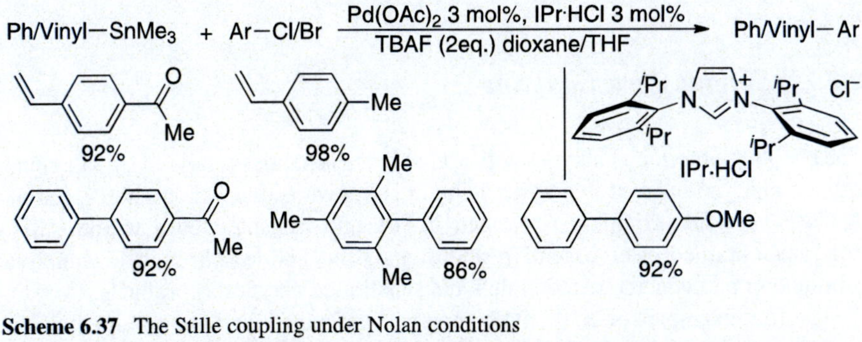

Scheme 6.37 The Stille coupling under Nolan conditions

Very recently, the coupling of tributyl(phenyl) and tributyl(vinyl) stannanes with cyclohexene derivatives was successfully carried out using a similar catalytic cocktail prepared from $Pd_2(dba)_3{\cdot}CHCl_3$, IPr·HCl and Cs_2CO_3 as base. Outstanding results in terms of yield were obtained (Scheme 6.38). The coupling of ethoxy- functionalised vinyl stannanes was only moderately successful [122].

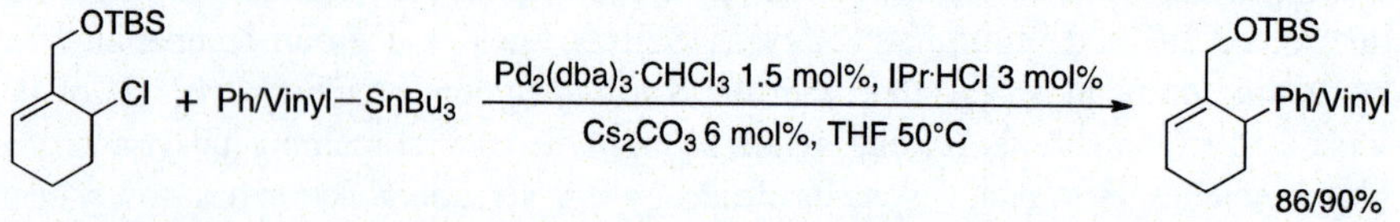

Scheme 6.38 Coupling between cyclic allyl chlorides and phenyl/vinyl stannanes

6.2.6 *Hiyama Coupling*

The C–C bond forming reaction of an organic halide with an organosilane, catalysed by nickel or palladium, is known as the Hiyama cross-coupling. Typically the C–Si bond needs to be activated by either electronegative substituents or by external fluoride anions.

In the context of NHC/metal catalysed cross-coupling reactions, the only example of a Hiyama reaction was reported by Nolan using an *in situ* protocol by mixing $Pd(OAc)_2$ and IPr·HCl for the formation of the catalyst. Activated aryl bromides and chlorides, such as 2-chloropyridine, were coupled with phenyl and vinyltrimethoxysilane in good yields [123] (Scheme 6.39).

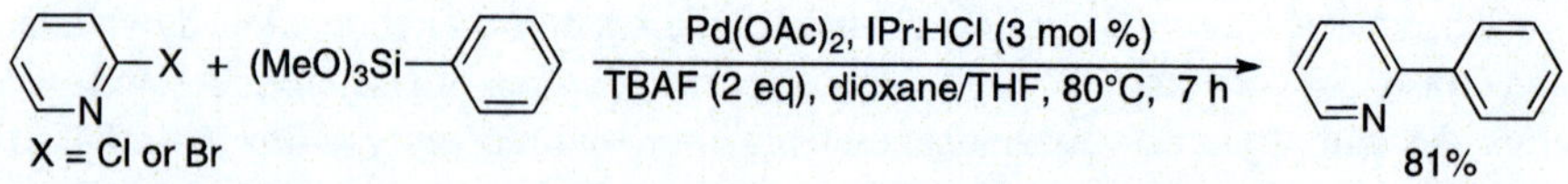

Scheme 6.39 Hiyama coupling between activated aryl bromides/chlorides and phenyl/vinyltrimethoxysilane

6.2.7 *Sonogashira Coupling*

The coupling of terminal alkynes with aryl or alkenyl halides catalysed by palladium and a copper co-catalyst in a basic medium is known as the Sonogashira reaction. A Cu(I)-acetylide complex is formed *in situ* and transmetallates to the Pd(II) complex obtained after oxidative addition of the halide. Through a reductive elimination pathway the reaction delivers substituted alkynes as products.

The first examples of NHC–Pd complexes applied to the Sonogashira reaction were reported to show a limited scope in the coupling of aryl iodides and activated aryl bromides with acetylene [23,33,52]. However, the use of *N*-carbamoyl-substituted heterocyclic carbene Pd(II) complexes expanded the use to alkyl-acetylenes and deactivated aryl iodides and bromides [124] (Scheme 6.40).

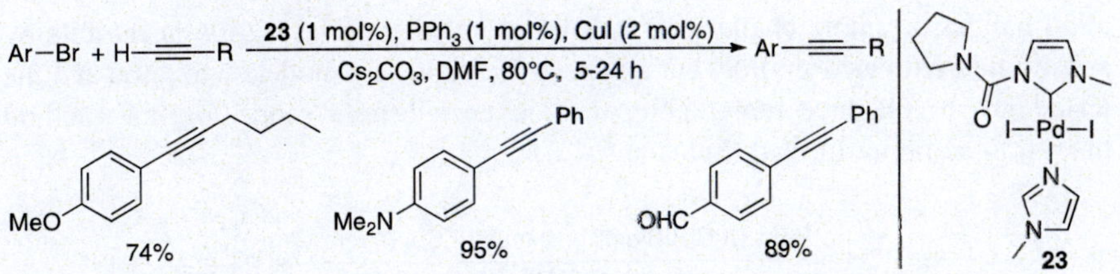

Scheme 6.40 Sonogashira coupling with *N*-carbamoyl-substituted NHC-Pd (II) complexes

The use of imidazolium salts for *in situ* catalyst formation was shown to be optimal for the coupling of TMS-protected alkynes even with sterically demanding aryl bromides and avoids the formation of homocoupling-derived products. For this reaction, Nolan reported that the activation of chlorobenzene by this catalytic system was possible in moderate yield [125] (Scheme 6.41).

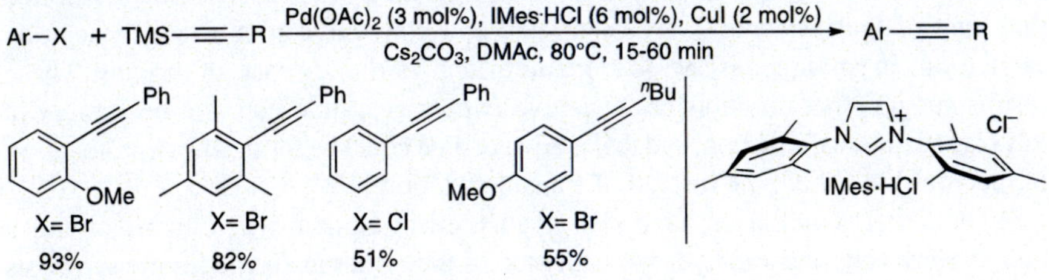

Scheme 6.41 Sonogashira coupling of TMS-protected alkynes

Related terminal alkyne substrates were later studied under copper-free and amine-free conditions, and the yield showed a remarkable dependence on the bulkiness of the *N*-substituents [126].

The favourable effect of steric demand was also confirmed by the experiments of Fu while studying the coupling of primary alkyl bromides. Good yields were achieved using IAd·HCl, under mild conditions and in the presence of CuI [127]. The results can be further improved using bulkier ligands such as IAd (bisdiamantyl-substituted NHCs) [128] (Scheme 6.42).

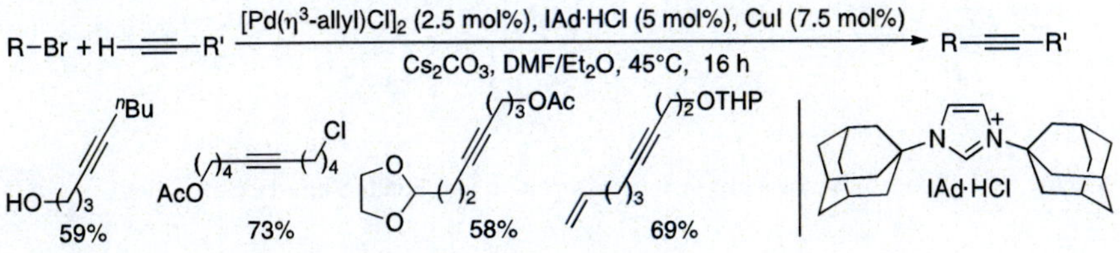

Scheme 6.42 Sonogashira coupling with alkyl bromides

The scope of this reaction was extended to secondary alkyl bromides with IBiox ligands that are both sterically hindered and flexible carbenes [129]. Secondary alkyl halides are more challenging substrates as they are less easily accessible by substitution with an acetylide, but although the yields are moderate to good and the tested group tolerance remarkable, the alkyne substrate scope for this method proved to be rather limited (Scheme 6.43).

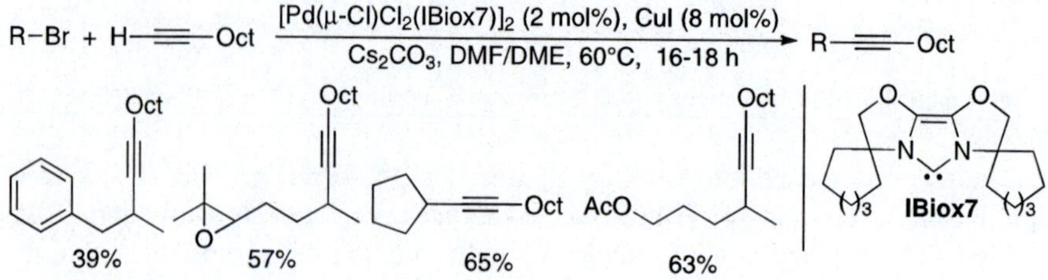

Scheme 6.43 Sonogashira coupling with secondary alkyl bromides

More recently, a study with di- and mono-carbene Pd(II) complexes has demonstrated that the Sonogashira coupling of activated and non-activated aryl iodides can be carried out in an aqueous, aerobic medium and in the absence of amines. These results suggest that the moisture-sensitive copper-acetylide may not be present in this particular transformation, and that a Pd-acetylide could be formed by deprotonation of the coordinated alkyne instead of transmetallation [130].

NHC–Pd(II) complexes have also been used in tandem reactions involving a Sonogashira coupling and hydroalkoxylation of the resulting alkyne for the synthesis of benzofurans [131] as well as sequential Heck-Sonogashira couplings [132] (Scheme 6.44).

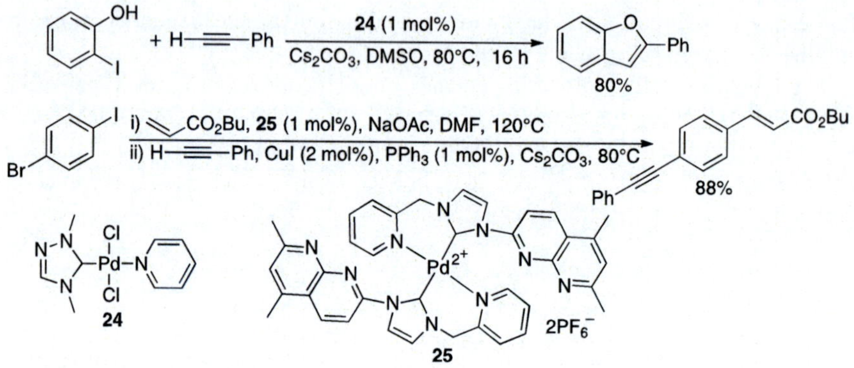

Scheme 6.44 Tandem Sonogashira-hydroalkoxylation and Sonogashira-Heck reactions

Concerning other metals, Sonogashira coupling products have also been observed in the reaction of Ag(I)-carbenes [133] and Au(I)-supported carbenes [134] in low to moderate yields, but only under harsh conditions (more than 100°C). In this regard, NHC based catalysts for Sonogashira reactions have been supported on different materials that include clays [135], polymers [136] and peptides [137].

6.3 Carbon–Heteroatom Bond Formation

6.3.1 C–N Bond Formation

The coupling between an aryl halide or triflate and an amine is known as the Buchwald-Hartwig amination [138]. Originally it was described using a tributyltin amine [139,140] and was thus considered to be a coupling reaction. Subsequently, tributyltin amine was replaced by a standard amine and a strong base. It is a reaction of great academic and industrial interest [11].

The mechanism involves a Pd(0) monocoordinate complex as the active species that undergoes oxidative addition to the aryl halide [141]. Thereafter, coordination of the amine to the palladium centre and deprotonation by the external base results in halide abstraction. After reductive elimination, the coupling product is obtained and the catalytic active species regenerated (Scheme 6.45).

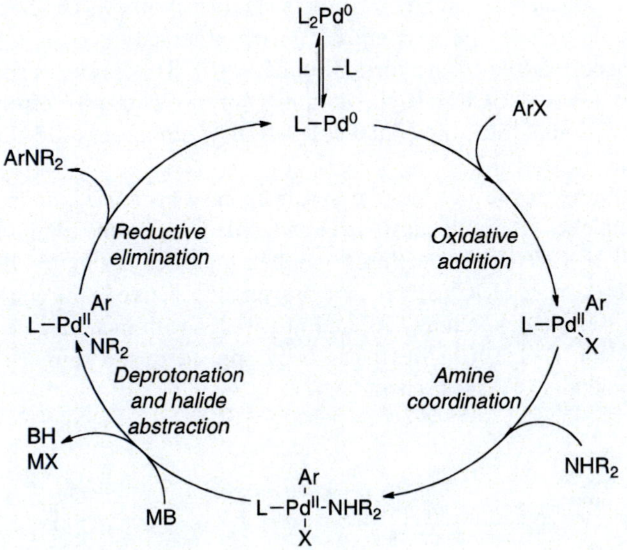

Scheme 6.45 Mechanism of the Buchwald-Hartwig amination

The first examples utilising *N*-heterocyclic carbenes as ligands in the Buchwald-Hartwig amination involved the *in situ* formation of the catalyst from the corresponding imidazolium salt and a Pd(0) source. Nolan reported IPr·HCl/$Pd_2(dba)_3$ as a catalytic system for the amination of aryl chlorides in excellent yields, using different types of amines, anilines, and also imines or indoles [142,143] (Scheme 6.46). Hartwig showed later that in some cases the reactions could be performed at room temperature and without anhydrous conditions even for aryl chlorides [144]. This was later shown for the less challenging bromides and iodides [145,146].

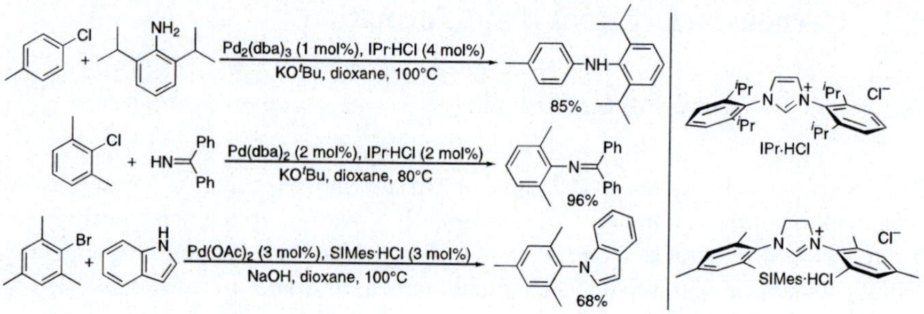

Scheme 6.46 Buchwald-Hartwig aminations with *in situ*-formed catalysts

The use of well-defined complexes has been widespread in this reaction, despite intriguing studies by Beller and others that have shown that *in situ* catalytic systems often give better yields in comparison to isolated carbene-Pd(0) complexes [147–149]. Since the mechanism consists of an oxidative addition on a Pd(0)-monocarbene species, efforts in catalyst synthesis have been directed towards Pd(II)-monocarbene complexes with other labile groups that can be easily released leading to the formation of Pd(0). This is the case for dimers of the type $[Pd(\mu\text{-}Cl)Cl(NHC)]_2$, a family of pre-catalysts effective under aerobic conditions [150], the [Pd(acac)Cl(NHC)] complexes [151] and related palladacycles [152–154].

Although other labile ligands like pyridines have been used, as in the case of PEPPSI complexes [155], the most active and widely applicable catalysts have been the $[Pd(\eta^3\text{-}C_3H_5)Cl(NHC)]$ complexes and their modified analogues. $[Pd(\eta^3\text{-}C_3H_5)Cl(IPr)]$ and $[Pd(\eta^3\text{-}C_3H_5)Cl(SIPr)]$ have proven to be active in the coupling of aryl chlorides at room temperature [156] and of triflates with moderate heating [157]. In addition $[Pd(\eta^3\text{-}C_3H_5)Cl(IMes)]$ has been applied to the formal synthesis of *cryptocarya* alkaloids [158] (Scheme 6.47).

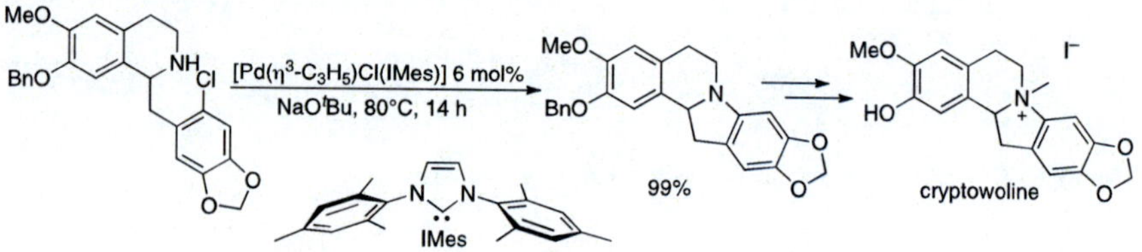

Scheme 6.47 NHC-Pd-Catalysed amination in the synthesis of cryptowoline

However, modification of the allyl fragment by substitution of one of the terminus positions has provided more active complexes by enabling a more facile activation step [159]. This allows the coupling of highly hindered amines with hindered aryl chlorides at room temperature and with low catalyst loadings [160] (Scheme 6.48).

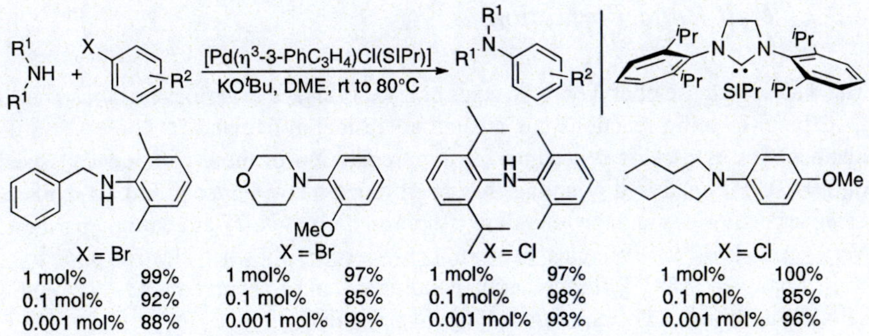

Scheme 6.48 Allyl-modified Pd-NHC complexes as improved catalysts in Buchwald-Hartwig aminations

Other related complexes that behave well at room temperature include dimeric $[Pd(alkyl)Cl(NHC)]_2$ [161] and [PdCl(Cp)(NHC)] [162].

However, the Buchwald-Hartwig reaction with NHCs as ligands is not limited to palladium. Nickel has also been successfully employed in this catalytic amination. *In situ* procedures have been described for the coupling of aryl chlorides [163] and tosylates [164] and, more interestingly, anisoles [165]. The use of well-defined Ni(0) catalysts has also been studied [166] (Scheme 6.49).

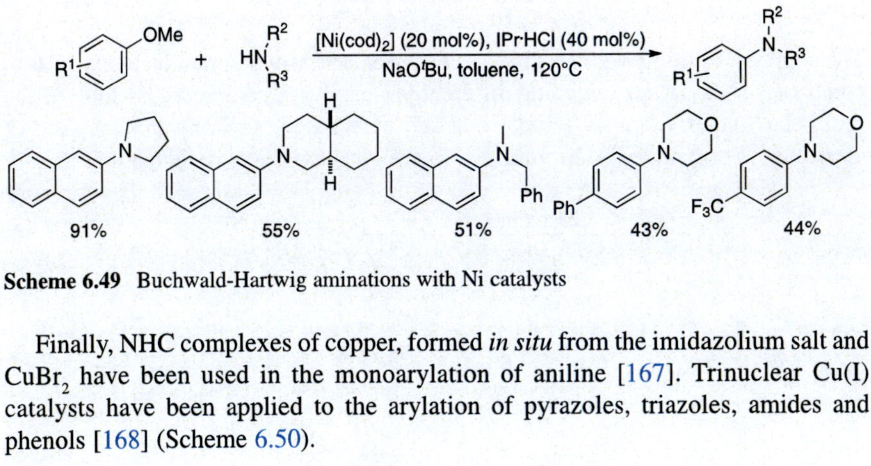

Scheme 6.49 Buchwald-Hartwig aminations with Ni catalysts

Finally, NHC complexes of copper, formed *in situ* from the imidazolium salt and $CuBr_2$ have been used in the monoarylation of aniline [167]. Trinuclear Cu(I) catalysts have been applied to the arylation of pyrazoles, triazoles, amides and phenols [168] (Scheme 6.50).

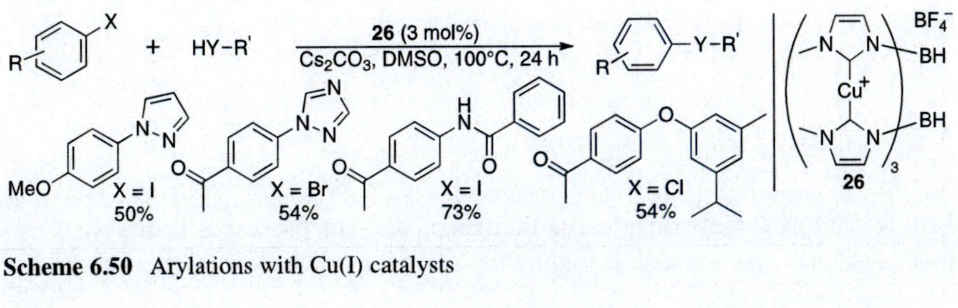

Scheme 6.50 Arylations with Cu(I) catalysts

6.3.2 C–B Bond Formation

The widespread use of aryl boronic acids or aryl boronates in various metal-catalysed C–C bond-forming reactions has created a substantial demand for these versatile nucleophiles. A general procedure for the preparation of these compounds, based on a NHC/Pd catalysed coupling, has been reported by Fürtsner and co-workers using aryl chlorides and the tetraalkoxy diboron derivative **27** as coupling partners. Very good yields were obtained in several cases especially when electron poor aryls were employed [169]. Milder reaction conditions can be achieved when diazonium salts are used instead of chlorides [170] (Scheme 6.51).

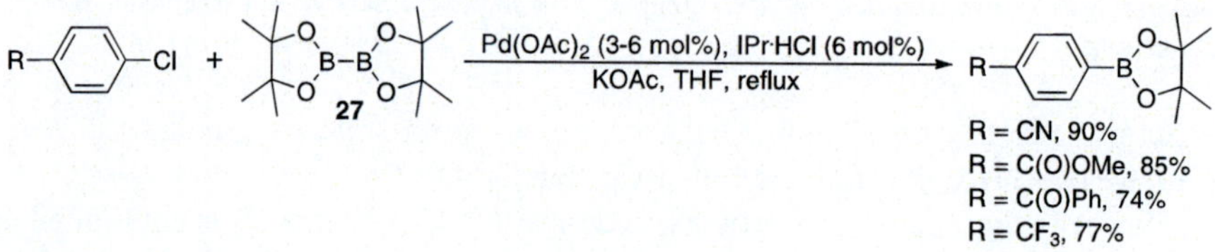

Scheme 6.51 Synthesis of boronic esters by Pd-coupling

6.3.3 C–S Bond Formation

The coupling of thiols with aryl halides has been recently reported using $Ni(NHC)_2$ complexes [171]. After screening different pre-catalysts, compound **28** showed the best behaviour in terms of activity and substrate scope, allowing the coupling of electron rich and poor aryl bromides with aryl or alkyl thiols (Scheme 6.52).

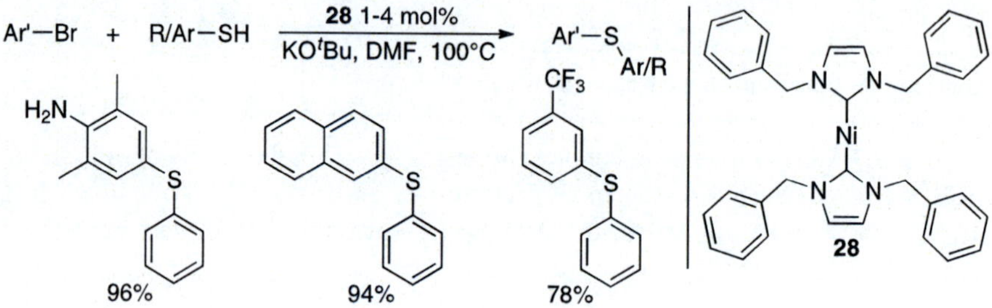

Scheme 6.52 Coupling of aryl halides and thiols

6.3.4 Other Cross Couplings

In this section some examples of multistep, one-pot processes using NHC/Pd-based catalytic systems will be highlighted. One remarkable example, although the yields are not uniformly excellent, is the synthesis of indoles by a sequential Sonogashira coupling, Buchwald-Hartwig amination and hydroamination reaction [172,173] (Scheme 6.53).

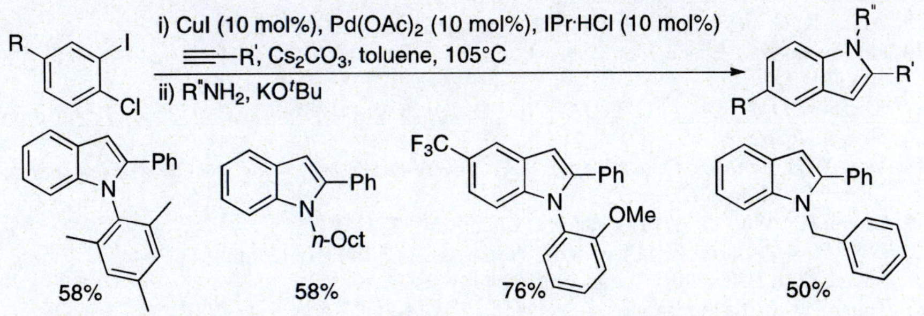

Scheme 6.53 Tandem couplings for the synthesis of indoles

Another example of a one-pot indole synthesis, which proceeds through a Heck carbonylation and a Suzuki coupling, is shown below. The reaction conditions are similar to the previous example; however microwave heating is employed [174] (Scheme 6.54).

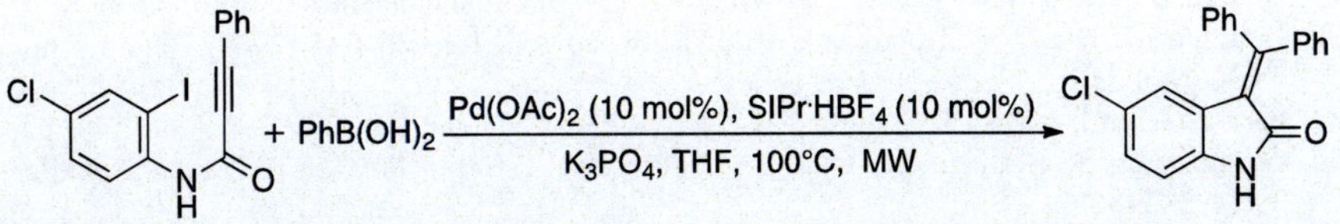

Scheme 6.54 Tandem Heck-Suzuki coupling

References

1. Negishi E (2007) Bull Chem Soc Jpn 80:233–257
2. Nicolaou KC, Bulger PG, Sarlah D (2005) Angew Chem Int Ed 44:4442–4489
3. McGuinness DS, Green MJ, Cavell KJ, Skelton BW, White AH (1998) J Organomet Chem 565:165–168
4. Huynh HV, Ho JHH, Neo TC, Koh LL (2005) J Organomet Chem 690:3854–3860
5. McGuinness DS, Cavell KJ (1999) Organometallics 18:1597–1605
6. de Lewis AK, Caddick S, Cloke FGN, Billingham NC, Hitchcock PB, Leonard J (2003) J Am Chem Soc 125:10066–10073
7. Tubaro C, Biffis A, Gonzato C, Zecca M, Basato M (2006) J Mol Catal A 248:93–98
8. Lebel H, Janes MK, Charette AB, Nolan SP (2004) J Am Chem Soc 126:5046–5047
9. Alcarazo M, Roseblade SJ, Cowley AR, Fernández R, Brown JM, Lassaletta JM (2005) J Am Chem Soc 127:3290–3291
10. Lee MT, Lee HM, Hu CH (2007) Organometallics 26:1317–1324

11. Bräse S, de Meijere A (2004) Cross-coupling of organic halides with alkenes: The Heck reaction. In: de Meijere A, Diederich F (eds) Metal-catalyzed cross-coupling reactions, 2nd edn. Wiley-VCH, Weinheim
12. Seyferth D (2009) Organometallics 28:1598–1605
13. Hadei N, Kantchev EAB, O'Brien CJ, Organ MG (2005) Org Lett 7:1991–1994
14. Farina V (2004) Adv Synth Catal 346:1553–1582
15. Albert K, Gisdakis P, Rösch N (1998) Organometallics 17:1608–1616
16. Herrmann WA, Elison M, Fischer J, Kocher C, Artus GRJ (1995) Angew Chem Int Ed 34:2371–2374
17. Khramov DM, Rosen EL, Er JAV, Vu PD, Lynch VM, Bielawski CW (2008) Tetrahedron 64:6853–6862
18. Yang W-H, Lee C-S, Pal S, Chen Y-N, Hwang W-S, Lin IJB, Wang J-C (2008) J Organomet Chem 693:3729–3740
19. Huynh HV, Wu J (2009) J Organomet Chem 694:323–331
20. Metallinos C, Barret FB, Chaytor JL, Heska MEA (2004) Org Lett 6:3641–3644
21. Shi M, Qian H-X (2005) Tetrahedron 61:4949–4955
22. Tubaro C, Biffis A, Basato M, Benetollo F, Cavell KJ, Ooi L-L (2005) Organometallics 24:4153–4158
23. Herrmann WA, Reisinger CP, Spiegler M (1998) J Organomet Chem 557:93–96
24. Lee HM, Lu CY, Chen CY, Chen WL, Lin HC, Chiu PL, Cheng PY (2004) Tetrahedron 60:5807–5825
25. Taige MA, Zeller A, Ahrens S, Goutal S, Herdtweck E, Strassner T (2007) J Organomet Chem 692:1519–1529
26. Magill AM, McGuinness DS, Cavell KJ, Britovsek GJP, Gibson VC, White AJP, Williams DJ, White AH, Skelton BW (2001) J Organomet Chem 617–618:546–560
27. Liu LJ, Wang F, Shi M (2009) Eur J Inorg Chem 1723–1728
28. Clyne DS, Jin J, Genest E, Gallucci JC, RajanBabu TV (2000) Org Lett 2:1125–1128
29. Nielsen DJ, Cavell KJ, Skelton BW, White AH (2006) Organometallics 25:4850–4856
30. Houghton J, Dyson G, Douthwaite RE, Whitwood AC, Kariuki BM (2007) Dalton Trans 28:3065–3073
31. Peris E, Loch JA, Mata J, Crabtree RH (2001) Chem Commun 201–202
32. Gründemann S, Albrecht M, Loch JA, Faller JW, Crabtree RH (2001) Organometallics 20:5485–5488
33. Loch JA, Albrecht M, Peris E, Mata J, Faller JW, Crabtree RH (2002) Organometallics 21:700–706
34. Nielsen DJ, Cavell KJ, Skelton BW, White AH (2002) Inorg Chim Acta 327:116–125
35. Nielsen DJ, Cavell KJ, Skelton BW, White AH (2006) Inorg Chim Acta 359:1855–1869
36. Herrmann WA, Böhm VPW, Gstöttmayr CWK, Grosche M, Reisinger CP, Westkamp T (2001) J Organomet Chem 617–618:616–628
37. Tsoureas N, Danopoulos AA, Tulloch AAD, Light ME (2003) Organometallics 22:4750–4758
38. Kremzov D, Seidel G, Lehmann CW, Fürstner A (2005) Chem Eur J 11:1833–1853
39. Lee HM, Zeng JY, Hu C-H, Lee M-T (2004) Inorg Chem 43:6822–6829
40. Yang C, Nolan SP (2001) Synlett 1539–1542
41. Andrus MB, Liu J (2006) Tetrahedron Lett 47:5811–5814
42. Xu L, Chen W, Xiao J (2000) Organometallics 19:1123–1127
43. Jin C-M, Twamley B, Shreeve JM (2005) Organometallics 24:3020–3023
44. Calò V, Nacci A, Monopoli A, Lopez L, di Cosmo A (2001) Tetrahedron 57:6071–6077
45. Zou G, Huang W, Xiao Y, Tang J (2006) New J Chem 30:803–809
46. Polshettiwar V, Hesemann P, Moreau JJE (2007) Tetrahedron Lett 48:5363–5366
47. Polshettiwar V, Varma RS (2008) Tetrahedron 64:4637–4643
48. Liu J, Zhao Y, Zhou Y, Li L, Zhang TY, Zhang H (2003) Org Biomol Chem 1:3227–3231
49. Yang C, Lee HM, Nolan SP (2001) Org Lett 3:1511–1514
50. Wang A-E, Xie J-H, Wang L-X, Zhou Q-L (2005) Tetrahedron 61:259–266

51. Tulloch AAD, Danopoulos AA, Tooze RP, Cafferkey SM, Kleinhenz S, Hursthouse MB (2000) Chem Commun 1247–1248
52. McGuinness DS, Cavell KJ (2000) Organometallics 19:741–748
53. Wang C-Y, Liu Y-H, Peng S-M, Chen J-T, Liu S-T (2007) J Organomet Chem 692:3976–3983
54. Zhang X, Xi Z, Liu A, Chen W (2008) Organometallics 27:4401–4406
55. Jahnke MC, Pape T, Hahn FE (2009) Eur J Inorg Chem 1960–1969
56. Zhang X, Liu A, Chen W (2008) Org Lett 10:3849–3852
57. Ye J, Chen W, Wang D (2008) Dalton Trans 30:4015–4022
58. Meyer D, Taige MA, Zeller A, Hohlfeld K, Ahrens S, Strassner T (2009) Organometallics 28:2142–2149
59. César V, Bellemin-Laponnaz S, Gade LH (2002) Organometallics 21:5204–5208
60. Lee HM, Chiu PL, Hu C-H, Lai C-L, Chou Y-C (2005) J Organomet Chem 690:403–414
61. Zhang T, Liu S, Shi M, Zhao M (2008) Synthesis 2819–2824
62. Iyer S, Jayanthi A (2003) Synlett 1125–1128
63. Kantchev EAB, Peh G-R, Zhang C, Ying JY (2008) Org Lett 10:3949–3952
64. Peh G-R, Kantchev EAB, Zhang C, Ying JY (2009) Org Biomol Chem 7:2110–2119
65. Frey GD, Schütz J, Herdtweck E, Herrmann WA (2005) Organometallics 24:4416–4426
66. Selvakumar K, Zapf A, Spannenberg A, Beller M (2002) Chem Eur J 8:3901–3906
67. Weck M, Jones CW (2007) Inorg Chem 46:1865–1875
68. Winkelmann O, Näther C, Lüning U (2008) J Organomet Chem 693:2784–2788
69. Caddick S, Kofie W (2002) Tetrahedron Lett 43:9347–9350
70. Caló V, Nacci A, Monopoli A, Spinelli M (2003) Eur J Org Chem 1382–1385
71. Calò V, Nacci A, Lopez L, Napola A (2001) Tetrahedron Lett 42:4701–4703
72. Touré BB, Lane BS, Sames D (2006) Org Lett 8:1979–1982
73. Zhong J, Xie J-H, Wang A-E, Zhang W, Zhou Q-L (2006) Synlett 1193–1196
74. Inamoto K, Kuroda J, Danjo T, Sakamoto T (2005) Synlett 1624–1626
75. Huang J, Nolan SP (1999) J Am Chem Soc 121:9889–9890
76. Frisch AC, Rataboul F, Zapf A, Beller M (2003) J Organomet Chem 687:403–409
77. Zhou Y, Xi Z, Chen W, Wang D (2008) Organometallics 27:5911–5920
78. Xi Z, Liu B, Chen W (2008) J Org Chem 73:3954–3957
79. Organ MG, Abdel-Hadi M, Avola S, Hadei N, Nasielski J, O'Brien CJ, Valente C (2007) Chem Eur J 13:150–157
80. Hamaguchi H, Uemura M, Yasui H, Yorimitsu H, Oshima K (2008) Chem Lett 37:1178–1179
81. Hartmann CE, Nolan SP, Cazin CSJ (2009) Organometallics 28:2915–2919
82. Hadei N, Kantchev EAB, O'Brien CJ, Organ MG (2005) Org Lett 7:3805–3807
83. Organ MG, Avola S, Dubovyk I, Hadei N, Kantchev EAB, O'Brien CJ, Valente C (2006) Chem Eur J 12:5749
84. Kantchev EAB, O'Brien CJ, Organ MG (2006) Aldrichimica Acta 39:97–110
85. O'keefe BM, Simmons N, Martin SF (2008) Org Lett 10:5301–5304
86. Xi Z, Zhou Y, Chen W (2008) J Org Chem 73:8497–8501
87. Miyaura N, Yamada K, Suzuki A (1979) Tetrahedron Lett 20:3437–3440. See also: Miyaura N, Suzuki A (1979) J Chem Soc Chem Commun 866–867
88. Herrmann WA, Reisinger CP, Spiegler M (1998) J Organomet Chem 557:93–96
89. Iyer S, Jayanthi A (2003) Synlett 1125–1128
90. Lee HM, Zeng JY, Hu CH, Lee MT (2004) Inorg Chem 43:6822–6829
91. Han Y, Huynh HV, Tan GK (2007) Organometallics 26:6581–6585
92. Kremzow D, Seidel G, Lehmann CW, Fürstner A (2005) Chem Eur J 11:1833–1853
93. Lee HM, Lu CY, Chen CY, Chen WL, Lin HC, Chiu PL, Cheng PY (2004) Tetrahedron 60:5807–5825
94. Ray L, Shaikh MM, Ghosh P (2007) Organometallics 26:958–964
95. McGuinness DS, Cavell KJ (2000) Organometallics 19:741–748
96. Zhang T, Wang W, Gu X, Shi M (2008) Organometallics 27:753–757
97. Han Y, Hong YT, Huynh HV (2008) J Organomet Chem 693:3159–3165

98. Organ MG, Çalimsiz S, Sayah M, Hoi KH, Lough AJ (2009) Angew Chem Int Ed 48: 2383–2387
99. Herrmann WA, Böhm VPW, Gstöttmayr CWK, Grosche M, Reisinger CP, Weskamp T (2001) J Organomet Chem 617–618:616–628
100. Zhang C, Huang J, Trudell ML, Nolan SP (1999) J Org Chem 64:3804–3805
101. Burstein C, Lehmann CW, Glorius F (2005) Tetrahedron 61:6207–6217
102. Song C, Ma Y, Chai Q, Ma C, Jiang W, Andrus MB (2005) Tetrahedron 61:7438–7446
103. Huang W, Guo J, Xiao Y, Zhu M, Zou G, Tang J (2005) Tetrahedron 61:9783–9790
104. Altenhoff G, Goddard R, Lehmann CW, Glorius F (2003) Angew Chem Int Ed 42:3690–3693
105. Marion N, Navarro O, Mei J, Stevens ED, Scott NM, Nolan SP (2006) J Am Chem Soc 128:4101–4111
106. O'Brien CJ, Kantchev EAB, Valente C, Hadei N, Chass GA, Lough A, Hopkinson AC, Organ MG (2006) Chem Eur J 12:4743–4748
107. Jin Z, Guo S, Qiu L, Song H, Fang J (2009) Adv Synth Catal 351:1575–1585
108. Diebolt O, Braunstein P, Nolan SP, Cazin CSJ (2008) Chem Commun 3190–3192
109. Diebolt O, Jurcík V, Correa da Costa R, Braunstein P, Cavallo L, Nolan SP, Slawin AMZ, Cazin CSJ (2010) Organometallics. 29:1443–1450
110. Glorius F, Altenhoff G, Goddard R, Lehmann CW (2002) Chem Commun 2704–2705
111. Altenhoff G, Goddard R, Lehmann CW, Glorius F (2004) J Am Chem Soc 126:15195–15201
112. Würtz S, Glorius F (2008) Acc Chem Res 41:1523–1533
113. Chiu PL, Lai CL, Chang CF, Hu CH, Lee HM (2005) Organometallics 24:6169–6178
114. Inamoto K, Kuroda J, Kwon E, Hiroya K, Doi T (2009) J Organomet Chem 694:389–396
115. Polshettiwar V, Varma RS (2008) Tetrahedron 64:4637–4643
116. Worm-Leonhard K, Meldal M (2008) Eur J Org Chem 26:5244–5253
117. Corma A, González-Arellano C, Iglesias M, Pérez-Ferrera S, Sánchez F (2007) Synlett 1771–1774
118. Fürtsner A, Leitner A (2001) Synlett 290–292
119. Arentsen K, Caddick S, Cloke FGN, Herring AP, Hitchcock PB (2004) Tetrahedron Lett 45:3511–3515
120. Weskamp T, Böhm VPW, Herrmann WA (1999) J Organomet Chem 585:348–352
121. Grasa GA, Nolan SP (2001) Org Lett 3:119–122
122. Yamazaki T, Sato Y (2008) Synthesis 2830–3834
123. Lee HM, Nolan SP (2000) Org Lett 2:2053–2055
124. Batey RA, Shen M, Lough AJ (2002) Org Lett 4:1411–1414
125. Yang C, Nolan SP (2002) Organometallics 21:1020–1022
126. Ma Y, Song C, Jiang W, Wu Q, Wang Y, Liu X, Andrus MB (2003) Org Lett 5:3317–3319
127. Eckhardt M, Fu GC (2003) J Am Chem Soc 125:13642–13643
128. Richter H, Schwertfeger H, Schreiner PR, Fröhlich R, Glorius F (2009) Synlett 193–197
129. Altenhoff G, Würtz S, Glorius F (2006) Tetrahedron Lett 47:2925–2928
130. Ray L, Barman S, Shaikh MM, Ghosh P (2008) Chem Eur J 14:6646–6655
131. Zanardi A, Mata JA, Peris E (2009) Organometallics 28:4335–4339
132. Zhang X, Liu A, Chen W (2008) Org Lett 10:3849–3852
133. Biffis A, Lobbia GG, Papini G, Pellei M, Santini C, Scattolin E, Tubaro C (2008) J Organomet Chem 693:3760–3766
134. Corma A, González-Arellano C, Iglesias M, Pérez-Ferreras S, Sánchez F (2007) Synlett 1771–1774
135. Mas-Marzá E, Segarra AM, Claver C, Peris E, Fernández E (2003) Tetrahedron Lett 44:6595–6599
136. Kim J-H, Lee D-H, Jun B-H, Lee Y-S (2007) Tetrahedron Lett 48:7079–7084
137. Worm-Leonhard K, Meldal M (2008) Eur J Org Chem 31:5244–5253

138. Jiang L, Buchwald SL (2004) Palladium-catalyzed aromatic carbon–nitrogen bond formation. In: de Meijere A, Diederich F (eds) Metal-catalyzed cross-coupling reactions, 2nd edn. Wiley-VCH, Weinheim
139. Paul F, Patt J, Hartwig JF (1994) J Am Chem Soc 116:5969–5970
140. Guram AS, Buchwald SL (1994) J Am Chem Soc 116:7901–7902
141. Green JC, Herbert BJ, Lonsdale R (2005) J Organomet Chem 690:6054–6067
142. Huang J, Grasa G, Nolan SP (1999) Org Lett 1:1307–1309
143. Grasa GA, Viciu MS, Huang J, Nolan SP (2001) J Org Chem 66:7729–7737
144. Stauffer SR, Lee S, Stambuli JP, Hauck SI, Hartwig JF (2000) Org Lett 2:1423–1426
145. Conesa Lerma I, Cawley MJ, Cloke FGN, Arentsen K, Scott JS, Pearson SE, Hayler J, Caddick S (2005) J Organomet Chem 690:5841–5848
146. Cheng J, Trudell ML (2001) Org Lett 3:1371–1374
147. Frisch AC, Zapf A, Briel O, Kayser B, Shaikh N, Beller M (2004) J Mol Catal A 214:231–239
148. Titcomb LR, Caddick S, Cloke FGN, Wilson DJ, McKerrecher D (2001) Chem Commun 1388–1389
149. Arentsen K, Caddick S, Cloke FGN (2005) Tetrahedron 61:9710–9715
150. Viciu MS, Kissling RM, Stevens ED, Nolan SP (2002) Org Lett 4:2229–2231
151. Navarro O, Marion N, Scott NM, González J, Amoroso D, Bell A, Nolan SP (2005) Tetrahedron 61:9716–9722
152. Viciu MS, Kelly RA III, Stevens ED, Naud F, Studer M, Nolan SP (2003) Org Lett 5:1479–1482
153. Li J, Cui M, Yu A, Wu Y (2007) J Organomet Chem 692:3732–3742
154. Broggi J, Clavier H, Nolan SP (2008) Organometallics 27:5525–5531
155. Organ MG, Abdel-Hadi M, Avola S, Dubovyk I, Hadei N, Kantchev EAB, O'Brien CJ, Sayah M, Valente C (2008) Chem Eur J 14:2443–2452
156. Viciu MS, Germaneau RF, Navarro-Fernández O, Stevens ED, Nolan SP (2002) Organometallics 21:5470–5472
157. Navarro O, Kaur H, Mahjoor P, Nolan SP (2004) J Org Chem 69:3173–3180
158. Cämmerer SS, Viciu MS, Stevens ED, Nolan SP (2003) Synlett 1871–1873
159. Marion N, Navarro O, Mei J, Stevens ED, Scott NM, Nolan SP (2006) J Am Chem Soc 128:4101–4111
160. Navarro O, Marion N, Mei J, Nolan SP (2006) Chem Eur J 12:5142–5148
161. Esposito O, Gois PMP, Lewis AKK, Caddick S, Cloke FGN, Hitchcock PB (2008) Organometallics 27:6411–6418
162. Jin Z, Guo S-X, Gu X-P, Qiu L-L, Song H-B, Fang J-X (2009) Adv Synth Catal 351:1575–1585
163. Chen C, Yang L-M (2007) J Org Chem 72:6324–6327
164. Gao C-Y, Yang L-M (2008) J Org Chem 73:1624–1627
165. Tobisu M, Shimasaki T, Chatani N (2009) Chem Lett 38:710–711
166. Matsubara K, Miyazaki S, Koga Y, Nibu Y, Hashimura T, Matsumoto T (2008) Organometallics 27:6020–6024
167. Haider J, Kunz K, Scholz U (2004) Adv Synth Catal 346:717–722
168. Tubaro C, Biffis A, Scattolin E, Basato M (2008) Tetrahedron 64:4187–4195
169. Fürstner A, Seidel G (2002) Org Lett 4:541
170. Ma Y, Song C, Jiang W, Xue G, Cannon JF, Wang X, Andrus MB (2003) Org Lett 5:4635
171. Zhang Y, Ngeow KC, Ying JY (2007) Org Lett 9:3495–3498
172. Ackermann L (2005) Org Lett 7:439–442
173. Kaspar LT, Ackermann L (2005) Tetrahedron 61:11311–11316
174. Cheung WS, Patch RJ, Player MR (2005) J Org Chem 70:3741

Chapter 7
N-Heterocyclic Carbene Complexes in Arylation Reactions other than Cross-couplings

Christophe Berini and Oscar Navarro

Abstract This chapter highlights the use of *N*-heterocyclic carbenes as supporting ligands in arylation reactions different than the more common cross-coupling reactions, including C-F bond activation, catalytic arylation, homocoupling, direct arylation and oxidative Heck reactions.

7.1 Introduction

Transition metal-catalysed reactions have emerged as powerful tools for carbon-carbon (C–C) bond formation [1]. Cross-coupling reactions (Suzuki-Miyaura, Mizoroki-Heck, Stille, etc.) are recognised to be extremely reliable, robust and versatile. However, some other catalysed arylation reactions have been studied and have been reported to be very efficient [2]. In recent years, *N*-heterocyclic carbenes (NHC) have been extensively studied and their use as ligands for transition-metal catalysis has allowed for the significant improvement of many reactions [3]. This chapter highlights the use of NHC-bearing complexes in those arylation reactions.

7.2 C–F Bond Activation

Organofluorine compounds have received particular attention from scientists because of their distinctive properties and their application in materials chemistry (liquid crystals, coatings, fabrics, coolants or artificial blood), drug synthesis (anti-cancer agents, antibiotics) and agrochemicals (herbicides, insecticides) [4]. Those special properties are due, among other reasons, to the high electronegativity of fluorine, which forms with carbon the strongest bond in organic chemistry [5].

O. Navarro (✉)
Department of Chemistry, University of Hawaii at Manoa, Honolulu, Hawaii 96822, USA
e-mail: oscarnf@hawaii.edu

C. S. J. Cazin (ed.), *N-Heterocyclic Carbenes in Transition Metal Catalysis and Organocatalysis*, Catalysis by Metal Complexes 32,
DOI 10.1007/978-90-481-2866-2_7,

The bond dissociation energy of fluoromethane is 115 kcal mol^{-1}, which is much higher than the other halides (C–Cl, C–Br and C–I, respectively 84, 72 and 58 kcal mol^{-1}) [6]. Due to its strength, the carbon-fluorine (C–F) bond is one of the most challenging bonds to activate [7]. A variety of C–F bond activation reactions have been carried out with different organometallic complexes [8]. Among them, nickel [9] and ruthenium complexes have proven to proceed selectively under mild conditions [10].

7.2.1 Nickel Complexes

In 1977, Fahey and Mahan described the first C–F oxidative addition of hexafluorobenzene (C_6F_6) to the zerovalent nickel complex [Ni(cod)(PEt_3)$_2$] [11]. The reaction was very slow and led, after several days at 30–35°C, to the formation of *trans*-[Ni(C_6F_5)(F)(PEt_3)$_2$] in very poor yield (7%). The use of [Ni(PEt_3)$_4$] for the same reaction by Perutz and co-workers 20 years later allowed for a better conversion but the reaction rate remained slow (4 weeks were necessary to obtain 48% of the desired product) [12].

The first use of NHC as ligands for the synthesis of a pentafluorophenyl fluoride nickel(II) complex through C–F activation was reported by Radius and co-workers. The authors synthesised a complex [NiL_n(NHC)$_2$] by adding a solution of [Ni(cod)$_2$] in toluene to a solution of I^{i}Pr (I^{i}Pr = *N,N'*-(di-*iso*-propyl)imidazol-2-ylidene) **1** [13]. The procedure afforded the dimeric complex [Ni_2(I^{i}Pr)$_4$(cod)] **2** in moderate yield (Scheme 7.1), which was characterised by NMR spectroscopy and X-ray analysis. The reaction of this complex with an equimolar amount of C_6F_6 proceeded rapidly in toluene and the reaction reached completion after being stirred overnight to yield the new complex *trans*-[Ni(C_6F_5)(F)(I^{i}Pr)$_2$] **3**. It is worth mentioning that this new complex could also be prepared on multigram scale in a one-pot procedure in better overall yield (56% *versus* 35%) [14].

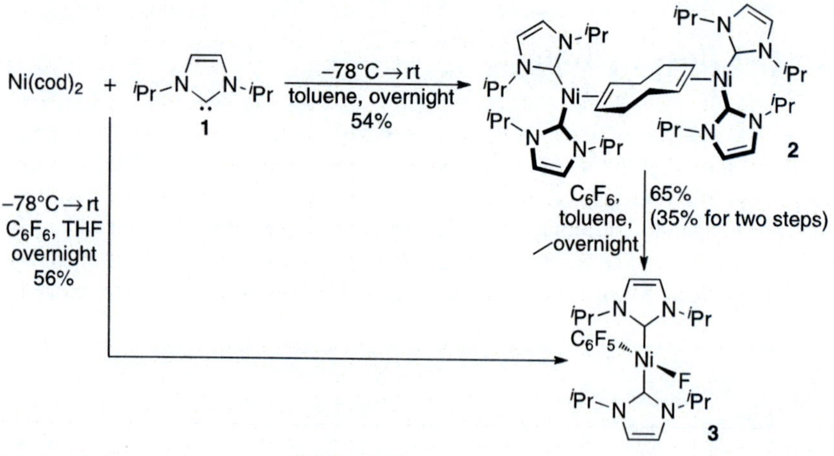

Scheme 7.1 Synthesis of *trans* [Ni(C_6F_5)(F)(IiPr)$_2$]

These conditions provided a major improvement in terms of yield but more particularly of timescale, and have been successfully applied to many different fluorinated arenes with high chemo- and regioselectivity [15]. Fluorinated heteroaromatics have also been used as substrates. Perutz and co-workers described the reaction of pentafluoropyridine (C_5F_5N) with [Ni(cod)$_2$] in the presence of PEt_3 to afford *trans*-[Ni(2-C_5F_4N)(F)(PEt_3)$_2$] **4** (the configuration was assigned by NMR analysis) (Scheme 7.2, pathway a) [12]. This reaction proceeded much faster than with hexafluorobenzene (2 h *versus* several weeks), suggesting that the nitrogen atom could play a role at accelerating the reaction rate. The cleavage at the C^2 position of C_5F_4N is in contrast with the observed activation on C^4 position with other metals like platinum or palladium [16]. This unusual regioselectivity has been attributed to an oxidative addition through a concerted reaction pathway [9]. The reaction of C_5F_5N with **2** at low temperature afforded *trans*-[Ni (4-C_5F_4N)(F)(PEt_3)$_2$] **5** as the sole product (Scheme 7.2, pathway b) [15].

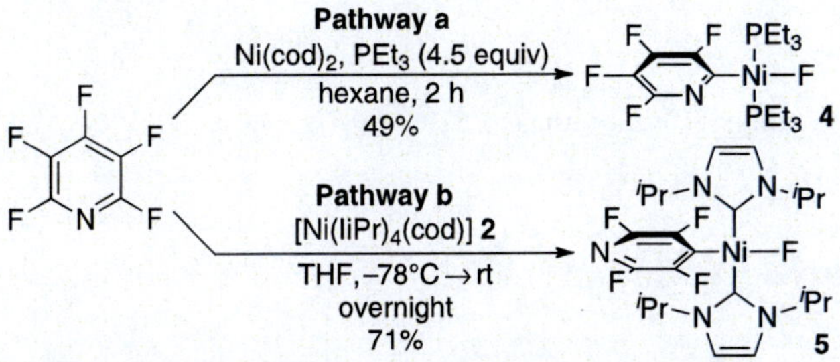

Scheme 7.2 Reactivity of pentafluoropyridine with nickel(0) complexes

The authors carried out theoretical calculations on this system as well as on the [Ni(PMe$_3$)$_2$] system to compare their reactivity with hexafluorobenzene. They found that the barrier for [Ni(IiPr)$_2$] is approximately 10 kJ/mol lower in energy than the corresponding barrier for the phosphine-bearing system. This value was in agreement with the different reactivity of both complexes but could not fully explain the large difference in reaction times. Density functional Theory (DFT) calculations also showed that the *trans* product is more stable than the *cis* product (total energies are respectively –130.9 and 91.1 kJ/mol), which was in agreement with the experimental values.

7.2.2 Ruthenium Complexes

Ruthenium has also been found to be very efficient promoting C–F bond activation under mild conditions [8]. In 1997, Perutz and co-workers reported C–F bond activation of C_6F_6 with [RuH$_2$(dmpe)$_2$] [10]. More recently, Whittlesey and co-workers described a similar dihydrido-ruthenium complex bearing an NHC, which achieved

C–F bond activation in some arene substrates (Scheme 7.3) [17]. The reaction of $[RuH_2(CO)(dppp)(ICy)]$ **6** complex with C_6F_6 and C_5F_5N at 120°C in the presence of Et_3SiH afforded the corresponding $[Ru(C_6F_5)H(CO)(ICy)(dppp)]$ **7**. and $[Ru(C_5F_4N)H(CO)(ICy)(dppp)]$ **8** complexes in 76% and 87% yield, respectively.

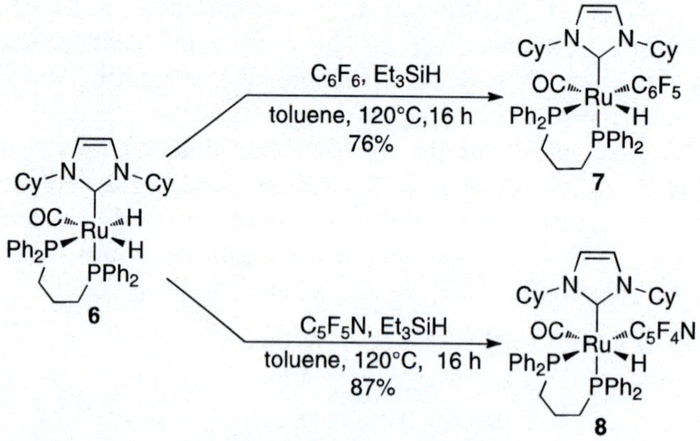

Scheme 7.3 C–F Bond activation using an NHC–Ru(II) complex

7.3 Catalytic Arylation

While the synthesis of functionalised secondary alcohols and amines can be achieved without catalyst by the addition of organolithium and organomagnesium reagents to C=N and C=O groups, these methods lack a significant functional group tolerance. In order to overcome this limitation and access to more functionalised compounds, the catalytic arylation of aldehydes and imines has been extensively studied [2].

7.3.1 Catalytic Aldehyde Arylation

Diaryl methanols are structural core units in a large number of biologically active compounds [18]. An increased attention has been paid to rhodium-catalysed methods to prepare these compounds [19]. In 1998, Miyaura and co-workers were the first to report the rhodium(I)-catalysed arylation of aldehydes with arylboronic acids in the presence of dppf **9** to obtain secondary alcohols in good to excellent yields (Table 7.1) [20]. Noticing that this reaction was very efficient using sterically demanding, strongly basic monodentate ligands, Fürstner and co-workers described an *in situ* generated catalytic system using $RuCl_3$ and IPr·HCl (IPr·HCl = *N*, *N*′-bis-(2,6-di-*iso*-propylphenyl)imidazolium chloride) **10** [21]. This procedure was highly efficient even with low catalyst loading, affording the desired arylation products five to eight times faster than with the previous phosphine system. A variety of boronic acids and aldehydes were successfully combined under these conditions.

Table 7.1 Rhodium-catalysed addition of phenylboronic acid to *p*-anisaldehyde

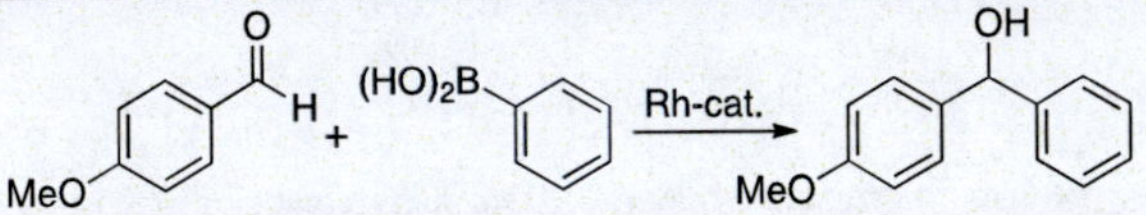

Rh source (mol%)	Ligand	Conditions	Temperature (°C)	Time (h)	Yield (%)
$[Rh(acac)(CO)_2]$ (3)	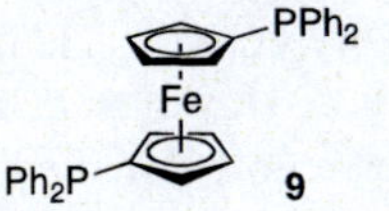 9	DME/H_2O 1:1	80	16	83
[Rh(acac)(coe)] (3)	9	DME/H_2O 4:1	80	16	85
[Rh(acac)(coe)] (3)	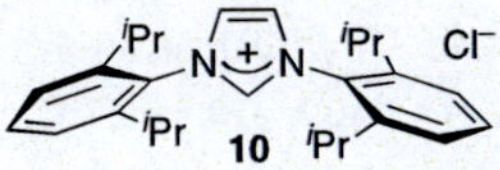 10	DME/H_2O 4:1 + NaOMe	80	1.5	77
$RhCl_3{\cdot}3H_2O$ (1)	10	DME/H_2O 4:1 + NaOMe	80	0.2	93

After these seminal studies, the use of NHC–Rh systems for the addition of aryl boronic acids to carbonyl compounds became a very fertile area and many groups have reported on variations of this catalytic system [22].

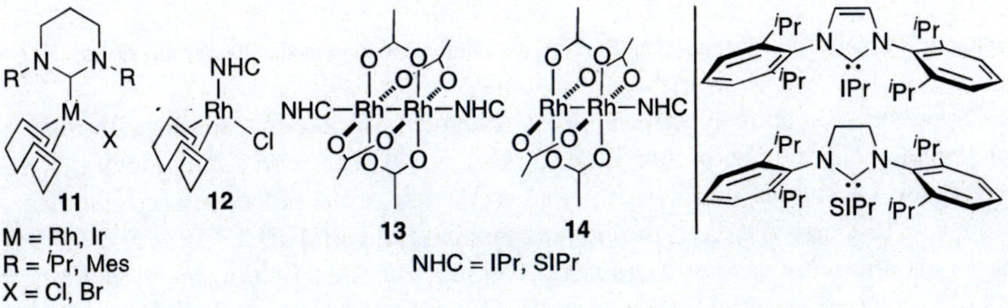

Fig. 7.1 NHC-Rh complexes for the catalytic arylation of aldehydes

Buchmeiser and co-workers prepared a series of NHC-bearing complexes of rhodium and iridium with the general structure **11** for this purpose, the latter metal being less reactive towards this reaction (Fig. 7.1) [23]. Over the last few years, the Çetinkaya group has synthesised a plethora of [RhCl(cod)(NHC)] complexes **12** gathering different scaffolds such as imidazolidinylidenes [24], perhydrobenzimidazolylidenes [24], tetrahydropyrimidinylidenes [25], and benzimidazolylidenes (Figs. 7.1 and 7.2) [26]. All these NHC-bearing complexes catalysed the arylation of aldehydes with arylboronic acids using low catalyst loadings (1 mol%) and under mild conditions (60 – 80°C, 0.4 – 10 h), to afford the desired diarylmethanol. Gois and co-workers have described the synthesis of symmetrical **13** and unsymmetrical **14** dirhodium(II) complexes (Fig. 7.1) and their use in this reaction [27], these complexes being particularly efficient for the conversion of aliphatic aldehydes. Finally, some examples of silica- and polymer-supported rhodium carbene complexes have shown to be very competent in this reaction [28].

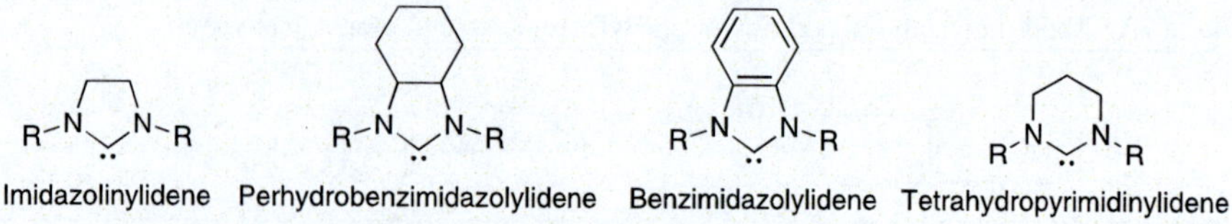

Fig. 7.2 General structure of some *N*-heterocyclic carbenes

The facile arylation of aldehydes with arylboronic acid has prompted the exploration of asymmetric versions of this reaction. However, this field has been scarcely explored and only few examples have been reported in the literature, with moderate results. The first diastereoselective example was described by Fürstner and co-workers. By reacting the Garner aldehyde **15** with phenylboronic acid under their set of experimental conditions (*i.e.* $RhCl_3{\cdot}3H_2O$, IPr·HCl) (Scheme 7.4) [21], the secondary alcohol was obtained in higher selectivity than that observed in the addition of phenylmagnesium bromide reported by Joullié (*de* = 94% versus 66%), with the *anti* isomer as the major compound [29].

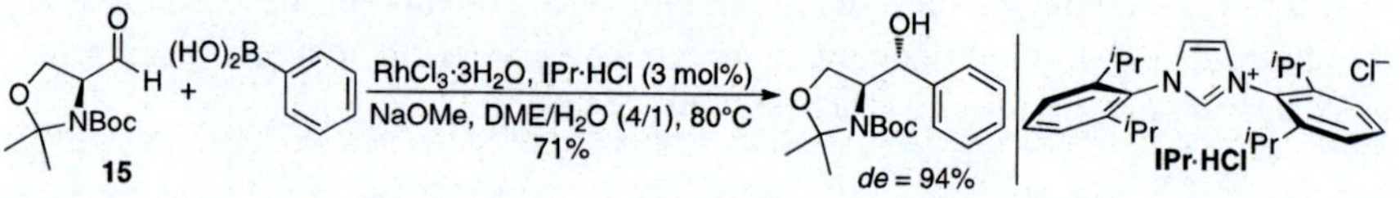

Scheme 7.4 (NHC)Rh-catalysed addition of phenylboronic acid to the Garner aldehyde

Some other enantioselective approaches have been attempted, still with moderate enantioselectivities, by making use of *in situ* systems containing a chiral NHC precursor. Luo and co-workers reported on the use of the bidentate chiral imidazolium salt **16**, derived from L-proline, in combination with $[Rh(\mu\text{-}Cl)(cod)]_2$, leading to an enantiometic excess of around 20% [30]. The use of chiral imidazolium salt **17** in combination with $[RhCl(CH_2{=}CH_2)_2]_2$ by Aoyama afforded slightly better *ee* (Fig. 7.3) [31]. So far, Bolm and co-workers have obtained the best enantioselectivities (up to 38% *ee*) for the catalytic addition of phenylboronic acid to aromatic aldehydes by using planar chiral imidazolium salts **18**, derived from paracyclophane, in combination with $[Rh(OAc)_2]_2$ [32].

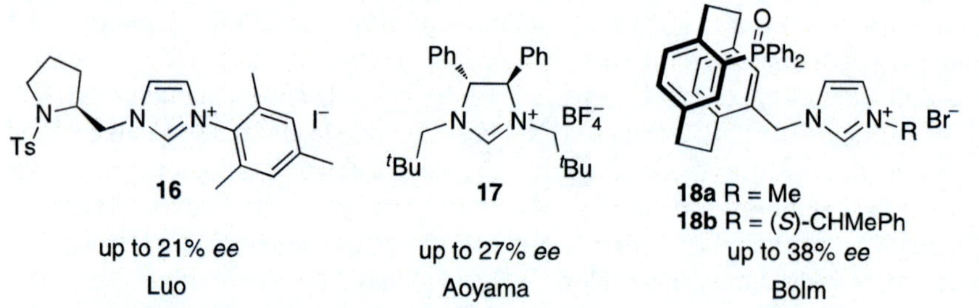

Fig. 7.3 NHC-Ligands for asymmetric catalysis with rhodium

Regarding the use of other metals for this transformation, Shirai and co-workers reported that a system constituted by palladium(II) complex $[Pd(\mu\text{-}Cl)(\eta^3\text{-allyl})]_2$ and thioether-imidazolium chloride **19** achieved the arylation of aldehydes with boronic acids [33] and potassium trifluoroborates in good to excellent yields (Scheme 7.5) [34]. More recently, Buffard and Itami showed that a $Ni(cod)_2$/IPr·HCl system could catalyse the reaction of arylboronate esters and inactivated aldehydes and ketones (Scheme 7.5) [35].

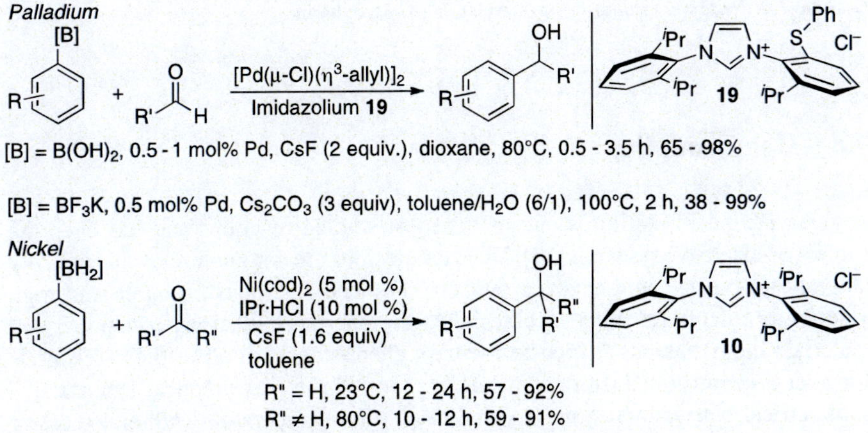

Scheme 7.5 NHC–Pd and NHC–Ni catalysed arylation of carbonyls compounds with boron reagents

7.3.2 Imine Arylation

The arylation of imines is an important transformation due to the fact that diarylamine scaffolds are important subunits of biologically active compounds [18]. Charette and co-workers have shown that (NHC)-Rh(I) complex (with NHC = *N*-mesityl-*N'*-benzoyliminoimidazolium ylide), prepared *in situ* from the corresponding transfer agent NHC–Ag, was catalytically active in the addition of phenylboronic acid to *p*-methylphenyl-*N*-phosphinoylimine [36]. The Sato group has recently reported on the arylation of *N*-phosphinoylarylimines and *N*-sulfonylarylimines using IAd·HCl and $[Rh(\mu\text{-}Cl)(cod)]_2$ [37].

So far, there is only one report describing the use of chiral NHC-metal complexes in catalytic asymmetric arylation of imines. This was achieved by using C_2-symmetric cationic NHC–Pd^{2+} diaquo complex **20** (Scheme 7.6) [38]. The arylation of a variety of *N*-tosylimines with different arylboronic acids was carried out under mild conditions. The presence of electron-withdrawing or electron-donating substituents on both partners did not seem to affect the reaction and the corresponding chiral diarylamines were obtained in good to excellent yields and high enantiomeric excess.

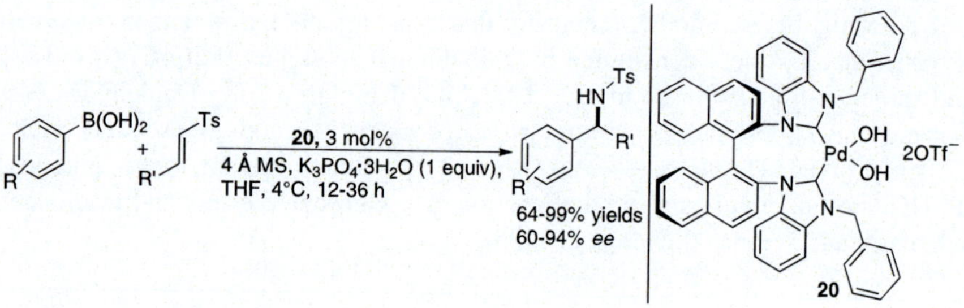

Scheme 7.6 Catalytic enantioselective arylation of *N*-tosylimines

7.4 Homocoupling

Aryl–aryl bond formation is one of the most studied reactions and has been the subject of extensive research [39]. Biaryl molecules are common scaffolds in many natural and biologically active products. The need for robust, reliable and high yielding protocols for biaryl synthesis for the pharmaceutical, agrochemical and materials industries has helped to increase chemists' attention in these reactions for over a century [40]. In fact, Ullmann reported the first practical synthesis of symmetrical biaryls by heating aryl halides at high temperature with an excess of copper [41, 42]. Unfortunately these relatively harsh conditions prevented the application of the Ullmann reaction in the synthesis of functionalised biaryls. Among transition metal-catalysed cross-coupling reactions which can be performed with different organometallic reagents and aryl halides and pseudo-halides [43], the palladium-catalysed Suzuki-Miyaura coupling of arylboronic acids or esters with aryl halides has emerged as a favourite [44]. This reaction has attracted much attention because these reagents are non-toxic, readily available and air and moisture stable. However, in order to drive the reaction to completion with aryl halides, many cross-coupling protocols use an excess of the organoboron reagent, and side reactions like the homocoupling [45] of arylboronic acid occur frequently. This side reaction is very interesting by itself as it allows for the synthesis of very useful scaffolds.

Many groups have reported on different protocols for the homocoupling of organoborons. For instance, Jackson and co-workers reported that *ligandless* palladium (II) acetate promoted the homocoupling reaction of arylboronic acids under oxygen atmosphere at room temperature [46]. The reaction proceeded slowly and only moderate yields of the symmetrical biaryl coupling products were obtained. Later, it was shown that the use of phosphine ligands enhanced the catalytic activity of $Pd(OAc)_2$ at elevated temperature [47]. Sometimes a side-product is formed in the presence of triarylphosphine, *via* an aryl-aryl interchange reaction between PPh_3 and a palladium complex intermediate, producing an unsymmetrical biaryl [48]. More recently, and to avoid the formation of this undesired compound

Yamamoto and co-workers reported a base- and ligand-free palladium (II) catalysed method, in methanol at room temperature and under air [49]. While the conversion of arylboronic acids bearing an electro-donating group was very efficient, the presence of an electro-withdrawing substituent led to lower conversions. To solve this problem and also prevent the fast formation of palladium black, Yamamoto and co-workers described a new approach where the reaction was catalysed by NHC-bearing complexes **21** and **22** in the presence of an oxidant (Fig. 7.4) [50]. The best results were obtained when complex **21b** was used in methanol at room temperature, in the presence of a small excess of *p*-benzoquinone.

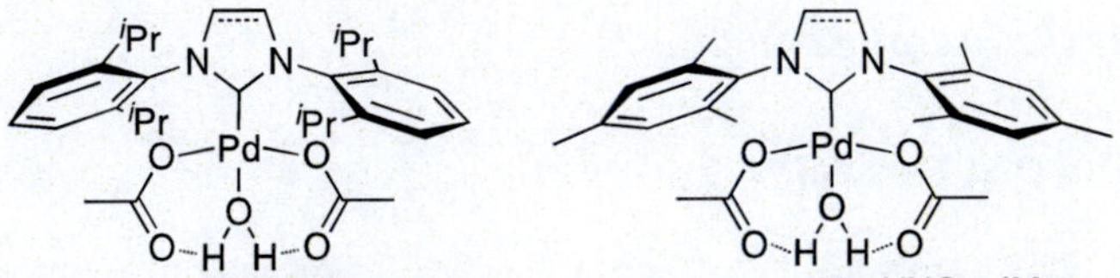

Fig. 7.4 (NHC)Pd(acetate)$_2$(H$_2$O) complexes

Very recently, well-defined complexes with general formula [PdCl(η^5-Cp)(NHC)] were synthesised and tested for the homocoupling of non-electrodeficient arylboronic acids at room temperature with good results (Scheme 7.7) [51]. This new class of catalysts were synthesised from commercially available NHC-palladium(II) chloride dimers and are air-stable.

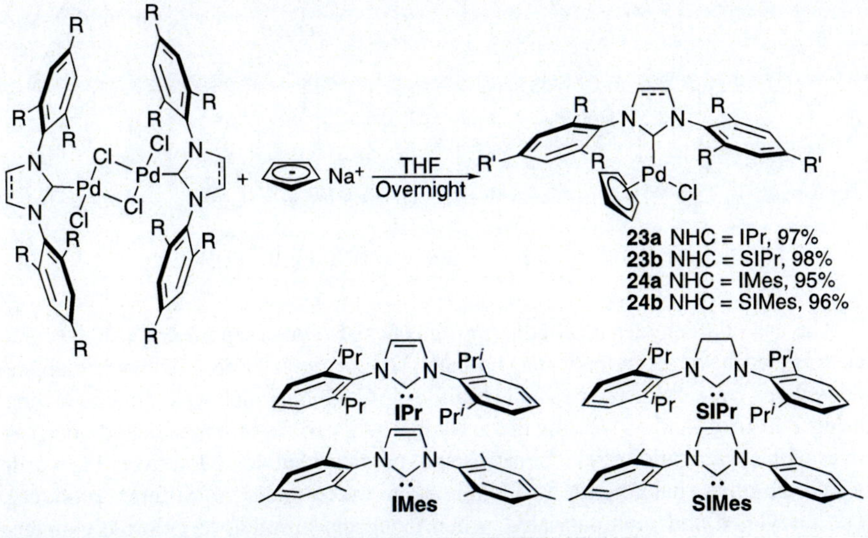

Scheme 7.7 Synthesis of well-defined [PdCl(η^5-Cp)(NHC)] complexes

7.5 Direct Arylation

While cross-coupling methods are extremely useful and of easy and common use, several drawbacks for these reactions could be pointed out, like the need of pre-functionalisation of both coupling partners and the use of the organometallic partner in stoichiometric amounts, with the corresponding generation of toxic waste. In recent years, alternative methods have emerged and the direct arylation through C–H bond activation probably constitutes the most attractive [52].

7.5.1 *Arylation of Arenes*

7.5.1.1 Directed Reaction

Even if the direct arylation reaction has become an attractive search area, the use of NHC as ligands for this reaction has not been intensively explored. The regioselectivity of the direct arylation reaction is governed either by the electronic effects of the arene, or more commonly by the presence of a directing group, limiting the degree of freedom in the system by coordinating a lone pair of electrons of the directing group (in N or O atoms) to the metal centre. The pyridine moiety has been used as directing group in ruthenium-catalysed arylation reactions [53]. The catalytic activity of bidentate (NHC)Ru(II) complexes **25** was evaluated in the direct arylation of 2-phenylpyridine with chlorobenzene (Scheme 7.8). These conditions led to the formation of di-*ortho*-arylated compounds as main products (up to 70%).

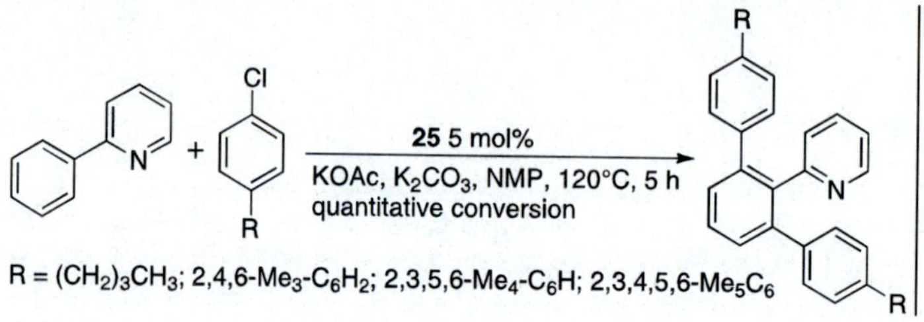

Scheme 7.8 Direct arylation of 2-phenylpyridine with NHC-Ru(II) complexes

The use of aldehydes as directing groups has also been reported. Çetinkaya and co-workers have described the Pd-catalysed *ortho*-arylation of benzaldehyde derivatives with aryl bromides and chlorides [54]. Good yields were obtained when using $Pd(OAc)_2$ and SIMes for the arylation of a variety of benzaldehyde derivatives with a wide range of electron-rich and -poor aryl halides. Moreover, it is worth mentioning that changing the aryl halide led to the formation of different products, *i.e.*, diarylated and mono-arylated, when using the bromide or chloride counterparts, respectively (Scheme 7.9).

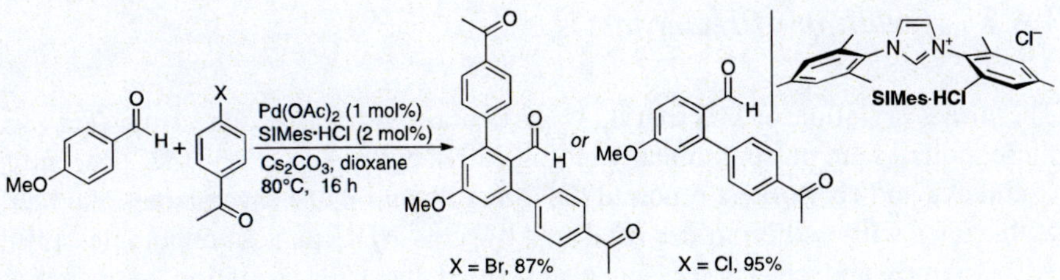

Scheme 7.9 Direct arylation using benzaldehyde as the directing group

7.5.1.2 Undirected Reaction

A series of new palladium complexes **26** bearing a phosphine-functionalised NHC ligand has been shown to be more efficient than typical catalytic systems (*i.e.* $Pd(OAc)_2$, PPh_3) for the reaction depicted in Scheme 7.10 [55]. In fact, these conditions afforded the 9-benzylidene-9*H*-fluorene in higher yields when iodobenzene was used, whereas the reaction of the less reactive phenyl bromide with diphenylacetylene was achieved for the first time affording the desired product in moderate to good yields.

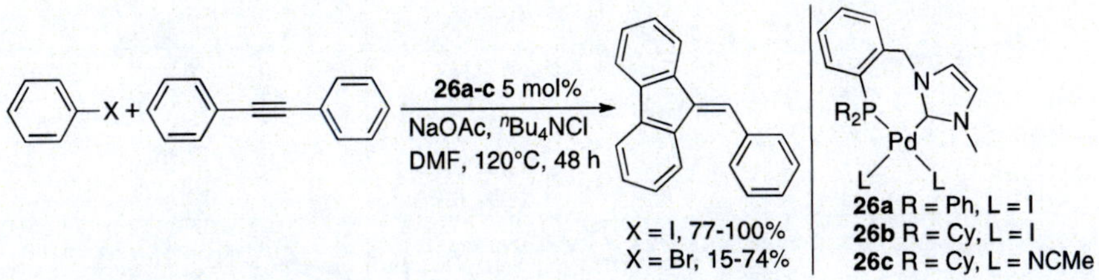

Scheme 7.10 Direct arylation of phenyl halides and diphenylacetylene

Fagnou and co-workers reported on the use of a palladium source in the presence of different phosphine ligands for the intramolecular direct arylation reaction of arenes with bromides [56]. Later, they discovered that new conditions employing palladium complex **27** promoted the direct arylation of a broad range of aryl chlorides to form six- and five-membered ring biaryls including different functionalities as ether, amine, amide and alkyl (Scheme 7.11) [57].

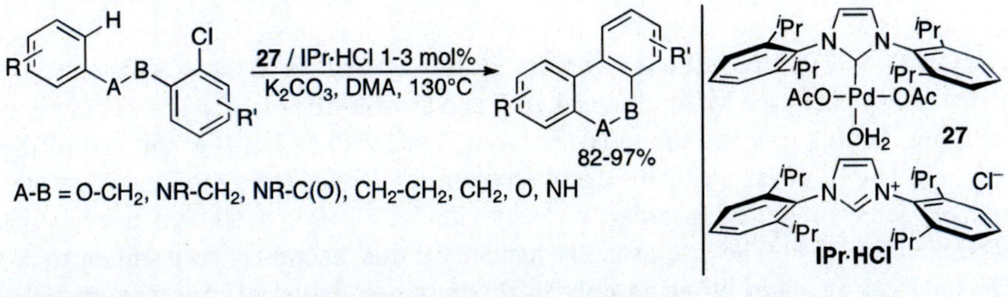

Scheme 7.11 Intramolecular direct arylation reactions with aryl chlorides

7.5.2 *Arylation of Heteroaryls*

The direct arylation of heteroaryls is particularly attractive due to the fact that these moieties are present in many biologically active compounds [58]. Recently, Çetinkaya and co-workers reported the direct arylation of benzoxazoles and benzothiazoles with aryl bromides catalysed by a bis-NHC-palladium complex [59]. Also, Sames and co-workers have described the C–H arylation of different SEM-protected heteroarenes, catalysed by NHC–Pd complex **28** (Scheme 7.12, pathway a) [60].

Sanford and co-workers reported on an alternative procedure for the same reaction. Their arylation strategy occurs under milder conditions (acetic acid at 25°C) with unprotected hetaroaryls (Scheme 7.12, pathway b) [61]. Instead of following a conventional Pd(0)/Pd(II) catalytic cycle, the authors proposed a Pd(II)/Pd(IV) catalytic cycle that used complex **29** and a diaryliodonium salt.

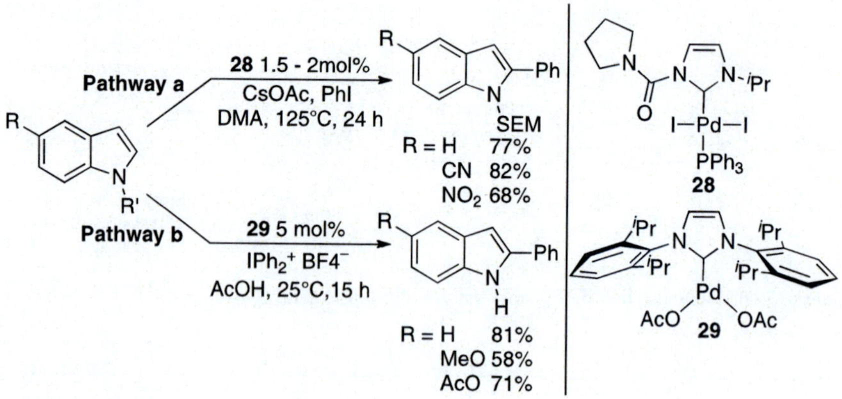

Scheme 7.12 Catalytic arylation of indoles

7.6 Oxidative Heck

The Oxidative Heck reaction is a very interesting alternative to the well-established palladium(0)-catalysed Mizoroki-Heck [62] and Suzuki-Miyaura [63] cross-coupling reactions. In this process, the Pd(II)-catalyst is reduced to Pd(0) at the end of the reaction. Therefore, an oxidising agent (copper, silver salt, benzoquinone or molecular oxygen) is required in order to regenerate the Pd(II) species that sustains the catalytic cycle [64]. The proposed mechanism for this reaction is very similar to the regular Heck reaction differing only in the transmetallation of an organometallic reagent with the Pd(II) catalyst and the oxidation of the resulting Pd(0) to the Pd(II) species in the last step of the cycle (Scheme 7.13) [65].

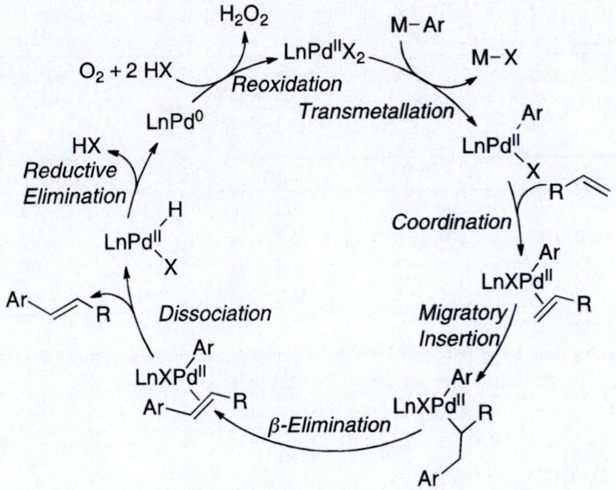

Scheme 7.13 Proposed mechanism of the oxidative Heck reaction with O_2 as oxidising agent

In 1975, Heck described the first non-catalytic example of this reaction, using organoboronic acids as precursors and stoichiometric amounts of palladium acetate [66]. Two decades later, a catalytic version was reported by Uemura and co-workers [67]. Since then, numerous systems for the oxidative Heck reaction have been reported [65, 68], including asymmetric versions [69]. Despite this renovated interest in this reaction and the fact that NHC were first used in the Mizoroki-Heck coupling [70], examples of the use of these ligands in oxidative Heck reactions is minimal. Only recently Jung and co-workers have reported on the preparation of air- and moisture-stable chiral palladium(II) dimers **30** bearing tridentate NHC ligands and their application in asymmetric oxidative Heck reactions [69]. The chiral ligands were obtained in good yields starting from optically active α-amino acids (Scheme 7.14). The coordination to palladium was achieved in two steps by transmetallation from the corresponding silver complexes. These chiral NHC–Pd(II) dimers were used for the coupling of aryl boronic acids and acyclic olefins. Although the yields were moderate (21–61%), both dimers provided excellent enantioselectivities (more than 90%), superior to those of existing methods (Table 7.2) [69].

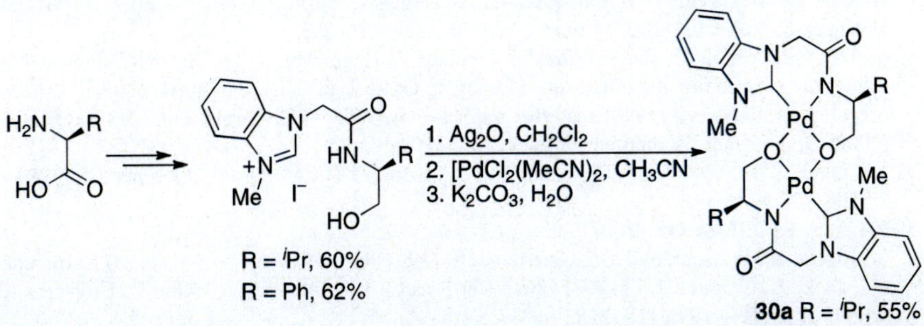

30a R = ^{i}Pr, 55%
30b R = Ph, 61%

Scheme 7.14 Synthesis of chiral NHC-Pd(II) complexes

Table 7.2 Chiral NHC-Pd(II) complexes in asymmetric oxidative Heck reaction

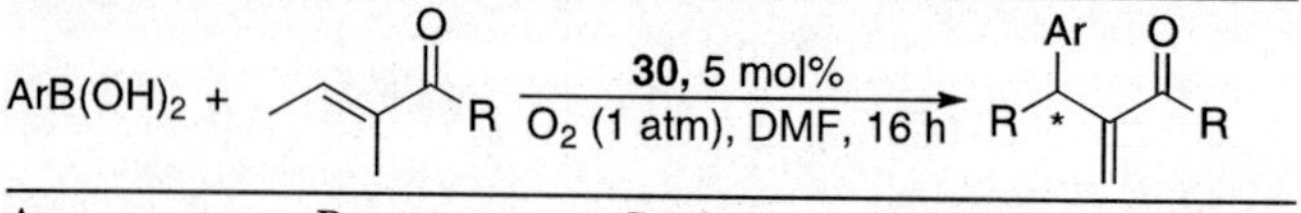

Ar	R	Catalyst	Yield[a] (%)	*ee*[b] (%)
Ph	OMe	**30a**	49	91
2-Naphthyl	OMe	**30a**	61	92
Ph	H	**30a**	31	98
	H	**30b**	32	90

[a]Isolated yields
[b]Determined by HPLC or ^{1}H NMR in the presence of chiral europium (III) tris[3-(heptafluoro-propylhydroxy- methylene)-(+)-camphorate]

7.7 Conclusion

As in the case of cross-coupling reactions, the examples of catalytic systems using NHC ligands in C–F bond activation and the arylation reactions described above prove a versatility and potential larger than that of ligandless and phosphine-based systems. The use of these ligands opens a new window of possibility in terms of substrate scope by activating classically "unreactive" bonds and sometimes acting with a regioselectivity different than that of phosphine-based systems. Although examples on the use of NHC systems in these reactions are currently limited, there is no doubt that more and exciting applications will appear in the near future.

References

1. (a) de Meijere A, Diederich F (2004) Metal-catalysed cross-coupling reactions. Wiley-VCH, Weinheim; (b) Beller M, Bolm C (2004) Transition metals for organic synthesis. Wiley-VCH, Weinheim
2. Ackermann L (2009) Modern arylation methods. Wiley-VCH, Weinheim
3. (a) Hermann WA (2002) Angew Chem Int Ed 41:1290–1309; (b) Nolan SP (2006) N-heterocyclic carbenes in synthesis. Wiley-VCH: Weinheim; (c) Glorius F (2007) N-heterocyclic carbenes in transition metal catalysis. Springer-Verlag, Berlin; (d) Díez-González S, Nolan SP (2007) Coord Chem Rev 251:874–883
4. (a) Hiyama T (2000) Organofluorine Compounds. Springer-Verlag, Berlin; (b) Chambers RD (2004) Fluorine in organic chemistry. Blackwell, Oxford; (c) Kirsch P (2004) Modern fluoro-organic chemistry: synthesis, reactivity, applications. Wiley-VCH, Weinheim (d) Uneyama K (2006) Organofluorine chemistry. Blackwell, Oxford
5. (a) Lemal DM (2004) J Org Chem 69:1–11 (b) O'Hagan D (2008) Chem Soc Rev 37:308–319
6. Blanksby SJ, Ellison GB (2003) Acc Chem Res 36:255–263
7. (a) Kiplinger JL, Richmond TG, Osterberg CE (1994) Chem Rev 94:373–431; (b) Burdeniuc J, Jedlicka B, Crabtree RH (1997) Chem Ber/Recueil 130:145–154; (c) Amii H, Uneyama K (2009) Chem Rev 109:2119–2183
8. Torrens H (2005) Coord Chem Rev 249:1957–1985
9. Braun T, Perutz RN (2002) Chem Commun 2749–2757

10. Whittlesey MK, Perutz RN, Moore MH (1996) Chem Commun 787–788
11. Fahey DR, Mahan JE (1977) J Am Chem Soc 99:2501–2508
12. Cronin L, Higgitt CL, Karch R, Perutz RN (1997) Organometallics 16:4920–4928
13. Schaub T, Radius U (2005) Chem Eur J 11:5024–5030
14. Schaub T, Backes M, Radius U (2008) Eur J Inorg Chem 2680–2690
15. Schaub T, Fischer P, Steffen A, Braun T, Radius U, Mix A (2008) J Am Chem Soc 130:9304–9317
16. Jasim NA, Perutz RN, Whitwood AC, Braun T, Izundu J, Neumann B, Rothfeld S, Stammler H-G (2004) Organometallics 23:6140–6149
17. Reade SP, Acton AL, Mahon MF, Martin TA, Whittlesey MK (2009) Eur J Inorg Chem 1774–1785
18. Schmidt F, Stemnler RT, Rudolph J, Bolm C (2006) Chem Soc Rev 35:454–470
19. Fagnou K, Lautens M (2003) Chem Rev 103:169–196
20. (a) Sakai M, Ueda M, Miyaura N (1998) Angew Chem Int Ed 37:3279–3281 (b) Ueda M, Miyaura N (2000) J Org Chem 65:4450–4452
21. Fürstner A, Krause H (2001) Adv Synth Catal 343:343–350
22. Díez-González S, Marion N, Nolan SP (2009) Chem Rev 109:3612–3676
23. (a) Imlinger N, Mayr M, Wang D, Wurst K, Buchmeiser MR (2004) Adv Synth Catal 346:1836–1843; (b) Imlinger N, Wurst K, Buchmeiser MR (2005) J Organomet Chem 690:4433–4440
24. (a) Özdemir I, Demir S, Çetinkaya B (2004) J Mol Catal A: Chem 215:45–48; (b) Yiğit M, Özdemir I, Çetinkaya E, Çetinkaya B (2005) Heteroat Chem 16:461–465; (c) Kilinçarslan R, Yiğit M, Özdemir I, Çetinkaya E, Çetinkaya B (2007) J Heterocycl Chem 44:69–73; (d) Özdemir I, Yiğit M, Çetinkaya E, Çetinkaya B (2006) Heterocycles 68:1371–1379; (e) Yiğit M, Özdemir I (2007) Transition Met Chem 32:536–540
25. Özdemir I, Demir S, Çetinkaya B, Çetinkaya E (2005) J Organomet Chem 690:5849–5855
26. Özdemir I, Gürbüz N, Gök Y (2005) Transition Met Chem 30:367–371
27. (a) Gois PMP, Trindade AF, Veiros LF, André V, Duarte MT, Afonso CAM, Caddick S, Cloke FGN (2007) Angew Chem Int Ed 46:5750–5753; (b) Trindade AF, Gois PMP, Veiros LF, André V, Duarte MT, Afonso CAM, Caddick S, Cloke FGN (2008) J Org Chem 73:4076–4086
28. (a) Özdemir I, Gürbüz N, Seçkin T, Çetinkaya B (2005) Appl Organometal Chem 19:633–638; (b) Yan C, Zeng X, Zhang W, Luo M (2006) J Organomet Chem 691:3391–3396
29. Williams L, Zhang Z, Shao F, Carroll PJ, Joullié MM (1996) Tetrahedron 52:11673–11674
30. Zhang W, Qin Y, Zhang S, Luo M (2005) Arkivoc 14:39–48
31. Arao T, Sato K, Kondo K, Aoyama T (2006) Chem Pharm Bull 54:1576–1581
32. Focken T, Rudolph J, Bolm C (2005) Synlett 429–436
33. Kuriyama M, Shimazawa R, Shirai R (2008) J Org Chem 73:1597–1600
34. Kuriyama M, Shimazawa R, Enomoto T, Shirai R (2008) J Org Chem 73:6939–6942
35. Bouffard J, Itami K (2009) Org Lett 11:4410–4413
36. Legault CY, Kendall C, Charette AB (2005) Chem Commun 3826–3828
37. Bakar Md A, Suzuki Y, Sato M (2008) Chem Pharm Bull 56:973–976
38. Ma G-N, Zhang T, Shi M (2009) Org Lett 11:875–878
39. Hassan J, Sévignon M, Gozzi C, Schulz E, Lemaire M (2002) Chem Rev 102:1359–1469
40. Stanforth SP (1998) Tetrahedron 54:263–303
41. Ullmann F, Bielecki J (1901) Chem Ber 34:2174–2185
42. For recent reviews of cross–coupling reactions with copper–based catalyst, see: (a) Lindley J (1984) Tetrahedron 40:1433–1456; (b) Ley SV, Thomas AW (2003) Angew Chem Int Ed 42:5400–5449; (c) Kunz K, Scholz U, Ganzer D (2003) Synlett 2428–2439; (d) Beletskaya IP, Cheprakov AV (2004) Coord Chem Rev 248:2337–2364
43. de Meijere A, Diederich F (2004) Metal-catalysed cross-coupling reactions, 2nd edn, vols 1 and 2. Wiley-VCH, Weinheim
44. (a) Miyaura N, Yamada K, Suzuki A (1979) Tetrahedron Lett 36:3437–3440; (b) Miyaura N, Suzuki A (1995) Chem Rev 95:2457–2483; (c) Kotha S, Lahiri K, Kashinath D (2002) Tetrahedron 58:9633–9695
45. Nelson TD, Crouch RD (2004) Org React 63:265–555

46. Smith KA, Campi EM, Jackson WR, Marcuccio S, Naeslund CGM, Deacon GB (1997) Synlett 131–132
47. Wong MS, Zhang XL (2001) Tetrahedron Lett 42:4087–4089
48. (a) O'Keefe DF, Dannock MC, Marcuccio SM (1992) Tetrahedron Lett 33:6679–6680; (b) Goodson FE, Wallow TI, Novak BM (1997) J Am Chem Soc 119:12441–12453
49. Yamamoto Y, Suzuki R, Hattori K, Nishiyama H (2006) Synlett 1027–1030
50. Yamamoto Y (2007) Synlett 1913–1916
51. Jin Z, Guo S-X, Gu X-P, Qiu L-L, Song H-B, Fang J-X (2009) Adv Synth Catal 351:1575–1585
52. (a) Alberico D, Scott ME, Lautens M (2007) Chem Rev 107:174–238; (b) McGlacken GP, Bateman LM (2009) Chem Soc Rev 38:2447–2464
53. (a) Özdemir I, Demir S, Çetinkaya B, Gourlaouen C, Maseras F, Bruneau C, Dixneuf PH (2008) J Am Chem Soc 130:1156–1157; (b) Özdemir I, Demir S, Gürbüz N, Çetinkaya B, Toupet L, Bruneau C, Dixneuf PH (2009) Eur J Inorg Chem 1942–1949
54. (a) Gürbüz N, Özdemir I, Çetinkaya B (2005) Tetrahedron Lett 46:2273–2277; (b) Doğan Ö, Gürbüz N, Özdemir I, Çetinkaya B (2009) J Heterocyclic Chem 46:186–190
55. Ho C-C, Chatterjee S, Wu T-L, Chan K-T, Chang Y-W, Hsiao T-H, Lee HM (2009) Organometallics 28:2837–2847
56. Campeau L-C, Parisien M, Leblanc M, Fagnou K (2004) J Am Chem Soc 126:9186–9187
57. Campeau L-C, Thansandote P, Fagnou K (2005) Org Lett 7:1857–1860
58. (a) Joullie MM, Berritt S, Forbeck EM (2008) Curr Opin Drug Discovery Dev 11:829–852; (b) Nicolaou KC, Chen JS (2008) Pure Appl Chem 80:727–742
59. Arslan H, Özdemir I, Vanderveer D, Demir S, Çetinkaya B (2009) J Coord Chem 62:2591–2599
60. Touré BB, Lane BS, Sames D (2006) Org Lett 8:1979–1982
61. Deprez NR, Kalyani D, Krause A, Sanford MS (2006) J Am Chem Soc 128:4972–4973
62. (a) Mizoroki T, Mori K, Ozaki A (1971) Bull Chem Soc Jpn 44:581; (b) Heck RF, Nolley JP Jr (1972) J Org Chem 37:2320–2322; (c) de Meijere A, Meyer FE (1994) Angew Chem Int Ed Engl 33:2379–2411
63. Varseev GN, Maier ME (2006) Angew Chem Int Ed 45:4767–4771
64. Stahl SS (2004) Angew Chem Int Ed 43:3400–3420
65. (a) Yoo KS, Yoon CH, Jung KW (2006) J Am Chem Soc 128:16384–16393; (b) Enquist P-A, Nilsson P, Sjöberg P, Larhed M (2006) J Org Chem 71:8779–8786
66. Dieck HA, Heck RF (1975) J Org Chem 40:1083–1090
67. Cho CS, Uemura S (1994) J Organomet Chem 465:85–92
68. For some recent articles, see: (a) Du X, Suguro M, Hirabayashi K, Mori A (2001) Org Lett 3:3313–3316; (b) Andappan MMS, Nilsson P, Larhed M (2003) Mol Diversity 7:97–106; (c) Jung YC, Mishra RK, Yoon CH, Jung KW (2003) Org Lett 5:2231–2234; (d) Andappan MMS, Nilsson P, Larhed M (2004) Chem Commun 218–219; (e) Enquist P-A, Lindh J, Nilsson P, Larhed M (2006) Green Chem 8:338–343; (f) Lindh J, Enquist P-A, Pilotti Å, Nilsson P, Larhed M (2007) J Org Chem 72:7957–7962; (g) Ruan J, Li X, Saidi O, Xiao J (2008) J Am Chem Soc 130:2424–2425; (h) Su Y, Jiao N (2009) Org Lett 11:2980–2983
69. (a) Akiyama K, Wakabayashi K, Mikami K (2005) Adv Synth Catal 347:1569–1575; (b) Yoo KS, Park CP, Yoon CH, Sakaguchi S, O'Neill J, Jung KW (2007) Org Lett 9:3933–3935; (c) Sakaguchi S, Yoo KS, O'Neill J, Lee JH, Stewart T, Jung KW (2008) Angew Chem Int Ed 47:9326–9329
70. Herrmann WA, Elison M, Fischer J, Köcher C, Artus GRJ (1995) Angew Chem Int Ed 34:2371–2374

Chapter 8
N-Heterocyclic Carbene Complexes in Dehalogenation Reactions

Elena Mas-Marzá, Michael J. Page, and Michael K. Whittlesey

Abstract Catalytic dehalogenation represents an underdeveloped transformation in M–NHC chemistry with a small number of reports detailing the reactivity of Co, Ru, Ni and Pd catalysts. *In situ* generated nickel and palladium NHC complexes catalyse the hydrodechlorination of aryl chlorides. Lower coordinate Ni complexes are proposed to operate in the hydrodefluorination of mono- and poly-fluorinated substrates. The single example of Ru–NHC catalysed hydrodefluorination of fully and partially fluorinated aromatic substrates is characterised by an unusual regioselectivity. The highly regioselective dehydrohalogenation of relatively unreactive alkyl halide substrates is achieved with a cobalt NHC catalyst.

8.1 Introduction

The dehalogenation of organic halides represents an important transformation in industry and environmental studies [1]. Due to the high environmental toxicity of a large number of chlorinated organics, as well as their oxidised combustion products, a wide variety of methods have been developed to neutralise these compounds *via* reductive methods. While a range of both transition metal and lanthanide complexes are known to catalyse the dehalogenation of organic halides, metal salts of Groups 8–10 in conjunction with phosphine ligands have received most attention [2]. In contrast, work with *N*-heterocyclic carbene ligands is in its infancy and this chapter aims to summarise the studies reported so far. The unsaturated and saturated *N*-aryl substituted imidazolium salts shown in Fig. 8.1 have been used to generate Ni, Pd and Co catalysts *in situ* for hydrodehalogenation and dehydrohalogenation. Isolated Ru complexes bearing IMes, SIMes, IPr and SIPr ligands have proved active for hydrodefluorination.

M. K. Whittlesey (✉)
Department of Chemistry, University of Bath, Claverton Down, Bath BA2 7AY, UK
e-mail: m.k.whittlesey@bath.ac.uk

C. S. J. Cazin (ed.), *N-Heterocyclic Carbenes in Transition Metal Catalysis and Organocatalysis*, Catalysis by Metal Complexes 32,
DOI 10.1007/978-90-481-2866-2_8, © Springer Science+Business Media B.V. 2011

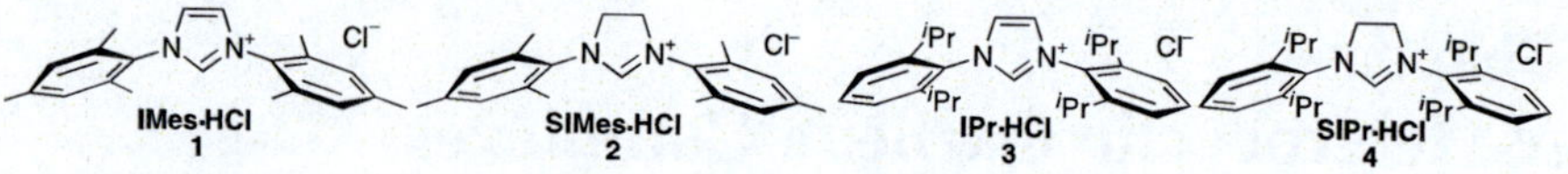

Fig. 8.1 Imidazolium salts commonly employed as NHC precursors

8.2 Nickel and Palladium

As most work has been done with complexes of the group 10 metals, these are discussed first. It is well known from C–C and C–N coupling reactions that Pd(0) and Ni(0) NHC complexes oxidatively add carbon-halogen bonds in a range of aryl halides. Hydrodehalogenation is often observed as a competing reaction pathway in catalytic C–C and C–N coupling [3,4], although only recent work by Fort with Ni [5,6] and Nolan with Pd [7–9] has specifically targeted the reaction.

Both Ni and Pd reactions are proposed to proceed *via* the general catalytic pathway shown in Scheme 8.1. Following the oxidative addition of a carbon-halogen bond to a coordinatively unsaturated zero valent metal centre (invariably formed *in situ*), displacement of the halide ligand by alkoxide and subsequent β-hydride elimination affords a Ni(II)/Pd(II) aryl-hydride complex, which reductively eliminates the dehalogenated product and regenerates $M(0)(NHC)_n$.

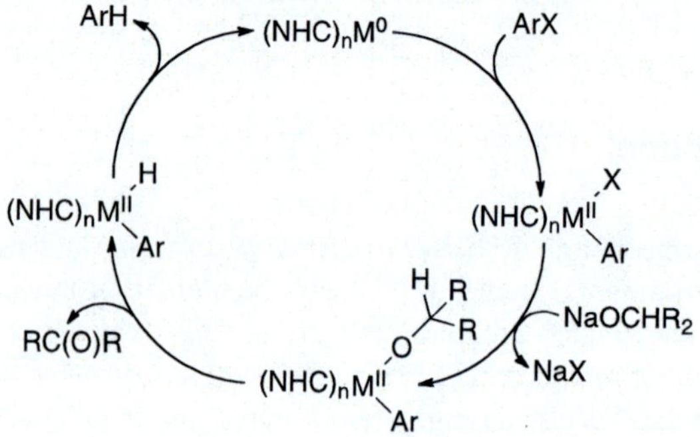

Scheme 8.1 General mechanistic cycle proposed for Ni and Pd catalysed hydrodehalogenation

The catalytic Ni system developed by Fort and co-workers was formed upon combining Ni(II) salts and two equivalents of imidazolium salt with a stoichiometric mixture of excess NaH and ^{i}PrOH [5]. The role of the *in situ* generated NaO^{i}Pr is to deprotonate the imidazolium salt to afford the free carbene, activate NaH to reduce Ni(II) to Ni(0), and to provide a source of hydrogen for the overall reduction of the aryl halide. Under optimised reaction conditions involving $Ni(acac)_2$ and **1**, the dehalogenation of chlorobenzene to benzene was achieved in 96% yield after 1 h at 65°C in THF. Poorer conversions were observed upon changing from *N*-aryl to either *N*-alkyl or pincer based imidazolium precursors, or upon changing the imidazolium counterion from chloride to tetrafluoroborate. The most effective alkoxide was found to be NaO^{i}Pr. The scope of this reaction towards a range of substituted aryl chlorides is summarised in Table 8.1.

Table 8.1 Comparison of Ni and Pd catalysed hydrodechlorination of aryl chlorides

Entry	Aryl halide	Yield/%[a] (time/h)			
		Ni(II)/**1**[b]	Pd(0)/**2**[c]	**5**[d]	**6**[e]
1	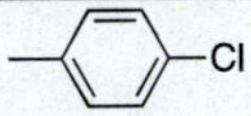	97 (1)	96 (1)	98 (1.75)	100 (1)
2	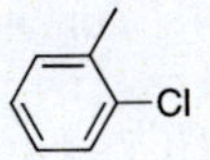	82 (1)	–	–	100 (1)
3	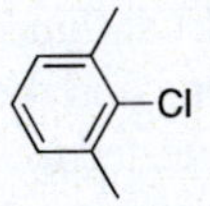	60 (1)	–	91 (1.75)	–
4	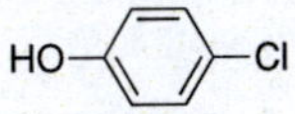	80 (1)	100 (1)	100 (1.75)	
5	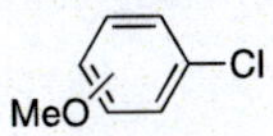	*o*- 81 (1) *m*- 100 (1) *p*- 87 (1)	*p*- 87 (1)	*o*- 95 (1.75) *m*- 99 (1.75) *p*- 98 (1.75)	*o, m, p*-100 (0.5)
6	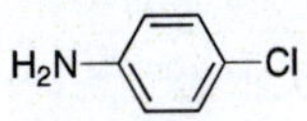	84 (1)	–	95 (1.75)	100 (1)
7	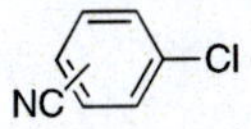	*o*- 86 (1) *m*- 87 (1)	*o*- 100 (16)[f] *p*- 21 (16)[f]	–	–
8	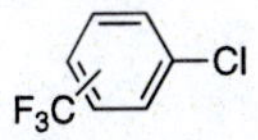	*o*- 96 (1) *m*- 96 (0.25) *p*- 84 (1)	–	*o*- 99 (1.75)	*o*- 100 (0.5) *p*- 90 (1.5)
9	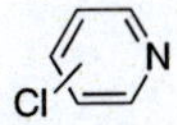	*o*- 43 (0.5) *m*- 99 (0.5) *p*- 97 (0.5)	*o*- 47 (1) *m*- 100 (1) *p*- 92 (1)	–	*m*- 100 (1)[g]
10[h]	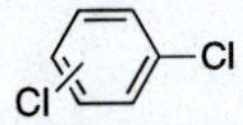	*o*- 91 (3) *m*- 95 (2.5)	*p*- 100 (1)	*p*- 97 (1.75)	–

[a]GC yields
[b]Aryl chloride (10 mmol), $Ni(acac)_2$ (3 mol%), IMes·HCl **1** (6 mol%), NaO^iPr (30 mmol), THF (12 mL), 65°C
[c]Aryl chloride (1.0 mmol), $Pd(dba)_2$ (2 mol%), SIMes·HCl **2** (2 mol%), KOMe (2 mmol), dioxane (3 mL), 100°C
[d]Aryl chloride (1 mmol), $[Pd(\eta^3\text{-}C_3H_5)Cl(IPr)]$ **5** (0.5 mol%), NaO^tBu (1.05 mmol), ^iPrOH (1.5 mL), 60°C
[e]Aryl chloride (1 mmol), $[PdCl\{\kappa^2\text{-}C,N\text{-}C_6H_4\text{-}2\text{-}(C_6H_4\text{-}2\text{-}NMe_2)\}(IPr)]$ **6** (1 mol%), KO^tBu (1.2 mmol), ^iPrOH (2 mL), 25°C
[f]Reactions performed at 25°C to minimise nucleophilic substitution side products
[g]Reaction at 60°C
[h]Twice the normal ratio of catalyst and base used

The data highlight the compatibility of the catalytic system for dehalogenation of both electron-rich (Table 8.1, entry 5) and electron-poor (Table 8.1, entry 8) substrates, sterically hindered *ortho*-substituted compounds, heteroaromatics (Table 8.1, entry 9), polyhalogenated species (Table 8.1, entry 10), and those bearing functional groups (Table 8.1, entry 4). Notable limitations of this system were encountered during the attempted reduction of the more activated substrates (e.g. 4-chlorobenzophenone and 2-chloropyridine), which gave products resulting from aromatic nucleophilic substitution by alkoxide.

The ease of dehalogenation of C_6H_5X by Ni(II)/ IMes·HCl **1**/NaO^{i}Pr decreased in the order I > Br > Cl >> F. Subsequent work showed that a 1:1 combination of Ni and NHC in the presence of $NaOCHEt_2$ resulted in enhanced reactivity towards aryl fluorides [6]. Again, the *N*-mesityl substituted ligand IMes·HCl **1** imparted the highest level of catalytic activity. Table 8.2 illustrates that hydrodefluorination is sensitive to both the nature of the substituents on the aromatic ring and the specific regioisomer. Thus, 2- or 4-fluorotoluene (Table 8.2, entry 2) proceeded to only 30% conversion after 15 h, whereas quantitative conversion of 2-fluoroanisole (Table 8.2, entry 3) and high conversion of 3-fluoropyridine (Table 8.2, entry 5) was achieved in only 2–3.5 h. The reactivity of 2-fluoropyridine was compromised by more efficient nucleophilic aromatic substitution.

Table 8.2 Ni–NHC catalysed hydrodefluorination of aryl fluorides[a]

Entry	Aryl fluoride	Yield/%[b] (time/h)
1	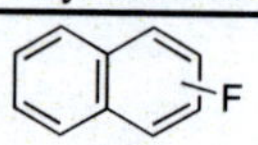	*o*- 92 (3) *m*- 100 (3)
2	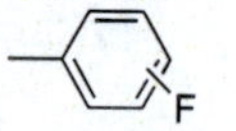	*m*- 30 (15) *p*- 30 (15)
3	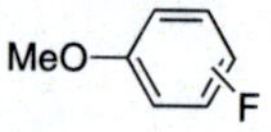	*o*- 100 (2.5) *m*- 75 (3.5) *p*- 46 (12)
4	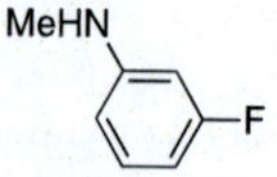	89 (23)[c]
5	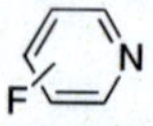	*o*- 0 (3.5) *m*- 100 (2)

[a]Aryl fluoride (10 mmol), $Ni(acac)_2$ (3 mol%), IMes·HCl **1** (3 mol%), $NaOCHEt_2$ (30 mmol), dioxane (3 mL), 100°C
[b]GC yields
[c]$NaOCHEt_2$ (40 mmol)

The exact nature of the catalytically active Ni species in these reactions is yet to be conclusively established. Hydrodechlorination proves optimal with a NHC:Ni ratio of 2:1 suggesting that 14-electron $Ni(NHC)_2$ is involved, whereas the 1:1 NHC:Ni ratio necessary for hydrodefluorination implies that it is the 12-electron mono-carbene adduct Ni(NHC) which is catalytically active [10]. Studies by Matsubara et al. revealed that treatment of $Ni(acac)_2$ with either one or two equivalents of IMes·HCl **1** or SIMes·HCl **2** in the presence of NaO^{t}Bu formed the mono-NHC complex $Ni(NHC)(acac)_2$ which, upon reduction with NaH in the presence or absence of carbene, formed $Ni(NHC)_2$ [11]. Density functional theory (DFT) calculations suggest that the strength of the Ni–NHC bond (*ca.* 50 kcal/mol) makes carbene loss unlikely, thereby casting doubt on formation of a 12-electron Ni(NHC) fragment under catalytic conditions. However, this species may be accessible, albeit quite transiently, as Sadighi's group have shown that reaction of $[Ni(\mu\text{-}Cl)(IPr)]_2$ with NaO*^t*Bu in benzene followed by treatment with excess bis(pinacol)diboron affords the dimer $[Ni(IPr)]_2$ which, on the basis of NMR evidence, appears to dissociate slowly in solution to the mono-carbene adduct $Ni(IPr)(\eta^6\text{-}C_6H_6)$ [12].

There has been a very brief report on the use of $[Ni(\eta^5\text{-}C_5H_5)Cl(NHC)]$/NaO*^i*Pr for the hydrodebromination of *p*-bromotoluene. As expected, the catalytic efficiency was carbene dependant (decreasing in the order IMes > SIMes > SIPr > IPr) although only a maximum yield of 40% was achieved. Interestingly, very similar catalytic efficiency was observed irrespective of whether the reaction was performed in refluxing THF at 65°C or in refluxing dioxane at 105°C [13].

Palladium NHC systems for the hydrodehalogenation of aryl chlorides and bromides and polyhalogenated aromatic substrates originate from about the same time as the first reports on Ni chemistry, and show many similarities. Initial efforts showed that the combination of $Pd(dba)_2$ (2 mol%), one equivalent of imidazolium chloride and KOMe produced an effective system for the reduction of 4-chlorotoluene, especially upon use of SIMes·HCl **2** (96% yield of toluene after 1 h at 100°C) [7]. Interestingly, higher ligand to metal ratios severely inhibited the catalysis with only 7% yield of toluene achieved in the same time in the presence of two equivalents of SIMes·HCl **2**. Variation of the metal source ($Pd(OAc)_2$, $Pd(C_6H_5CN)_2Cl_2$), alkoxide (NaOMe, KO*^t*Bu, NaOH/*sec*-BuOH) or imidazolium salt (IMes·HCl **1**, IPr·HCl **3**, IAd·HCl, ICy·HCl) all failed to give a more active catalyst.

Table 8.1 shows the reactivity of selected aryl chlorides towards the mixture of $Pd(dba)_2$/ SIMes·HCl **2**/KOMe. In general, good activity is observed for a range of both electron rich and neutral substrates, although a consequence of the high reaction temperatures used in this system is that nucleophilic aromatic substitution becomes a competitive reaction for substrates bearing strongly electron withdrawing groups. This can be minimised by lowering the reaction temperature (Table 8.1, entry 7) and reducing the amount of KOMe, although longer reaction times are required. Aryl bromides were found to be reduced much more effectively than aryl chlorides while aryl fluorides were unreactive; this allowed for the selective reduction of the bromo and chloro groups in 2-bromochlorobenzene and 2-fluorochlorobenzene, respectively. Unlike the Ni(0)/NHC systems described above, the reduction of sterically hindered substrates was inhibited.

Latter studies by the Nolan group employed the air stable allyl precursor $[Pd(\eta^3\text{-}C_3H_5)Cl(IPr)]$ (**5**) (Fig. 8.2) in *^i*PrOH in the presence of base in an effort to provide a simple and well-defined catalytic system for hydrodechlorination [8]. Using one equivalent of NaO*^t*Bu and only 0.5 mol% $[Pd(\eta^3\text{-}C_3H_5)Cl(IPr)]$ **5** in *^i*PrOH at 60°C, a similar range of substrates could be reduced in excellent yield in < 2 h (Table 8.1). Even the sterically crowded 2-chloroxylene could be reduced in 91% yield under these conditions (entry 3). The $[Pd(\eta^3\text{-}C_3H_5)Cl(IPr)]$ catalyst was also investigated under microwave assisted reaction conditions. Impressively, a catalyst loading of 0.025 mol% with one equivalent of NaO*^t*Bu in *^i*PrOH allowed the quantitative hydrodechlorination of several aryl chlorides in only 120 s at 120°C. The presence of the NHC ligand proved paramount to high activity as well as stability, as efforts to employ just [Pd(μ-Cl)(η^3-C_3H_5)]$_2$ at 1 mol% loading as the catalyst gave only a 45% conversion of 4-chlorotoluene to toluene and concomitant formation of a large amount of metallic palladium.

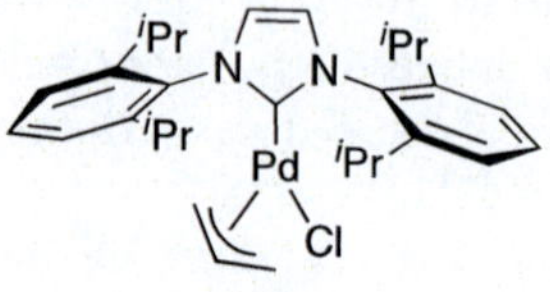

5

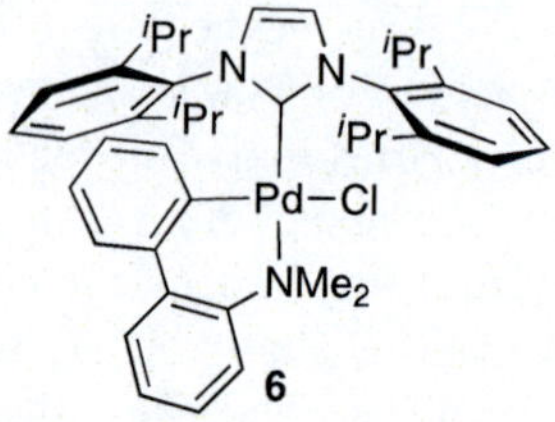

6

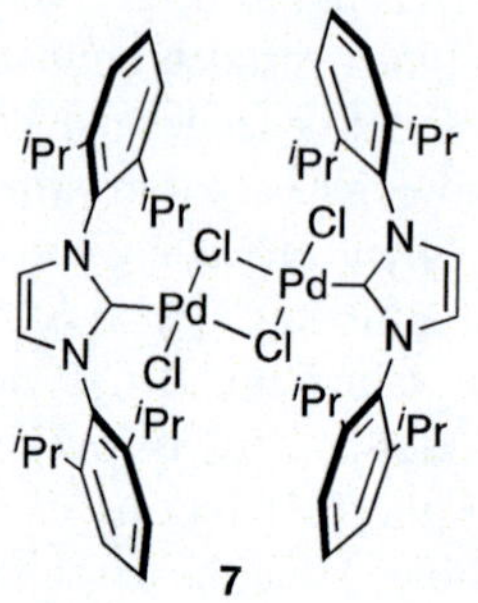

7

Fig. 8.2 Structures of Pd NHC dehalogenation catalysts

Very recently, efforts have turned towards the use of [Pd(η^3-C_3H_5)Cl(IPr)] **5** for the hydrodechlorination of polychlorinated phenyl substrates [14]. In these cases, [Pd(η^3-C_3H_5)Cl(IPr)] **5** proved to be less active than the non-allyl containing dimer [Pd(μ-Cl)Cl(IPr)]$_2$ **7** for the complete hydrodechlorination of 1,2,4,5-tetrachlorobenzene to benzene, with conversions of 40% and 95% respectively found at 80°C and 0.02 mol% catalyst loading in the presence of KO^{t}Bu/^{i}PrOH.

The potential of the dimeric system **7** is perhaps best exemplified by its ability to bring about the total hydrodechlorination of decachlorobiphenyl, a member of the notorious family of polychlorinated biphenyls or PCBs (Persistent Organic Pollutants, POPs), in the presence of the inexpensive base NaOH (Table 8.3).

Table 8.3 Pd–NHC catalysed hydrodehalogenation of polychlorinated phenyls and biphenyls[a]

$ArCl_n$	**7** mol%	Conv.[b] (yield)/%	TON (per Cl)	$ArCl_n$	**7** (mol%)	Conv.[b] (yield)/%	TON (per Cl)
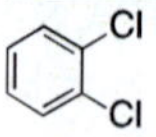	0.004	100[c]	25 000	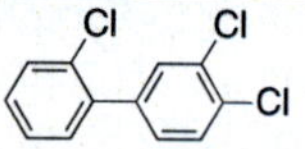	0.02	>99 (91)	7 500
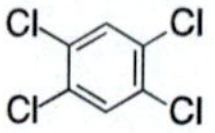	0.02	100	10 000	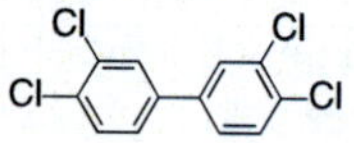	0.02	>99 (95)	10 000
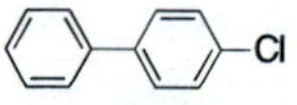	0.02	>99 (92)	2 500	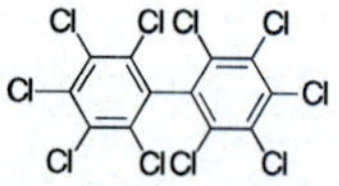	0.5	>99 (84)	1 000

[a]$ArCl_n$ (0.04–1 mmol), ^{i}PrOH (2–3 mL), NaOH (10% excess with respect to Cl), 24 h, 80°C
[b]Conversion to fully dechlorinated product
[c]Isolated yield, minimum average of two reactions; Base is KO^{t}Bu; 60°C

Nolan has reported that another well-defined system, the palladacycle **6** (Fig. 8.2), is an active hydrodehalogenation catalyst in the presence of 1.2 equivalents of KO^{t}Bu (Table 8.1) [6]. A particular feature of this system is that reactions could be run at room temperature in technical grade ^{i}PrOH. Interestingly, the reduction of electron rich substrates (Table 8.1, entry 5) was faster than that of electron-poor derivatives (Table 8.1, entries 8 and 9). As oxidative addition would be expected to be faster for aryl halides bearing electron-withdrawing substituents, this suggests that oxidative addition is not the rate-determining step in the catalytic pathway. Indeed, the significant difference in reaction rate between 2- and 4-trifluoromethylchlorobenzene (Table 8.1, entry 8) implies that it is reductive elimination of the substrate that is rate limiting.

8.3 Cobalt

The only example of dehalogenation by a group 9 metal NHC complex has been reported by Yorimitsu and Oshima and involves the Co catalysed dehydrohalogenation of alkyl halides [15]. The combination of $CoCl_2$ and IMes·HCl **1** or IPr·HCl **3** with $Me_2PhSiCH_2MgCl$ generates a catalytically active species capable of performing the highly regioselective dehydrobromination of 2-bromodecane to 1-dodecene at room temperature (Scheme 8.2). A more catalytically efficient intermediate is formed using IMes·HCl **1**, which yields 1-dodecene in 84% yield in 45 min compared to only 36% with IPr·HCl **3** (greater than half of the starting material is left unreacted in this case). Despite the limitations of the bulkier carbene, both ligands

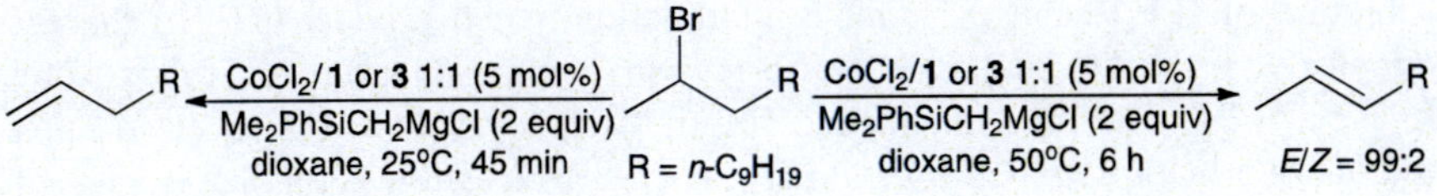

Scheme 8.2 Cobalt catalysed dehalogenation

perform significantly better than either alkyl or aryl substituted monodentate phosphines, or the chelating aryl phosphine dppe.

Dehydrohalogenation of both 2-iodo and 2-chlorododecane proved possible with the $CoCl_2$/ IMes·HCl **1**/$Me_2PhSiCH_2MgCl_2$ mixture, although while the former goes in 87% conversion after only 15 min, the latter is <15% complete after close to 4 h. The effectiveness of the dehydrobromination reaction upon variation of the R group on the alkyl halide was also investigated with little impact on the catalytic activity noted upon incorporation of either functional groups ($CH_2OC(O)Ph$, $(CH_2)_2OC_6H_4$-*p*-Cl, $(CH_2)_2N(CH_2Ph)Ts$) or sterically bulky substituents ($CH(n\text{-}C_5H_{11})_2$, CH_2-1-naphthyl).

As shown in Scheme 8.2, increasing the temperature to 50°C resulted in the selective formation of (*E*)-2-alkenes. While the formation of 1-dodecene was observed within minutes, isomerisation to 2-dodecene is slow, accounting for the timescale of hours required for reaction.

8.4 Ruthenium

N-aryl substituted NHCs have been used to afford ruthenium complexes that catalyse the hydrodefluorination of a range of fully and partially fluorinated aromatic substrates [16]. The catalytic activity revolves around the ability to interconvert the 16-electron hydride fluoride species [Ru(H)(F)(CO)(NHC)(PPh_3)] (**8a–d**) and 18-electron dihydride complexes [$Ru(H)_2$(CO)(NHC)$(PPh_3)_2$] (**9a–d**), where NHC = IMes, SIMes, IPr or SIPr. As shown in Scheme 8.3, treatment of the former with an alkylsilane in the presence of PPh_3 gives the corresponding dihydride complexes (along with R_3SiF, which provides the driving force for this process), which can subsequently undergo C–F bond activation. Thus, for example, hexafluorobenzene is converted to pentafluorobenzene with the reformation of **8a–d**.

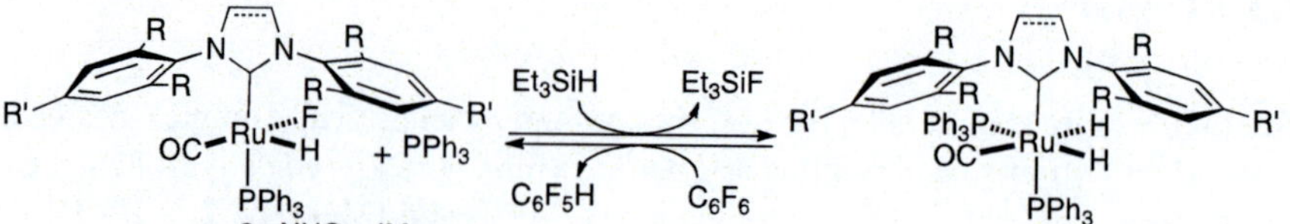

8a NHC = IMes
8b NHC = SIMes
8c NHC = IPr
8d NHC = SIPr

Scheme 8.3 Ruthenium NHC catalysed hydrodefluorination

Under catalytic conditions (10 mol% **9a–d**, 70°C, 20 h), C_6F_6 was converted to a mixture of C_6F_5H and the double hydrodefluorinated product $C_6F_4H_2$; the IPr complex **9c** proved to be the most efficient with over seven turnovers, followed by **9b** > **9a** > **9d**. The most interesting feature of this reaction was revealed upon hydrodefluorination of C_6F_5H, which gave 98% selectivity for 1,2-isomer of $C_6F_4H_2$. This regioselectivity contrasts with a few other non-NHC, non-Ru based hydrodefluorination catalysts, which afford exclusively the 1,4-isomer [17,18].

Attempts to perform hydrodefluorination on other polyfluoroarenes using **9c** met with mixed success (Table 8.4). Pentafluoropyridine proved to be highly active with a turnover number of 13.6, although the distribution of products revealed up to three hydrodefluorination reactions (Table 8.4, entry 4). Substrates with a lower fluorine content, such as $C_6F_5CH_3$ and 1,2- or 1,4-$C_6F_4H_2$ were unreactive.

A kinetic study of the hydrodefluorination of C_6F_5H in the presence of Et_3SiH indicated a first-order dependence on both [fluoroarene] and [ruthenium precursor] and a zero-order dependence on the concentration of alkylsilane, implying that the rate-limiting step in the catalytic cycle involves activation of the fluoroarene. The regioselectivity for hydrodefluorination of partially fluorinated substrates such as C_6F_5H has been accounted for by an initial C–H bond activation as shown in the proposed mechanism outlined in Scheme 8.4. C–H coordination of fluoroarene to the 16-electron dihydride species [Ru(H)$_2$(CO)(NHC)(PPh$_3$)] and subsequent metathesis provides a route to [Ru(C$_6$F$_5$)H(CO)(NHC)(PPh$_3$)], which upon ring-to-metal β-F transfer affords a tetrafluorobenzyne hydride fluoride species. Metal-to-ring hydrogen transfer then generates a σ Ru-*ortho*-C$_6$F$_4$H complex, which upon reaction with H$_2$ then eliminates the observed 1,2-hydrodefluorination product. Further metathesis with alkylsilane completes the cycle.

Table 8.4 Catalytic hydrodefluorination of polyfluoroarenes by **9 c**[a]

Entry	Aryl fluoride	Products (yield/%)[b]
1	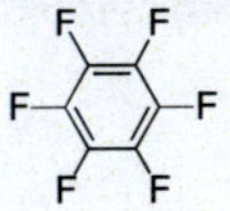	C_6F_5H (32.2) 1,2-$C_6F_4H_2$ (19.9) 1,4-$C_6F_4H_2$ (0.9)
2	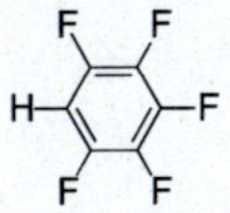	1,2-$C_6F_4H_2$ (67.7) 1,4-$C_6F_4H_2$ (1.6)
3	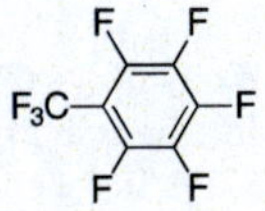	2,3,5,6-$C_6F_4HCF_3$ (14.1) 2,3,4,5-$C_6F_4HCF_3$ (3.1) 2,3,6-$C_6F_3H_2CF_3$ (3.3)
4	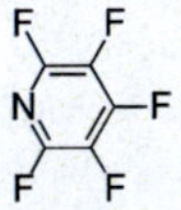	2,3,4,5-C_5F_4HN (23.8) 2,3,5,6-C_5F_4HN (15.9) 3,4,5-$C_5F_3H_2N$ (35.1) 2,3,5-$C_5F_3H_2N$ (8.2) 3,4-$C_5F_2H_3N$ (3.2)

[a]Fluoroarene (0.1 M), **9c** (10 mol%), Et_3SiH (0.2 M), THF, 70°C, 19.45 h
[b]Yields determined by NMR spectroscopy

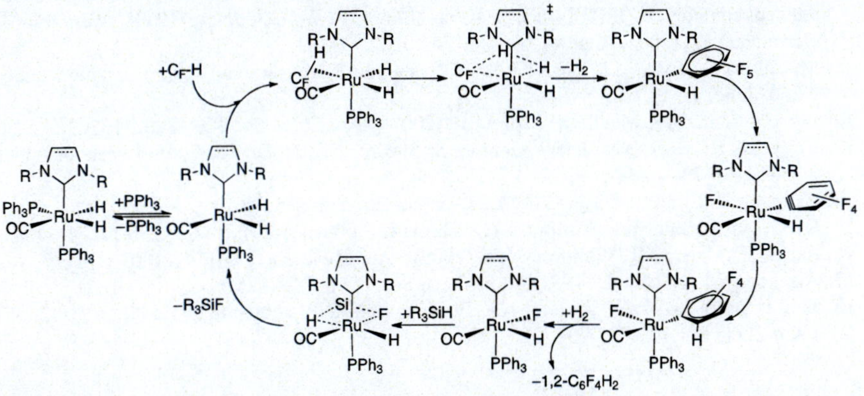

Scheme 8.4 Proposed mechanistic cycle to account for the conversion of C_6F_5H to 1,2-$C_6F_4H_2$

8.5 Conclusions

Despite a paucity of studies, this overview suggests that use of M–NHC complexes have considerable promise for catalytic dehalogenation reactions. *N*-aryl substituted carbenes have proven to be the most effective types of ligands and have generated catalysts that react under mild conditions, with high efficiency and at low catalyst loading. The ever increasing push for 'Green Chemistry' is likely to ensure that new dehalogenation catalysts will be developed and that further work in this field will be forthcoming.

Acknowledgements The authors would like to thank the Spanish Ministerio de Ciencia e Innovación (EMM) and the EPSRC (MJP) for financial support.

References

1. Hudlicky M (1991) In: Trost BM, Fleming I (eds) Comprehensive organic synthesis, vol 8. Pergamon, Oxford, pp 895–922
2. Alonso F, Beletskaya IP, Yus M (2002) Chem Rev 102:4009–4091
3. Gradel B, Brenner E, Schneider R, Fort Y (2001) Tetrahedron Lett 42:5689–5692
4. Maerten E, Sauthier M, Mortreux A, Castanet Y (2006) Tetrahedron 63:682–689
5. Desmarets C, Kuhl S, Schneider R, Fort Y (2002) Organometallics 21:1554–1559
6. Kuhl S, Schneider R, Fort Y (2003) Adv Synth Cat 345:341–344
7. Viciu MS, Grasa GA, Nolan SP (2001) Organometallics 20:3607–3612
8. Navarro O, Kaur H, Mahjoor P, Nolan SP (2004) J Org Chem 69:3173–3180
9. Navarro O, Marion N, Oonishii Y, Kelly RA III, Nolan SP (2006) J Org Chem 71:685–692
10. It is worth noting that Radius and co-workers have shown that $\{[Ni(I^iPr_2)_2]_2(\mu\text{-COD})\}$ and $[Ni(I^iPr_2)_2(\eta^2\text{-}C_2H_4)]$ (I^iPr_2 = *N, N'*-(di-*iso*-propyl) imidazol-2-ylidene), both effective sources of $Ni(NHC)_2$, will bring about the stoichiometric activation of aromatic C–F bonds (Schaub T, Fischer P, Steffen A, Braun T, Radius U, Mix A (2008) J Am Chem Soc 130:9304–9317). In contrast, Herrmann has reported that oxidative addition of aryl fluorides to the N-aryl substituted complex $Ni(IPr)_2$ is not possible. Böhm VPW, Gstöttmayr CWK, Weskamp T, Herrmann, WA (2001) Angew Chem Int Ed 40:3387–3389
11. Matsubara K, Miyazaki S, Koga Y, Nibu Y, Hashimura T, Matsumoto T (2008) Organometallics 27:6020–6024
12. Lee CH, Laitar DS, Mueller P, Sadighi JP (2007) J Am Chem Soc 129:13802–13803
13. Kelly RA III, Scott NM, Díez-González S, Stevens ED, Nolan SP (2005) Organometallics 24:3442–3447
14. Akzinnay S, Bisaro F, Cazin CSJ (2009) Chem Commun 5752–5753
15. Kobayashi T, Ohmiya H, Yorimitsu H, Oshima K (2008) J Am Chem Soc 130:11276–11277
16. Reade SP, Mahon MF, Whittlesey MK (2009) J Am Chem Soc 131:1847–1861
17. Aizenberg M, Milstein D (1995) J Am Chem Soc 117:8674–8675
18. Vela J, Smith JM, Yu Y, Ketterer NA, Flaschenriem CJ, Lachicotte RJ, Holland PL (2005) J Am Chem Soc 127:7857–7870

Chapter 9
N-Heterocyclic Carbene Complexes in Reactions Involving Carbon Monoxide

Matthew Jeletic and Adam Veige

Abstract This chapter focuses on carbon monoxide as a reagent in M–NHC catalysed reactions. The most important and popular of these reactions is hydroformylation. Unfortunately, uncertainty exists as to the identity of the active catalyst and whether the NHC is bound to the catalyst in a number of the reported reactions. Mixed bidentate NHC complexes and cobalt-based complexes provide for better stability of the catalyst. Catalysts used for hydroaminomethylation and carbonylation reactions show promise to rival traditional phosphine-based catalysts. Reports of decarbonylation are scarce, but the potential strength of the M–NHC bond is conducive to the harsh conditions required. This report will highlight, where appropriate, the potential benefits of exchanging traditional phosphorous ligands with *N*-heterocyclic carbenes as well as cases where the role of the NHC might need re-evaluation. A review by the author on this topic has recently appeared [1].

9.1 Hydroformylation

Hydroformylation represents the most industrially important homogeneous catalysed reaction by volume [2, 3]. The petrochemical, agrochemical and pharmaceutical industries are particularly interested in this transformation. The reaction uses *syngas* ($CO{:}H_2$ mix) and a catalyst, commonly rhodium or platinum, to transform an olefin into an aldehyde (Scheme 9.1) [4].

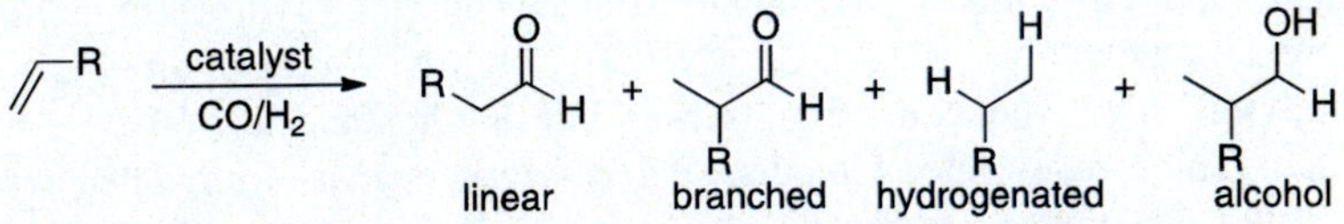

Scheme 9.1 General hydroformylation reaction showing all possible product types

A. Veige (✉)
Department of Chemistry, Center for Catalysis, University of Florida, P.O. Box 117200, Gainesville, FL, USA
e-mail: veige@chem.ufl.edu

C. S. J. Cazin (ed.), *N-Heterocyclic Carbenes in Transition Metal Catalysis and Organocatalysis*, Catalysis by Metal Complexes 32,
DOI 10.1007/978-90-481-2866-2_9, © Springer Science+Business Media B.V. 2011

Attaining high selectivity of the desired product remains the major obstacle in hydroformylation. Therefore, controlling the regioselectivity (branched:linear ratio, b:l), chemoselectivity (aldehyde, alcohol, hydrogenated product), and possibly enantioselectivity (in the branched isomer) becomes a delicate balancing act of the reaction condition variables. Early hydroformylation catalysts involved cobalt complexes modified with CO ligands, but were not very selective. Introduction of phosphorous ligands, specifically bulky trialkylphosphines, greatly improved the selectivity as well as operating conditions. However, excess PR_3 is necessary to compensate for deleterious PR_3 substitution (with CO), PR_3 oxidation, and P–C bond cleavage [5–8].

The high concentration of CO under typical hydroformylation conditions causes ligand competition between CO and PR_3. Thus, avoiding the formation of the highly active species **A** (Scheme 9.2) often requires excess PR_3. Pruett and Smith demonstrated an increase from 31% to 89% in the selectivity of linear aldehyde by increasing the equivalents of $P(OPh)_3$ from 0 to 60 during hydroformylation of 1-pentene [9]. The goal in hydroformylation becomes manipulating the equilibrium such that the predominant species is **D** by varying CO concentration, ligand concentration and PR_3 identity (Scheme 9.2).

$$\underset{\textbf{A}}{HRh(CO)_4} \underset{CO}{\overset{PR_3}{\rightleftharpoons}} \underset{\textbf{B}}{HRh(CO)_3PR_3} \underset{CO}{\overset{PR_3}{\rightleftharpoons}} \underset{\textbf{C}}{HRh(CO)_2(PR_3)_2} \underset{CO}{\overset{PR_3}{\rightleftharpoons}} \underset{\textbf{D}}{HRh(CO)(PR_3)_3} \underset{CO}{\overset{PR_3}{\rightleftharpoons}} \underset{\textbf{E}}{HRh(PR_3)_4}$$

Scheme 9.2 PR_3/CO competing intermediates in Rh-modified hydroformylation

Replacement of the phosphorous ligand with an NHC is a logical next step toward stabilising the **D**-type intermediate due to the σ-donor strength of the NHC. Thus, choosing the correct NHC should allow for high selectivities without excess ligand.

9.1.1 *Hydroformylation Using Monodentate NHC Complexes*

Herrmann and co-workers first demonstrated catalytic hydroformylation of 1-hexane using the catalyst [RhCl(COD)(NHC)] (NHC=*N,N′*-dimethylimidazolidin-2-ylidine) [10]. The product distribution was 33.8% *n*-heptanal and 66.2% 2-methylhexanal. Three years later Crudden and co-workers hydroformylated styrene (60°C, 1000 psi syngas, benzene) using two pre-catalysts (**1** and **2**) synthesised from $[RhCl(PPh_3)_3]$ (Scheme 9.3) by adding the NHC IMes [11].

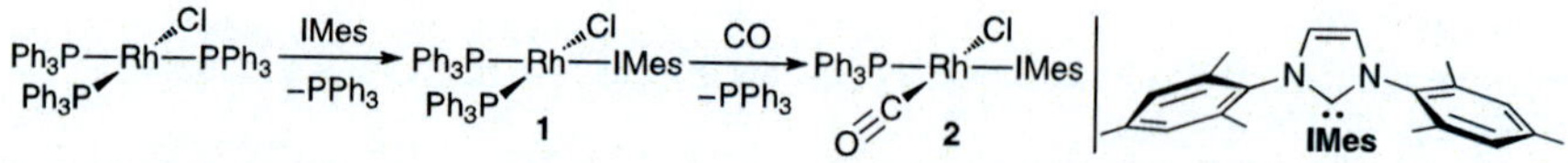

Scheme 9.3 Formation of Crudden's pre-catalysts for styrene hydroformylation

Substitution of PR_3 for CO during the conversion of **1** into **2** epitomises the need for excess PR_3 during hydroformylation conditions. Presumably **1** converts into **2** under hydroformylation conditions (CO atmosphere), therefore **1** and **2** should perform uniformly in hydroformylation reactions. However, **1** is twice as reactive as **2** ($TOF_1 = 7$, $TOF_2 = 4$). Surprisingly, the activity of **2** equalled **1** with the addition of at least one equivalent of PPh_3. This raises the question whether the NHC is actually bound during catalysis. McGuinness and Cavell and co-workers demonstrated a facile catalyst decomposition *via* NHC elimination to form the acyl imidazolium salt when using CO as a reagent in 2000 [12]. Displacement of the NHC ligand during hydroformylation could lead to the formation of $[RhCl(CO)(PPh_3)_2]$, but when tested under the same conditions, the b:l ratio was different (88:12 compared to 96:4 for **2**). The role the additional phosphine plays remains unclear, but regardless of the conditions, the pre-catalyst, or substrate used, the b:l ratio is consistently 95:5 (±3). This particular b:l ratio, concomitant with those of other pre-catalysts, suggests $HRh(CO)_4$ (**A**) as another possible active catalyst [2]. A later report by Crudden and co-workers describing the clean displacement of IMes from complex **1** further complicates the question of the active catalyst identity [13]. This lends credence to the theory that NHC ligands are not substitutionally inert. Upon treating complexes **1** or **2** with one equivalent of PPh_3, $[RhCl(PPh_3)_3]$ cleanly forms in dichloroethane. In hydroformylation studies performed by Crudden, the most efficient reactions always have at least one equivalent of PPh_3.

Crudden and co-workers explored hydroformylation with the saturated NHC complexes $[RhCl(CO)(PR_3)(SIMes)]$ (**3-R**; R = Ph, OPh, *p*-OMePh, *p*-FPh, DBP, furyl, Cy, *o*-tolyl) [14]. Complexes **1** and **3-Ph** allow for a direct comparison of two different NHCs while retaining the same phosphine. The catalyst **3-Ph** yielded a TOF (14) *ca* twice that of **1** (8) and the b:l ratios of the resulting styrenic aldehydes were statistically identical, 95:5 (±3), but again one equivalent of PPh_3 was added. This rate difference suggests the active species retains the NHC ligand, but this conclusion presumes linear kinetics operates [15]. A greater spread of steric and electronically adjusted NHCs, as well as a full kinetic profile during catalysis, will provide conclusive evidence. Crudden presents clear evidence for the importance of PR_3 during the rate-determining step. There was a one order of magnitude difference in TOF between the electron-withdrawing substituted **3-*p*-FPh** catalyst (31) and the bulky **3-*o*-tolyl** derivative (3). Switching the *para*-substituent to the electron-donating -OMe (**3-*p*-OMePh**) resulted in a less reactive species ($TOF_{p\text{-FPh}} = 31$ compared to $TOF_{p\text{-OMePh}} = 13$). The b:l ratio dropped to 65:35 when using **3-OPh**, but most of the b:l ratios were statistically identical when an equivalent of PPh_3 is present during hydroformylation.

Crudden and co-workers drastically increased the TOF (212–515 compared to 3–31) for hydroformylation of styrene and its derivatives using the carboxylato complex *cis*-$[Rh(OAc)(IPr)(CO)_2]$ **4** [16]. Complex **4** was converted to *trans*-$[Rh(OAc)(CO)(IPr)_2]$ **5** by adding an additional equivalent of IPr. Similar to previous reactions, they added an equivalent of phosphine to the reaction (0.5 mol% **4**, benzene, 80°C, 1,000 psi). Likewise the b:l ratios fell to 95:5 (±3) with the exception of the *p*-methoxy derivative (90:10). Hydroformylation of linear alkenes was also tested. Complex **4** was highly active (TOF = 345–725) but not very selective (l:b = 2.3–3). When measured at 1 and 19 h intervals, selectivity did not change, indicating isomerisation did not occur. The TOFs for linear alkenes using **5** were low (1–3 h^{-1}) and were attributed to small amounts of **5** decomposing into **4**, implying **5** is inactive.

Catalyst **4** is a square-planar, catalytically-active rhodium carboxylato complex bearing an NHC ligand. The analogous acetylacetonate complexes [Rh(acac)(CO)(IPr)] **6** and [Rh(acac)(CO)(IMes)] **7** do not exhibit hydroformylation activity towards 1-hexene until the addition of PPh_3 or $P(O\text{-}2,4\text{-}{}^tBu_2\text{–}Ph)_3$ [17]. Otto and co-workers demonstrated *via* high pressure ^{31}P NMR techniques the cleavage of the Rh–NHC bond in **7** under hydroformylation conditions. Otto concluded the resultant activity is identical to that using $[Rh(acac)(CO)_2]$ with PPh_3, and no benefit was realised by the preinstalled NHC ligand. This again raises concerns as to whether these catalysts retain the NHC ligand. Crudden admits they cannot rule out NHC-free catalytically active species, but also argues their yields and selectivities do not match those for $[Rh(acac)(CO)(PPh_3)]/PPh_3$ mixtures reported by Trzeciak and co-workers [18].

Following the work of Crudden, Trzeciak and co-workers prepared several new complexes of the form [RhX(COD)(NHC)] (X = Cl, Br, I, SCN and NHC = 1-butyl-3-methylimidazolidin-2-ylidene or 1,3-diethoxymethylimidazolidin-2-ylidene) for the hydroformylation of 1-hexene with $P(OPh)_3$ as an additive [19]. The catalysts were formed *in situ* and the authors, like Crudden, found that activity was only turned on in the presence of $P(OPh)_3$. They found four equivalents of $P(OPh)_3$ yields the best l:b ratio (7.9). $[RhH(CO)\{P(OPh)_3\}]$ provided a substantially lower ratio (1.3) than the catalyst, indicating the NHC is both intact and responsible for the increase in l:b ratio. The authors also used the NHC as an additive for $[RhH(CO)\{P(OPh_3)\}_3]$ (reverse experiment), and observed no significant increase in l:b ratio (1.4). They also found that different PR_3 groups affected l:b ratios and conversions to aldehyde. To further demonstrate that the NHC was attached during catalysis, the authors performed detailed ^{31}P NMR mechanistic studies and determined the main active catalyst is $[RhH(CO)(NHC)\{P(OPh_3)\}_2]$. They also postulated the high linear selectivity is due to the square pyramidal geometry of the complex containing the NHC in contrast to the typical trigonal bipyramidal geometry of hydroformylation catalysts.

Buchmeiser and Nuyken [20], Weberskirch [21] and co-workers, examined the hydroformylation of 1-octene with a series of compounds of the formula [RhX(COD)(NHC)] (Fig. 9.1). The choice of halogen does not affect the reactivity, implying an active catalyst of the general formula $[RhH(CO)_3(NHC)]$. Also, significant differences were observed in the initial TOF between catalysts (**8**,**9**) and (**10**,**11**) bearing *N*-^{i}Pr and *N*-Mes groups, respectively. Complex **11** exhibited a TOF of 1 480 h^{-1}, whereas under identical conditions, **9** only turned over at a frequency of

520 h^{-1} (T = 100°C, P = 50 bar). The decreased electron donating capacity of the NHC in complexes **14** and **15** provided even higher TOFs (2 410 and 3 540, respectively). Complex **13** decomposed during the reaction to an unknown brown solid, which the authors attribute to the destabilisation of the Rh–NHC bond due to bulky adamantyl groups. None of the catalysts were particularly selective for the formation of aldehyde and the l:b ratios were low (*ca.* 0.5) because of competing isomerisation during hydroformylation.

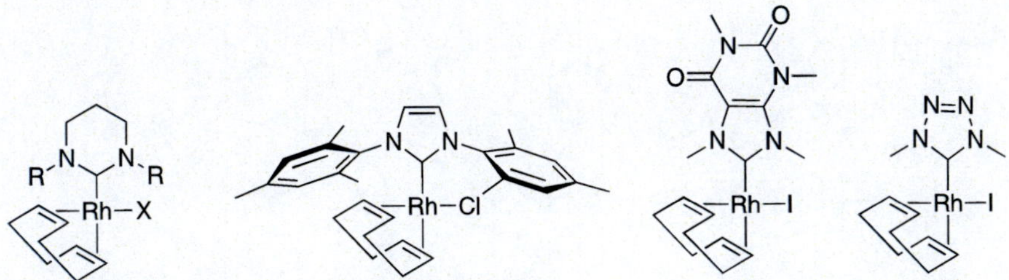

8 X = Cl, R = ^{i}Pr
9 X = Br, R = ^{i}Pr
10 X = Cl, R = Mes
11 X = Br, R = Mes

12 X = Cl, R = Mes
13 X = Cl, R = Ad

14

15

Fig. 9.1 Series of monodentate complexes of the form [RhX(COD)(NHC)]

9.1.2 Hydroformylation Using bis-κ²–NHC Complexes

Keeping the NHC ligand bound to the metal centre during catalysis can be problematic, and therefore a multidentate NHC ligand may pose a solution. Peris [22] and Veige [23, 24] prepared multimetallic and *bis*-κ²–NHC complexes, and tested them for hydroformylation activity (Fig. 9.2). Hydroformylation of styrene with complexes **16–18** led to high conversions and similar b:l ratios (0.1–1 mol%, 50°C, 80–100 bar, solvent = toluene or $CHCl_3$). After the reaction, the authors noted the presence of benzimidazolium salt in solution, suggesting reductive elimination of the NHC had occurred. Several key pieces of evidence signaled decomposition to a common species during hydroformylation: sigmoidal kinetic profile, positive mercury test, and the observation of rhodium starting material, $[Rh(nbd)_2]BF_4$, provided the same b:l ratio (95 ± 3) as the NHC complexes **16–18**. Furthermore, enantiopure versions of **16–18** yielded no enantioselectivity. The authors concluded the true identity of the catalyst is most likely $HRh(CO)_4$, and thus showed even di-NHC ligands are susceptible to reductive elimination from metal centres during hydroformylation. Complex **19** yielded the same b:l ratio (95 ± 3) in the styrenic derivatives used, with TOFs of 11–15 (0.25 mol% **19**, 40°C, 50 bar). The authors attempted to determine the nature of the catalytic species *via* high pressure ^{13}C NMR. While **19** keeps the NHC bound after 3 days at 40°C and 30 atm CO/H_2 in $CDCl_3$, this on its own is not sufficient evidence to claim with certainty the NHC is bound during catalysis. No substrate was present [23] and the conditions used during hydroformylation were harsher. Complex **19** provided low TOFs for linear alkenes and **20** showed zero activity toward hydroformylation of styrene even under harsh conditions (80°C, 80 atm).

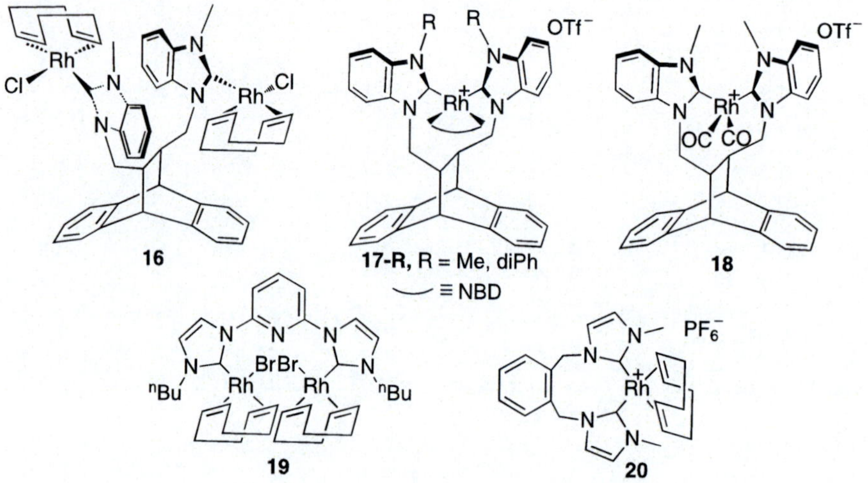

Fig. 9.2 Complexes containing di-NHC ligands used in hydroformylation

9.1.3 *Hydroformylation Using Mixed Multidentate NHC Complexes*

Another approach to stabilising active Rh(I) hydroformylation catalysts involves using hemilabile ligands. Two such complexes have been synthesised (Fig. 9.3) and tested for hydroformylation activity. Green and co-workers prepared the stable non-cyclic imino NHC complex **21** [25]. Hydroformylation of 1-octene with **21** yielded >99% conversion to aldehyde, but produced non-spectacular l:b ratios (0.8–1.9) at pressures greater than 10 bar or temperatures greater than 60°C. Under optimal conditions (60°C, 10 bar) the l:b ratio was 2.5, but only 30% of the olefin was converted to aldehyde. An unremarkable l:b ratio (2.5) was also obtained with the mixed Cp–NHC complex **22** [26]. In addition, selection of the appropriate aldehyde was problematic because significant alkene isomerisation (followed by hydroformylation) occurred.

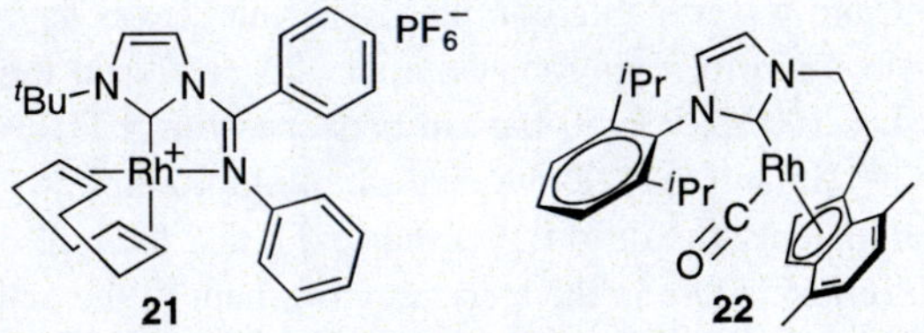

Fig. 9.3 Rhodium complexes with mixed bidentate NHC ligands

9.1.4 Hydroformylation with Thione/NHC Complexes

Raubenheimer and co-workers pursued the hydroformylation of 1-hexene using Rh(I) precursors with particular focus on the potential activity when thione ligands (C=S, Fig. 9.4) were employed [27]. The authors prepared numerous complexes by varying the combinations of ligands (Fig. 9.4). Several of the combinations included either monodentate NHC–R (R = Me, ^{i}Pr) or bidentate NHC–NHC ligands. The composition of Rh catalysts containing at least one NHC ligand combination follows: [RhCl(COD)(NHC–^{i}Pr)], [RhCl(COD)(NHC–Me)], *cis*-[RhCl(CO)$_2$(NHC–^{i}Pr)], *trans*-[RhCl(CO)(NHC–Me)$_2$], [Rh(COD)(NHC–NHC)]PF_6, [Rh(CO)$_2$(NHC–NHC)]PF_6, [RhCl(CO)(C=S)(NHC–^{i}Pr)], [Rh(COD)(C=S)(NHC–^{i}Pr)]BF_4, *cis*-[Rh(CO)$_2$(C=S)(NHC–^{i}Pr)]BF_4, [RhCl(CO)(NHC–^{i}Pr){P(thiazole)$_3$}], [RhCl(CO)(NHC–^{i}Pr)(PPh$_3$)], and [Rh(CO)$_2$(Hbbtm)]BF_4 (Hbbtm = *bis*-{benzothiazol-2-yl}methane). The authors presented hydroformylation results of 1-hexene obtained under a variety of conditions, but there was no apparent benefit to including the NHC ligand in the pre-catalyst. The percent conversion, yield of aldehyde, and b:l ratio were all similar to those obtained by simply using [RhCl(PPh$_3$)$_3$] (b:l = 1.3).

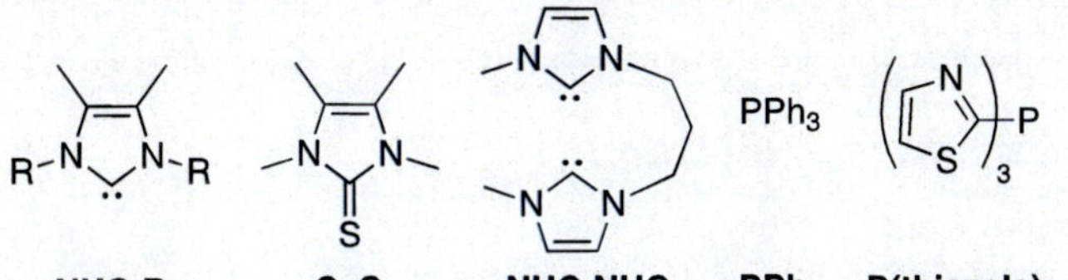

Fig. 9.4 1-Hexene hydroformylation ligands

9.1.5 Hydroformylation Using Immobilised M–NHC Complexes

In 2004, Weberskirch and co-workers tried a new approach by synthesising [RhBr(COD)(NHC)] **23** (NHC = 1-(2′-hydroxyethyl)-3-methylimidazolidine-2-ylidene) [28]. Subsequently, attaching the unsymmetrical, monohydroxy-functionalised NHC by ester groups to an amphiphilic, water-soluble polymer support (ps) created $\mathbf{24}_{ps}$, an air and water stable complex. The intent was to recycle $\mathbf{24}_{ps}$ after performing biphasic hydroformylation. Complex **23** catalysed the hydroformylation of 1-octene (T = 100°C, P = 50 bar) in benzene with a TOF of 2400 h^{-1}, but with poor selectivity (l:b = 0.67). However, when employing $\mathbf{24}_{ps}$ under biphasic conditions (water/1-octene) the ratio increased to 2.6 and the initial TOF was 1100 h^{-1}, indicating a definite change in the steric environment of the active metal centre once supported. When the catalyst was recycled twice the TOF increased to 2320 h^{-1} which was nearly identical to the unbound catalyst. However the l:b ratio declined to 1.2. Slow catalyst activation explained the pronounced increase in TOF, requiring approximately 4 h to fully activate each metal centre.

9.1.6 Hydroformylation Using Co–NHC Complexes

Co–NHC complexes receive considerably less attention than Rh–NHC complexes, despite the fact that NHCs may actually be better suited to cobalt. That is, basic phosphines work well with Co whereas less basic phosphines and phosphites provide the best Rh catalysts [2]. Van Rensburg and co-workers synthesised the dimer $[Co(CO)_3(IMes)]_2$ (**25**) and combined it with 1-octene and syngas (H_2:CO = 2:1) at 60 bar and 170°C [29, 30]. The result was a dark brown oily precipitate identified as the imidazolium salt $[IMesH]^+[Co(CO)_4]^-$.

Llewellyn, Green, and Cowley isolated the Co–H complex $[CoH(CO)_3(IMes)]$ **26**, a relatively stable complex under inert conditions [31]. The authors examined the hydroformylation activity of 1-octene with Co-hydride complex **26**. With 8 atm of syngas (H_2/CO) at 50°C for 17 h and 1 mol% **26**, the conversion to aldehyde products was 47% with a l:b of 0.78. However, 83% of the product was the internal aldehyde 2-methyl-octanal, indicating isomerisation competed with hydroformylation and the rate of isomerisation occurred faster than hydroformylation.

9.1.7 Hydroaminomethylation

Hydroaminomethylation allows for the direct synthesis of key pharmaceuticals, agrochemicals, and natural products (Scheme 9.4) [32]. Synthesis of 3,3-diarylpropylamines is an excellent application of hydroaminomethylation [33, 34].

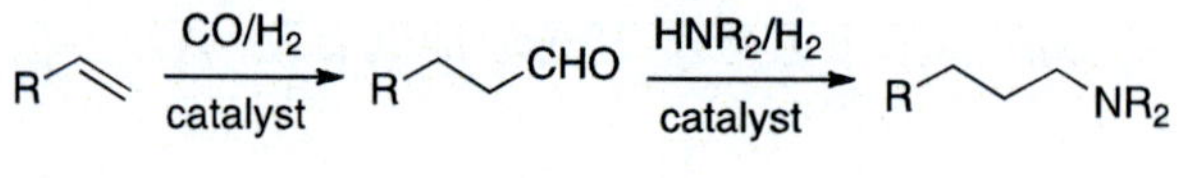

Scheme 9.4 General scheme for hydroaminomethylation

Hydroaminomethylation is hydroformylation coupled to amine–aldehyde condensation followed by hydrogenation. The hydroformylation reaction establishes the regiochemistry, and the hydrogenation step determines the chemoselectivity between amine and eneamine products. Beller and co-workers first reported hydroaminomethylation with the series of [RhCl(COD)(NHC)] pre-catalysts in Fig. 9.5 [34, 35]. Hydroaminomethylation of 1- and 2-pentene with piperidine using **27** led to results that varied dramatically with changes in H_2/CO pressure, temperature, and solvent. The best result obtained converted 99% of 1-pentene into 99% amine with a l:b ratio of 1.6. $[Rh(\mu\text{-}Cl)(COD)]_2$ catalyses the reaction more quickly, but without selectivity, and decomposes to Rh(0) during catalysis.

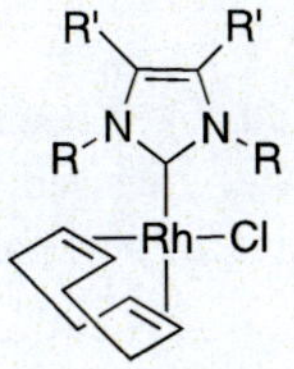

27 R = Mes, R' = H
28 R = Mes, R' = Me
29 R = Mes, R' = Cl
30 R = 2,6-^{i}Pr$_2$-C$_6$H$_4$, R' = H
31 R = 2,6-^{i}Pr$_2$-C$_6$H$_4$, R' = Me

Fig. 9.5 Monodentate NHC catalyst for hydroaminomethylation

The authors optimised conditions for the general reaction of 1,1-diphenylethylene and piperidine (Scheme 9.5). They obtained the highest TOF (288 h^{-1}) and all linear product for this specific reaction when using complex **31** and 5:1 H_2:CO at 125°C for 24 h. An important note here is that the sterics of the substrate, 1,1-diarylethylenes, are responsible for generation of only linear products instead of the catalyst. With a one-pot method, the authors procured the active pharmaceuticals prozapine, fendilline, milverine, and diisopromine in 85%, 91%, 35% and 88% yield, respectively. The catalyst activities compare well to the established Rh–Xantphos system [33].

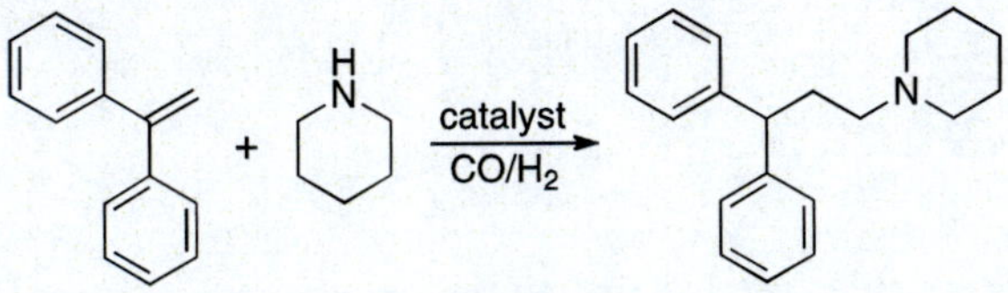

Scheme 9.5 Hydroaminomethylation of 1,1-diphenylethylene with piperidine

9.2 Carbonylation Reactions

Carbonylation reactions encompass a diverse set of transformations used to synthesise many important high-value fine chemicals, synthetic intermediates and materials such as polycarbonates [36]. Palladium catalysts modified with PR_3 ligands facilitate these reactions. However, carbonylation often requires harsh conditions, especially for less reactive C–X bonds, thereby promoting catalyst degradation *via* P–C bond cleavage. The strength of the NHC bond may demonstrate the utility of

M–NHC catalysts in this area. Metal catalysed carbonylation also provides an alternative synthetic route to the production of materials that traditionally require highly toxic precursors, like phosgene. This section discusses carbonylation of aryl halides, oxidative carbonylation of phenolic and amino compounds, carbonylation of aryl diazonium ions, alcohol carbonylation, carbonylative amidation, and copolymerisation of ethylene and CO.

9.2.1 *Carbonylation of Aryl Halides*

A number of Pd or Cu catalysts bearing NHC ligands have been prepared for carbonylation of aryl halides. Nacci and co-workers synthesised the benzothiazole carbene ligated Pd complex **32** (Fig. 9.6) and tested it for aryl halide carbonylation activity (Scheme 9.6) [37].

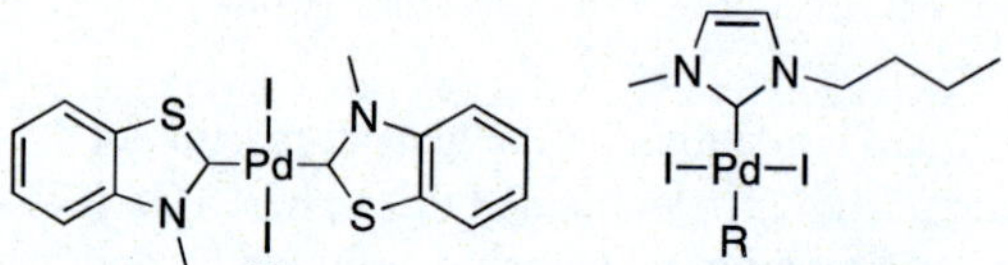

32

33 R = PPh_3
34 R = pyridine
35 R = aniline
36 R = NHC

Fig. 9.6 Palladium carbonylation catalysts

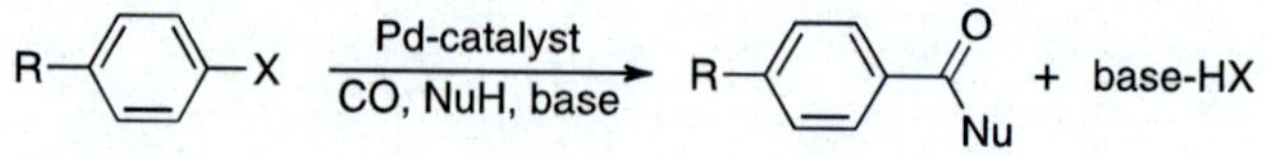

Scheme 9.6 General scheme for aryl halide carbonylation

The complex is stable to high temperatures, oxygen and moisture. Carbonylating iodobenzene with **32** led to high yields (P_{CO} = 1 atm, T = 80°C). Carbonylation of less active aryl bromides and chlorides generated moderate to high yields (T = 130–140°C), but only in the presence of additives (tetrabutylammonium bromide and PPh_3). This again brings into question the nature of the active catalyst. That is, the authors achieved the best results when PPh_3 was present during catalysis, suggesting the possible substitution of the benzothiazole carbene. However, when using just $Pd(OAc)_2$ and PPh_3, the system loses significant reactivity after the third cycle in recycling experiments, and Pd(0) forms. In contrast, there is only a 10% drop in product yield and no formation of Pd(0) after seven catalytic cycles with **32**. This suggests at least one benzothioazole remains bound during catalysis. The authors also wanted to explore the effects of doing catalysis in ionic liquids (tetrabutylammonium bromide and NHC based liquids). A few other reports have appeared describing the Pd catalysed methoxycarbonylation of aryl and alkyl iodides in ionic liquids. The ionic liquids used are imidazolium salts, and several groups propose Pd–NHC species as the active catalyst, though conclusive evidence remains elusive [38–40].

Xia and co-workers synthesised a number of Pd–NHC complexes (**33**, **34**, **36**) for carbonylative Suzuki reactions (Fig. 9.6) [41]. Various aryl iodides were carbonylatively coupled (P = 1 atm) with either phenylboronic acid or sodium tetraphenylborate. All the complexes were highly active, but **33** provided the best results with >76% selectivity for ketone in all the reactions. Xia followed this work with the double carbonylation of various aryl iodides with several secondary amines using the catalysts [CuX(IMes)] (**37-X**) and [Cu(IPr)X] (**38-X**) (X = I, Br, Cl) (3 MPa, 100°C, 10 h) (Scheme 9.7) [42].

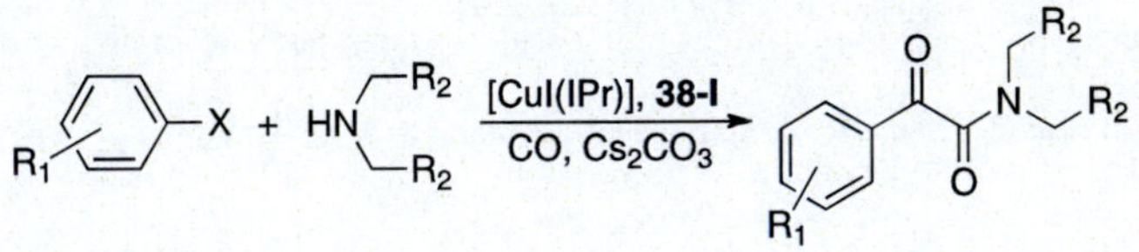

Scheme 9.7 General scheme for double carbonylation of aryl iodides with secondary amines

During optimisation experiments, complex **38-I** yielded the best results, though the NHC and halide had little effect on the conversion. The catalyst was tolerant to different amines (morpholine, piperidine, pyrolidine, and diethylamine) and R groups (Me, Et, Br, Cl, OMe, MeCO, NO_2) on the aryl iodide giving yields ranging from 68–93%. Curiously, to achieve good conversions, the reaction required excess IPr. This led the authors to explore double carbonylation with $[Cu(IPr)_2]BF_4$ **39**. The conversion dropped by 50% using **39**, but upon adding 10 mol% NaI to the reaction, the activity was restored. This suggests the active catalyst is a *bis*-NHC complex and the nature of the counterion is important.

9.2.2 Oxidative Carbonylation of Amino Compounds

Oxidative carbonylation generates a number of important compounds and materials such as ureas, carbamates, 2-oxazolidinones, and aromatic polycarbonates. The [CuX(IPr)] complexes **38-X** (X = Cl, Br, I) were tested as catalysts for the oxidative carbonylation of amino alcohols by Xia and co-workers [43]. Complex **38-I** is the first catalyst to selectively prepare ureas, carbamates, and 2-oxazolidinones without any additives. The important findings were the identity of the counterion and that the presence of the NHC ligand influenced the conversions. 2-Oxazolidinones were formed from primary amino alcohols in 86–96% yield. Complex **38-I** also catalysed the oxidative carbonylation of primary amines to ureas and carbamates. *n*-Propylamine, *n*-butylamine, and *t*-butylamine were transformed into the corresponding ureas, and when using CH_3OH, the corresponding carbamates were isolated in yields ranging from 86% to 98%. Complex **38-I** does not convert aromatic amines, however.

Following this publication, the authors tested a series of Pd–NHC complexes (**33–36**) for the oxidative carbonylation of amino compounds (Scheme 9.8) [44, 45]. These complexes catalysed the oxidative carbonylation of amino compounds selectively to the ureas with good conversion and very high TOFs. Unlike the Cu–NHC catalyst **38-X**, the palladium complexes catalysed the oxidative carbonylation of a variety of aromatic amines. For example, **35** converted 4-Me–C_6H_4–NH_2, 4-Cl–C_6H_4–NH_2, 2,4-Me_2–C_6H_3–NH_2, 2,6-Me_2–C_6H_3–NH_2, and 4-Ac–C_6H_3–NH_2 to the corresponding ureas with very high TOFs (>6000) in 1 h at 150°C, in 99%, 87%, 85%, 72%, and 60% isolated yields, respectively ($P_{CO/O2}$ = 3.2/0.8 MPa).

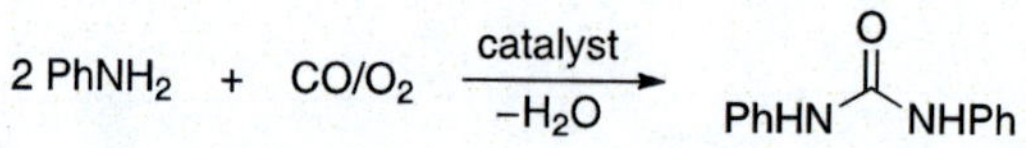

Scheme 9.8 General scheme for oxidative carbonylation of phenylamine with CO/O_2

9.2.3 *Oxidative Carbonylation of Phenolic Compounds*

Sugiyama and co-workers explored the synthesis of diphenyl carbonate (DPC), traditionally made from phosgene, using the catalyst **39-R** [46]. The authors directly carbonylated phenol to form DPC and then converted it to polycarbonates (PCs), without phosgene. The complicated catalyst system required an inorganic redox active co-catalyst, alkylammonium halide, 3 Å molecular sieves and the NHC-catalyst **39-R** (Scheme 9.9). Complex **39-^{t}Bu** along with Ce(TMHD)$_4$ (TMHD = 2,2,6,6-tetramethyl-3,5-heptanedionate), hydroquinione, ^{n}Bu$_4$NBr, and 3 Å molecular sieves imparted the best results by providing DPC in 45% yield with a TOF of 50.7 (mol-DPC/mol-Pd h).

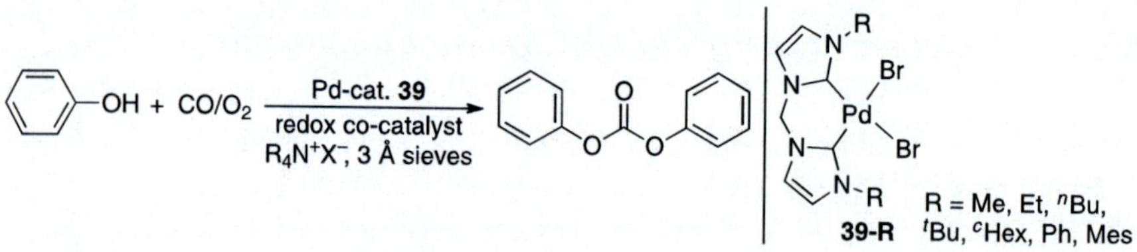

Scheme 9.9 Oxidative carbonylation of phenol with CO/O_2 with catalyst **39**

The authors also successfully catalysed the formation of PC with **39-R** *via* oxidative carbonylation of bisphenol A [47]. Under similar conditions with **39-^{t}Bu**, a PC with M_w = 24000 and M_n = 9400 was obtained in 80% yield. The polymer obtained has the highest molecular weight synthesised directly from PC *via* oxidative carbonylation of bisphenol A, and compared favourably to commercially available PC.

The properties and yield of the polymer product were correlated to the NHC identity, providing clear evidence that the NHC ligand was bound and influenced the reaction. Smaller R groups (Me, Et) on **39-R** provided low molecular weights, yields, and detectable amounts of impurity. Sugiyama only examined the influence of sterics on the formation of PC, but the initial success inspired Tanaka and co-workers to extend this application by tethering NHC ligands to styrene beads [48].

9.2.4 Methanol Carbonylation

Methanol carbonylation is an important process for the production of acetic acid, accounting for approximately 75% of worldwide supply. Monsanto first patented $[RhI_2(CO)_2]^-$ for producing methanol, and later BP Chemicals introduced a new process based upon an iridium catalyst for the purpose of increasing catalyst lifetime [49]. This provides a significant avenue for catalyst improvement with NHC ligands. Cole-Hamilton, Danopoulos and co-workers were successful in the catalytic carbonylation of methanol with **22** [26]. Monitoring the reaction kinetics determined both the rate of catalysis and if catalyst decomposition occurred (linear kinetics signaled stable catalysts). Under the same conditions, T = 150°C and P_{CO} = 27 bar, complex **22** is slightly less active than $[RhI_2(CO)_2]^-$, and equivalent to a phosphane catalyst with similar electronics $[Rh(Cp^{Me4}\text{–}PEt_2)CO]$ (Rates = 1.5, 1.7 and 1.5 M h^{-1}, respectively). However, an additional phosphane ligand was required which is susceptible to detrimental reactions with MeI, thus, complex **22** holds an advantage over the Rh–phosphane catalysts.

9.2.5 Copolymerisation of Alkenes and CO

Herrmann and co-workers examined the copolymerisation of CO and ethylene with **40a/b** (Fig. 9.7) [50]. The optimal conditions for copolymerisation were 5:2

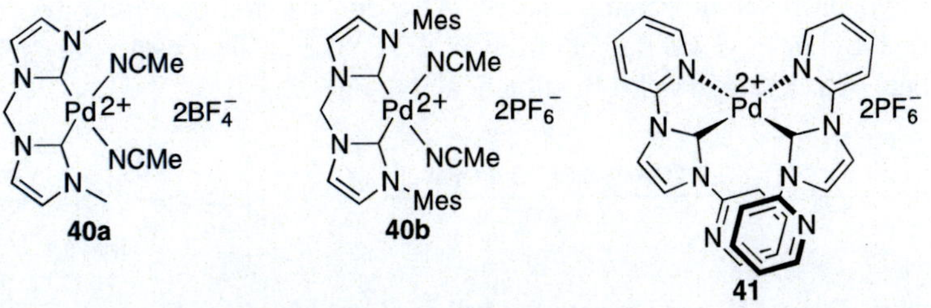

Fig. 9.7 Pd–NHC alkene/CO copolymerisation catalysts

ethylene:CO (70 bar total), T = 50°C and three equivalents of *p*-benzoquinone (Scheme 9.10).

$$n\,\mathrm{CO} + n\,\mathrm{CH_2{=}CH_2} \xrightarrow[\mathrm{MeOH}]{\mathbf{40a\text{-}b}} \mathrm{H{-}(CH_2CH_2CO)_n{-}OCH_3}$$

Scheme 9.10 Copolymerisation of CO and alkenes with catalyst **40a** or **40b**

The ethylene/CO ratio and equivalents of oxidant heavily influenced the yield of polymer. Under these conditions, catalyst **40b** performed better than **40a** by providing 810 and 324 $g_{polymer}/g_{Pd}$, respectively, in high purity. The authors were unable to determine the identity of the active catalyst but noted that some of the Pd metal centres were inactive. The estimated average molecular weight of the copolymer is higher than the weights reported by Drent's analogous PR_3 based catalysts, which are the highest to date [51]. However, the catalysts were inactive for CO/propene and CO/styrene copolymerisation.

Since this initial report, there is only one other report for M–NHC catalysed copolymerisation of CO/alkenes [52]. Lin and co-workers synthesised the *bis*-NHC complex dication **41**, that copolymerises CO and norbornene. The copolymer is synthesised in 87% yield by employing 0.5 mol% **41**, and 750 psi CO gas after 3 days at 60°C. The polymer formed contains ~37 repeat units and M_w = 4660 and M_n = 3790.

9.2.6 *Carbonylation of Aryl Diazonium Ions to Make Ketones*

Carbonylative coupling of aryl diazonium ions and aryl boronic acids supplies a route to access unsymmetrical ketones. A major drawback to this method involves the formation of significant amounts of biaryl coupled product. Andrus and co-workers successfully coupled aryl boronic acids to aryl diazonium ions with Pd–NHC catalysts (Scheme 9.11) [53]. The Pd–NHC catalyst was formed *in situ*, but the authors demonstrated the active catalyst has the NHC bound by showing $Pd(OAc)_2$ by itself led to >99% biaryl formation. After conducting optimisation experiments, the authors chose 100°C, 1 atm CO, 5 h, and dioxane as the catalytic conditions. Strangely, other conditions (rt, 10 atm CO) achieved better conversion and ketone selectivity (90% ketone, 0% biaryl compared to 71%:6%). Regardless, electron-rich diazonium ions provided yields of ketone ranging from 50–82%, and electron-poor *p*-bromo, *p*-nitro, and 2-naphthyl gave yields ranging from 81–90%. The catalyst was highly selective for ketone, with <12% biaryl formation in all but one case.

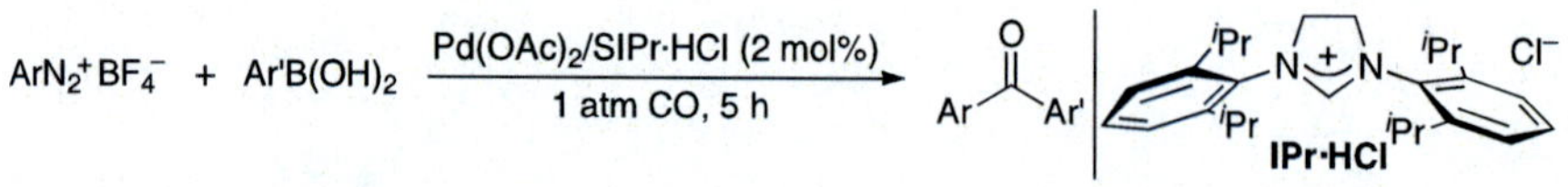

Scheme 9.11 General scheme for catalytic coupling of aryl boronic acids with diazonium ions using Pd-SIPr to make ketones

9.2.7 *Carbonylative Amidation*

Carbonylative coupling marks an efficient method for installing a variety of functional groups including ketones and amides. Andrus and co-workers again successfully coupled diazonium ions, CO, and organoboron compounds in the presence of a Pd(NHC) catalyst and NH_3 (Scheme 9.12) [54]. The catalyst was formed *in situ* from $Pd(OAc)_2$ and SIPr·HCl. During optimisation experiments the authors found CO pressure and order of addition of the reagents had a pronounced effect on product distribution (amide, biaryl product, biaryl amine or reverse amide). The yields ranged from 64% to 89% with aryl and alkyl boronic acids, and with potassium phenyltrifluoroborate. When using the organoboron compounds, phenylpinacolatoborane or *o*-anisyl diazonium ion, the yields diminished by 10–20%.

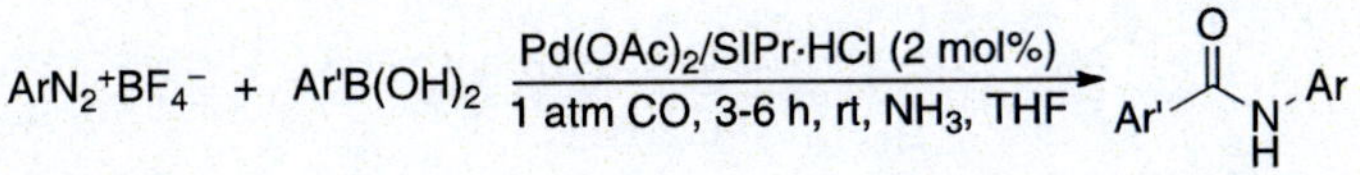

Scheme 9.12 General scheme for catalytic coupling of aryl boronic acids with diazonium ions using Pd/SIPr to produce amides

9.3 Decarbonylation Reactions

Catalytic decarbonylation is significantly more difficult to achieve than the reverse process. The process involves activating either a C–H or C–C bond, hence generally requires high temperatures [55, 56]. In the case of C–C bond activation, only activated or highly strained substrates are used. The harsh conditions should be well suited to the strong M–NHC bond. To date, there are only three reports on decarbonylation with NHC complexes. Murakami and co-workers first reported decarbonlyation of cyclobutanones with [RhCl(COD)(NHC)] (NHC = 1,3,4,5-tetramethylimidazol-2-ylidene), **42**, in 2006 (Fig. 9.8) [57]. The reaction required high

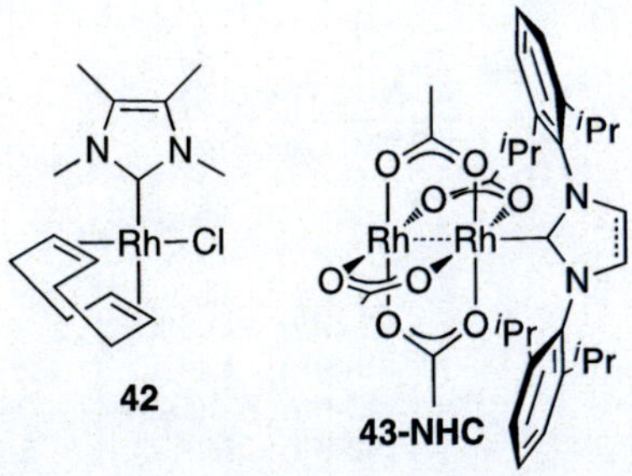

Fig. 9.8 Rh–NHC decarbonylation catalysts

temperatures (150°C) and they achieved yields above 80% for a number of cyclobutanones. Complex **42** failed to decarbonylate linear ketones and less strained cycloalkanones, even when using stoichiometric amounts of **42**. The authors demonstrated interesting chemoselectivity in a cyclobutanone containing a linear aldehyde group; complex **42** selectively decarbonylated only the ketone while [RhCl(PPh_3)$_3$] only decarbonylated the aldehyde (Scheme 9.13). Ligand choice is obviously responsible for the selectivity, providing opportunity for selective decarbonylation applications.

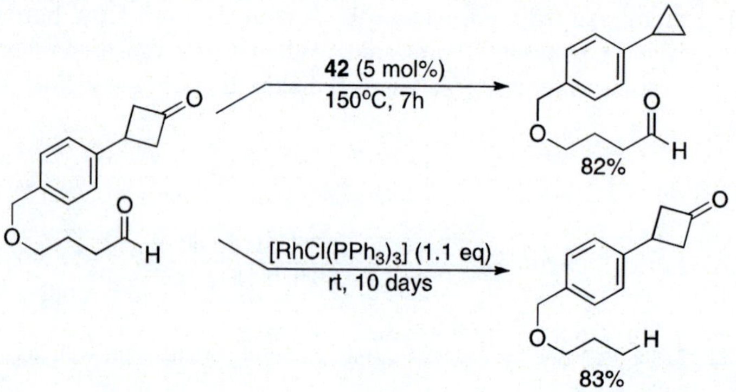

Scheme 9.13 Selective decarbonylation of a cyclobutanone to form the corresponding ketone with PR_3 ligands or the aldehyde with NHC ligands

Afonso, Gois, and co-workers followed this report with an unexpected decarbonylation of diazo-acetamides (Scheme 9.14) using **43-NHC** (Fig. 9.8) [58]. The reaction generated three different products, with low selectivity for the decarbonylated product. The authors tested other substrates with different R groups and bulk at the amine position but found no correlation to the amount of decarbonylation product formed. However, **43-IPr** was more selective than **43-SIPr** for the decarbonylation product. The authors attributed the decarbonylation to the axial coordination of the NHC ligand to the dirhodium (II) complexes.

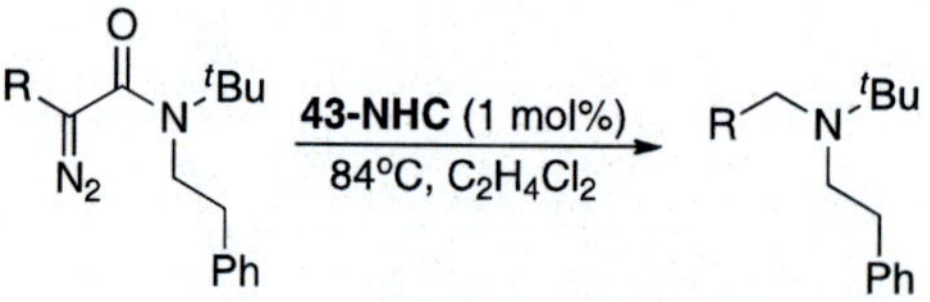

Cat.	R	Yield (%)
43-SIPr	PO(OEt)$_2$	9
43-SIPr	MeCO	29
43-IPr	PO(OEt)$_2$	25
43-IPr	MeCO	36

Scheme 9.14 Decarbonylation of diazo-acetamides with catalyst **43-NHC**

Andrus and Liu exploited a Pd(NHC) decarbonylative Heck coupling reaction in the total synthesis of resveratrol [59]. The catalyst was formed *in situ* with $Pd(OAc)_2$ and IPr·HCl.

9.4 Concluding Remarks

Metal–NHC complexes are competent catalysts in a number of reactions involving CO as a reagent. Hydroaminomethylation of vinylic arenes led to a number of important pharmaceuticals. Carbonylation of amino and phenolic compounds eliminated phosgene as a reagent in the synthesis of ureas, 2-oxazolidinones, carbamates and aromatic PCs. In addition, NHC ligands compete with or surpass phosphine ligands in a number of these reports. Some instances, particularly decarbonylation reactions, help validate the concept that NHCs provide more robust catalysts because they are stronger σ donors. However, the idea that NHCs are not labile during reactions is not absolute, especially in hydroformylation reactions where the results are mixed.

Several groups present hydroformylation results where the role of the NHC during catalysis is ambiguous. Hydroformylation with the aim of isolating the branched isomer complicates this problem because unmodified pre-catalysts provide excellent selectivity for that regioisomer, especially in styrenic derivatives. By that same token, hydroformylation of linear alkenes producing low selectivity may suggest loss of the NHC. With this in mind, one would think taking advantage of the chelate effect would make a more stable catalyst; unfortunately even di-NHC ligands are susceptible to reductive elimination. Mixed di-NHCs, however, show promise for staying bound to the metal centre. Cobalt based catalysts also may be a route to stabilising the M–NHC bond. Certainly, mounting evidence exists to suggest one should be both cautious and critical of future claims involving the beneficial effects of NHC ligands in hydroformylation. Computational mechanistic work comparing the activity of NHC and PR_3 ligands in metal catalysed hydroformylation exists but there is no accompanying experimental work [60].

Catalysis by nature is a kinetic phenomenon. As such, TOFs are irrelevant for proving whether an NHC is bound or not without the corresponding kinetic profile demonstrating linear kinetics [15]. If phosphines are used, whether used as additives or directly attached to the metal centre, detailed mechanistic studies should be done to determine whether the NHC is replaced by a phosphine. Comparison of results to presumed authentic active catalysts or testing the stability under mock hydroformylation conditions without substrate do not indicate whether the NHC is attached. Though Halpern's axiom [61, 62] may prevent one from identifying the active catalyst, making conclusions from a combination of fundamental and secondary experiments will undoubtedly clarify the role of the NHC during hydroformylation [15].

References

1. Veige AS (2008) Polyhedron 27:3177–3189
2. van Leeuwen PWNM, Carmen C (2002) Rhodium catalysed hydroformylation. Kluwer, New York
3. Bizzari S, Blagoev M, Kishi A (2006) Chemical economics handbook. SRI Consulting, Menlo Park
4. Cotton FA, Wilkinson G, Murillo C, Bochmann M (1999) Advanced inorganic chemistry. Wiley, New York
5. Abatjoglou AG, Bryant DR (1984) Organometallics 3:932–934
6. Lewin M, Aizenshtat Z, Blum J (1980) J Organomet Chem 184:255–261
7. Garrou PE (1985) Chem Rev 85:171–185
8. Herrmann WA, Brossmer C, Öfele K, Beller M, Fischer H (1995) J Organomet Chem 491:C1–C4
9. Pruett RL, Smith JA (1969) J Org Chem 34:327–330
10. Herrmann WA, Fischer J, Elison M, Köcher C (1997) (Hoechst AG), DE 4447068.1 A1 (1994); EP0719758 A1; US 5.703.269
11. Chen AC, Ren L, Decken A, Crudden CM (2000) Organometallics 19:3459–3461
12. McGuinness DS, Cavell KJ (2000) Organometallics 19:4918–4920
13. Allen DP, Crudden CM, Calhoun LA, Wang RY (2004) J Organomet Chem 689:3203–3209
14. Chen AC, Allen DP, Crudden CM, Wang RY, Decken A (2005) Can J Chem 83:943–957
15. Widegren JA, Finke RG (2003) J Mol Cat A: Chem 198:317–341
16. Praetorius JM, Kotyk MW, Webb JD, Wang RY, Crudden CM (2007) Organometallics 26:1057–1061
17. Datt MS, Nair JJ, Otto S (2005) J Organomet Chem 690:3422–3426
18. Mieczynska E, Trzeciak AM, Ziolkowski JJ (1992) J Mol Cat 73:1–8
19. Wojciech G, Trzeciak AM, Ziolkowski JJ (2008) Organometallics 27:4131–4138
20. Bortenschlager M, Mayr M, Nuyken O, Buchmeiser MR (2005) J Mol Cat A: Chem 233:67–71
21. Bortenschlager M, Schutz J, von Preysing D, Nuyken O, Herrmann WA, Weberskirch R (2005) J Organomet Chem 690:6233–6237
22. Poyatos M, Uriz P, Mata JA, Claver C, Fernandez E, Peris E (2003) Organometallics 22:440–444
23. Jeletic MS, Jan MT, Ghiviriga I, Abboud KA, Veige AS (2009) Dalton Trans 2764–2776
24. Lowry RJ, Jan MT, Abboud KA, Ghiviriga I, Veige AS (2010) Polyhedron 29:553–563
25. Dastgir S, Coleman KS, Cowley AR, Green MLH (2006) Organometallics 25:300–306
26. Downing SP, Pogorzelee PG, Danopoulos AA, Cole-Hamilton DJ (2009) Eur J Inorg Chem 1816–1824
27. Neveling A, Julius GR, Cronje S, Esterhuysen C, Raubenheimer HG (2005) Dalton Trans 181–192
28. Zarka MT, Bortenschlager M, Wurst K, Nuyken O, Weberskirch R (2004) Organometallics 23:4817–4820
29. van Rensburg H, Tooze RP, Foster DF, Slawin AMZ (2004) Inorg Chem 43:2468–2470
30. van Rensburg H, Tooze RP, Foster DF, Otto S (2007) Inorg Chem 46:1963–1965
31. Llewellyn SA, Green MLH, Cowley AR (2006) Dalton Trans 4164–4168
32. EliBracht P, Barfacker L, Buss C, Hollmann C, Kitsos-Rzychon BE, Kranemann CL, Rische T, Roggenbuck R, Schmidt A (1999) Chem Rev 99:3329–3366
33. Ahmed M, Seayad AM, Jackstell R, Beller M (2003) J Am Chem Soc 125:10311–10318
34. Seayad AM, Selvakumar K, Ahmed M, Beller M (2003) Tetrahedron Lett 44:1679–1683
35. Ahmed M, Buch C, Routaboul L, Jackstell R, Klein H, Spannenberg A, Beller M (2007) Chem Eur J 13:1594–1601
36. Kollar L (2008) Modern carbonylation methods. Wiley-VCH, Weinheim
37. Calo V, Giannoccaro P, Nacci A, Monopoli A (2002) J Organomet Chem 645:152–157
38. Mizushima E, Hayashi T, Tanaka M (2001) Green Chem 3:76–79

39. Fukuyama T, Inouye T, Ryu I (2007) J Organomet Chem 692:685–690
40. Zawartka W, Trzeciak AM, Ziolkowski JJ, Lis T, Ciunik Z, Pernak J (2006) Adv Synth Cat 348:1689–1698
41. Zheng S, Xu L, Xia C (2007) Appl Organomet Chem 21:772–776
42. Liu J, Zhang R, Wang S, Sun W, Xia C (2009) Org Lett 6:1321–1324
43. Zheng S, Li F, Liu J, Xia C (2007) Tetrahedron Lett 48:5883–5886
44. Zheng S, Peng X, Liu J, Sun W, Xia C (2007) Helv Chim Acta 90:1471–1476
45. Zheng S, Peng X, Liu J, Sun W, Xia C (2007) Chin J Chem 25:1065–1068
46. Okuyama K, Sugiyama J, Nagahata R, Asai M, Ueda M, Takeuchi K (2003) J Mol Cat A: Chem 203:21–27
47. Okuyama K, Sugiyama J, Nagahata R, Asai M, Ueda M, Takeuchi K (2003) Macromolecules 36:6953–6955
48. Okuyama K, Sugiyama J, Nagahata R, Asai M, Ueda M, Takeuchi K (2003) Green Chem 5:563–566
49. Layman P (1996) C&EN July 1, pp 7–8
50. Gardiner MG, Herrmann WA, Reisinger CP, Schwarz J, Spiegler M (1999) J Organomet Chem 572:239–247
51. Drent E, Budzelaar PHM (1996) Chem Rev 96:663–681
52. Chen JCC, Lin IJB (2000) Organometallics 19:5113–5121
53. Andrus MB, Ma Y, Zang Y, Song C (2002) Tetrahedron Lett 43:9137–9140
54. Ma Y, Song C, Chai Q, Ma C, Andrus MB (2003) Synthesis 2886–2889
55. Doughty DH, Pignolet LH (1983) Homogenous catalysis with metal phosphine complexes. Plenum, New York, , pp 343–375, Chapter 11
56. Crabtree RH (2005) The organometallic chemistry of the transition metals. Wiley, New Jersey
57. Matsudu T, Shigeno M, Murakami M (2006) Chem Lett 35:288–289
58. Gomes LFR, Trindade AF, Candeias NR, Gois PMP, Afonso CAM (2008) Tetrahedron Lett 49:7372–7375
59. Andrus MB, Liu J (2006) Tetrahedron Lett 47:5811–5814
60. Sparta M, Borve KJ, Jensen VR (2007) J Am Chem Soc 129:8487–8499
61. Halpern J (1982) Science 217:401–407
62. Chan ASC, Pluth JJ, Halpern J (1980) J Am Chem Soc 102:5952–5954

Chapter 10
N-Heterocyclic Carbene Complexes in Oxidation Reactions

Václav Jurčík and Catherine S. J. Cazin

Abstract This chapter describes applications of *N*-heterocyclic carbenes (NHCs) in oxidation chemistry. The strong σ-donation capabilities of the NHCs allow an efficient stabilisation of metal centres in high oxidation states, while high metal-NHC bond dissociation energies suppress their oxidative decomposition. These properties make NHCs ideal ligands for oxidation processes. The first part of this chapter is dedicated to the reactivity of NHC-metal complexes towards molecular oxygen whilst the second half highlights all oxidation reactions catalysed by such complexes. These include oxidation of alcohols and olefins, oxidative cyclisations, hydrations of alkynes and nitriles, oxidative cleavage of alkenes and the oxidation of methane.

10.1 Introduction

In recent years, *N*-heterocyclic carbenes (NHCs) have emerged as a viable alternative to *N*- and *O*-donor ligands for transition metal catalysed oxidations [1–4]. This is due to the unique electronic and steric properties of NHCs (see Chapter 1, Sections 1.2 and 1.3 for detailed discussion) that affect all individual steps of the catalytic manifold and bring a remarkable stability to the complexes under oxidation conditions. This stability can be attributed to the high dissociation energy of the M–NHC bond that prevents catalyst decomposition, the strong σ-donor ability that stabilises higher oxidation states and the remarkable stability of the NHC towards oxygen. This hence leads to catalytic complexes stable under oxidation conditions, and that can activate O_2, as illustrated in the following section.

C. S. J. Cazin (✉)
EaStCHEM School of Chemistry, University of St Andrews,
St Andrews, KY16 9ST, UK
e-mail: cc111@st-andrews.ac.uk

C. S. J. Cazin (ed.), *N-Heterocyclic Carbenes in Transition Metal Catalysis and Organocatalysis*, Catalysis by Metal Complexes 32,
DOI 10.1007/978-90-481-2866-2_10,

10.2 Reactions of NHC-Complexes with O_2

The ability of transition metal complexes to activate molecular oxygen is a fundamental step enabling catalytic aerobic oxidations, which represents one of the most important challenges in oxidation catalysis. In this regard, computational and experimental studies have been undertaken to assess the behaviour of NHC-complexes in the presence of molecular oxygen. Pioneering work by Stahl and co-workers examined homoleptic NHC complexes of palladium. In early work, the authors showed that $[Pd(IMes)_2]$ **1a** reacts readily with O_2 at −78°C to form the corresponding η^2-peroxo-complex **2a** (Scheme 10.1). Upon reaction with acetic acid, **2a** evolves into the hydroperoxo-complex **4a** [5], a species frequently proposed as an intermediate in oxidation catalysis, [6] that could be isolated and characterised due to the stabilisation imparted by the IMes ligand. Subsequent studies showed that **4a** can also be prepared by reaction of the hydride **5a** with O_2 [7]. Surprisingly, the authors found that the presence of exogenous carboxylic acid catalyses the oxidation of complexes of type **5**. This prompted computational [8] and experimental [9] studies on the reaction of O_2 with the hydride. These results suggest that the preferred pathway for the oxidation of **5a** to form **4a** proceeds *via* reductive elimination of the carboxylic acid, leading to **1**, which then reacts with O_2 to form **2**, that reacts with the RCOOH liberated during the first step to form **4a**.

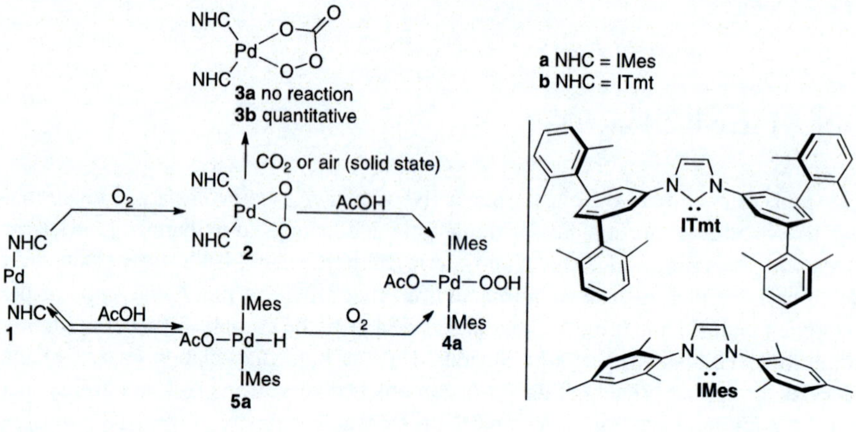

Scheme 10.1 The O_2 and CO_2 fixation to $[Pd(NHC)_2]$ complexes (NHC = IMes or ITmt)

In a similar fashion, the homoleptic complex $[Pd(ITmt)_2]$ **1b** readily reacts with O_2 to form the corresponding peroxo-complex **2b** (Scheme 10.1). This complex, upon exposure to CO_2, leads to the peroxo-carbonate complex **3b** [10]. Under the same reaction conditions, the formation of **3a** does not occur, presumably due to the larger steric hindrance of the IMes ligand.

Computational studies performed to shed light on the reactivity of Pd(0) complexes towards molecular oxygen showed that O_2 coordination is considerably more favoured with complexes bearing NHC rather than for those supported by tertiary phosphines (Table 10.1) [11]. This result can be rationalised by considering the stronger σ-donor character of the NHC ligands relative to tertiary phosphines. The strong dependence on the polarity of the reaction medium in the reaction of O_2 with the $[PdL_2]$ complexes appears crucial. Furthermore, it also brings into question the important possibility of reversibility of O_2 binding in oxidation catalysis.

Table 10.1 DFT calculations: reaction of O_2 with $[PdL_2]$ complexes [11]

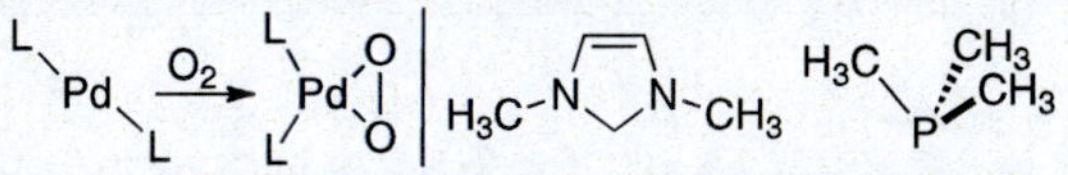

		G^0 (kcal/mol)	
Medium	Dielectric constant	IMe	PMe_3
Gas phase	1.0	+3.6	+10.3
Toluene	2.379	−5.7	−2.3
THF	7.58	−12.0	−3.1
MeCN	36.64	−14.9	−2.8

The results obtained from the computational studies prompted the authors to re-evaluate the reversibility of O_2 binding to $[Pd(IMes)_2]$. It was found that heating $[Pd(\eta^2-O_2)(IMes)_2]$, in the solid state, *in vacuo* for 1 week at 80°C resulted in a 31% recovery of $[Pd(IMes)_2]$ [11]. In contrast, the phosphine complex $[Pd(\eta^2-O_2)(PPh^tBu_2)_2]$ released O_2 almost quantitatively under similar conditions.

Another study on O_2 binding to NHC complexes, that combined experiments and DFT (density functional theory) calculations was recently reported on a ruthenium system. This study shows the reversible binding of oxygen to the tetra-NHC complex $[Ru(NHC)_4H)][BAr_4^F]$ **6** ($BAr_4^F = B\{(3,5\text{-}CF_3)_2C_6H_3\}_4$), which leads to complex **7** (Scheme 10.2) [12]. Unexpectedly, the chemical shift of the hydride ligand undergoes a large downfield shift upon coordination to O_2 (from −41.2 ppm for **6** to +4.8 ppm for **7**). Both the reversibility of O_2 coordination and the positive hydride chemical shift were predicted and rationalised by the DFT studies.

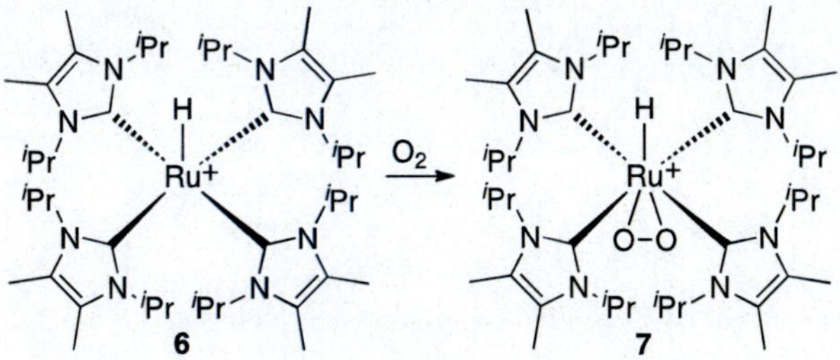

6 **7**

Scheme 10.2 Reaction of O_2 with $[Ru(NHC)_4H)][BAr_4^F]$ **6** [12]

An additional study on the reaction of molecular oxygen with an NHC complex is depicted in Scheme 10.3 [13]. The tri-NHC cobalt complex **9** is obtained by reaction of $TIMEN^{xyl}$ **8** (TIMEN = tris[2-(3-alkylimidazol-2-ylidene)ethyl]amine) with

[$CoCl(PPh_3)_3$]. Reaction of [Co(TIMEN^{xyl})]Cl **9** with oxygen in the presence of $NaBPh_4$ leads to the formation of the peroxo-complex [Co(η^2-O_2)(TIMEN^{xyl})] BPh_4 **10**, which is a rare example of a side-on η^2-peroxo cobalt complex (the majority of Co-O_2-adducts are η^1-O_2-complexes, *i.e.* end-on). The authors also showed that **10** is capable of converting molecular oxygen to benzoylchloride.

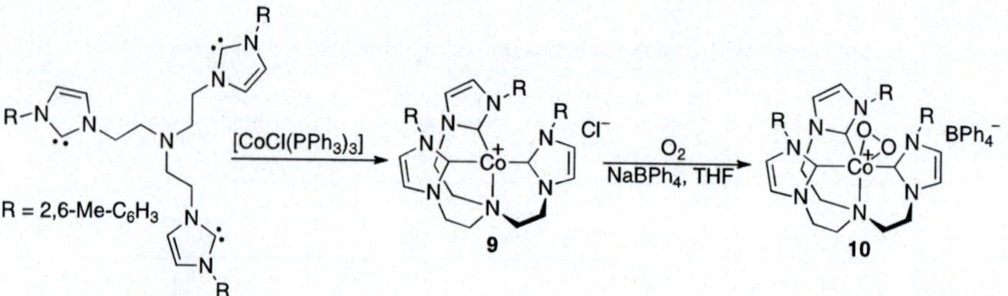

8
TIMEN^{xyl}

Scheme 10.3 Preparation of [Co(η^2-O_2)(TIMEN^{xyl})]BPh_4 [13]

A series of complexes of type [Ni(η^3-allyl)Cl(NHC)] highlight the important influence of the NHC on the reactivity of the resulting complex towards O_2. It was shown that O_2-activation is disfavoured when the rotation around the Ni–C_{NHC} bond is restricted. On the other hand, with complexes displaying free rotation around the Ni–C_{NHC} bond, the complexes react cleanly with O_2. The overall reaction results in the oxidation of the allyl group and the formation of hydroxy-bridged dimers (Scheme 10.4) [14,15].

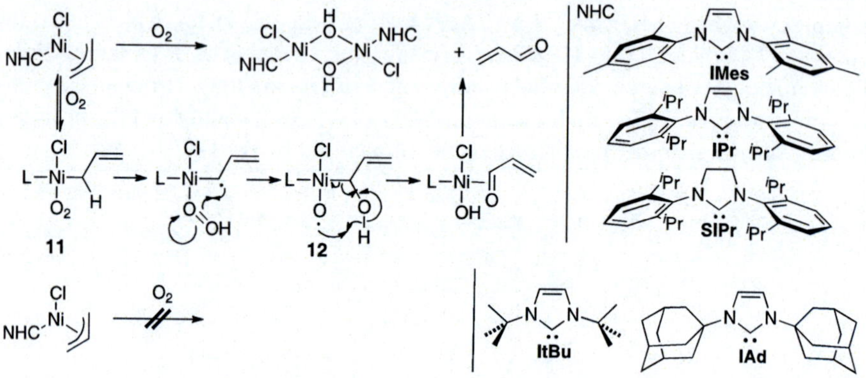

Scheme 10.4 Reaction of NHC–Ni–allyl complexes with O_2 [14,15]

The proposed mechanism consists of O_2 binding to Ni, leading to an activated complex which undergoes decomposition *via* formation of a Ni(III)-peroxide intermediate **11**, in which an oxygen radical abstracts a hydrogen from the neighbouring allyl group. Intramolecular hydroxylation to alcohol **12** followed by hydrogen transfer leads to nickel hydroxide dimer with concomitant liberation of the aldehyde.

All mechanistic studies on O_2 activation by NHC complexes are of great interest (particularly when computations and experiments are combined) as they help rationalise catalytic behaviour, they allow catalyst improvement and ultimately they might allow rational catalyst design.

10.3 Oxidation Reactions Catalysed by Metal–NHC Complexes

10.3.1 Oxidation of Alcohols

The oxidation of alcohols to the corresponding carbonyl compounds is one of the key reactions in organic synthesis and numerous methods have been developed over the years to accomplish this transformation [16]. A general mechanism for Pd-catalysed aerobic oxidation is shown below (Scheme 10.5).

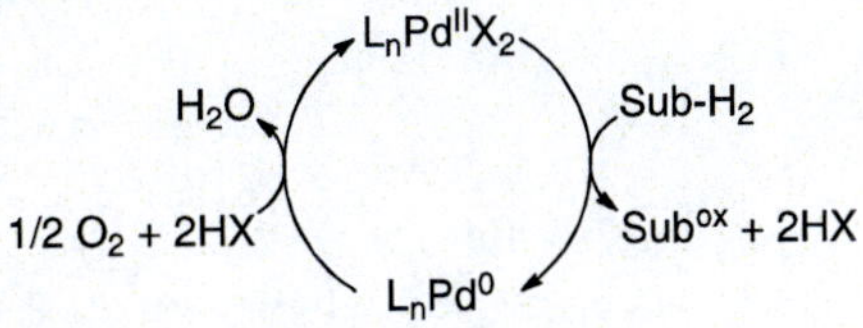

Scheme 10.5 Simplified mechanism for Pd-catalysed oxidation

10.3.1.1 Aerobic Oxidation

Despite a significant number of metals having been used in oxidation reactions, palladium has been recognised as the most promising. From the myriads of Pd-based systems for oxidation of alcohols [17], systems developed by Uemura ($Pd(OAc)_2$/pyridine) [18–20] and Sheldon ($Pd(OAc)_2$/phenanthroline) [21] are nowadays recognised as benchmarks. Sigman and co-workers reported that complex **13** (Scheme 10.6) promotes the aerobic oxidation of alcohols [22–24]. In the proposed mechanism, the catalyst enters the cycle upon loss of water and binds the alcohol substrate. Intramolecular deprotonation and loss of acetic acid lead to the palladium alkoxide **14**, which undergoes β-hydride elimination to give the hydride **15**. The hydride reductively eliminates acetic acid producing a Pd–NHC species that is oxidised by dioxygen to give η^2-peroxo complex **16**. Protonation by two equivalents of acetic acid leads to the elimination of H_2O_2 and regeneration of the catalyst (Scheme 10.6).

Interestingly, the scope of the reaction using this catalyst can be extended to oxidative kinetic resolution of secondary alcohols by using (-)-sparteine as a base (Table 10.2) [25]. The best enantiomeric excess of the alcohol was obtained when a chiral enantiopure base and an achiral catalyst were used. The use of chiral enantiopure catalyst bearing ligand **17** led to low enantioselectivity.

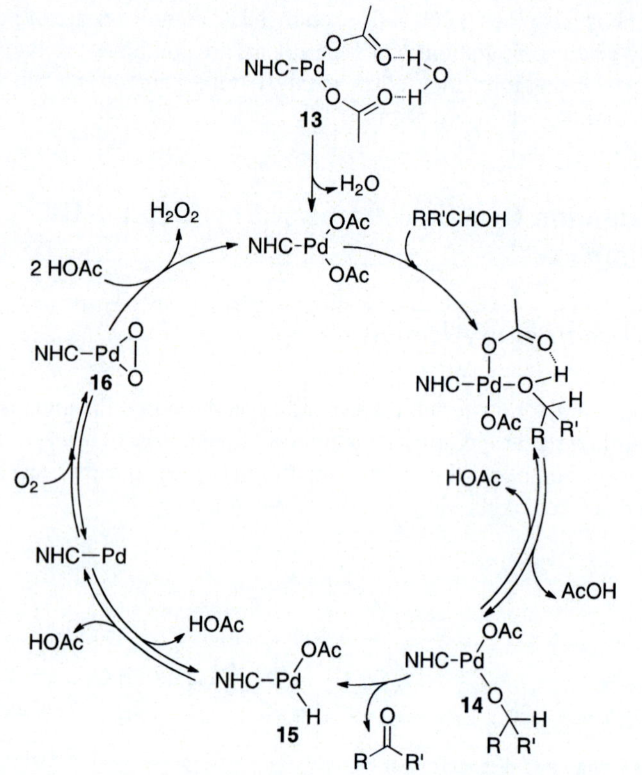

Scheme 10.6 Mechanism of aerobic oxidation catalysed by complex **13** [23]

Table 10.2 Oxidative kinetic resolution of alcohols using (-)-sparteine [25]

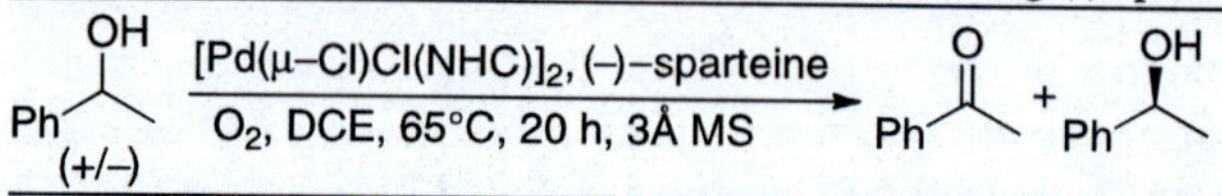

Cat. loading (mol%)	NHC	Additive (mol%)	Conv. (%) (% *ee*)	k_{rel}
1.5	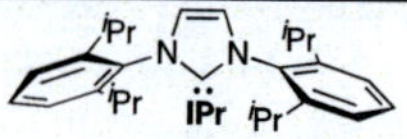 IPr	(-)-sparteine (15)	65 (96)	11.6
2.5	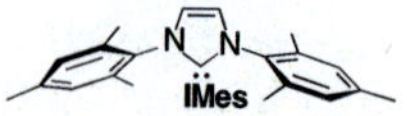 IMes	(-)-sparteine (20)	32 (35)	6.1
2.5	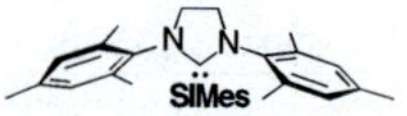 SIMes	(-)-sparteine (20)	45 (54)	6.4
2.5	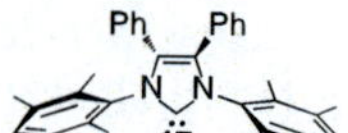 17	(-)-sparteine (20)	35 (42)	11.8
2.5		AgOAc (10.5)	34 (10)	1.6

Axially chiral Pd–NHC complexes reported by Shi and co-workers [26–28] have shown high selectivity in the oxidative kinetic resolution of alcohols without the need of addition of a chiral base. Enantiomeric excesses of up to 99% were obtained (Scheme 10.7).

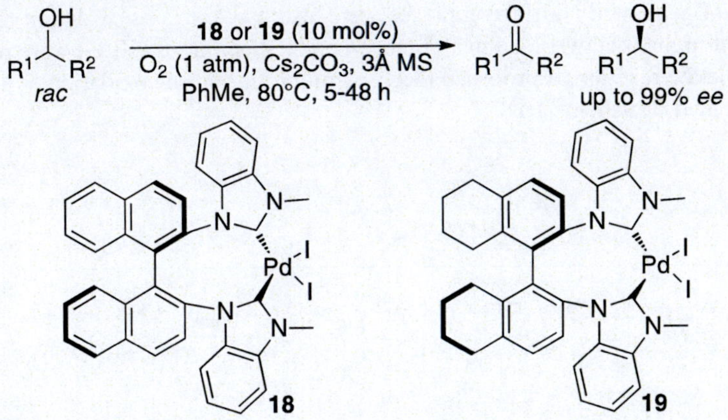

Scheme 10.7 Oxidative kinetic resolution of alcohols using chiral bis-NHC ligands [26–28]

10.3.1.2 Anaerobic Oxidation

The prevailing tendency in the development of systems for alcohol oxidation makes use of oxygen, which as a reagent could represent a potential fire/explosion risk when used in conjunction with flammable solvents. An appealing alternative is the use of chlorobenzene as an oxidant [29,30]. An example of this anaerobic oxidation catalysed by NHC–Pd or NHC–Ni complexes was recently reported by the Navarro group [31]. Being inspired by the dehalogenation reactions [32] that produce acetone when 2-propanol is used as a hydride source, this methodology was expanded to the oxidation of alcohols. [Pd(η^3-2-PhC_3H_4)Cl(IPr)] **22** was found to be the most efficient catalyst for this transformation. Reactions typically proceed at room temperature. While a broad range of secondary alcohols could be oxidised in high yields, the methodology cannot be applied to most primary alcohols (Table 10.3).

Table 10.3 Anaerobic oxidation of alcohols [31]

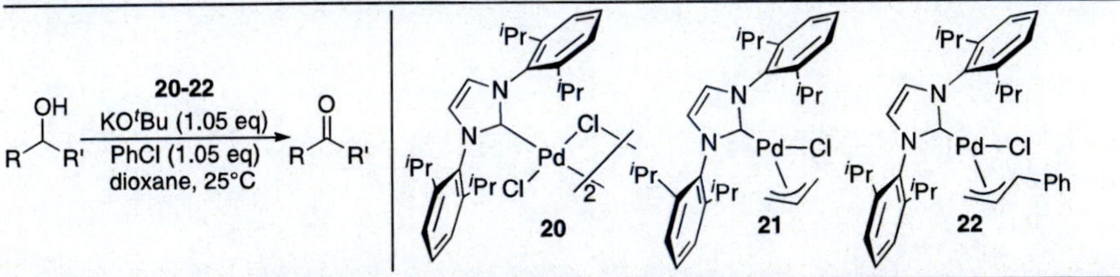

Catalyst	Loading (mol%)	R	R'	Time (h)	Yield (%)
20	0.5	C_6H_5	Et	24	45
21	0.5	Ph	Et	24	90
22	1	Ph	Et	11	91
22	1	4-MeOC_6H_4	Et	14	90
22	1	Ph	^{t}Bu	12	91
22	1	Me	Hex	20	71
22	1	H	Dec	120	<5

In situ generated Ni–IPr complexes were also active in this oxidation reaction, however higher catalyst loadings (5 mol%) and temperatures (60°C) were required to enable the reaction. A proposed mechanism for the aerobic oxidation of alcohols in presented in Scheme 10.8.

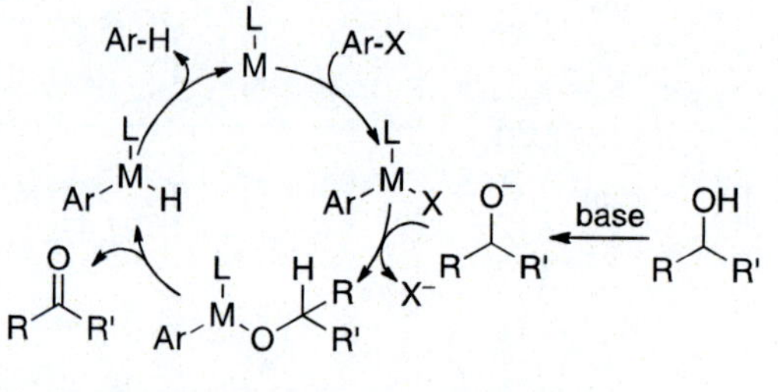

Scheme 10.8 Mechanism of anaerobic oxidation of alcohols [31]

10.3.1.3 Oppenauer Oxidation

The Oppenauer oxidation makes use of ketones (typically acetone) or alkenes as hydrogen acceptors and this absence of a strong oxidising agent allows to overcome some potential NHC oxidative instability. Reactions consist of an equilibrium between an alcohol and its oxidised form (Scheme 10.9).

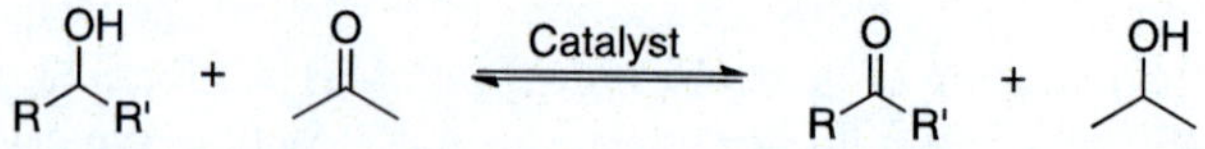

Scheme 10.9 Oppenauer oxidation

Yamaguchi and co-workers have reported the synthesis of various Ir–NHC complexes starting from [Ir(μ-Cl)ClCp*]$_2$ **23** (Scheme 10.10) and their application in Oppenauer-type reactions [33,34].

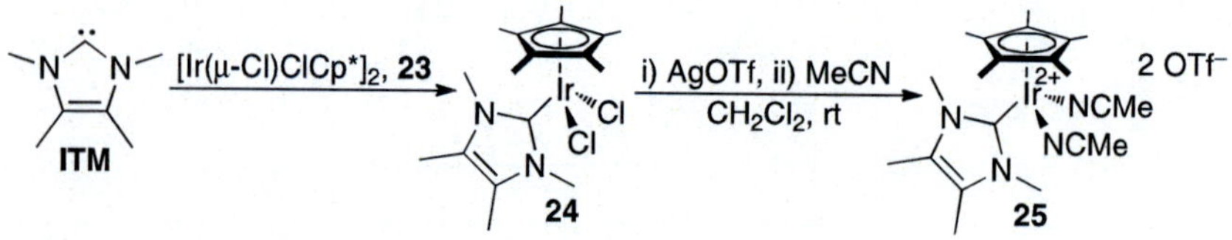

Scheme 10.10 Preparation of Ir-NHC complexes [34]

The use of NHCs as ancillary ligands was shown to be highly beneficial, providing that a cationic Ir-species was present (Table 10.4, compare entries 1–4).

Interestingly, when analogous phosphine adducts were tested, very poor catalytic activity was obtained (Table 10.4, entries 6 and 7). On the other hand, the NHC-cationic complex $[IrCp^*(ITM)(NCMe)_2][OTf]_2$ **25** proved highly efficient for the Oppenauer-type oxidation of a large range of secondary as well as primary alcohols in acetone (e.g., Table 10.4, entries 4, 5, 8–10).

Table 10.4 Ir-catalysed Oppenauer oxidation [34]

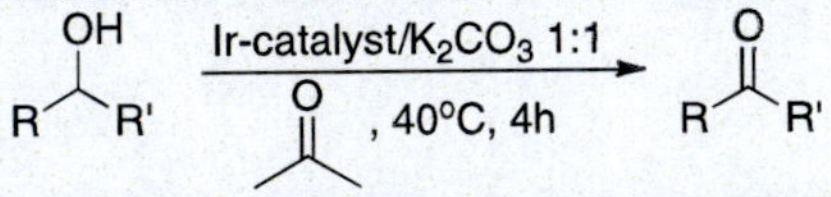

Entries	Catalyst (load. mol% Ir)	Product	Conv. (Yield) (%)	TON
1	$[Ir(\mu\text{-}Cl)ClCp^*]_2$ **23** (0.1)		48(43)	430
2	$[IrCl_2Cp^*(ITM)]$ **24** (0.1)		3(1)	10
3	$[IrCl_2Cp^*(ITM)]$ **24** (0.1) + AgOTf[a]		95(92)	920
4	$[IrCp^*(ITM)(NCMe)_2][OTf]_2$ **25** (0.1)	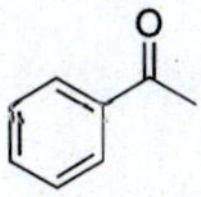	95(95)	950
5	$[IrCp^*(ITM)(NCMe)_2][OTf]_2$ **25** (0.025)		81(80)	3200
6	$[IrCp^*(P^nBu_3)(NCMe)_2][OTf]_2$ (0.1)		8(1)	10
7	$[IrCp^*(PPh_3)(NCMe)_2][OTf]_2$ (0.1)		8(1)	10
8	$[IrCp^*(ITM)(NCMe)_2][OTf]_2$ **25** (0.1)	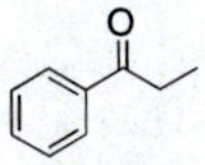	93(91)	910
9	$[IrCp^*(ITM)(NCMe)_2][OTf]_2$ **25** (0.1)	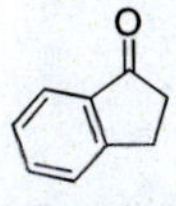	91(81)	810
10	$[IrCp^*(ITM)(NCMe)_2][OTf]_2$ **25** (0.5)	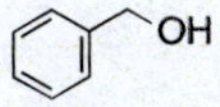	89(86)	172

[a]0.045 mmol added

Ruthenium NHC dihydride complex **26** was found to exhibit interesting reversible hydrogenation/dehydrogenation activity (Scheme 10.11) [35,36]. When excess acetone was used as a hydrogen acceptor, dehydrogenation of several alcohols was achieved (Table 10.5).

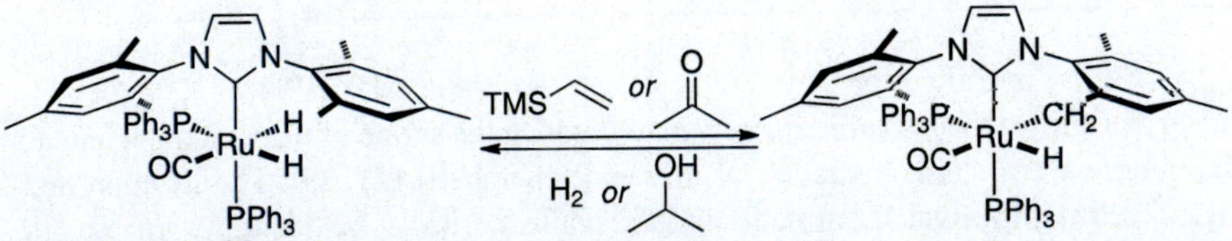

26

Scheme 10.11 Hydrogenation/dehydrogenation equilibrium involving complex **26**

Table 10.5 Dehydrogenation of alcohols catalysed by $[Ru(H)_2(IMes)(CO)(PPh_3)_2]$, **26** [36]

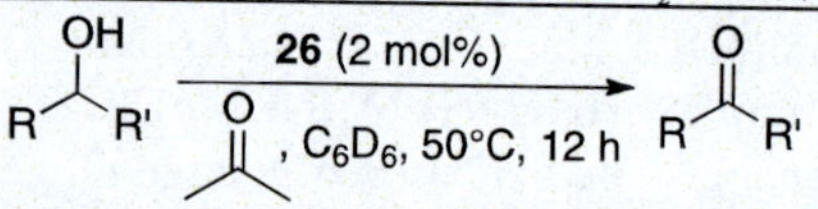

Product	Conv. (%)	TON	Product	Conv. (%)	TON
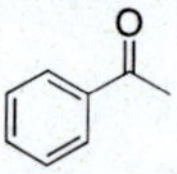	88	44	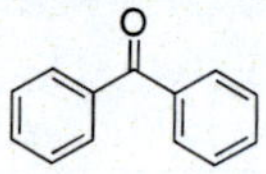	30	15
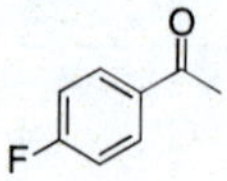	87	44	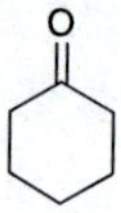	47	24
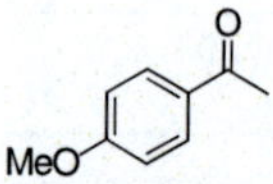	96	48			

10.3.2 *Alkyne and Nitrile Hydration*

The hydration of C–C triple bonds represents one of the most atom economical and environmentally friendly oxidation reactions [37]. Recently, Nolan and co-workers reported the cationic $[Au(IPr)][SbF_6]$ system, which was generated *in situ* from [AuCl(IPr)] and $AgSbF_6$. The catalyst system showed remarkable activity in the hydration of a large range of alkynes, at Au loadings as low as 10 ppm (typically 50–100 ppm), under acid-free conditions (Table 10.6) [38].

Table 10.6 $[AuCl(IPr)]/AgSbF_6$ catalysed hydration of alkynes [38]

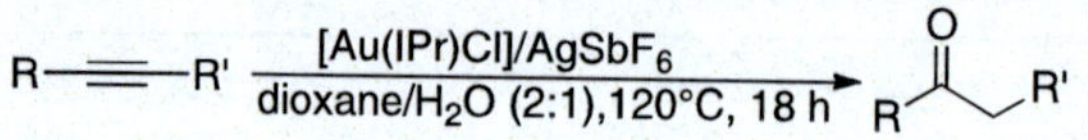

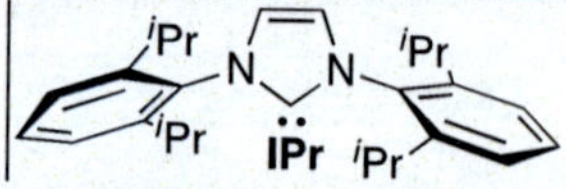

R	R'	Cat. load. (ppm)	Yield (%)	TON
Ph	Ph	1000	77	770
^{t}Bu	Me	100	72	7200
4-MeOPh	Me	100	88	8800
^{n}Pr	^{n}Bu	100	95	9500
Ph	Me	50	85	17000
^{n}Pent	Me	50	100	20000
Et	Et	10	84	84000

A slight modification of the catalyst to $[Au(NTf_2)(IPr)]$ [NTf_2 = bis (trifluoromethylsulfonyl)amide] allowed to broaden the scope of this reaction genre to the microwave-assisted hydration of nitriles (Scheme 10.12) [39]. The reaction proceeds well with activated aromatic nitriles, while sterically hindered, heteroaromatic and aliphatic nitriles require longer reaction times and higher catalyst loading but proceed as well to product. Although the precise mechanism is not known, the authors suggest that the catalytically active species might be a gold-hydroxide complex.

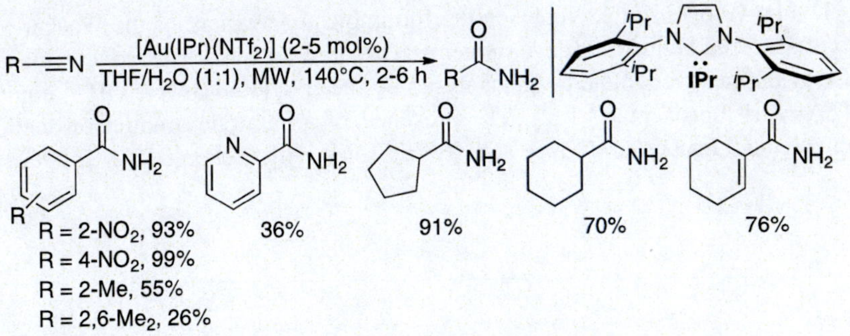

Scheme 10.12 [Au(IPr)(NTf_2)] catalysed hydration of nitriles [39]

10.3.3 *Wacker Type Reactions and Oxidative Cyclisations*

The conversion of ethylene to acetaldehyde using a soluble palladium complex, developed in the late 1950s, was one of the early applications of homogeneous catalysis and the first organo-palladium reaction practised on an industrial scale [40]. Typically this reaction requires stoichiometric amounts of $CuCl_2$ under aerobic conditions. The use of copper represents not only an environmental issue, but often limits the scope of ligands that can be used in conjunction with Pd.

Pd($OTf)_2$(IPr) generated *in situ* from [Pd(μ-Cl)(Cl)(IPr)]$_2$ and AgOTf was reported to catalyse the copper-free Wacker-type oxidation of styrene derivatives using *tert*-butyl hydroperoxide (TBHP) as the oxidant (Table 10.7) [41]. Reaction conditions minimised oxidative cleavage of styrene, which is a common side-reaction in Wacker-type oxidations. However, when *trans*-stilbene was used as a substrate, a significant amount of oxidative cleavage occurred.

Table 10.7 Copper-free Wacker oxidation of substituted styrenes [41]

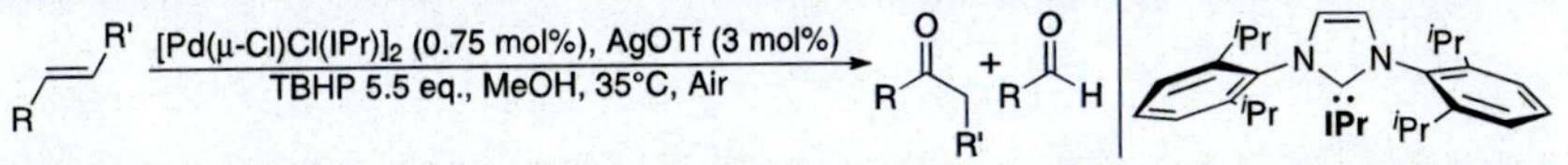

R	R'	Time (h)	Yield (%)	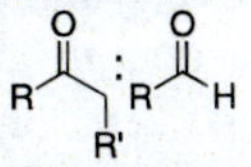
C_6H_5	H	24	75	>130:1
2-Me-C_6H_4	H	48	79	36:1
3-Me-C_6H_4	H	32	83	22:1
4-Me-C_6H_4	H	16	86	22:1
2,4,6-$Me_3C_6H_2$	H	24	71	>150:1
Ph	Ph	48	42	42:35
Ph	H	24	75	>130:1

Oxidative cyclisations representing intramolecular variant of the Wacker reaction have seen significant developments. The intramolecular oxidative cyclisation of tosylamines was found to be catalysed by the $[Pd(TFA)_2(IMes)(OH_2)]$ complex (TFA = trifluoroacetate) [42]. The presence of a catalytic amount of acetic or benzoic acid leads to improved activity and selectivity (Scheme 10.13).

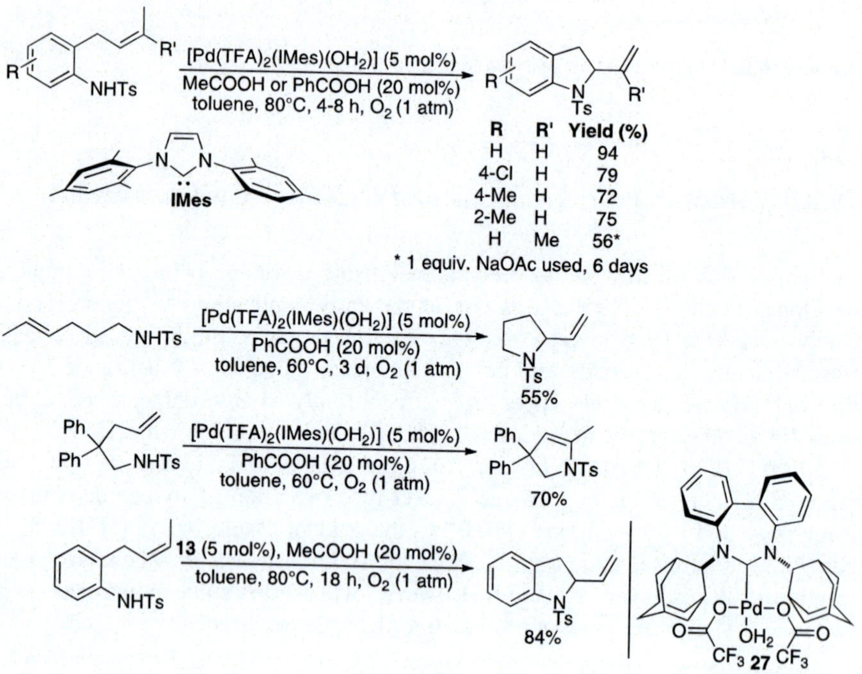

Scheme 10.13 Aza-Wacker cyclisation [42]

Asymmetric variants of these reactions are highly interesting since they provide access to chiral heterocycles. A recent comprehensive study by Stahl and co-workers reports the synthesis of various enantiopure $[Pd(\mu\text{-Cl})Cl(NHC)]_2$ complexes and their application in asymmetric aza-Wacker cyclisations. The reactions generally proceed with low yields or enantioselectivity [43]. The best enantioselectivity (63%) was achieved using complex **28** (Table 10.8).

Table 10.8 Enantioselective aza-Wacker cyclisations [43]

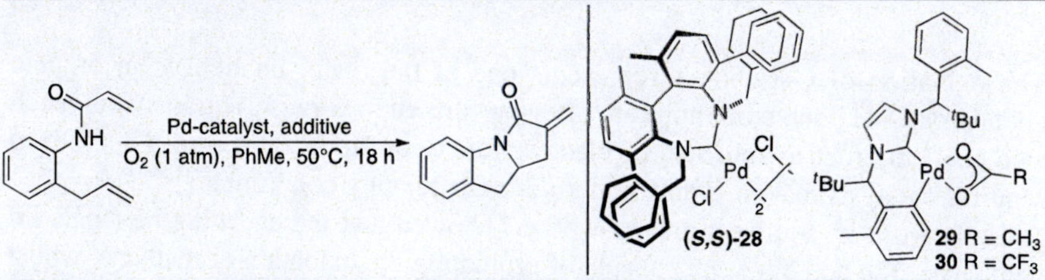

Catalyst	Loading (mol%)	Additive	Yield (%)	*ee* (%)
(*S*,*S*)-**28**	10	AgTFA (40 mol%), $^{i}Pr_2NEt$	24	27
(*S*,*S*)-**28**	10	AgTFA (40 mol%), $^{i}Pr_2NEt$ 3Å MS	35	63
29	5	–	57	2
29	5	Na_2CO_3 (2 equiv.)	66	7
30	5	Na_2CO_3 (2 equiv.)	13	4

10.3.4 Oxidative Cleavage of Alkenes

The oxidative cleavage of alkenes is a common reaction usually achieved by ozonolysis or the use of potassium permanganate. An example of NHC-coordinated Ru complex (**31**) capable of catalysing the oxidative cleavage of alkenes was reported by Peris and co-workers (Table 10.9) [44]. Despite a relatively limited substrate scope, this reaction reveals an intriguing reactivity of ruthenium and will surely see further elaboration.

Table 10.9 Ru bis-NHC catalysed oxidative cleavage of alkenes [44]

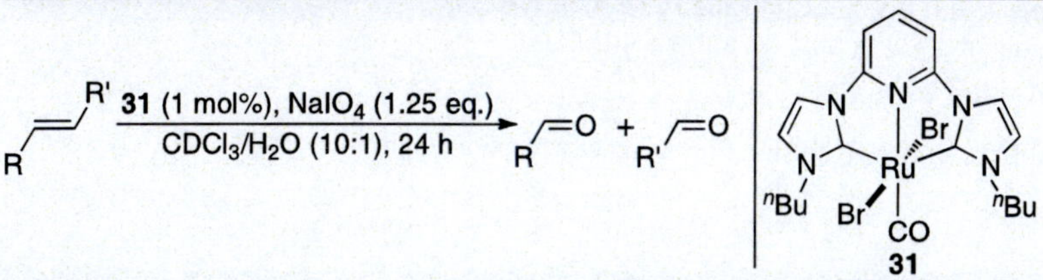

Substrate	Product	Yield (%)	Substrate	Product	Yield (%)
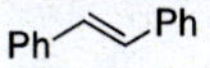	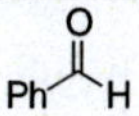	58	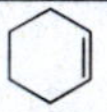	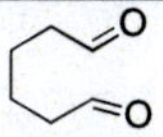	72
	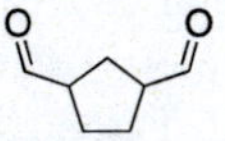	>99	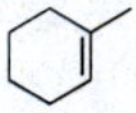	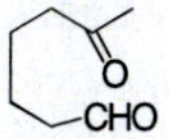	>99
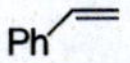	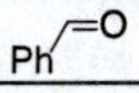	23			

10.3.5 Oxidation of Methane

The oxidation of C–H bonds represents one of the major challenges in organic chemistry. C–H bond oxidations are often performed in strongly acidic media (such as oleum) in order to enhance the electrophilicity of the metal centre. This indeed requires catalysts that are stable under very challenging conditions.

Early work by Strassner and co-workers showed that the chelating bis-NHC Pd complexes **32a** and **32c** were capable of promoting the oxidation of methane, whilst the iodo-analogues **32b** and **32d** were inactive under the same reaction condition [45]. Indeed, in a mixture of TFA and TFAA, in the presence of potassium peroxodisulfate under 20–30 bar of methane, trifluoroactic acid methyl ester is produced, using **32a** or **32c** as catalyst (Scheme 10.14). In a more recent work, the authors disclosed the use of pyrimidine-NHC Pd complexes for the same reaction. A slightly better catalytic activity was obtained with the unexpected cationic complex **34** [46].

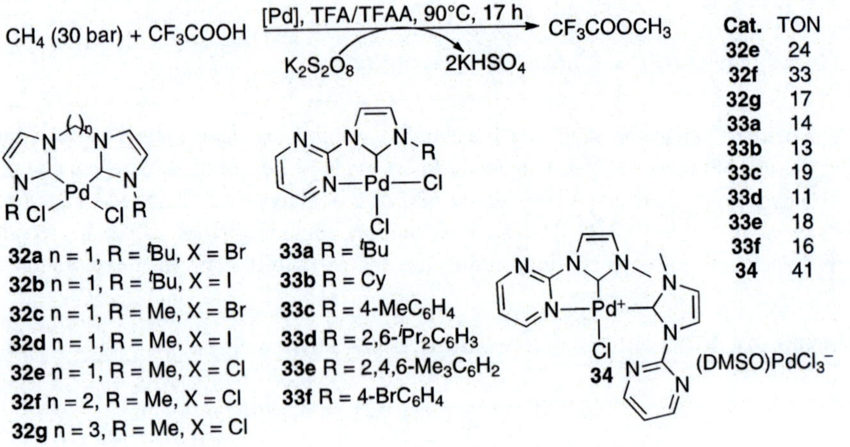

Scheme 10.14 Pd–NHC systems for the catalytic oxidation of methane [45,46]

10.4 Conclusions

Thus far, the described applications of NHC-coordinated complexes clearly illustrate the real potential of this ligand family in oxidation chemistry. Their strong σ-donation properties allow them to stabilise transition metal centres in high oxidation states and the relatively strong metal-NHC bond dissociation energy helps prevent oxidative decomposition. Numerous intermediates in metal-mediated oxidation reactions have been isolated due in part to the stabilisation brought by the use of NHC ligands. Despite the fact that most oxidation catalytic systems make use of palladium, examples employing other transition metal-NHC complexes are emerging. In the coming years, we can expect further developments and uses of NHC ligands in oxidation catalysis, especially in the important area of asymmetric catalysis. This area is in its infancy and the unique ligand properties displayed by NHCs should facilitate advances in oxidation catalysis.

References

1. Rogers MM, Stahl SS (2007) In: Glorius F (ed) N-Heterocyclic carbenes in transition metal catalysis. Springer, Berlin/Heidelberg. Top Organomet Chem 21:21–46
2. Strassner T (2007) In: Meyer F, Limberg C (eds) Organometallic oxidation catalysis. Springer, Berlin/Heidelberg. Top Organomet Chem 22:125–148
3. Schultz MJ, Sigman MS (2006) In: Nolan SP (ed) N-Heterocyclic carbenes in synthesis. Wiley-VCH, Weinheim, pp 103–118
4. Popp BV, Stahl SS (2007) In: Meyer F, Limberg C (eds) Organometallic oxidation catalysis. Springer, Berlin/Heidelberg. Top Organomet Chem 22:149–189
5. Konnick MM, Guzei IA, Stahl SS (2004) J Am Chem Soc 126:10212–10213
6. Stahl SS (2004) Angew Chem Int Ed 43:3400–3420
7. Konnick MM, Gandhi BA, Guzei IA, Stahl SS (2006) Angew Chem Int Ed 45:2904–2907
8. Popp BV, Stahl SS (2007) J Am Chem Soc 129:4410–4422
9. Konnick MM, Stahl SS (2008) J Am Chem Soc 130:5753–5762
10. Yamashita M, Goto K, Kawashima T (2005) J Am Chem Soc 127:7294–7295
11. Popp BV, Wendlandt JE, Landis CR, Stahl SS (2007) Angew Chem Int Ed 46:601–604
12. Häller LJL, Mas-Marza E, Moreno A, Lowe JP, Macgregor SA, Mahon MF, Pregosin PS, Whittlesey MK (2009) J Am Chem Soc 131:9618–9619
13. Hu X, Castro-Rodriguez I, Meyer K (2004) J Am Chem Soc 126:13464–13473
14. Dibble BR, Sigman MS (2006) Inorg Chem 45:8430–8441
15. Dibble BR, Sigman MS (2003) J Am Chem Soc 125:872–873
16. Larock RC (1999) Comprehensive organic transformations, 2nd ed. VCH, New York, pp 1234–1250
17. Muzart J (2003) Tetrahedron 59:5789–5816
18. Nishimura T, Onoue T, Ohe K, Uemura S (1999) J Org Chem 64:6750–6755
19. Nishimura T, Maeda Y, Kakiuchi N, Uemura S (2000) J Chem Soc Perkin Trans 1:1915–1918
20. Nishimura T, Kakiuchi N, Onoue T, Ohe K, Uemura S (2000) J Chem Soc Perkin Trans 1:4301–4305
21. Ten Brink GJ, Arends IWCE, Sheldon RA (2000) Science 287:1636–1639
22. Jensen DR, Schultz MJ, Mueller JA, Sigman MS (2003) Angew Chem Int Ed 42:3810–3813
23. Mueller JA, Goller CP, Sigman MS (2004) J Am Chem Soc 126:9724–9734
24. Schultz MJ, Hamilton SS, Jensen DR, Sigman MS (2005) J Org Chem 70:3343–3352
25. Jensen DR, Sigman MS (2003) Org Lett 5:63–65
26. Chen T, Jiang JJ, Xu Q, Shi M (2007) Org Lett 9:865–868
27. Liu L, Wang F, Shi M (2009) Organometallics 28:4416–4420
28. Liu SJ, Liu L, Shi M (2009) Appl Organomet Chem 23:183–190
29. Guram AS, Bei X, Turner HW (2003) Org Lett 5:2485–2487
30. Bei X, Hagemeyer A, Volpe A, Saxton R, Turner H, Guram AS (2004) J Org Chem 69:8626–8633
31. Brayton DF, Mocka C, Berini C, Navarro O (2009) Org Lett 11:4244–4247
32. Navarro O, Kaur H, Mahjoor P, Nolan SP (2004) J Org Chem 69:3173–3180
33. Hanasaka F, Fujita K-I, Yamaguchi R (2004) Organometallics 23:1490–1492

34. Hanasaka F, Fujita K-I, Yamaguchi R (2005) Organometallics 24:3422–3433
35. Edwards MG, Jazzar RFR, Paine BM, Shermer DJ, Whittlesey MK, Williams JMJ, Edney DD (2004) Chem Commun 90–91
36. Burling S, Whittlesey MK, Williams JMJ (2005) Adv Synth Catal 347:591–594
37. Hintermann L, Labonne A (2007) Synthesis 1121–1150
38. Marion N, Ramon RS, Nolan SP (2009) J Am Chem Soc 131:448–449
39. Ramon SR, Marion N, Nolan SP (2009) Chem Eur J 15:8695–8697
40. Smidt J, Hafner W, Jira R, Sedlmeier J, Sieber R, Rüttinger R, Kojer H (1959) Angew Chem 71:176–182
41. Cornell CN, Sigman MS (2005) J Am Chem Soc 127:2796–2797
42. Rogers MM, Wendledt JE, Guzei IA, Stahl SS (2006) Org Lett 8:2257–2260
43. Scarborough CC, Bergant A, Sazama GT, Guzei IA, Spencer LC, Stahl SS (2009) Tetrahedron 65:5084–5092
44. Poyatos M, Mata JA, Falomir E, Crabtree RH, Peris E (2003) Organometallics 22:1 110–1114
45. Muehlhofer M, Strassner T, Herrmann WA (2002) Angew Chem Int Ed 41:1745–1747
46. Meyer D, Taige MA, Zeller A, Hohlfeld K, Ahrens S, Strassner T (2009) Organometallics 28:2142–2149

Chapter 11
N-Heterocyclic Carbene Complexes in other Transition Metal Mediated Reactions

Tracy D. Nixon and Jonathan M. J. Williams

Abstract This chapter describes reactions involving NHC-transition metal complexes that have not been considered in previous chapters. The reactions are treated in six sections, presenting borrowing hydrogen reactions where C–C and C–N bonds are formed from alcohols in the first section. Then dehydration reactions either with or without coupling are discussed. The dynamic kinetic resolution of alcohols using a combination of NHC–TM catalysed racemisation coupled with enzyme-catalysed resolution is described. The next section considers the emerging area of dehydrogenation reactions, followed by a section discussing the isomerisation of alkene-containing substrates. The final section details hydrogen/deuterium exchange reactions within aryl and alkyl substrates.

11.1 Borrowing Hydrogen

The temporary removal of hydrogen from alcohols provides access to the more electrophilic aldehydes and ketones. In addition, suitable aldehyde and ketone intermediates can undergo deprotonation to form nucleophilic enolates. The borrowing hydrogen approach exploits the greater reactivity of carbonyl compounds over alcohols by using the temporary metal-catalysed removal of hydrogen from the relatively inert alcohol substrates [1]. The formation of C–C bonds and C–N bonds has been achieved using this strategy and NHC–TM complexes have been successfully used to catalyse these reactions.

The coupling of a secondary alcohol **1** with a primary alcohol **2** is achieved by the temporary removal of H_2 from each substrate which generates the ketone **3** and aldehyde **4** intermediates. A crossed aldol condensation occurs under the reaction conditions by the enolate derived from ketone **3** undergoing nucleophilic addition

J. M. J. Williams (✉)
Department of Chemistry, University of Bath, Claverton Down, Bath BA2 7AY, UK
e-mail: j.m.j.williams@bath.ac.uk

C. S. J. Cazin (ed.), *N-Heterocyclic Carbenes in Transition Metal Catalysis and Organocatalysis*, Catalysis by Metal Complexes 32,
DOI 10.1007/978-90-481-2866-2_11,

to aldehyde **4**, which after loss of water leads to the α,β-unsaturated ketone **5**. Return of the two equivalents of hydrogen reduces the carbonyl and alkenyl groups providing the coupled alcohol product **6** (Scheme 11.1).

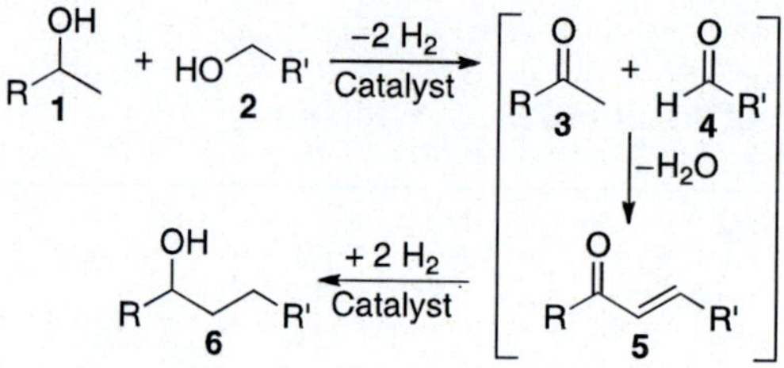

Scheme 11.1 Coupling of a secondary alcohol with a primary alcohol

Various secondary alcohols and primary alcohols have been coupled, with a common example involving the reaction of 1-phenylethanol **7** with benzyl alcohol **8** to generate the coupled product **9** (Scheme 11.2). Reactions are generally run in toluene at reflux with 1 mol% of a suitable catalyst. Potassium hydroxide is required in order to promote the aldol coupling step and also facilitates the exchange of the alcohol/alkoxide with the halide in the complex. Peris and co-workers have used the ruthenium Janus-head complex **10** [2], along with a range of other NHC ligands complexed to [$RuCl_2$(*p*-cymene)], including complex **11** [3]. More basic NHC ligands were found to lead to more reactive catalysts.

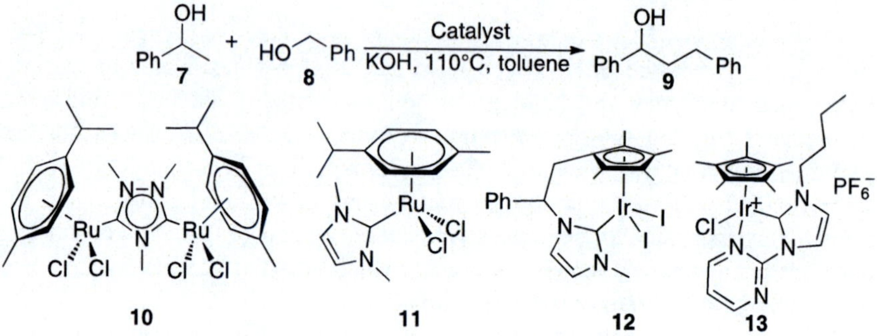

Scheme 11.2 Ru and Ir NHC complexes used in the coupling of alcohols **7** and **8**

Iridium NHC complexes have also been used to catalyse the coupling of alcohols **7** and **8**. Complexes **12** [4] and **13** [5], containing chelating NHC ligands, are also effective for catalysing the same reaction.

The ruthenium NHC complex **14** readily donates hydrogen to suitable acceptors and forms the C–H activated complex **15** [6]. The reversible nature of the C–H activation is established by the return of the hydrogen to restore the original complex **14** (Scheme 11.3). Catalyst **14** has been shown to be effective for borrowing hydrogen reactions for the indirect Wittig reaction on alcohols [7]. Complex **14** undergoes activation by loss of hydrogen to vinylsilane to give complex **15**. It is believed that benzyl alcohol **8** is dehydrogenated to give benzaldehyde, and that Wittig reaction on the aldehyde with phosphonium ylide **16** leads to alkene formation, with return of the borrowed hydrogen giving product **17**.

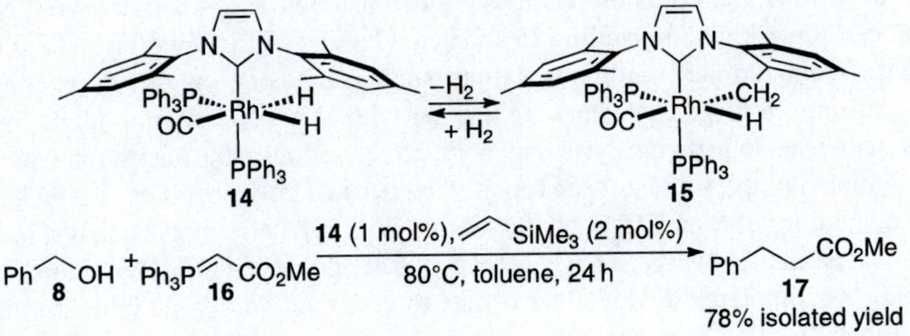

Scheme 11.3 Indirect Wittig reaction of an alcohol

Whittlesey, Williams and co-workers further developed the catalytic indirect Wittig reaction and found that the more electron-rich NHC present in complex **18** provided a more reactive catalyst [8]. Catalyst **18** was used to convert benzyl alcohol **8** and phosphonium ylide **19** into the product **20** under slightly milder reaction conditions and in a shorter time than in previous work (Scheme 11.4). Other C–C bond-forming reactions from alcohols using a borrowing hydrogen approach have been reported, with Peris and co-workers using Ir-NHC complexes for the C-3 alkylation of indoles with alcohols [9].

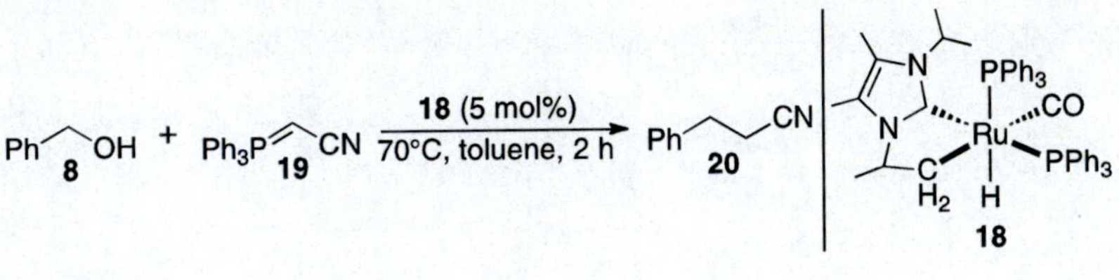

Scheme 11.4 An improved catalyst for indirect Wittig reactions on an alcohol

Borrowing hydrogen reactions have also been developed for the alkylation of amines by alcohols according to Scheme 11.5. Temporary removal of hydrogen from an alcohol **21** leads to the formation of an aldehyde **22**. Aldehydes readily condense with primary amines to form the corresponding imine **23**. The catalyst then returns the hydrogen, which gives the alkylated product **24**. The use of alcohols as alkylating agents for amines avoids the use of traditional alkylating agents such as alkyl halides, which can be toxic and mutagenic.

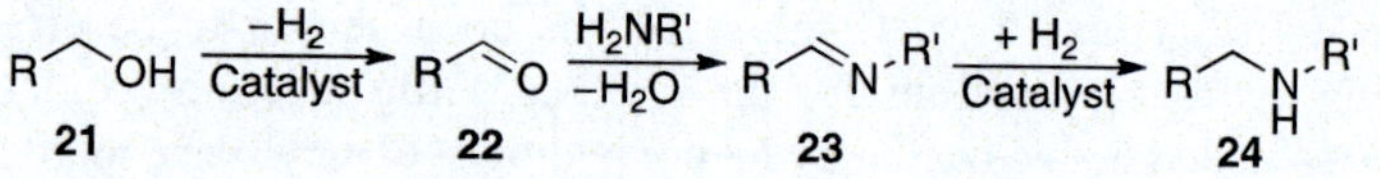

Scheme 11.5 Alkylation of amines by alcohols

Typical amine alkylation reactions involve the coupling of primary alcohols such as benzyl alcohol **8** or secondary alcohols such as 1-phenylethanol **7** with primary amines including aniline **25** and benzylamine **27** (Scheme 11.6). Less common are amine/amine coupling reactions, such as the coupling of benzylamine **27** with aniline **25** to give *N*-benzylaniline **26** with loss of ammonia. NHC ligands have been used to form catalysts which are effective for amine alkylation reactions. The iridium complex **29** has been reported by Liu and co-workers to give 96% yield in the coupling of **8** and **25** to give *N*-benzylaniline **26** at 1 mol% catalyst loading [10]. Peris and co-workers have activated a range of iridium NHC complexes, including compound **30** with silver triflate to give over 95% yield in the formation of amines **26** and **28** by alcohol/amine coupling [11]. At higher temperature, the same complex was found to be effective for coupling amines **27** and **25** to give *N*-benzylaniline **26**. Complex **13**, which was successful for C–C bond-formation, was shown by Crabtree and co-workers to be equally effective for amine alkylation, including the formation of *N*-benzylaniline **26** in 98% conversion [5]. Complex **30** has also been used by Marr, Stephens and co-workers for the alkylation of aniline **25** with 1,3-propanediol and closely related catalysts were used in the amination reactions of waste glycerol from biodiesel production [12].

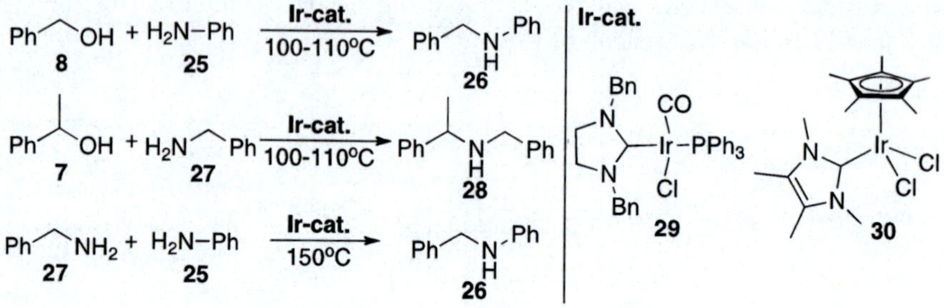

Scheme 11.6 Alkylation of amines by alcohols or amines

Madsen and co-workers have reported an important extension to the amine alkylation chemistry, in which oxidation takes place to give the amide product [13]. A ruthenium NHC complex is formed *in situ* by the reaction of [$RuCl_2$(cod)] with a phosphine and an imidazolium salt in the presence of base. Rather than returning the borrowed hydrogen, the catalyst expels two equivalents of H_2. For example, alcohol **31** and benzylamine **27** undergo an oxidative coupling to give amide **32** in good isolated yield (Scheme 11.7).

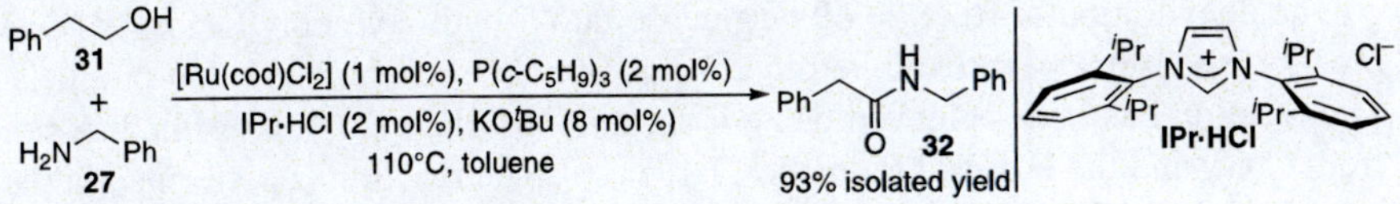

Scheme 11.7 Oxidative coupling of an amine and alcohol

11.2 Dehydration

Dehydration reactions catalysed by NHC complexes have been reported where a new C–C bond is formed. Peris has used $[Ir(OTf)_2Cp^*(NHC)]$ complexes including compound **35** to benzylate arenes with alcohols and other reagents [14]. For example, the dehydrative C–C coupling of benzyl alcohol **8** with toluene **33** is catalysed by 0.1 mol% of **35** to give a mixture of benzylated products **34** (Scheme 11.8).

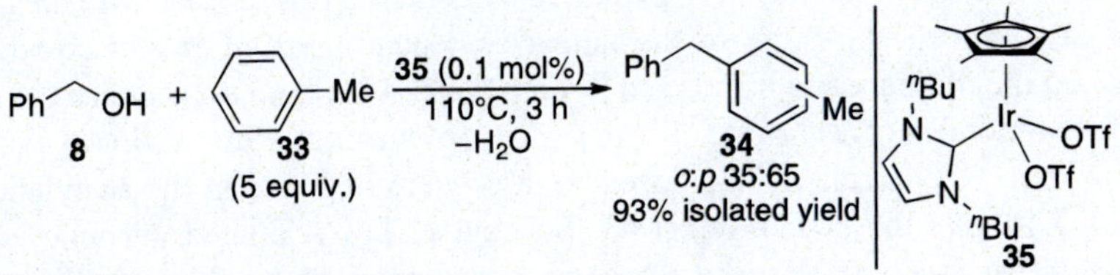

Scheme 11.8 Dehydrative C–C coupling

Another interesting example of dehydrative C–C coupling involves the alkylation of benzimidazole **36** with allyl alcohol **37**, which is catalysed by complex **39** [15]. The reaction is believed to proceed by alkene complex formation with the allyl alcohol **37** with loss of water from the NH proton of the NHC ligand and OH of the allyl alcohol to give an intermediate π-allyl complex. The initially formed 2-allylbenzimidazole isomerises to a mixture of the internal alkenes **38** (Scheme 11.9).

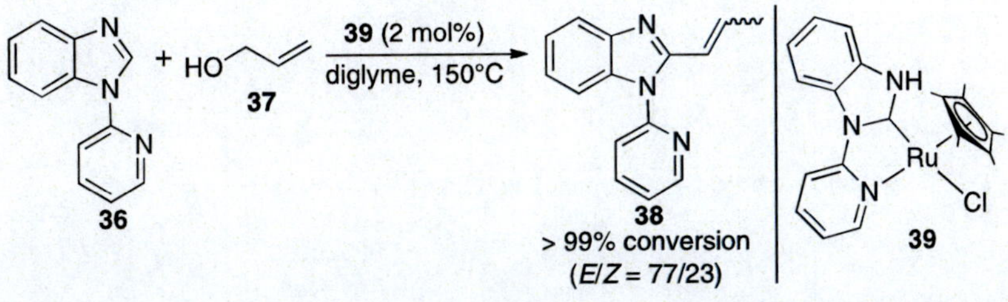

Scheme 11.9 Dehydrative coupling of a benzimidazole with allyl alcohol

The dehydration of fructose **40** or glucose into 5-hydroxymethylfurfural **41** is a process which has been exploited to convert biomass into higher value products. The reaction has been achieved using a chromium NHC complex, formed *in situ* from $CrCl_2$ and the NHC **42** (Scheme 11.10) [16]. The reaction is performed in the ionic liquid $BMIM^+Cl^-$ (1-butyl-3-methylimidazolium chloride).

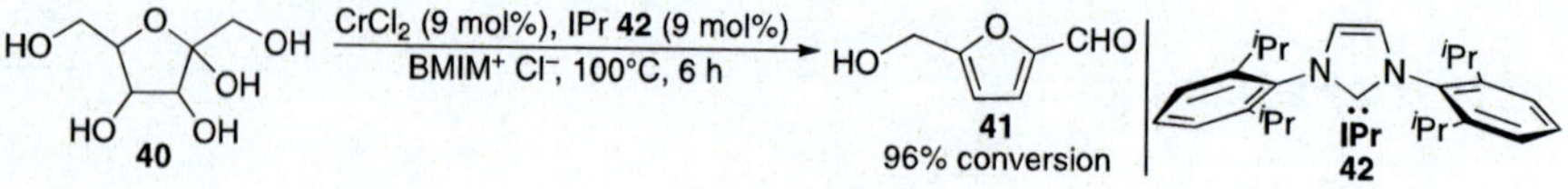

Scheme 11.10 Dehydration of fructose to give 5-hydroxymethylfurfural

11.3 Dynamic Kinetic Resolution

The ability of enzymes to achieve the selective esterification of one enantiomer of an alcohol over the other has been exploited by coupling this process with the *in situ* metal-catalysed racemisation of the unreactive enantiomer. Marr and co-workers have used the rhodium and iridium NHC complexes **44** and **45** to racemise the unreacted enantiomer of substrate **7** [17]. In combination with a lipase enzyme (Novozyme 435), excellent enantioselectivities were obtained in the acetylation of alcohol **7** to give the ester product **43** (Scheme 11.11). A related dynamic kinetic resolution has been reported by Corberán and Peris [18]. In their chemistry, the aldehyde **46** is readily racemised and the iridium NHC catalyst **35** catalyses the reversible reduction of aldehyde **46** to give an alcohol which is acylated by an enzyme to give the ester **47** in reasonable enantiomeric excess.

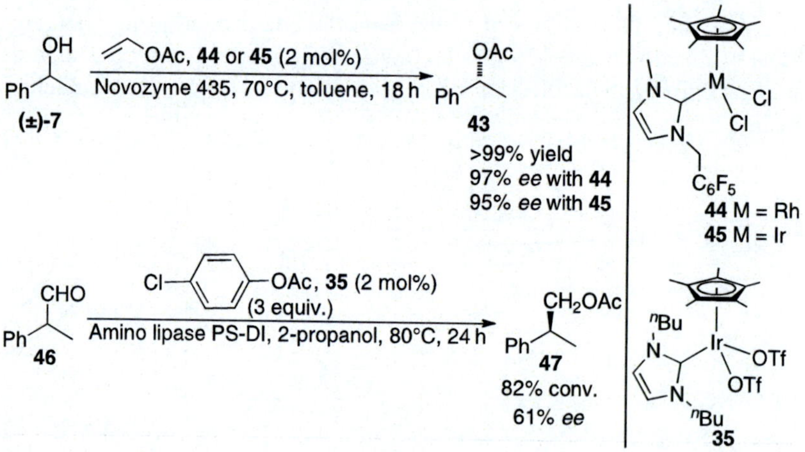

Scheme 11.11 Dynamic kinetic resolution reactions using enzyme/metal combinations

11.4 Dehydrogenation

The ease of dehydrogenation of organic substrates clearly depends on the nature of the functional groups which are present. Whilst the dehydrogenation of alcohols to give ketones can be relatively straightforward, the dehydrogenation of hydrocarbons is a significant challenge. Cole-Hamilton, Santini and co-workers have used rhodium and iridium NHC complexes, including the cationic compound **51** for the dehydrogenation of cyclooctadiene **48** [19]. Isomerisation and dehydrogenation were both observed, although the presence of the NHC ligand was not found to be beneficial for these reactions (Scheme 11.12).

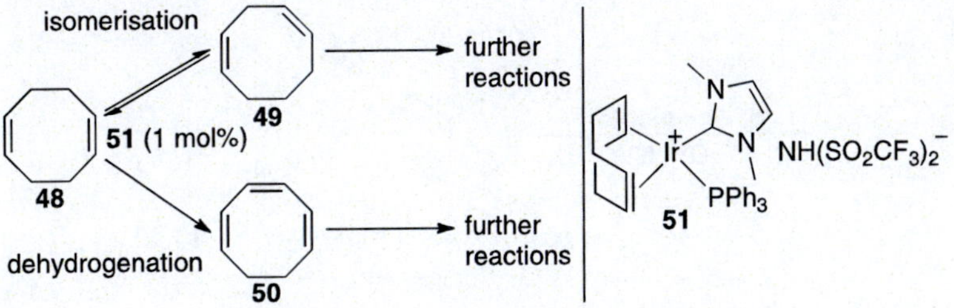

Scheme 11.12 Isomerisation and dehydrogenation of 1,5-cyclooctadiene

There has been considerable recent interest in the use of ammonia–borane ($H_3N{\cdot}BH_3$) for hydrogen storage. Dehydrogenation of $H_3N{\cdot}BH_3$ initially gives $H_2N{-}BH_2$, which can then undergo further reaction. Baker and co-workers have developed a catalyst derived from $Ni(cod)_2$ and NHC **52** for this reaction (Scheme 11.13) [20]. It is the first example of the use of a first row transition metal catalyst for the dehydrogenation of ammonia–borane. Work considering the details of the mechanism of the nickel-catalysed dehydrogenation of ammonia–borane has been reported [21–23].

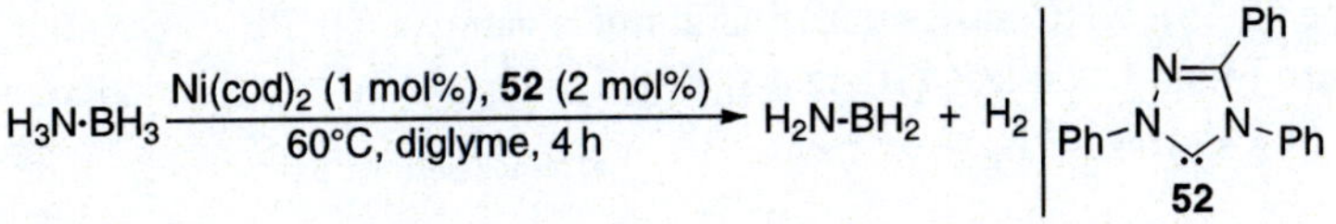

Scheme 11.13 Catalytic dehydrogenation of ammonia–borane

11.5 Isomerisation

The most common isomerisation reactions catalysed by transition metals are those involving the isomerisation of alkenes. Faller, Crabtree and co-workers have reported that the iridium *bis*-NHC complex **55** is effective for the isomerisation of allylbenzene **53** into the conjugated alkene **54** as the major product (Scheme 11.14) [24]. Alkene isomerisation has been reported using ruthenium NHC complexes that are more commonly used for alkene metathesis reactions. For example, Hanessian and co-workers have developed conditions where the Grubbs second generation catalyst **58** catalyses the isomerisation of terminal alkenes in preference to metathesis reactions [25]. There have been many reports where unwanted alkene isomerisation has been seen as an unwanted byproduct in alkene metathesis reactions [26,27], along with studies to identify the origin of the intermediates involved [28,29]. Nolan and Prunet have developed conditions where alkene isomerisation can be avoided [30], and Grubbs and co-workers have addressed the same problem [31].

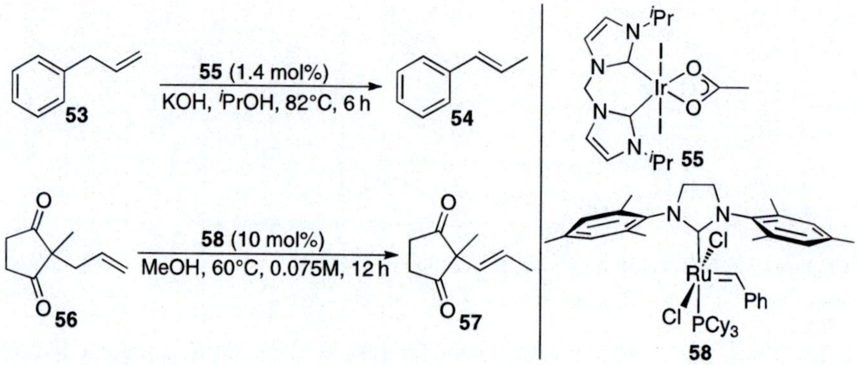

Scheme 11.14 Alkene isomerisation reactions

The isomerisation of allylic alcohols to saturated ketones usually has a strong thermodynamic driving force. The ruthenium NHC complex **62** has been used to catalyse the isomerisation of allylic alcohol **59** which gives ketone **60** as the principal product along with some of the reduction product **61** [32]. The catalyst was water-soluble and the aqueous phase could be re-used for several runs (Scheme 11.15). NHC analogues of Crabtree's catalyst, [Ir(PCy$_3$)(pyridine)(cod)]PF$_6$, were found to be less efficient for the isomerisation of allylic alcohols than the parent catalyst [33].

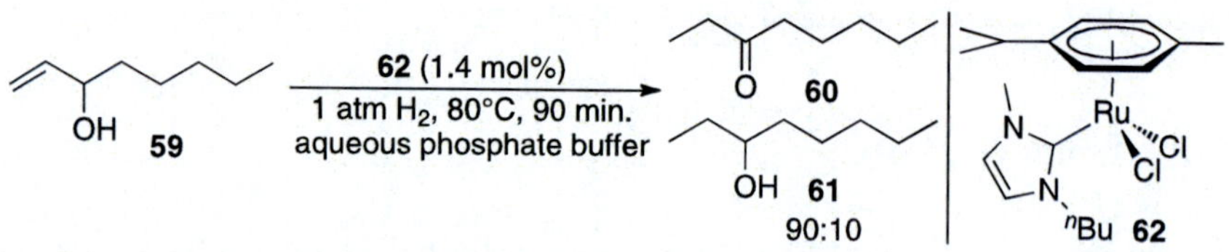

Scheme 11.15 Isomerisation of allylic alcohols to ketones.

11.6 H–D Exchange

The ability to exchange hydrogen atoms for deuterium has a number of applications, most notably in making the identification of drug metabolites easier. Iridium NHC complex **65** has been reported by Kerr and co-workers to catalyse H–D exchange on various aromatic substrates [34]. For example, the deuteration of acetophenone **63** was catalysed by complex **65** to give dideuterioacetophenone **64**. In this case, the catalytic activity was much greater than using similar iridium complexes without the NHC ligand. Powell and co-workers have used the related catalyst **66** for H–D exchange, and have also seen improved activity when the NHC ligand was present [35]. The iridium catalyst **67** is one of several reported by Peris and co-workers for H–D exchange of aromatic and aliphatic C–H bonds [36]. The palladium NHC complex **68** also has good activity for a range of H–D exchange reactions [37] (Scheme 11.16).

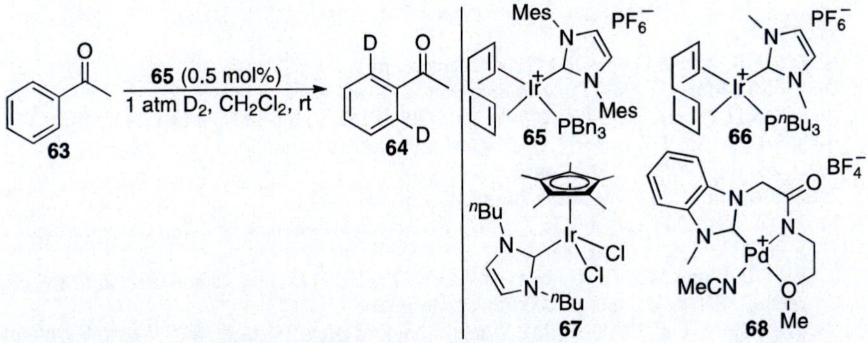

Scheme 11.16 H–D exchange reaction of acetophenone

References

1. Nixon TD, Whittlesey MK, Williams JMJ (2009) Dalton Trans 753–762
2. Viciano M, Sanaú M, Peris E (2007) Organometallics 26:6050–6054
3. Prades A, Viciano M, Sanaú M, Peris E (2008) Organometallics 27:4254–4259
4. Pontes da Costa A, Viciano M, Sanaú M, Merino S, Tejeda J, Peris E, Royo B (2008) Organometallics 27:1305–1309
5. Gnanamgari D, Sauer ELO, Schley ND, Butler C, Incarvito CD, Crabtree RH (2009) Organometallics 28:321–325
6. Burling S, Whittlesey MK, Williams JMJ (2005) Adv Synth Catal 347:591–594
7. Edwards MG, Jazzar RFR, Paine BM, Shermer DJ, Whittlesey MK, Williams JMJ, Edney DD (2004) Chem Commun 90–91
8. Burling S, Paine BM, Nama D, Brown VS, Mahon MF, Prior TJ, Pregosin PS, Whittlesey MK, Williams JMJ (2007) J Am Chem Soc 129:1987–1995
9. Zanardi A, Corberán R, Mata JA, Peris E (2008) Organometallics 27:3570–3576
10. Chang Y-H, Fu C-F, Liu Y-H, Peng S-M, Chen J-T, Liu S-T (2009) Dalton Trans 861–867

11. Prades A, Corberán R, Poyatos M, Peris E (2008) Chem Eur J 14:11474–11479
12. Liu S, Rebros M, Stephens G, Marr AC (2009) Chem Commun 2308–2310
13. Nordstrøm LU, Vogt H, Madsen R (2008) J Am Chem Soc 130:17672–17673
14. Prades A, Corberán R, Poyatos M, Peris E (2009) Chem Eur J 15:4610–4613
15. Araki K, Kuwata S, Ikariya T (2008) Organometallics 27:2176–2178
16. Yong G, Zhang Y, Ying JY (2008) Angew Chem Int Ed 47:9345–9348
17. Marr AC, Pollock CL, Saunders GC (2007) Organometallics 26:3283–3285
18. Corberán R, Peris E (2008) Organometallics 27:1954–1958
19. Hintermair U, Gutel T, Slawin AMZ, Cole-Hamilton DJ, Santini CC, Chauvin Y (2008) J Organomet Chem 693:2407–2414
20. Keaton RJ, Blacquiere JM, Baker RT (2007) J Am Chem Soc 129:1844–1845
21. Zimmerman PM, Paul A, Zhang Z, Musgrave CB (2009) Angew Chem Int Ed 48:2201–2205
22. Yang X, Hall MB (2008) J Am Chem Soc 130:1798–1799
23. Yang X, Hall MB (2009) J Organomet Chem 694:2831–2838
24. Albrecht M, Miecznikowski JR, Samuel A, Faller JW, Crabtree RH (2002) Organometallics 21:3596–3604
25. Hanessian S, Giroux S, Larsson A (2006) Org Lett 8:5481–5484
26. Arisawa M, Terada Y, Takahashi K, Nakagawa M, Nishida A (2006) J Org Chem 71:4255–4261
27. Lehman SE Jr, Schwendeman JE, O'Donnell PM, Wagener KB (2003) Inorg Chim Acta 354:190–198
28. Hong SH, Day MW, Grubbs RH (2004) J Am Chem Soc 126:7414–7415
29. Dinger MB, Mol JC (2003) Eur J Inorg Chem 2827–2833
30. Bourgeois D, Pancrazi A, Nolan SP, Prunet J (2002) J Organomet Chem 643–644:247–252
31. Hong SH, Sanders DP, Lee CW, Grubbs RH (2005) J Am Chem Soc 127:17160–17161
32. Fekete M, Joó F (2006) Catal Commun 7:783–786
33. Mantilli L, Mazet C (2009) Tetrahedron Lett 50:4141–4144
34. Brown JA, Irvine S, Kennedy AR, Kerr WJ, Andersson S, Nilsson GN (2008) Chem Commun 1115–1117
35. Powell ME, Elmore CS, Dorff PN, Heys JR (2007) J Label Compd Radiopharm 50:523–525
36. Corberán R, Sanaú M, Peris E (2006) J Am Chem Soc 128:3974–3979
37. Lee JH, Yoo KS, Park CP, Olsen JM, Sakaguchi S, Prakash GKS, Mathew T, Jung KW (2009) Adv Synth Catal 351:563–568

Chapter 12
N-Heterocyclic Carbenes in Organocatalysis

Craig D. Campbell, Kenneth B. Ling, and Andrew D. Smith

Abstract The use of *N*-heterocyclic carbenes (NHCs) to promote organocatalytic transformations has rapidly expanded in recent years, building upon the classic use of these compounds to generate acyl anion equivalents from aldehydes. This chapter gives an overview of the recent progress made in this area, describing the use of NHCs to generate synthetic intermediates from a range of readily accessible starting materials and recent developments in their reactivity. The reaction of an NHC can result in the formation of a range of d^1, d^2 and d^3 synthons (acyl anion, azolium enol or enolate, and azolium homoenolate intermediates) or an electrophilic a^1 acylazolium species, with typical processes that proceed through each of these intermediates described. The use of NHCs to participate in stereoselective reaction processes within each of these areas is also described, with an emphasis upon a mechanistic understanding of these processes given where appropriate.

12.1 Introduction

The ability of *N*-heterocyclic carbenes (NHCs) to promote a remarkably diverse series of organocatalytic transformations has been identified in recent years, with a series of excellent reviews documenting progress in this area [1]. The aim of this chapter is twofold; firstly to describe the ability of NHCs to generate synthetically useful intermediates and their reactivity, and secondly to review the use of NHCs in stereoselective reaction processes. In order to categorise the recent advances in reactivity involving NHC mediated transformations, Section 12.2 classifies reaction processes by the type of intermediate through which C–C, C–H or C–heteroatom bond-forming reactions proceed. Within this framework, four main synthetic intermediates are

A. D. Smith (✉)
EaStCHEM School of Chemistry, University of St Andrews,
St Andrews, KY16 9ST, UK
e-mail: ads10@st-andrews.ac.uk

C. S. J. Cazin (ed.), *N-Heterocyclic Carbenes in Transition Metal Catalysis and Organocatalysis*, Catalysis by Metal Complexes 32,
DOI 10.1007/978-90-481-2866-2_12, © Springer Science+Business Media B.V. 2011

accessed: three nucleophilic species **1–3**, formally resulting in the formation of d^1, d^2 and d^3 synthons [2] (acyl anion, azolium enol or enolate, and azolium homoenolate intermediates, respectively) and an electrophilic acylazolium species **4** (Fig. 12.1).

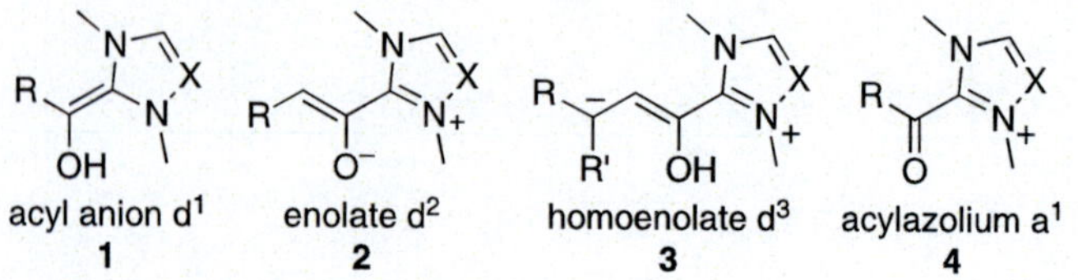

Fig. 12.1 Synthetic intermediates generated with NHCs

In Section 12.3 stereoselective reactions involving chiral NHCs have been classified in the same manner. Reactions that proceed through a number of these intermediates have been categorised according to the first asymmetric step in the transformation, with a miscellaneous section covering other reaction types.

12.2 Reactivity of NHC Derived Synthons

12.2.1 Generation of Acyl Anion Equivalents (d^1 Synthons) from Aldehydes

The simplest and most direct manner to generate acyl anion equivalents is through reaction of an NHC with an aldehyde, generating an enamine species **8**, commonly referred to as a "Breslow intermediate". Subsequent reaction with an electrophile, classically using aldehydes or enones, generates the benzoin and Stetter products **10** and **11** respectively (Scheme 12.1).

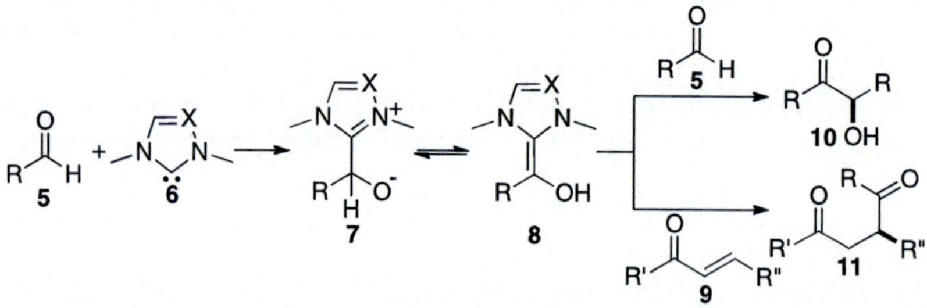

Scheme 12.1 Generation of acyl anion equivalents with NHCs

Alternative methods to access the same "Breslow intermediate" using NHCs utilising acyl silanes [3] or α-keto-acids have been developed [4], although these processes have not been utilised in asymmetric transformations to date.

12.2.2 Generation of Homoenolates, Enolates and Acylazoliums from Enals

Treatment of enals with NHCs allows the generation of nucleophilic (homoenolate and enolate) and electrophilic (acylazolium) species. Initial addition of an NHC to an enal **12** generates hydroxyazolium **13**, that can either be oxidised *in situ* to generate α,β-unsaturated acylazolium **19**, or, after proton transfer, generates intermediate **14** that acts as a homoenolate equivalent. Azolium enol or enolate species can also be accessed through variation of the reaction conditions and electrophilic partner in these reactions; selective β-protonation or tautomerisation of the homoenolate can generate an intermediate **16** with enolate reactivity. Protonation of enolate **16** generates an acylazolium **17**, with nucleophilic attack giving the desired product and regenerating the NHC (Scheme 12.2).

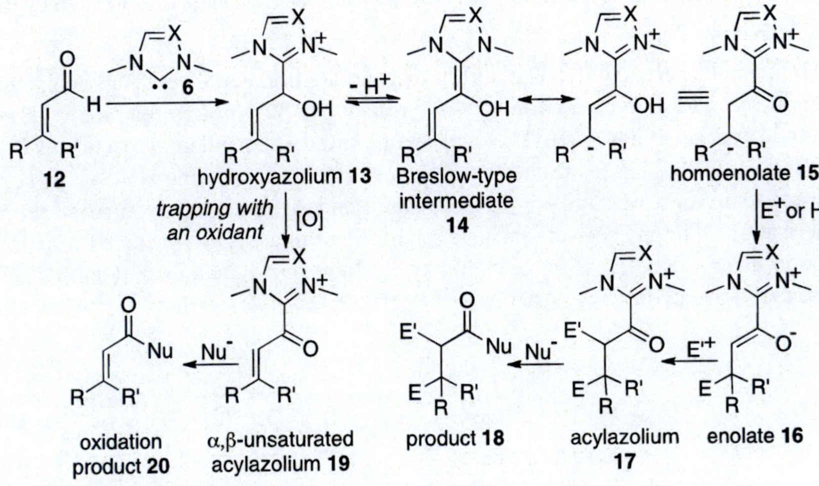

Scheme 12.2 Modes of reactivity of enals with NHCs

Representative examples of the synthetic utility of these reaction processes, all of which have been utilised in asymmetric transformations, are covered below.

1. Oxidation Processes: Scheidt and co-workers have employed cascade oxidation pathways from allylic or propargylic alcohols to afford unsaturated ester products **25**. *In situ* oxidation of an unsaturated alcohol **21** to the enal **22** using MnO_2,

followed by NHC addition affords hydroxyazolium **23** that is oxidised to acylazolium **24** and trapped with an alcohol. The oxidation of the hydroxyazolium intermediate is assumed to occur considerably faster than deprotonation to afford the Breslow-type intermediate in this process, as no homoenolate products are observed under these conditions (Scheme 12.3) [5].

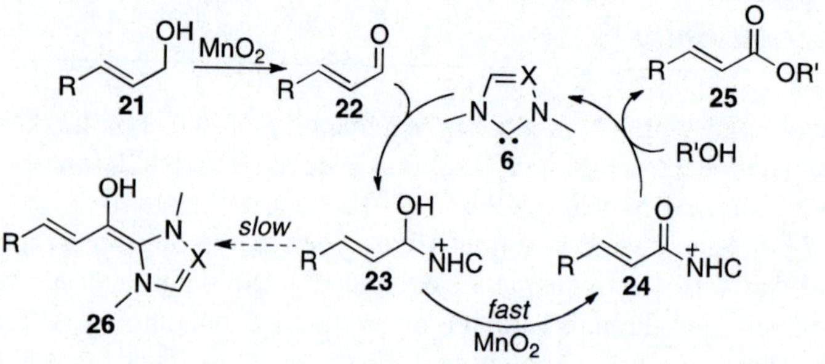

Scheme 12.3 Cascade oxidation pathway by hydroxyazolium trapping

Related oxidation processes have been reported that allow the generation of esters directly from aryl aldehydes [6] and the hydroacylation of α-keto esters with aldehydes [7].

2. Homoenolate Reactivity: The ability to generate homoenolates from enals and its application to the preparation of γ-butyrolactones **30**, through reaction with an aldehyde or aryl trifluoromethyl ketone, was reported independently by Glorius [8], and Bode and Burstein [9] (Scheme 12.4). A sterically demanding NHC catalyst is required to promote reactivity at the d^3 terminus and to prevent competitive benzoin dimerisation. Nair and co-workers have reported a similar spiro-γ-lactone formation reaction using cyclic 1,2-diones, including cyclohexane-1,2-dione and substituted isatin derivatives [10].

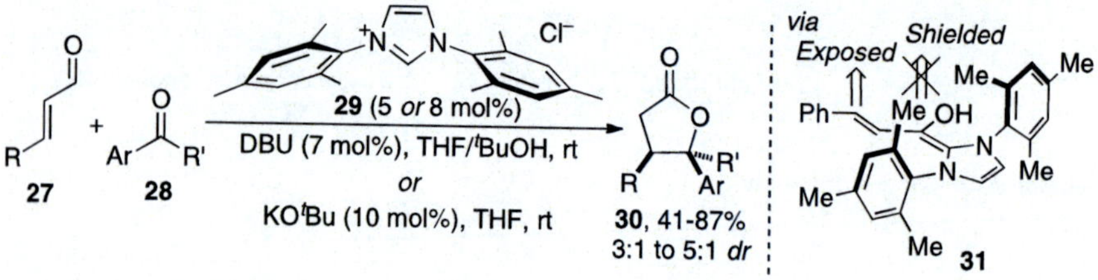

Scheme 12.4 γ-Butyrolactone formation with homoenolates

Bode and co-workers have used NHCs to form γ-butyrolactams **34** from enals **27** and saccharin-derived cyclic sulfonylimines **32**. A range of β-alkyl and β-aryl substituted enals, and a variety of substituted imines, are tolerated in this reaction, generating lactams in good yield with moderate to good diastereoselectivity, with a general preference for the *cis*-diastereoisomer [11]. Whilst this product outcome can be explained *via* homoenolate attack on the imine, the authors also propose an alternative mechanism involving an ene reaction of the hydroxyazolium derived from NHC addition to the aldehyde *via* transition state **35**, which would deliver products with the observed relative stereochemistry (Scheme 12.5). Either of these mechanisms may be invoked to explain this type of transformation, however all such reactions will be classified as occurring *via* a formal homoenolate pathway unless noted.

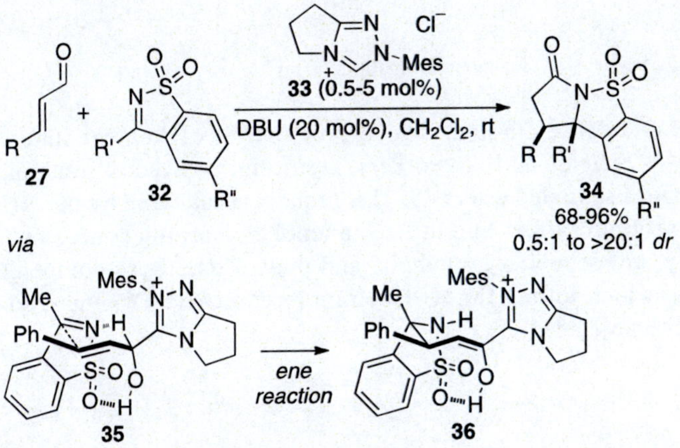

Scheme 12.5 Lactamisation and postulated alternative ene reaction pathway

Scheidt and Chan have shown that NHC promoted homoenolate formation and addition to azomethine imines **37** generates pyridazinones **41** with high diastereoselectivity, *via* a proposed highly organised transition state **40** due to a key hydrogen bonding interaction (Scheme 12.6) [12].

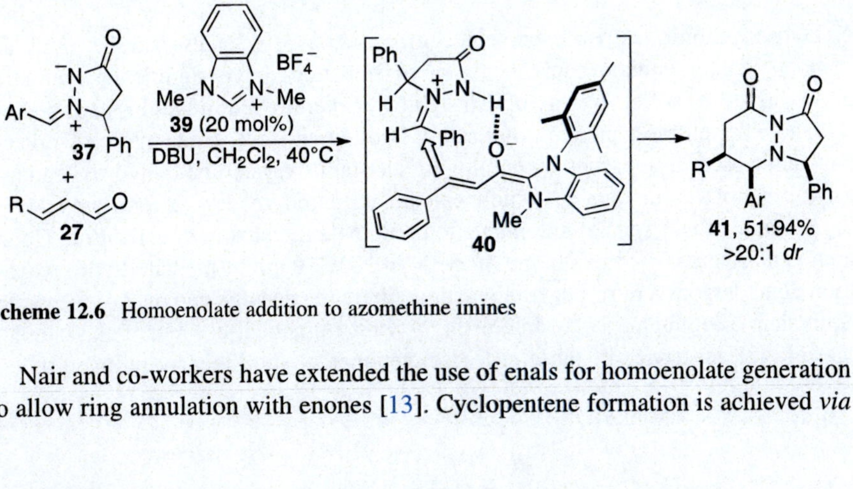

Scheme 12.6 Homoenolate addition to azomethine imines

Nair and co-workers have extended the use of enals for homoenolate generation to allow ring annulation with enones [13]. Cyclopentene formation is achieved *via* homoenolate attack upon the enone **42**, generating β-lactone **43** after cyclisation, which undergoes spontaneous decarboxylation to afford the *trans*-substituted cyclopentene products **44** in good to excellent yield. (Scheme 12.7)

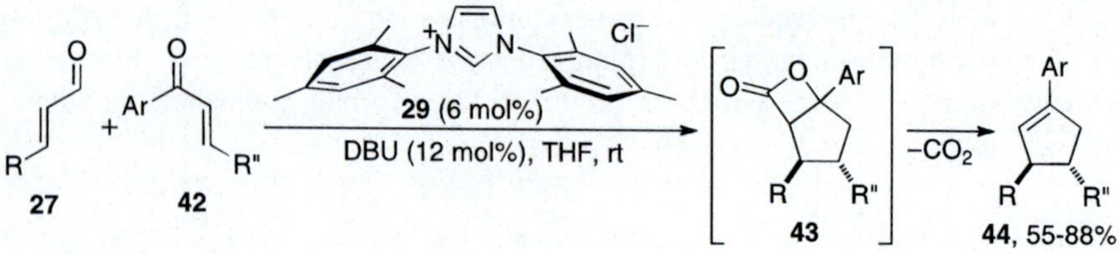

Scheme 12.7 Scope of cyclopentene formation

3. Homoenolate Protonation: The β-protonation of homoenolates has been observed by Scheidt and co-workers, resulting in a redox transformation of enals to afford saturated esters **48**. This process is catalysed by the NHC derived from imidazolium salt **46** and utilises phenol as a proton source [14]. A range of primary and secondary alcohols, and phenol itself, are competent nucleophiles with which to trap the acylazolium intermediate **47** generated by protonation (Scheme 12.8).

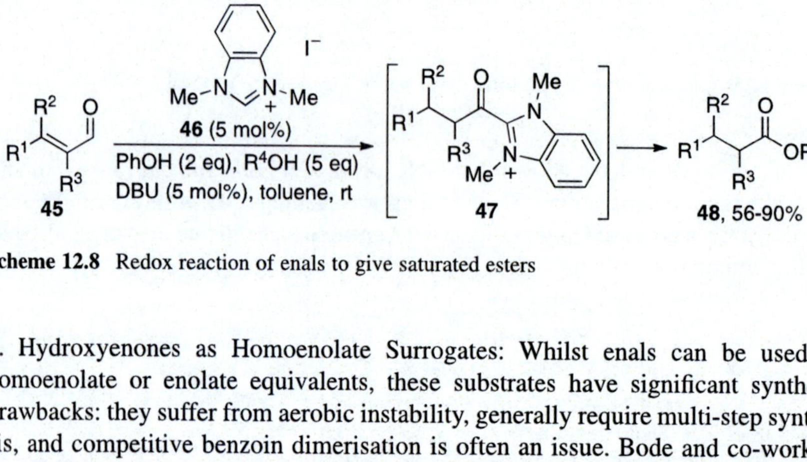

Scheme 12.8 Redox reaction of enals to give saturated esters

4. Hydroxyenones as Homoenolate Surrogates: Whilst enals can be used as homoenolate or enolate equivalents, these substrates have significant synthetic drawbacks: they suffer from aerobic instability, generally require multi-step synthesis, and competitive benzoin dimerisation is often an issue. Bode and co-workers have addressed these issues by employing air-stable, crystalline α-hydroxyenones as homoenolate surrogates, which can be prepared readily in one step [15]. Significantly, these species are unable to participate in benzoin dimerisation. Upon treatment of the α-hydroxyenone **49** with an NHC, the intermediate hydroxyazolium **50** undergoes a retro-benzoin reaction, liberating acetone and the homoenolate equivalent **52** (Scheme 12.9). Interestingly, whilst practical to prepare and handle, the relative reactivity of these α-hydroxyenones is significantly reduced when

compared to their enal counterparts. Competition experiments show that hydroxyenones are approximately five to six times less reactive than the corresponding enals in cyclopentene formation.

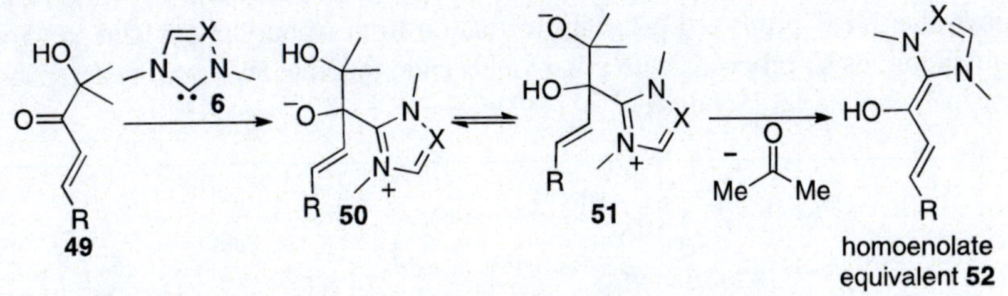

Scheme 12.9 Generation of homoenolate equivalents from α-hydroxyenones

Bode and co-workers have demonstrated the application of such substrates in a number of reactions, including cyclopentene ring formation and γ-lactamisation [15]. For example, γ-lactamisation to give **56** with saccharin-derived sulfonylimines **54** was achieved from α-hydroxyenone **53** with moderate to excellent stereocontrol (Scheme 12.10).

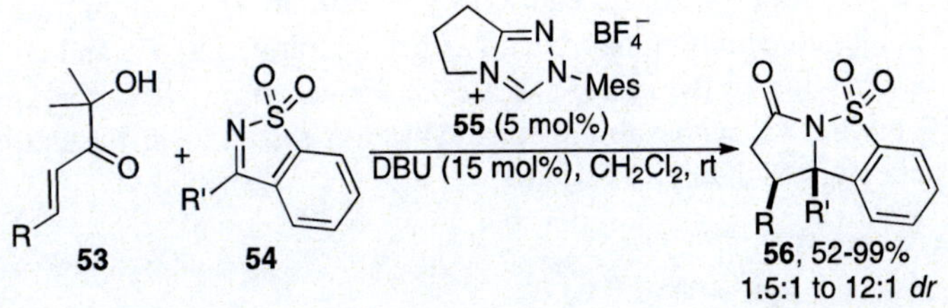

Scheme 12.10 γ-Lactam formation using α-hydroxyenones

Hydroxyenones have also been used in catalytic amide formation, although 1,2,4-triazole is required as a co-catalyst. Assumed protonation of the Breslow intermediate and tautomerisation generates an acylazolium intermediate, which is trapped by triazole, releasing the NHC and generating the acyltriazole **60** that is the active acylating agent for the amine (Scheme 12.11) [16].

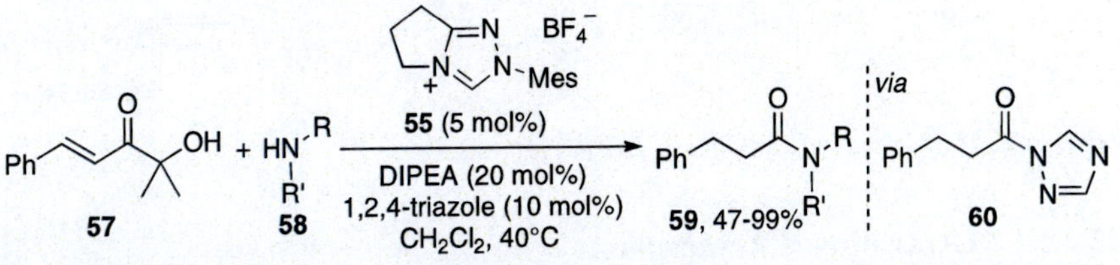

Scheme 12.11 Redox amide formation using hydroxyenones

12.2.3 *Direct Formation of Azolium Enolates*

Azolium enolates such as **65** can be generated directly through addition of NHCs to symmetrical or unsymmetrical ketenes. For example, Smith and co-workers have shown that NHC promoted β-lactam formation from isobutylphenylketene **61** and *N*-tosyl imines **63** proceeds with good yields and moderate levels of diastereoselectivity *via* enolate **65** (Scheme 12.12) [17].

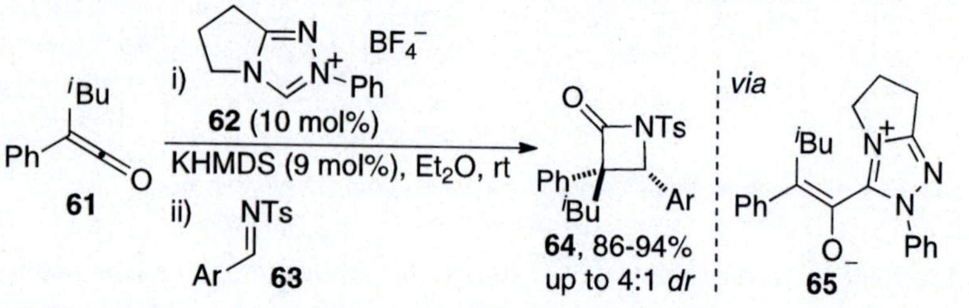

Scheme 12.12 β-Lactam formation promoted by NHCs

Ye and co-workers have shown that NHC **67** can catalyse the aza-Morita-Baylis-Hillman reaction of enones **66** and *N*-tosyl imines **63**, presumably *via* initial NHC conjugate addition to the enone to generate an azolium enolate **68** [18]. A related conjugate addition approach has been exploited by Fu and co-workers, with tautomerisation of the initial enolate **72** derived from NHC conjugate addition to **70** giving **73**, with subsequent cyclisation resulting in the umpolung of Michael acceptors (Scheme 12.13) [19].

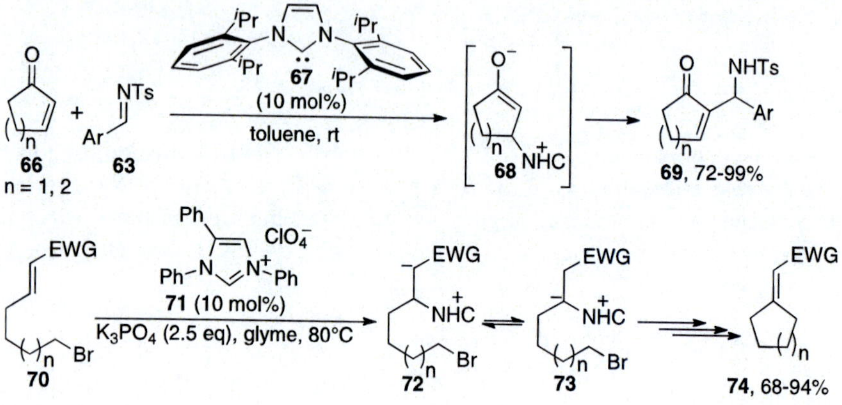

Scheme 12.13 Alternative azolium enolates in synthesis

12.2.3.1 Generation of Acylazoliums

Acylazolium species can be prepared through a number of discrete reaction pathways. The first catalytic reactions known to proceed through such an intermediate were the transesterification processes published independently by the Nolan [20] and Hedrick groups (Scheme 12.14) [21, 22].

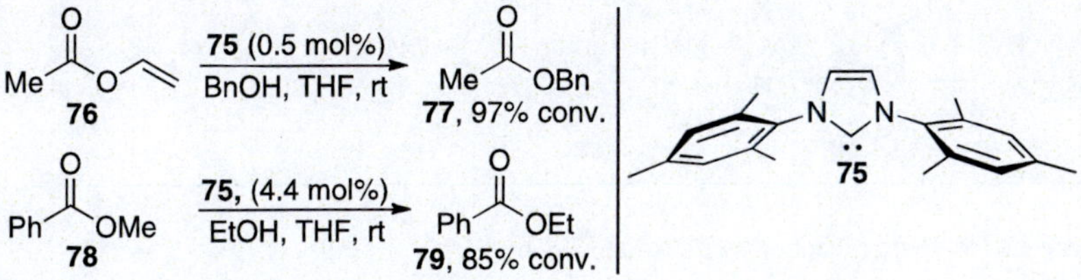

Scheme 12.14 Acyl transfer using NHCs

This work has been extended to transesterification with secondary alcohols [23], and of phosphonate esters [24]. Movassaghi and co-workers have demonstrated that NHCs effectively catalyse the amidation of esters with amino alcohols, although an alternative mechanism involving the NHC acting as a Brønsted base, resulting in nucleophilic activation of the alcohol for an initial transesterification event, followed by rapid *O*- to *N*-acyl transfer, has been proposed [25, 26].

Smith and co-workers have employed NHC **81** to catalyse the *O*- to *C*-carboxyl transfer of a range of oxazolyl carbonates **80**, forming **82** with the generation of a C–C bond at a quaternary centre with good catalytic efficiency [27]. This transformation presumably proceeds *via* the generation of an intermediate carboxyazolium species, and has been utilised as a component of domino multi-component reactions [28], as well as the rearrangement of indolyl and benzofuranyl carbonates (Scheme 12.15) [29].

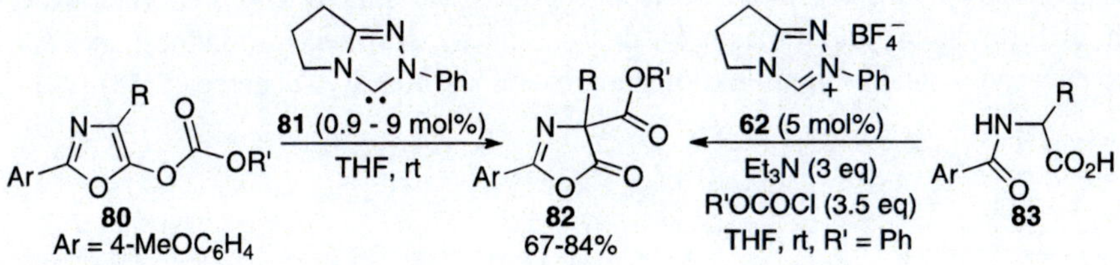

Scheme 12.15 *O*- to *C*-carboxyl transfer using NHCs

12.2.3.2 Enol to Acylazolium Tautomerisation – Redox Transformations

A range of α-functionalised aldehydes have been used to generate acylazolium species *via* the corresponding enol intermediate. For example, addition of an NHC to an α-halo aldehyde **84** presumably generates the Breslow species **86**, with elimination of HX to afford the enol **87**. Subsequent *in situ* tautomerisation generates the acylazolium species **88** that can be trapped by a nucleophile to afford the product **89** that has formally undergone a redox reaction (Scheme 12.16).

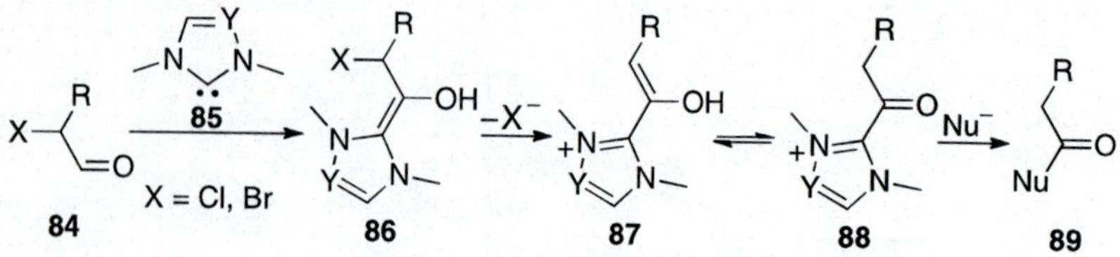

Scheme 12.16 Redox reactions with α-halo aldehydes

This approach has been applied by Rovis and co-workers to the formation of saturated esters from α-haloaldehydes and alcohols. A range of aliphatic alcohols, phenol and aniline proved competent nucleophiles (Scheme 12.17) [30].

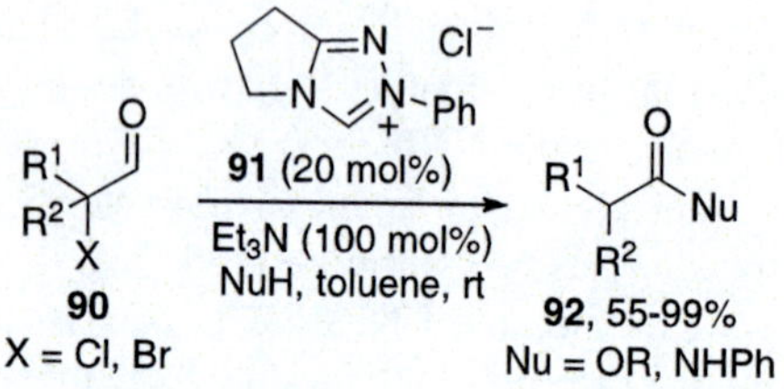

Scheme 12.17 Acyl transfer using α-haloaldehydes

A conceptually similar protocol was reported simultaneously by Bode and co-workers, employing α,β-epoxy aldehydes **93** as acylazolium precursors. With these substrates, good levels of diastereoselectivity of the β-hydroxyester products **95** are obtained resulting from stereoselective protonation of the enol intermediate [31]. A related ring opening reaction catalysed by NHCs was subsequently reported, using chiral formylketocyclopropanes **96** that can be transformed into ketoesters **97** in excellent yield and without loss of stereochemical integrity (Scheme 12.18) [32].

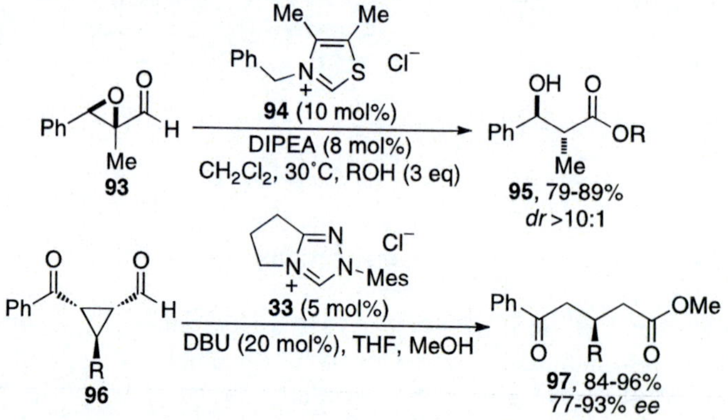

Scheme 12.18 Other representative redox transformations

Related redox transformations allow the conversion of ynals to α,β-unsaturated esters [33], as well as the ring expansion of formyl β-lactams [34], oxacycloalkane-2-carboxaldehydes [35], and 2-acyl-1-formylcyclopropanes [36]. Further developments allow the synthesis of amides from a range of α-functionalised compounds, but require an additive (imidazole, HOAt or HOBt) for efficient amidation [37].

12.3 Asymmetric Reactions Involving Acyl Anion Equivalents

12.3.1 NHC-Catalysed Benzoin Reactions

The use of NHCs to catalyse the benzoin condensation of aldehydes was first reported by Ugai in 1943 [38], with the now generally accepted mechanism for this transformation first proposed by Breslow in 1958 [39].

12.3.1.1 Asymmetric Benzoin Reactions: Thiazolium Salt Pre-Catalysts

The first asymmetric benzoin reactions were reported by Sheehan and Hannemann using chiral thiazolium salt pre-catalyst **100** of unknown absolute configuration [40]. Low yields and enantioselectivities were obtained, and although a wide range of thiazolium salt pre-catalysts have since been studied, of which **101–105** are representative, the enantioselectivities obtained for the condensation of benzaldehyde using thiazolium pre-catalysts are generally poor (Scheme 12.19) [41].

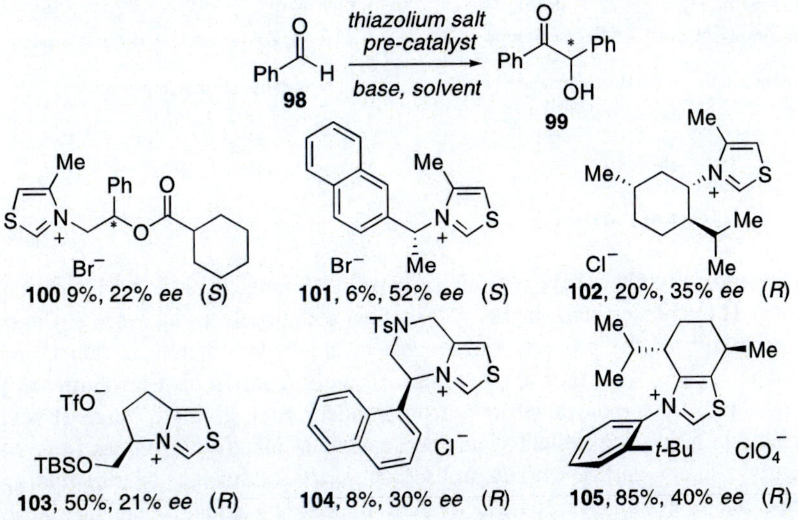

Scheme 12.19 Chiral thiazolium pre-catalysts for the asymmetric benzoin reaction

12.3.1.2 Triazolium Salt Pre-Catalysts

Following their studies of stable triazolin-5-ylidine NHC catalysts [42], Enders and co-workers studied chiral triazolium salt pre-catalysts for the asymmetric benzoin reaction. The NHC derived from **108** produced acyloin products **107** in moderate to good yields and up to 84% *ee*, although the asymmetric benzoin reaction of electron-deficient aldehydes remained a challenge [43]. In 1997, Knight and Leeper developed the bicyclic triazolium salt **109**, derived from (*S*)-phenylalanine. Good enantioselectivities (up to 83% *ee*) were obtained for a range of aromatic aldehydes, although the isolated yields were modest, even with high catalyst loadings (30 mol%) [44]. In 2002, Enders and co-workers reported triazolium salt **110**, derived from (*S*)-*tert*-leucine, as an effective catalyst for the asymmetric benzoin reaction. Excellent enantioselectivities (up to 99% *ee*) were observed for a range of aromatic aldehydes, with electron-rich aldehydes generally giving higher enantioselectivities (but lower yields) than electron-deficient aldehydes (Scheme 12.20) [45, 46].

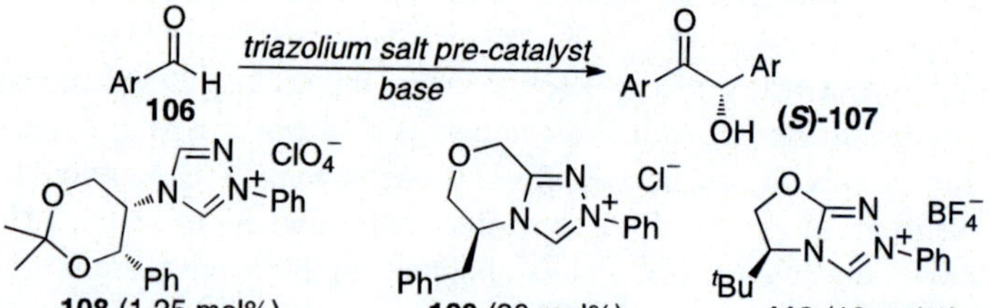

Ar	**108** (1.25 mol%) K_2CO_3, THF, rt Yield (%)	*ee*	**109** (30 mol%) Et_3N, MeOH, rt Yield (%)	*ee*	**110** (10 mol%) KO^{t}Bu, THF, rt Yield (%)	*ee*
Ph	66	75 (*ent*)	45	80	83	90
4-MeC_6H_4	28	84 (*ent*)	38	83	16	93
4-ClC_6H_4	51	29 (*ent*)	11	76	80	64

Scheme 12.20 Triazolium pre-catalysts for the asymmetric benzoin condensation

12.3.1.3 Recent Advances

Enders and co-workers have recently shown that (*S*)-pyroglutamic acid derived pre-catalyst **111** accesses benzoin **107** in excellent enantioselectivity (95% *ee*), though the generality of the process with respect to aldehyde structure is rather limited (Scheme 12.21) [47]. You and co-workers recently reported bis-triazolium salt pre-catalyst **113**, for the asymmetric benzoin condensation, giving a range of acyloin products in good to excellent yields and excellent enantioselectivities [48]. More recently, Connon and co-workers utilised chiral triazolium salts bearing hydrogen-bond donating substituents. NHCs derived from secondary amide-containing triazolium salts **112** give acyloin products in up to 62% *ee*, while the corresponding *N*-methylated catalyst provided benzoin **107** with only 13% *ee* [49].

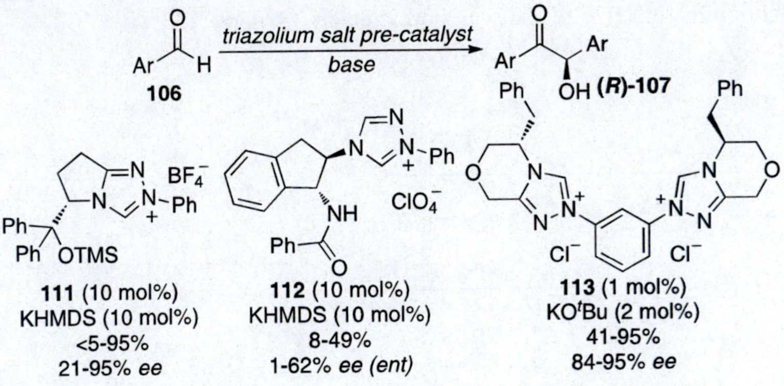

Scheme 12.21 Triazolium pre-catalysts for the asymmetric benzoin condensation

12.3.1.4 Asymmetric Intramolecular Cross Benzoin Reaction

Attempted intermolecular cross-benzoin reactions typically generate a thermodynamically controlled mixture of products [50], although several groups including Enders [51], Suzuki [52] and You [53] have utilised catalysts **116–118** for the intramolecular crossed benzoin of keto-aldehydes (Scheme 12.22).

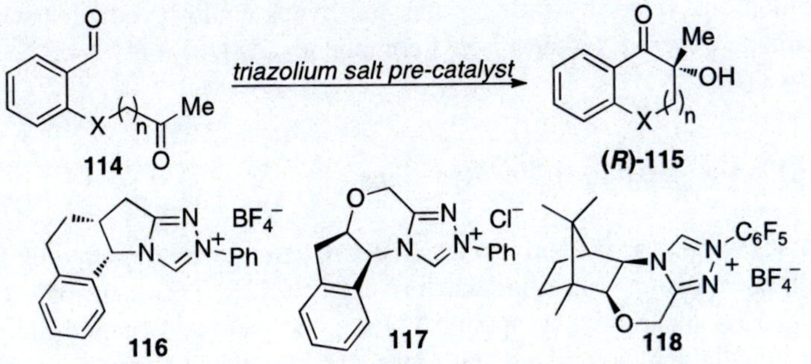

X	n	116 Yield (%)	116 *ee*	117 Yield (%)	117 *ee*	118 Yield (%)	118 *ee*
CH_2	0	95	74	69	60 (*ent*)	87	48
CH_2	1	93	94	70	96 (*ent*)	-	-
O	1	70	90	-	-	93	84

Scheme 12.22 Asymmetric intramolecular cross-benzoin reactions

Suzuki and co-workers recently applied the asymmetric intramolecular benzoin reaction to the synthesis of the homoisoflavonoid (+)-sappanone B **122** [54]. The authors found that triazolium salt pre-catalyst **120** gave the best results for the cyclisation of keto-aldehyde **119**, with electron-withdrawing *N*-aryl substituents on the NHC important in order to suppress competing aldol pathways. The synthesis of (+)-sappanone B from benzoin product **121** was achieved in two steps by sequential deprotection of the aryl methyl ethers (Scheme 12.23).

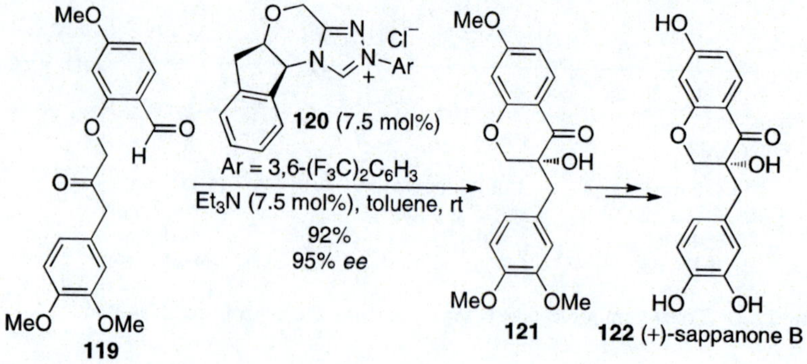

Scheme 12.23 Synthesis of sappanone B using NHC-mediated catalysis

12.3.2 NHC-Catalysed Stetter Reactions

In 1976, Stetter extended the synthetic utility of the Breslow intermediate (**1**) as an acyl anion equivalent by showing that aldehydes could be coupled with Michael acceptors to generate 1,4-dicarbonyl compounds [55].

12.3.2.1 Asymmetric Stetter Reactions

The first asymmetric intramolecular Stetter reactions were reported by Enders and co-workers utilising triazolium salt pre-catalyst **125**. Treatment of substrate **123** generated 1,4-dicarbonyl compound **124** in good yield and enantioselectivity [56]. These salicylaldehyde-derived substrates **123** have since become the standard test substrates for the development of new catalysts for the asymmetric intramolecular Stetter reaction. Bach and co-workers have achieved moderate enantioselectivities using axially-chiral thiazolium pre-catalyst **126** [41], whilst Miller and co-workers have developed peptidic thiazolium pre-catalyst **127** [57]. In 2005, Rovis and co-workers showed that the NHCs derived from triazolium salts **128–130** were excellent catalysts for the asymmetric intramolecular Stetter reaction of a wide range of substrates, giving typically excellent yields and enantioselectivities [58]. The *N*-pentafluorophenyl catalyst **129** currently represents the state of the art in asymmetric Stetter reactions (Scheme 12.24) [59].

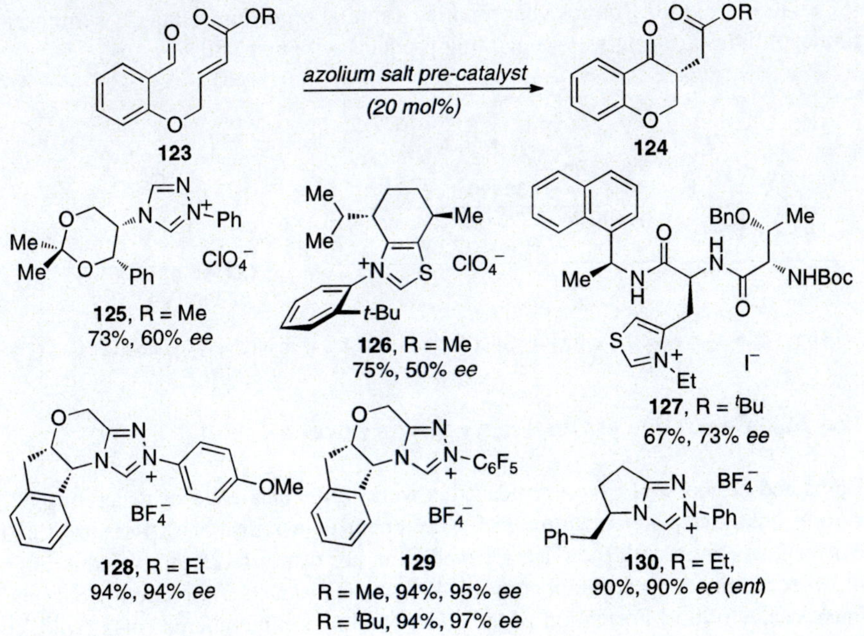

Scheme 12.24 Asymmetric intramolecular Stetter reactions

Rovis and co-workers have also shown that pre-catalyst **129** is competent with a wide range of Michael acceptors including α,β-unsaturated aldehydes, amides, nitriles, esters, thioesters, vinylphosphonates and vinylphosphine oxides (Scheme 12.25) [58, 60].

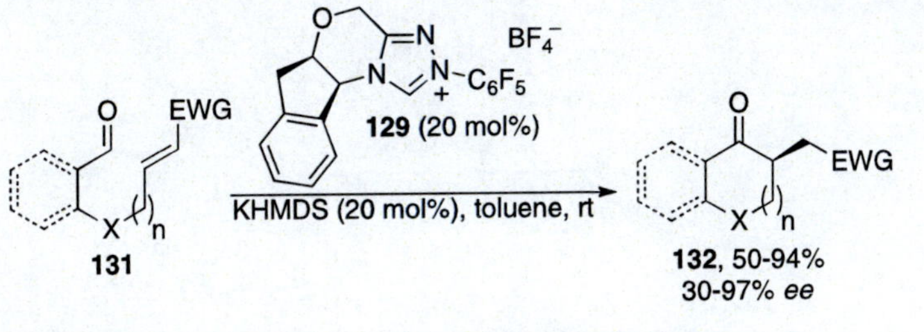

Scheme 12.25 Scope of the asymmetric intramolecular Stetter reaction

The intramolecular asymmetric Stetter reaction of aliphatic aldehydes is generally more difficult to achieve due to the presence of acidic α-protons. Rovis and co-workers have demonstrated that the NHC derived from pre-catalyst **130** promotes the intramolecular Stetter cyclisation with enoate and alkylidene malonate Michael acceptors **133**. Cyclopentanones are generally accessed in excellent yields and enantioselectivities, however cyclohexanones are obtained in significantly lower yields unless very electron-deficient Michael acceptors are employed

(Scheme 12.26) [58]. Tomioka has reported a similar procedure using C2-symmetric imidazolinium pre-catalysts, generating products with up to 80% *ee* [61].

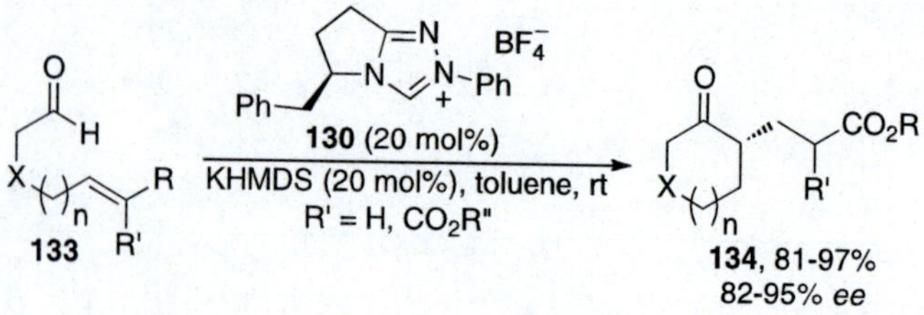

Scheme 12.26 Asymmetric intramolecular Stetter reactions with aliphatic aldehydes

12.3.2.2 Generation of Quaternary Stereocentres

Rovis and co-workers have extended this work to the challenge of generating all-carbon quaternary stereocentres [62]. Treatment of a range of α,β-disubstituted Michael acceptors with the NHC derived from pre-catalyst **129** generated a range of benzofuranone, benzothiophenone and indanone products in excellent yields and enantioselectivities. Formation of the corresponding six-membered rings typically occurred in lower yield but excellent *ee* (Scheme 12.27). Aliphatic aldehydes were also suitable substrates, generating the corresponding Stetter products in good yields and enantioselectivities. Notably, while a sulfide tether (X = S) was not tolerated, switching to an electron-withdrawing sulfone tether (X = SO_2) restored good reactivity [63].

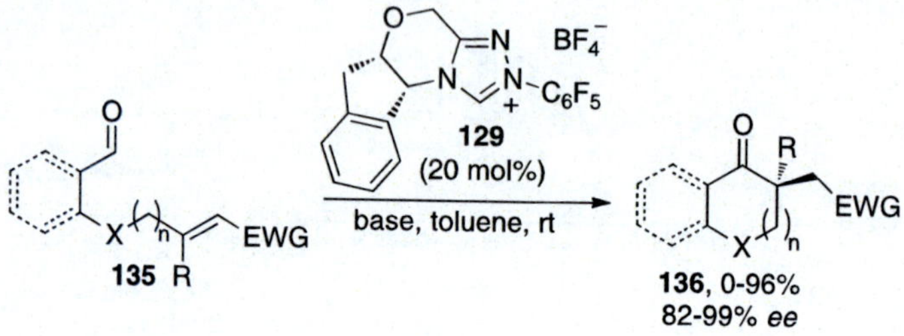

Scheme 12.27 Generation of quaternary stereocentres

Rovis and co-workers further extended the scope of the reaction to the enantio- and diastereoselective cyclisation of α,β-disubstituted Michael acceptors **137**. The high diastereoselectivity of the process relies on selective protonation of the resultant enolate after conjugate addition. It was found that HMDS (formed during deprotonation of the triazolium salt pre-catalyst) was detrimental to the diastereoselectivity of the process, and so under optimised conditions the free NHC **138** was used (with HMDS being removed in high vacuum after deprotonation) (Scheme 12.28) [64].

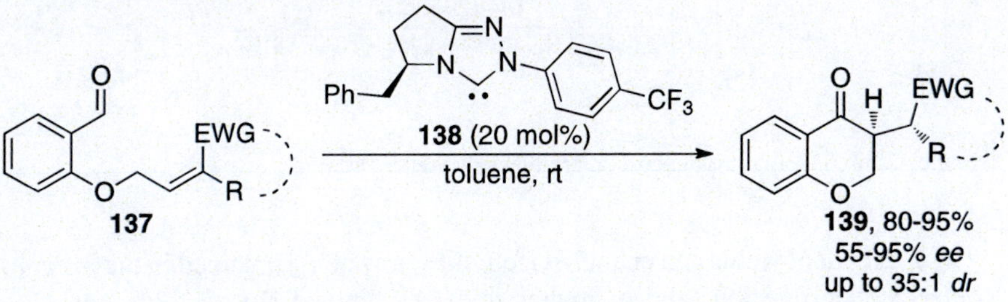

Scheme 12.28 Generation of contiguous stereocentres in the Stetter reaction

12.3.2.3 Synthesis of Hydrobenzofuranones

Rovis and co-workers have applied the asymmetric intramolecular Stetter reaction to the desymmetrisation of cyclohexadienones **140**, generating a quaternary stereocentre and forming hydrobenzofuranones **141** in excellent yields and enantioselectivities. Substitution at the two, four and six-positions is tolerated, and even substitution at the three-position is accommodated (Scheme 12.29) [65].

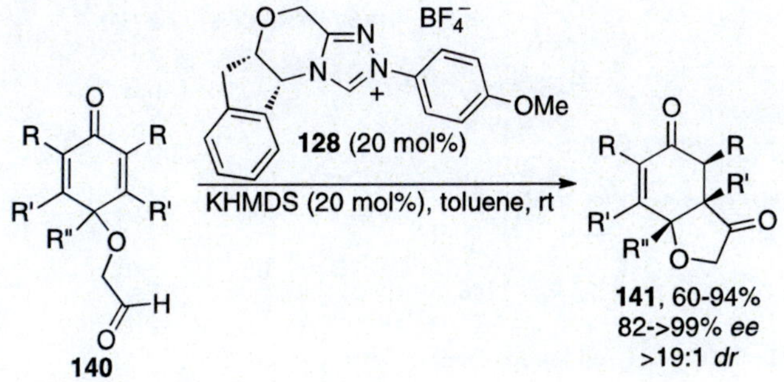

Scheme 12.29 Synthesis of hydrobenzofuranones *via* desymmetrisation

12.3.2.4 Intermolecular Stetter Reactions

The first attempted asymmetric intermolecular Stetter reaction was reported by Enders and co-workers who showed in 1989 that reaction of *n*-butanal **142** with chalcone **143** in the presence of the NHC derived from thiazolium salt **144** generated Stetter product **145** in 39% *ee* but only 4% yield (Scheme 12.30) [66].

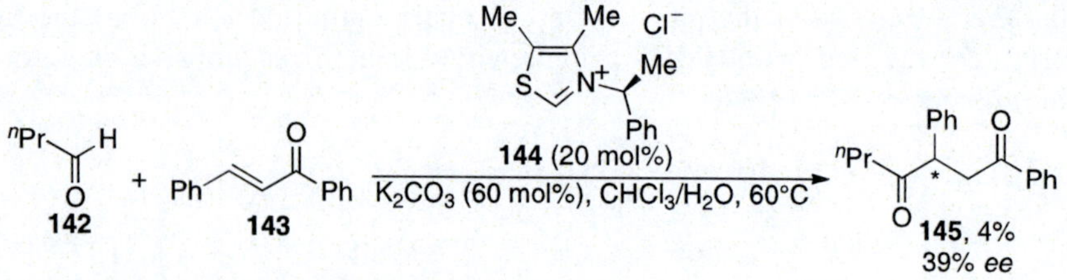

Scheme 12.30 The first asymmetric intermolecular Stetter reaction

With excellent yields and enantioselectivities now being obtained in the intramolecular Stetter reaction, and a greater understanding of the factors controlling enantioselectivity in NHC-mediated processes, research is increasingly returning to the original problem of intermolecular Stetter reactions. Recent work by Enders and co-workers has shown that *N*-benzyl-substituted triazolium salt pre-catalyst **148** is effective for promoting the Stetter reaction of aromatic aldehydes with chalcones. Remarkably, the corresponding *N*-phenyl-substituted triazolium salt **147** (which is effective in the asymmetric benzoin reaction) does not catalyse the Stetter reaction, highlighting the dramatic effect that subtle changes in the NHC structure can have upon reactivity (Scheme 12.31) [67].

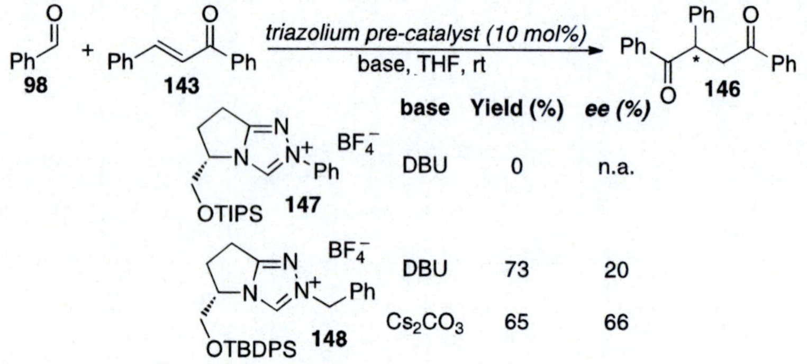

Scheme 12.31 NHC-catalysed intermolecular Stetter reaction

Rovis and co-workers have recently shown that the NHC derived from triazolium pre-catalyst **151** catalyses the intermolecular Stetter reaction of glyoxamides **149** with alkylidene malonates **150**. Enantioselectivities of up to 91% have been obtained with typically good to excellent yields of the corresponding Stetter products (Scheme 12.32) [68].

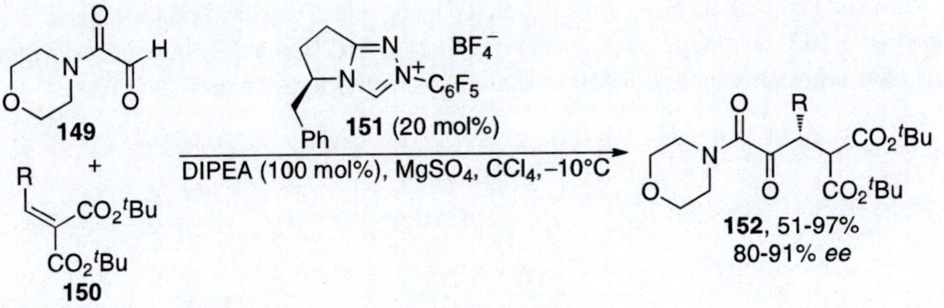

Scheme 12.32 Intermolecular Stetter reactions of glyoxamides

Rovis and co-workers have also extended the intermolecular Stetter reaction to include nitroalkenes as the electrophilic component. Fluorinated triazolium pre-catalyst **155** was effective in catalysing the reaction of a variety of heteroaromatic aldehydes **153** with nitroalkenes **154** to generate β-nitroketones in excellent yields and enantioselectivities. The authors propose that stereoelectronically induced conformational effects on the catalyst skeleton are key to the high selectivities observed with fluorinated catalyst **155** (Scheme 12.33) [69].

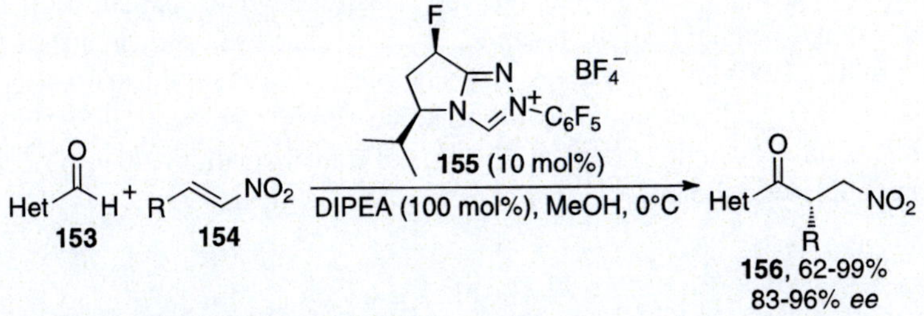

Scheme 12.33 Intermolecular Stetter reaction of nitroalkenes

12.3.3 *Asymmetric Reactions Involving Homoenolates*

12.3.3.1 Enals as Homoenolate Equivalents

As noted in Section 12.2.2, homoenolates can be accessed from enals. Glorius and co-workers have used pre-catalyst **159** to prepare γ-butyrolactones from enals enantioselectively, though *ee*s of up to only 25% were obtained (Scheme 12.34) [8].

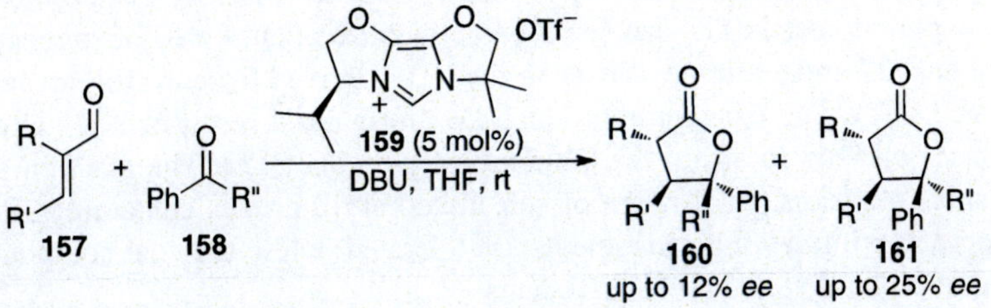

Scheme 12.34 Asymmetric γ-butyrolactone formation

You and co-workers have demonstrated enantioselective γ-lactone formations using glyoxalate **163**, achieving up to 78% *ee* with the NHC derived from chiral triazolium salt **164**, although with low levels of diastereoselectivity (Scheme 12.35) [70].

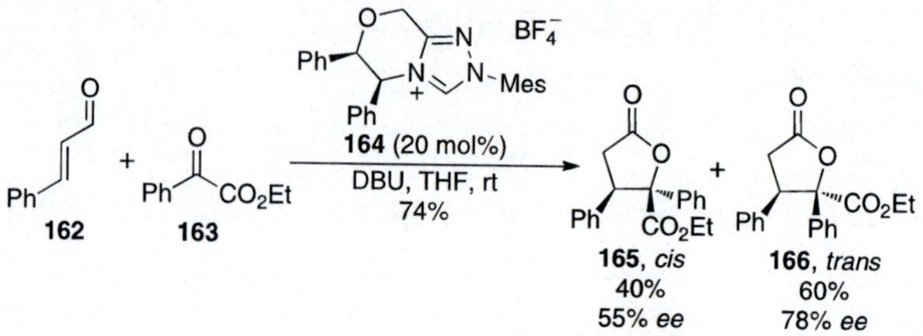

Scheme 12.35 Asymmetric γ-butyrolactone synthesis *via* glyoxalates

Bode and co-workers have extended the synthetic utility of homoenolates to the formation of enantiomerically enriched *N*-protected γ-butyrolactams **169** from saccharin-derived cyclic sulfonylimines **167**. While racemic products have been prepared from a range of β-alkyl and β-aryl substituted enals and substituted imines, only a single example of an asymmetric variant has been shown, affording the lactam product **169** with good levels of enantioselectivity and diastereoselectivity (Scheme 12.36) [71]. As noted in the racemic series (see Section 12.2.2), two mechanisms have been proposed for this type of transformation, either by addition of a homoenolate to the imine or *via* an ene-type mechanism.

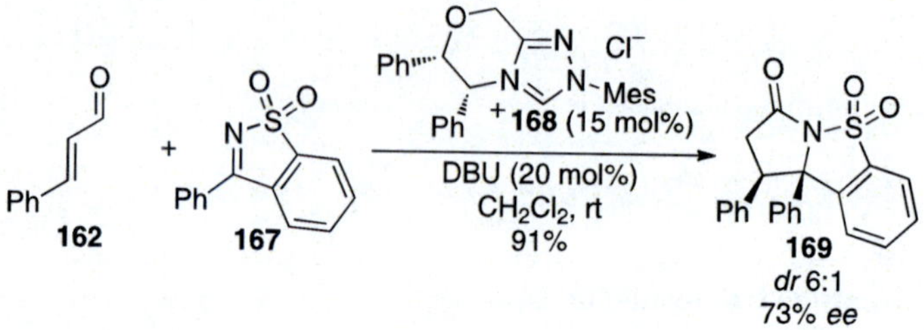

Scheme 12.36 Asymmetric γ-lactamisation

This homoenolate methodology has been extended to the use of nitrones **170** as electrophiles [72]. Scheidt and co-workers have shown that enantiomerically enriched γ-amino esters **172** can be prepared with excellent levels of stereocontrol from an enal **27** and a nitrone **170** using the NHC derived from triazolium salt **164** (Scheme 12.37). The oxazinone product **171**, formally a result of a [3+3] cycloaddition, is cleaved to afford the γ-amino ester product **172**. The reaction shows broad substrate scope, as a range of substituted aryl nitrones containing electron donating and withdrawing substituents are tolerated, while the enal component is tolerant of both alkyl and aryl substituents.

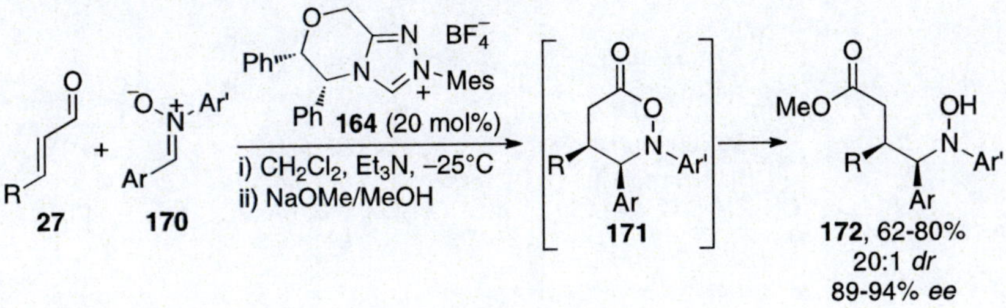

Scheme 12.37 Asymmetric homoenolate addition to nitrones

Nair and co-workers have demonstrated NHC-catalysed formation of spirocyclic diketones **173** from α,β-unsaturated aldehydes **174** and substituted dibenzylidine-cyclopentanones **175**. Where chalcones and dibenzylidene cyclohexanones give only cyclopentene products (as a result of β-lactone formation then decarboxylation), cyclopentanones **175** give only the spirocyclic diketone products **173** [73]. Of particular note is the formation of an all-carbon quaternary centre and the excellent level of diastereoselectivity observed in the reaction. An asymmetric variant of this reaction has been demonstrated by Bode using chiral imidazolium salt **176**, obtaining the desymmetrised product with good diastereo- and enantioselectivity, though in modest yield (Scheme 12.38) [74].

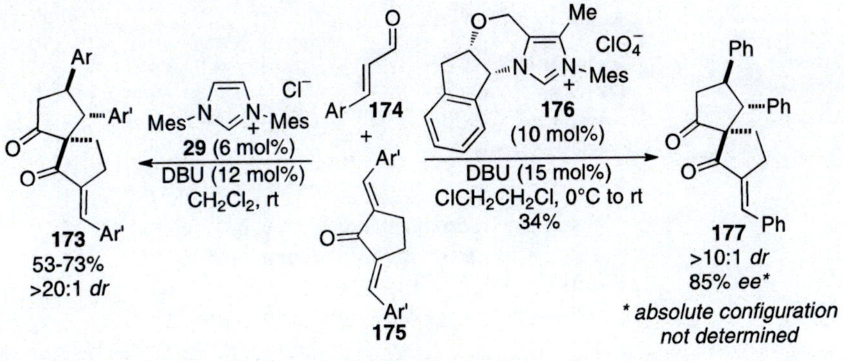

Scheme 12.38 Spirocyclopentanone formation

A formal [3+2] cycloaddition reaction with homoenolates has also been realised with nitrogen-based electrophiles such as *N*-acyl-*N*'-aryldiazenes **180**. Pyrazolidinones **178** can be prepared from enals **27** and acyldiazenes **180**, as demonstrated by Scheidt and Chan [75]. An example of the asymmetric variant demonstrates excellent levels of enantioselectivity in this reaction (90% *ee*) (Scheme 12.39).

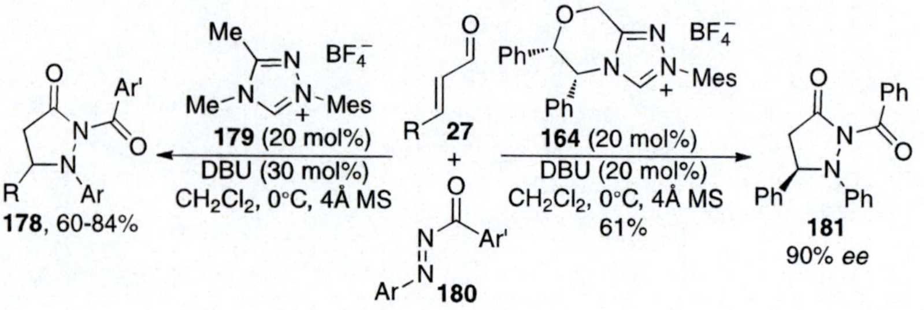

Scheme 12.39 Pyrazolidinone formation

In addition to the use of homoenolates to access C–C or C–heteroatom bond formation, Scheidt and co-workers have investigated enantioselective homoenolate protonation using chiral triazolium salt **183** as the NHC pre-catalyst. The reaction gives moderate levels of enantioselectivity in both polar and non-polar solvents, but a competing *in situ* oxidation process is more pronounced in THF than in toluene. Experimental and computational studies have allowed the authors to attribute this solvent dependence to the ability of the solvent to stabilise the charged intermediates resulting from the oxidative pathway. As an illustration, treatment of enal **182** with the chiral NHC derived from **183** gave varying amounts of the redox product **184** and oxidation product **185** dependent on the nature of the solvent, but with similar levels of enantioselectivity (Scheme 12.40) [76].

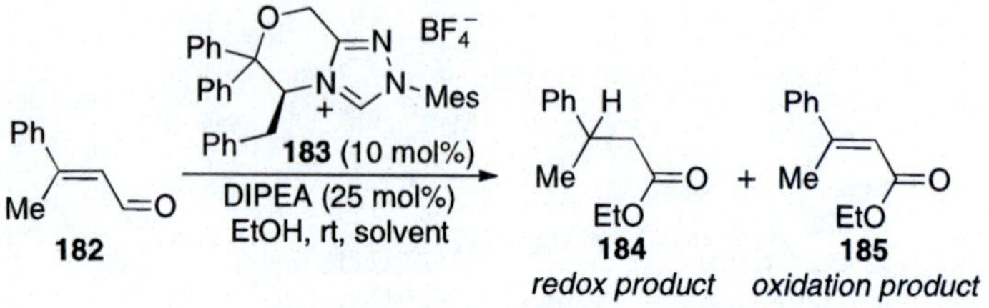

solvent	redox:oxidation product ratio	% yield (combined)	*ee* (%)
THF	3:1	58	55
toluene	13:1	81	53

Scheme 12.40 Enantioselective homoenolate protonation *versus* oxidation

NHC-catalysed homoenolate generation has been applied by Bode and Struble in the formal synthesis of the natural product salinosporamide A [77]. The key step in the synthesis is a late-stage NHC-catalysed intramolecular lactonisation step of intermediate **186**. When this reaction was attempted with an achiral triazolium-derived NHC, a 4:1 diastereomeric ratio of products was obtained in preference for the *undesired* product **189**. In order to circumvent this, chiral triazolium salt **187** was employed, giving an approximately 1:1 mixture of desired:undesired diastereoisomers (Scheme 12.41).

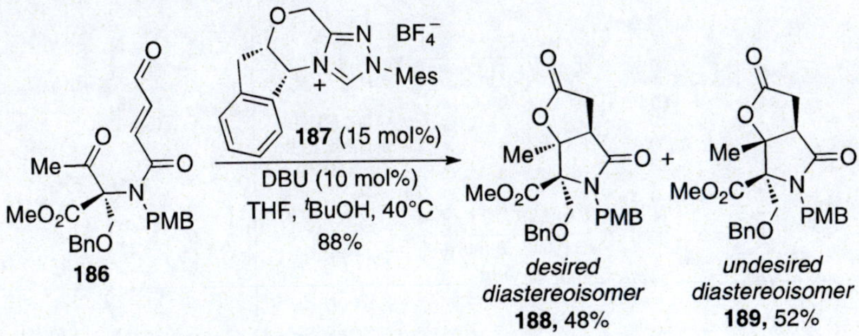

Scheme 12.41 Homoenolate methodology in natural product synthesis

12.3.3.2 α-Hydroxyenones as Homoenolate Surrogates

As discussed in Section 12.2.2, homoenolates can be accessed from α-hydroxy-enones, and an asymmetric synthesis of cyclopentenes has been achieved using this strategy with enones (Scheme 12.42) [15].

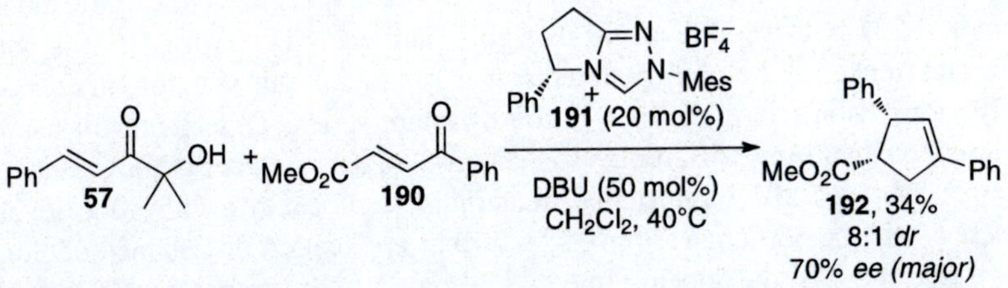

Scheme 12.42 Asymmetric cyclopentene formation

12.3.4 Asymmetric Reactions Involving Azolium Enolates

12.3.4.1 Enolate Generation from Ketenes

The direct generation of an enolate intermediate through the addition of a chiral NHC to a ketene was independently investigated by the Ye and Smith groups, who both showed the utility of this reaction for the asymmetric synthesis of β-lactams. Ye and co-workers showed that the pyroglutamic acid derived NHC pre-catalyst **195** gave good diastereocontrol (up to 99:1 *dr*) and excellent levels of enantiocontrol (up to 99% *ee*) for the preparation of β-lactams **196** using a range of unsymmetrical alkylarylketenes **193** and *N*-Boc imines **194** [78]. Smith and co-workers showed that good catalytic activity and reasonable levels of enantioselectivity (up to 74% *ee*) could be observed in the reaction of diphenylketene **197** with *N*-tosyl imines **63** employing the NHCs from either imidazolidinium or triazolium pre-catalysts **198** or **199**, with crystallisation generating products with excellent *ee* (up to >99% *ee*, Scheme 12.43) [17].

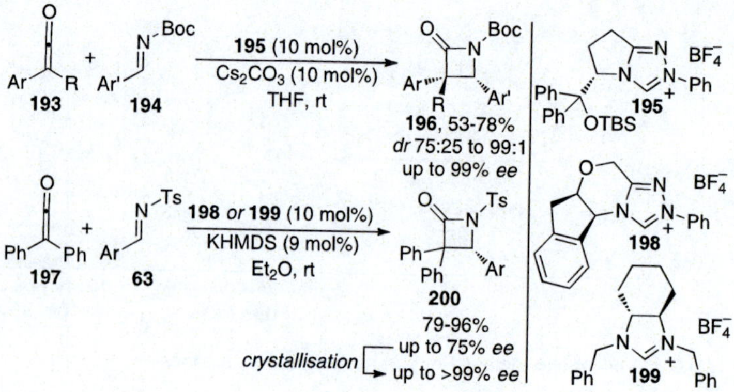

Scheme 12.43 Asymmetric β-lactam synthesis using chiral NHCs

Ye and co-workers have extended this methodology to a range of enantioselective formal [2+2] and [4+2] cycloaddition processes involving unsymmetrical ketenes and a range of electrophiles. All of these processes utilise structurally related NHCs derived from pyroglutamic acid. For example, the formal asymmetric [2+2] reaction of alkylarylketenes **193** with 2-oxoaldehydes **202** generates the corresponding β-lactones **201** with good levels of diastereo- and enantiocontrol [79]; it is notable that a preference for the *anti*-diastereoisomer is noted in this reaction, in contrast to the *syn*-stereoisomeric preference in the β-lactam series. This chemistry has been extended to encompass trifluoromethylketones **206** as the electrophilic partner, generating β-lactones **205** with vicinal quaternary stereocentres with good levels of diastereo- and enantio-control [80]. The further application of this methodology to utilise diazenes as the electrophile has allowed both formal [2+2] and [4+2] cycloaddition processes to be developed by changing the nature of the *N*-substituent of the diazene. For example, NHC **195** promoted cycloaddition of ketenes **193** with *N*,*N*-dibenzoyldiazenes **203** generate the formal [4+2] products in excellent *ee* [81], whereas diazenedicarboxylates **209** give access to the aza-β-lactam [2+2] cycloaddition products **210** in similarly good *ee* (Scheme 12.44) [82].

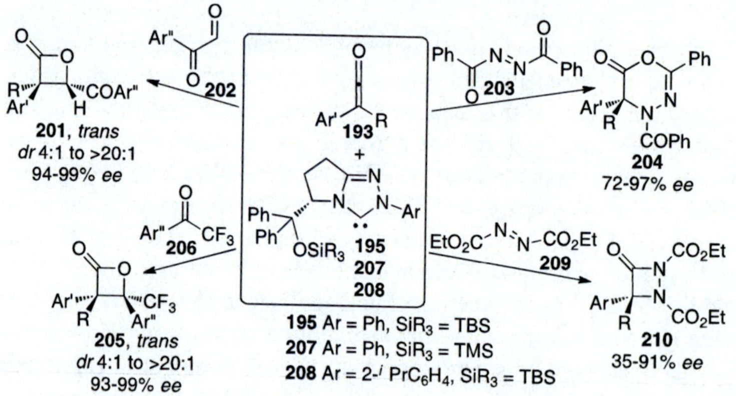

Scheme 12.44 Applications of ketenes in [2+2] and [4+2] cycloadditions

An interesting switch in enantioselectivity is observed with variation of the *N*-substituent of the NHC used to catalyse the [4+2] addition of ketenes and *N*-aryl-*N*-benzoyldiazenes. For example, *N*-phenyl-substituted NHC **212** (Ar = Ph) gave (*R*)-**215** in 84% *ee* in the reaction with ethylphenylketene and *N*-phenyl-*N*-benzoyldiazene **213**, while *N*-mesityl-substituted NHC **214** (Ar = Mes) gave (*S*)-**211** in 76% yield and 96% *ee* (Scheme 12.45) [81].

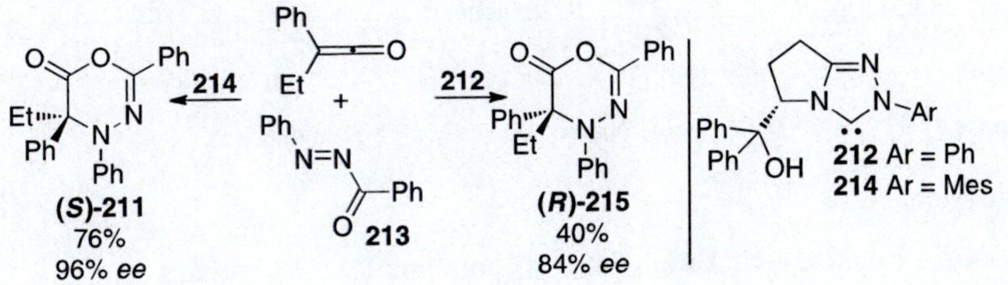

Scheme 12.45 Effect of NHC *N*-aryl substituent on enantioselectivity

Further work by the Ye group has shown that NHCs derived from pre-catalyst **215** can also promote the asymmetric dimerisation of alkylarylketenes **193** to generate alkylidene β-lactones **216** in good diastereo- and enantio-selectivity [83]. The asymmetric [4+2] addition of enones and alkylarylketenes to generate δ-lactones **218** in high *ee* has also been accomplished [84], as has the asymmetric esterification of alkylarylketenes to give esters **217** using benzhydrol, which is assumed to proceed *via* a Lewis-base mediated mechanism (Scheme 12.46) [85].

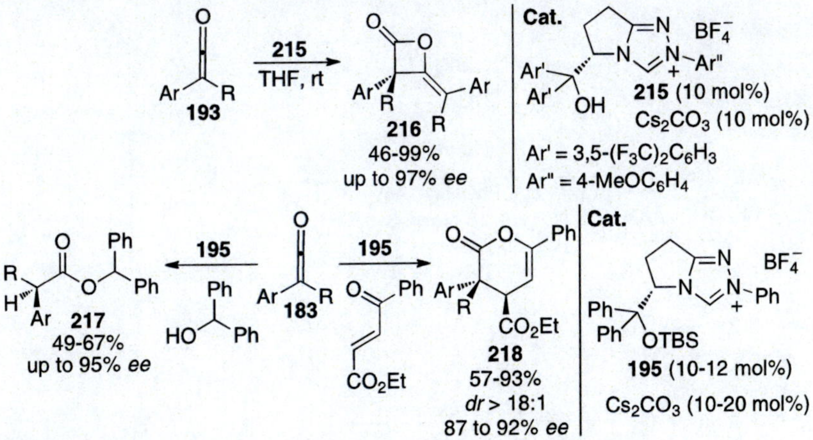

Scheme 12.46 Further applications of ketenes as enolate equivalents

Within the latter area, Smith and co-workers have shown that asymmetric NHC promoted esterifications of arylalkylketenes **193** can also be achieved with

2-phenylphenol using the oxazolidinone derived triazolium pre-catalyst **220** (up to 84% *ee*, Scheme 12.47) [86].

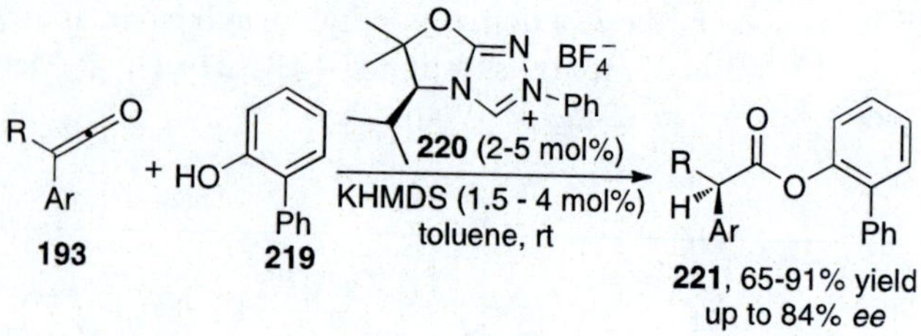

Scheme 12.47 Asymmetric esterification of ketenes with 2-phenylphenol

12.3.4.2 Enolate Generation from α-Functionalised Aldehydes

Whilst the addition of a chiral NHC to a ketene generates a chiral azolium enolate directly, a number of alternative strategies have been developed that allow asymmetric reactions to proceed *via* an enol or enolate intermediate. For example, Rovis and co-workers have shown that chiral azolium enolate species **225** can be generated from α,α-dihaloaldehydes **222**, with enantioselective protonation and subsequent esterification generating α-chloroesters **224** in excellent *ee* (84–93% *ee*). Notably, in this process a bulky acidic phenol **223** is used as a buffer alongside an excess of an alternative phenolic component to minimise product epimerisation (Scheme 12.48). An extension of this approach allows the synthesis of enantiomerically enriched α-chloro-amides (80% *ee*) [87].

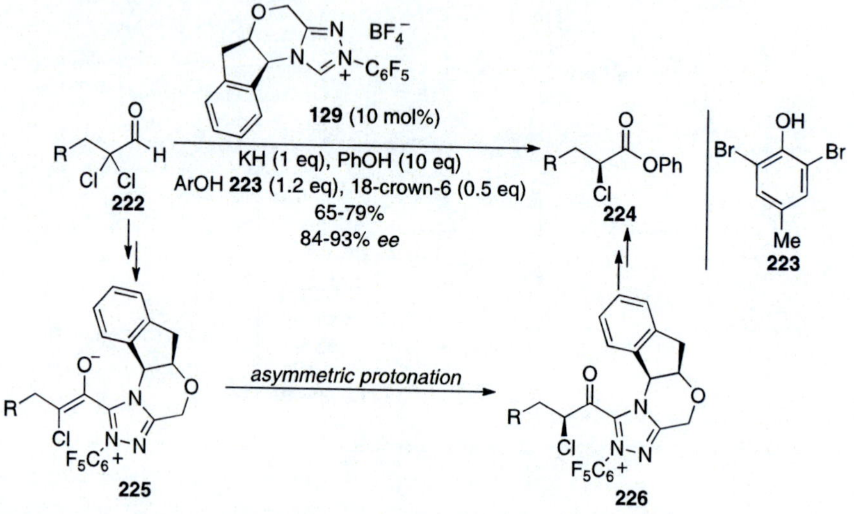

Scheme 12.48 Asymmetric protodehalogenation

Bode and co-workers have used the NHC derived from pre-catalyst **227** to generate an enolate **228** from a range of enals, with subsequent hetero-Diels-Alder type reaction with an α,β-unsaturated imine **229** generating dihydropyridinones **230** in excellent diastereo- and enantio-selectivities (Scheme 12.49) [88]. This methodology has recently been applied to the synthesis of amino functionalised 3,4-dihydropyranones in high *ee*, with a stepwise, rather than a concerted, process thought to be operating [89].

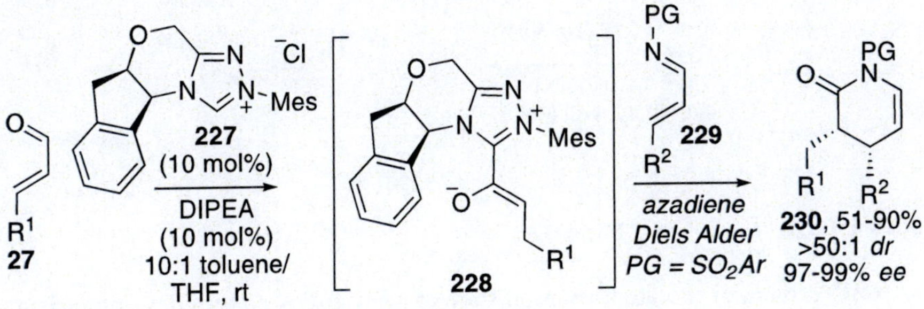

Scheme 12.49 Asymmetric dihydropyridone formation

Further studies by Bode and co-workers have shown that enolate formation from α-chloroaldehydes and subsequent reaction with 4-oxo-enoates or unsaturated α-ketoesters **232** generates dihydropyranones **233** in excellent diastereo- and enantio-selectivities, and with impressively low catalyst loadings [90]. This work has been extended to the generation of enolate equivalents from bisulfite adducts of α-haloaldehydes **234** under aqueous conditions (Scheme 12.50) [91].

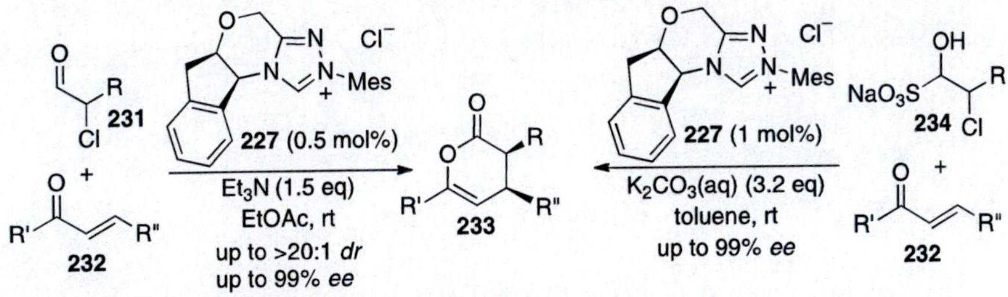

Scheme 12.50 Asymmetric dihydropyranone formation

Scheidt and co-workers have also accessed enolate equivalents from enals to furnish cyclopentanes **236** asymmetrically. Formation of the enolate equivalent from enals **235** with the NHC, followed by an intramolecular Michael reaction and *O*-acylation, gives the lactone products **236**, which are readily opened by either alcohols or amines to generate functionalised cyclopentane derivatives **237** in excellent *ee*.

This reaction is equally amenable to enals with both aliphatic and aromatic β-substituents, although the formation of substituted cyclohexanes (from analogous enals) proceeds with reduced enantioselectivity (Scheme 12.51) [92]. You and co-workers have shown that the same reaction is also promoted by triazolium salts derived from camphor in excellent enantioselectivity (95–99% *ee*) [93]).

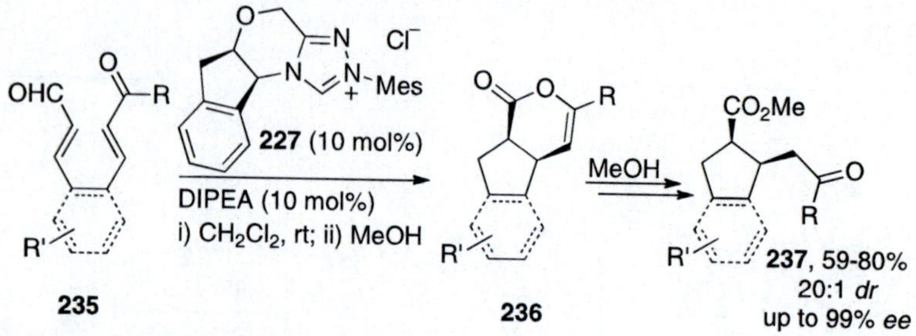

Scheme 12.51 Intramolecular Michael reactions

NHC-promoted enolate formation from an enal, followed by a desymmetrising aldol event to generate β-lactones and loss of CO_2, has been exploited by Scheidt and co-workers to generate functionalised cyclopentenes **240** in high *ee* from enal substrates **238** (Scheme 12.52) [94]. Interestingly, the use of alkyl ketones in this reaction manifold allows the isolation of the β-lactone intermediates; with acyclic diketones, β-lactones **239** are formed with the R group *anti*- to the tertiary alkoxide, while with cyclic diketones the β-lactone products have the R group with a *syn* relationship to the alkoxide [95].

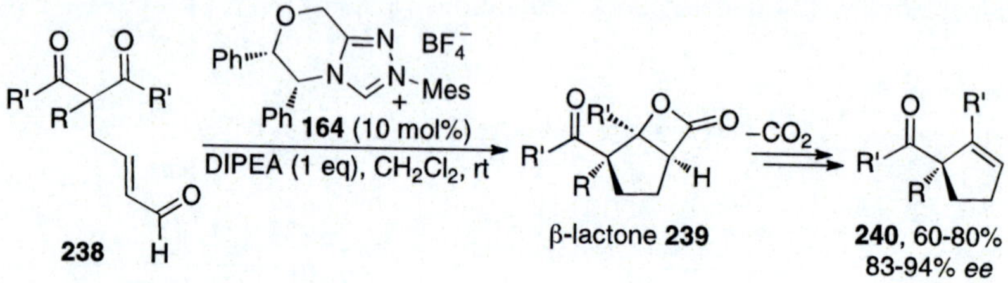

Scheme 12.52 Desymmetrisation of 1,3-diketones

An alternative method of generating enolates by conjugate addition has been applied to aza Baylis-Hillman reactions with low enantiocontrol (up to 44% *ee*) [96].

12.3.5 *Asymmetric Reactions Involving Acylazolium Species*

Considerable effort into utilising the Lewis basic properties of NHCs to catalyse acyl transfer reactions has been made as noted in Section 12.2.4.1. This concept has been extended to the use of enantiomerically pure NHCs to facilitate asymmetric

O-acylations. For example, Marouka and co-workers have employed chiral C_2-symmetric imidazolium salt **242** and enol ester **243** in the kinetic resolutions of alkyl-aryl carbinols **241** with excellent levels of enantioselectivity (Scheme 12.53) [97]. A small number of alkyl-alkenyl carbinols are also resolved with high levels of selectivity (selectivity values up to 22). Suzuki et al. have also shown that an analogous α-methylbenzyl substituted imidazolium salt can resolve alkyl-aryl carbinols using vinyl acetate [98].

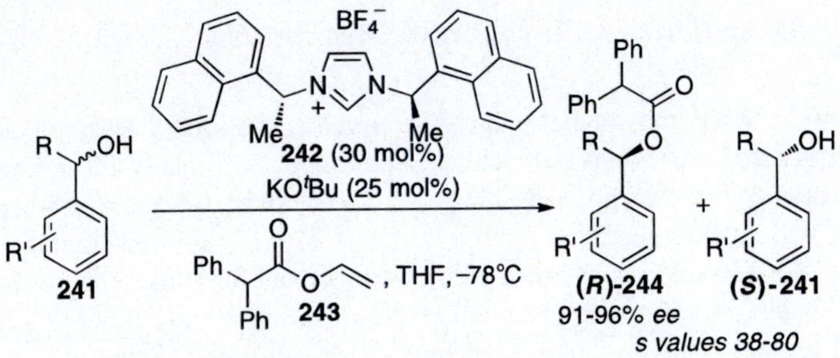

Scheme 12.53 Example of kinetic resolution of alkyl-aryl alcohols

Rovis and co-workers have extended the application of redox transformations to generate chiral acylazolium species from α-haloaldehydes **245** and the NHC derived from pre-catalyst **247**, allowing the desymmetrisation of *meso*-hydrobenzoin **246** to give ester **248** in good yield and enantioselectivity (Scheme 12.54) [30].

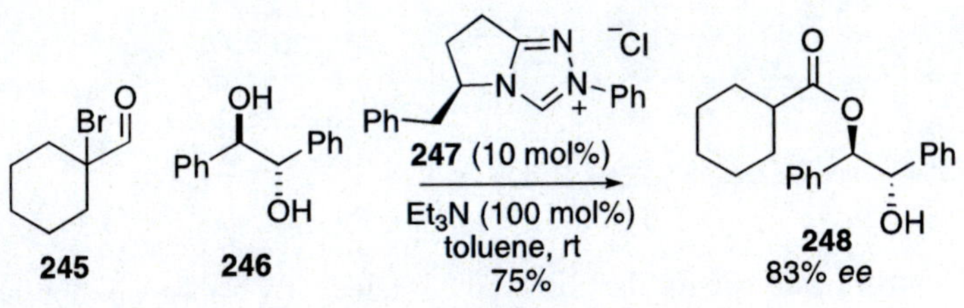

Scheme 12.54 Desymmetrisation using α-haloaldehydes

Scheidt and co-workers have used their *in situ* hydroxyazolium oxidation strategy to allow the desymmetrisation of diol **249** using chiral triazolium salt **187**, giving mono-ester **250** in 80% *ee* (Scheme 12.55) [99].

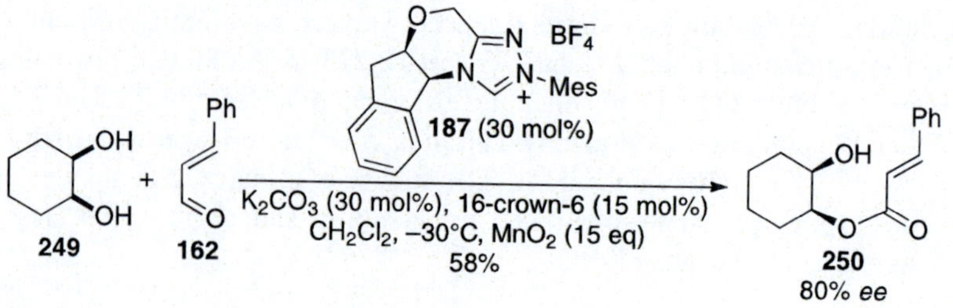

Scheme 12.55 Desymmetrisation *via* oxidation pathway

12.3.6 Miscellaneous Asymmetric Reactions

Bode and He have shown that chalcone-derived (acyclic) α,β-unsaturated sulfonylketimines **251** can act as competent electrophiles with enals **27**, giving highly enantio-enriched β-lactam products **252** [100]. The scope and generality of the reaction has been widely examined and a range of enals **27** and sulfonylketimines **251** are tolerated. Notably, alkyl-substituted enals, acrolein and 3,3-dimethylacrolein are all suitable substrates, all giving excellent levels of enantio- and diastereo-selectivity, although the acroleins give only moderate yields (45–50%) (Scheme 12.56). This reaction is postulated to follow a different pathway to the related work by the same authors using sulfonylaldimines **229**, which generates dihydropyridone products **230**.

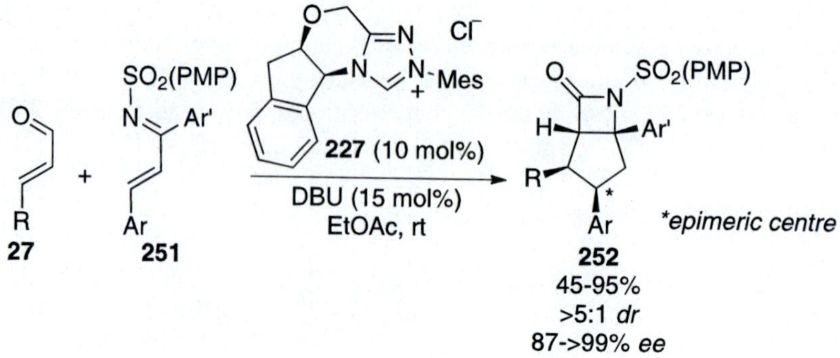

Scheme 12.56 Enantioselective γ-lactamisation

A mechanistic rationale for the observed *cis*-selectivity has been proposed based on preorganisation of the Breslow-type intermediate and imine through hydrogen bonding **253**, with an aza-benzoin oxy-Cope process proposed. Reaction *via* a boat transition state delivers the observed *cis*-stereochemistry of the product (Scheme 12.57). Related work by Nair and co-workers (using enones **42** in place of α,β-unsaturated sulfonylimines **251**, see Section 12.2.2) generates β-lactones **43** with *trans*-ring substituents, while the β-lactam products **252** possess a *cis*-stereochemical relationship.

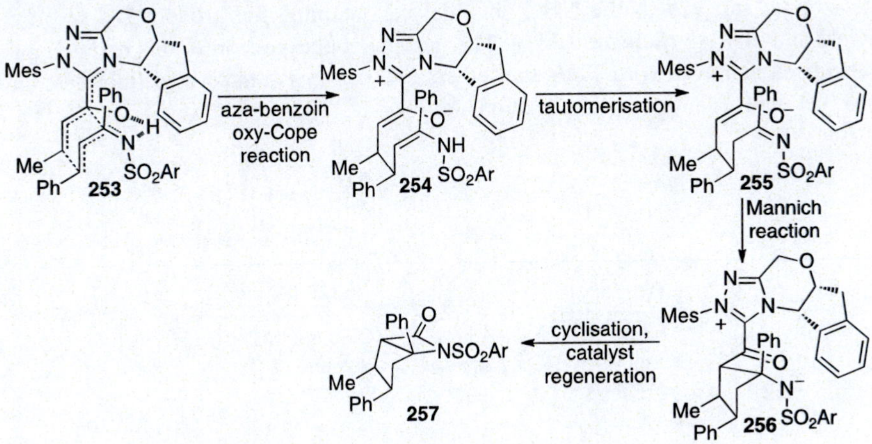

Scheme 12.57 Rationale for enantio- and diastereoselective β-lactam formation

Bode and Kaeobamrung have demonstrated an analogous reaction using α-hydroxyenones **259** as the electrophile [101]. Interestingly, triazolium-derived NHCs give β-lactones **258** and imidazolium-derived NHCs give γ-lactones **261**, both with excellent levels of enantioselectivity. The authors postulate a similar benzoin oxy-Cope mechanism to account for the high level of stereoselectivity in the process. The observed divergence in behaviour has been attributed to the leaving group ability of the NHC (Scheme 12.58).

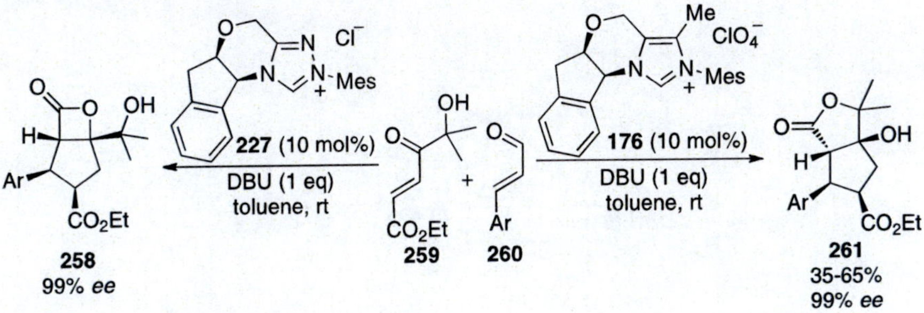

Scheme 12.58 Divergence between triazolium- and imidazolium-derived NHCs in lactone formation with hydroxyenones

Recent developments by Lupton and co-workers have shown alternative uses for NHCs in organocatalysis through trapping of α,β-unsaturated azolium intermediates [102]. Treatment of enol esters **263** with an imidazolium-derived NHC gives pyranones **262** in good yield. Mechanistically, it is proposed that nucleophilic addition of the NHC to the enol ester generates an α,β-unsaturated azolium intermediate and enolate, which recombine in a conjugate manner to afford pyranones **262**. An enantioselective variant of this reaction has been demonstrated using chiral triazolium salt **129** as the NHC pre-catalyst, obtaining the product **264** in good yield and 50% *ee* (Scheme 12.59). The intermolecular version of this reaction has also been achieved with TMS enol ethers and α,β-unsaturated acyl fluorides, but not yet in an enantioselective fashion.

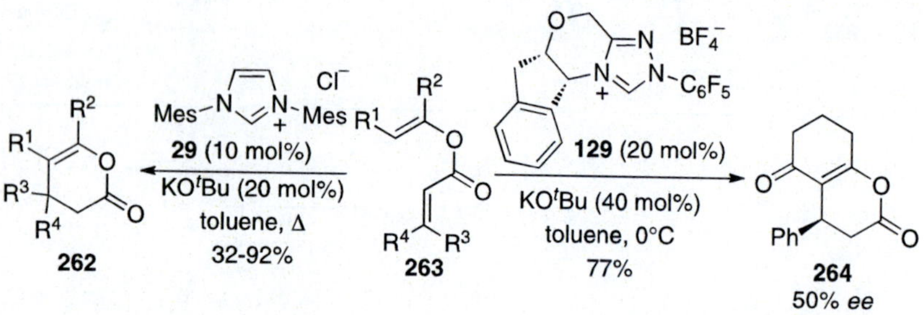

Scheme 12.59 Pyranone formation

You and co-workers have demonstrated a further application of NHCs in the kinetic resolution of formyl β-lactams (±)-**265** [103]. Upon treatment with a chiral NHC, the Breslow-type intermediate is formed, followed by ring-opening of the β-lactam moiety, with subsequent trapping of the acylazolium intermediate leading to the enantio-enriched succinimide product **266** and resolved formyl β-lactam (which is reduced to its alcohol **267**). The authors note that when R" = H, the products undergo racemisation readily, and this is a possible explanation for the lower levels of enantioselectivity observed in the succinimide products **266** (Scheme 12.60).

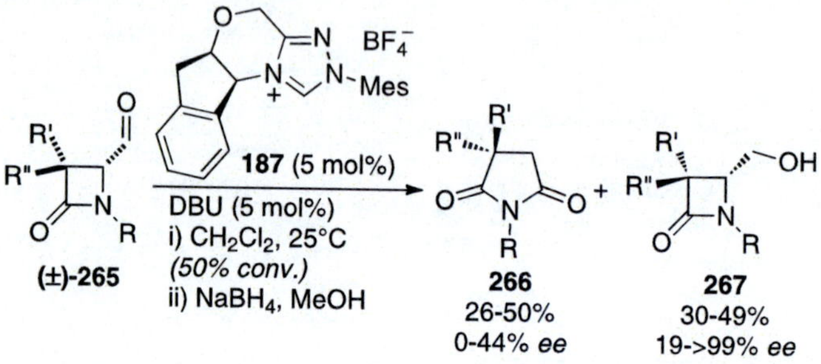

Scheme 12.60 Kinetic resolution of formyl β-lactams

12.4 Conclusions

Major advances in the application of NHCs in organocatalysis have been achieved, and this arena has become a focus of considerable research. The use of chiral NHCs has allowed access to highly enantioselective organocatalytic transformations and the breadth and depth of reactivity that can be accessed is ever expanding.

References

1. For recent reviews of the ability of NHCs to catalyse organocatalytic processes see Enders D, Niemeier O, Henseler A (2007) Chem Rev 107:5606–5655; Marion N, Díez-González S, Nolan, SP (2007) Angew Chem Int Ed 46:2988–3000
2. Seebach D (1979) Angew Chem Int Ed 18:239–258
3. Mattson AE, Scheidt KA (2004) Org Lett 6:4363–4366; Bharadwaj AR, Scheidt KA (2004) Org Lett 6:2465–2468; Mattson AE, Bharadwaj AR, Scheidt KA (2004) J Am Chem Soc 126:2314–2315
4. Stetter H, Lorenz G (1985) Chem Ber 118:1115–1125; Myers MC, Bharadwaj AR, Milgram, BC, Scheidt KA (2005) J Am Chem Soc 127:14675–14680
5. Maki EM, Chan A, Phillips EM, Scheidt KA (2007) Org Lett 9:371–374
6. Noonan C, Baragwanath L, Connon SJ (2008) Tetrahedron Lett 49:4003–4006; Guin J, Sarkar S De, Grimme S, Studer A (2008) Angew Chem Int Ed 47:8727–8730
7. Chan A, Scheidt KA (2006) J Am Chem Soc 128:4588–4589; Maki BE, Chan A, Phillips EM, Scheidt KA (2009) Tetrahedron 65:3102–3109
8. Burstein C, Glorius F (2004) Angew Chem Int Ed 43:6205–6208
9. Sohn SS, Rosen EL, Bode JW (2004) J Am Chem Soc 126:14370–14371
10. Nair V, Vellalath S, Poonoth M, Mohan R, Suresh E (2006) Org Lett 8:507–509
11. Rommel M, Fukuzumi T, Bode JW (2008) J Am Chem Soc 130:17266–17267
12. Chan A, Scheidt KA (2007) J Am Chem Soc 129:5334–5335
13. Nair V, Vellalath S, Poonoth M, Suresh E (2006) J Am Chem Soc 128:8736–8737
14. Chan A, Scheidt KA (2005) Org Lett 7:905–908
15. Chiang PC, Rommel M, Bode JW (2009) J Am Chem Soc 131:8714–8718
16. Chiang PC, Kim Y, Bode, JW (2009) Chem Commun 4566–4568
17. Duguet N, Campbell CD, Slawin AMZ, Smith AD (2008) Org Biomol Chem 6:1108–1113
18. He L, Jian T-Y, Ye S (2007) J Org Chem 72:7466–7468
19. Fischer C, Smith SW, Powell DA, Fu GC (2006) J Am Chem Soc 128:1472–1473
20. Grasa GA, Kissling RM, Nolan SP (2002) Org Lett 4:3583–3586
21. Nyce GW, Lamboy JA, Connor EF, Waymouth RM, Hedrick JL (2002) Org Lett 4:3587–3590
22. For the pioneering stoichiometric transfer of a carboxyl unit from 2-alkoxycarbonylimidazolium salts to benzyl alcohol in the presence of DABCO see Bakhtiar C, Smith EH (1994) J Chem Soc Perkin Trans 1:239–243
23. Grasa GA, Guveli T, Singh R, Nolan SP (2003) J Org Chem 68:2812–2819
24. Singh R, Nolan SP (2005) Chem Commun 5456–5458
25. Movassaghi M, Schmidt MA (2005) Org Lett 7:2453–2456
26. Schmidt MA, Müller P, Movassaghi M (2008) Tetrahedron Lett 49:4316–4318
27. Thomson JE, Rix K, Smith AD (2006) Org Lett 8:3785–3789; Thomson JE, Campbell CD, Concellón C, Duguet N, Rix K, Slawin AMZ, Smith AD (2008) J Org Chem 73:2784–2791
28. Campbell CD, Duguet N, Gallagher KA, Thomson JE, Lindsay AG, O'Donoghue A, Smith AD (2008) Chem Commun 3528–3530
29. Thomson JE, Kyle AF, Concellón C, Gallagher KA, Lenden P, Morrill LC, Miller AJ, Joannesse C, Slawin AMZ, Smith AD (2008) Synthesis 2805–2818
30. Reynolds NT, Read de Alaniz J, Rovis T (2004) J Am Chem Soc 126:9518–9519
31. Chow KY-K, Bode JW (2004) J Am Chem Soc 126:8126–8127
32. Sohn SS, Bode JW (2006) Angew Chem Int Ed 45:6021–6024
33. Zeitler K (2006) Org Lett 8:637–640
34. Li G-Q, Li Y, Dai L-X, You S-L (2007) Org Lett 9:3519–3521; Alcaide B, Almendros P, Cabrero G, Ruiz MP (2007) Chem Commun 4788–4790
35. Wang L, Thai K, Gravel M (2009) Org Lett 11:891–893
36. Li G-Q, Dai L-X, You S-L (2009) Org Lett 11:1623–1625

37. Vora H, Rovis T (2007) J Am Chem Soc 129:13796–13797; Bode JW, Sohn SS (2007) J Am Chem Soc 129:13798–13799
38. Ugai T, Tanaka S, Dokawa S (1943) J Pharm Soc Jpn 63:296–300
39. Breslow R (1958) J Am Chem Soc 80:3719–3726
40. Sheehan JC, Hunneman DH (1966) J Am Chem Soc 88:3666–3667
41. (a) Sheehan JC, Hara T (1974) J Org Chem 39:1196–1199; (b) Tagaki, W, Tamura Y, Yano Y (1980) Bull Chem Soc Jpn 53:478–480; (c) Martí J, Castells J, López-Calahorra, F (1993) Tetrahedron Lett 34:521–524; (d) Yamashita K, Sasaki S-I, Osaki T, Nango M, Tsuda K (1995) Tetrahedron Lett 36:4817–4820; (e) Knight RL, Leeper F (1997) Tetrahedron Lett 38:3611–3614; (f) Pesch J, Harms K, Bach T (2004) Eur J Org Chem 2025–2035; (g) Dvorak CA, Rawal VH (1998) Tetrahedron Lett 39:2925–2928
42. (a) Enders D, Breuer K, Raabe G, Runsink J, Teles JH, Melder J-P, Ebel K, Brode S (1995) Angew Chem Int Ed 34:1021–1023; (b) Enders D, Breuer K, Kallfass U, Balansiefer T (2003) Synthesis 1292–1295; (c) Melder J-P, Ebel K, Schneider R, Gehrer E, Harder W, Brode S, Enders D, Breuer K, Teles JH, Raabe G (1996) Helv Chim Acta 79:61–83
43. Enders D, Breuer K, Teles JH (1996) Helv Chim Acta 79:1217–1221
44. Knight RL, Leeper FJ (1998) J Chem Soc Perkin Trans 1:1891–1893
45. Enders D, Kallfass U (2002) Angew Chem Int Ed 41:1743–1745
46. For a computational study on the origin of the enantioselectivity in this reaction, see: Dudding T, Houk KN (2004) Proc Natl Acad Sci USA 101:5770–5775
47. Enders D, Han J (2008) Tetrahedron: Asymm 19:1367–1371
48. Ma Y, Wei S, Wu J, Yang F, Liu B, Lan J, Yang S, You J (2008) Adv Synth Catal 350:2645–2651
49. O'Toole S, Connon SJ (2009) Org Biomol Chem 7:3584–3593
50. Stetter H, Dämkes G (1977) Synthesis 403–404
51. (a) Enders D, Niemeier O, Balensiefer T (2006) Angew Chem Int Ed 45:1463–1467; (b) Enders D, Niemeier O, Raabe G (2006) Synlett 2431–2434
52. Takikawa H, Hachisu Y, Bode JW, Suzuki K (2006) Angew Chem Int Ed 45:3492
53. Li Y, Feng Z, You S-L (2008) Chem Commun 2263–2265
54. Takikawa H, Suzuki K (2007) Org Lett 9:2713–2716
55. Stetter H (1976) Angew Chem Int Ed 15:639–647
56. Enders D, Breuer K, Runsink J, Teles JH (1996) Helv Chim Acta 79:1899–1902
57. Mennen SM, Blank JT, Tran-Dubé MB, Imbriglio JE, Miller SJ (2005) Chem Commun 195–197
58. (a) Kerr MS, Read de Alaniz J, Rovis T (2002) J Am Chem Soc 124:10298–10299 (b) Read de Alaniz J, Kerr MS, Moore JL, Rovis T (2008) J Org Chem 73:2033–2040
59. Kerr MS, Rovis T (2003) Synlett 1934–1936
60. Cullen SC, Rovis T (2008) Org Lett 10:3141–3144
61. Matsumoto Y, Tomioka K (2006) Tetrahedron Lett 47:5843–5846
62. Kerr MS, Rovis T (2004) J Am Chem Soc 126:8876–8877
63. Moore JL, Kerr MS, Rovis T (2006) Tetrahedron 62:11477–11482
64. Read de Alaniz J, Rovis T (2005) J Am Chem Soc 127:6284–6289
65. (a) Liu Q, Rovis T (2006) J Am Chem Soc 128:2552–2553; (b) Liu Q, Rovis T (2007) Org Process Res Dev 11:598–604
66. Enders D (1993) Stereoselective synthesis. Springer, Heidelberg, Germany, p 63
67. Enders D, Han J, Henseler A (2008) Chem Commun 3989–3991
68. Liu Q, Perreault S, Rovis T (2008) J Am Chem Soc 130:14066–14067
69. DiRocco DA, Oberg KM, Dalton DM, Rovis T (2009) J Am Chem Soc 131:10872–10874
70. Li Y, Zhuo ZA, He H, You S-L (2008) Adv Synth Catal 350:1885–1890
71. Rommel M, Fukuzumi T, Bode JW (2008) J Am Chem Soc 130:17266–17267
72. Phillips EM, Reynolds TE, Scheidt KA (2008) J Am Chem Soc 130:2416–2417
73. Nair V, Babu BP, Vellalath S, Suresh E (2008) Chem Commun 747–749
74. Struble JR, Kaeobamrung J, Bode JW (2008) Org Lett 10:957–960

75. Chan A, Scheidt KA (2008) J Am Chem Soc 130:2740–2741
76. Maki BE, Chan A, Scheidt KA (2008) Synthesis 1306–1315; Maki BE, Patterson EV, Cramer CJ, Scheidt KA (2009) Org Lett 11:3942–3945
77. Struble JR, Bode JW (2009) Tetrahedron 65:4957–4967
78. Zhang YR, He L, Wu X, Shao PL, Ye S (2008) Org Lett 10:277–280
79. He L, Lv H, Zhang Y-R, Ye S (2008) J Org Chem 73:8101–8103
80. Wang X-N, Lv H, Huang X-L, Ye S (2009) Org Lett 11:4029–4031
81. Huang X-L, He L, Shao LP-L, Ye S (2009) Angew Chem Int Ed 48:192–195
82. Huang X-L, Shao P-L, Lv H, Ye S (2009) J Org Chem 74:7585–7587
83. Lv H, Zhang Y-R, Huang X-L, Ye S (2008) Adv Synth Catal 350:2715–2718
84. Zhang Y-R, Lv H, Zhou D, Ye S (2008) Chem Eur J 14:8473–8476
85. Wang X-N, Lv H, Huang X-L, Ye S (2009) Org Biomol Chem 7:346–350
86. Concellon C, Duguet N, Smith AD (2009) Adv Synth Cat 351:3001–3009
87. Reynolds NT, Rovis T (2005) J Am Chem Soc 127:16406–16407
88. He M, Struble JR, Bode JW (2006) J Am Chem Soc 128:8418–8420
89. Kobayashi S, Kinoshita T, Uehara H, Sudo T, Ryu I (2009) Org Lett 11:3934–3937
90. He M, Uc GJ, Bode JW (2006) J Am Chem Soc 128:15088–15089
91. He M, Beahm BJ, Bode JW (2008) Org Lett 10:3817–3820
92. Phillips EM, Wadamoto M, Chan A, Scheidt KA (2007) Angew Chem Int Ed 46:3107–3110
93. Li Y, Wang X-Q, Zheng C, You S-L (2009) Chem Commun 5823–5825
94. Wadamoto M, Phillips EM, Reynolds TE, Scheidt KA (2007) J Am Chem Soc 129:10098–10099
95. Phillips EM, Wadamoto M, Scheidt KA (2009) Synthesis 687–690
96. He L, Zhang Y-R, Huang X-L, Ye S (2008) Synthesis 2825–2829
97. Kano T, Sasaki K, Maruoka K (2005) Org Lett 7:1347–1349
98. Suzuki Y, Yamauchi K, Maramatsu K, Sato M (2004) Chem Commun 2770–2771; Suzuki Y, Muramatsu K, Yamauchi K, Morie Y, Sato M (2006) Tetrahedron 62:302–310
99. Maki BE, Chan A, Phillips EM, Scheidt KA (2009) Tetrahedron 65:3102–3109
100. He M, Bode JW (2008) J Am Chem Soc 130:418–419
101. Kaeobamrung J, Bode JW (2009) Org Lett 11:677–680
102. Ryan SJ, Candish L, Lupton DW (2009) J Am Chem Soc 131:14176–14177
103. Li G-Q, Li Y, Dai L-X, You S-L (2008) Adv Synth Catal 350:1258–1262

Chapter 13
N-Heterocyclic Carbene Complexes: Decomposition Pathways

Kingsley J. Cavell and Adrien T. Normand

Abstract *N*-heterocyclic carbenes, and their derivatives, are powerful donor ligands that form strong M–NHC bonds, particularly with late transition metals. Interest in these species has largely focused on their application as ligands in catalysis, in which they have shown some promise. However, they have not had the impact originally anticipated; NHCs undergo a number of processes that lead to catalyst deactivation and in the following chapter the range of deactivation reactions that have been reported will be discussed. By far the most detrimental deactivation process uncovered to date is reductive elimination, in which the NHC is lost as an imidazolium salt and the metal centre is reduced. This decomposition pathway is discussed in some detail and its impact on catalysis is put in context. Nevertheless, as apparent from some of the following discussion, the opportunity exists for an imaginative approach to the problem of catalyst decomposition, and it is possible to develop novel synthetic chemistry utilising these reaction pathways.

13.1 Reductive Elimination in Group 10 Metals NHC Complexes

13.1.1 Introduction

Developments in NHC chemistry have been considerable, and an exponential increase in publications has occurred since the mid 1990s. Numerous reviews have appeared providing comprehensive experimental and theoretical descriptions of advances in the field [1–13], and very recently a volume of *Coordination Chemistry Reviews* and complete issues of *Dalton Transactions* and *European Journal of Inorganic Chemistry* have been dedicated to NHC chemistry [14]. An excellent review on the topic of catalyst decomposition was published in 2004 [7], and consequently the present chapter will

K. J. Cavell (✉)
School of Chemistry, Cardiff University, Main Building, Park Place, CF10 3AT Cardiff, UK
e-mail: Cavellkj@cf.ac.uk

C. S. J. Cazin (ed.), *N-Heterocyclic Carbenes in Transition Metal Catalysis and Organocatalysis*, Catalysis by Metal Complexes 32,
DOI 10.1007/978-90-481-2866-2_13, © Springer Science+Business Media B.V. 2011

mostly focus on the latest developments in the field. Studies of *N*-heterocyclic carbenes (NHCs) have principally focused on their application as ligands in organometallic chemistry and catalysis. They are powerful donor ligands, forming strong metal-NHC bonds, and have properties that make them potentially useful in a wide range of catalytic reactions. The study of NHCs can be considered a mature field, as clearly exemplified by the current book (*N-Heterocyclic Carbenes in Transition Metal Catalysis and Organocatalysis*). Nevertheless, NHCs have not been universally applicable as ligands in catalysis, and it soon became apparent that fundamental "problems" exist in the chemistry of these ligands. In recent years, a number of studies have revealed some unexpected "decomposition" pathways and implications for catalysis; NHCs are reactive species and undergo a range of reactions, some considered as decomposition processes, and therefore undesirable, and some intriguing and potentially desirable reactions. The two aspects overlap, and consequently in considering "decomposition" or "deactivation" pathways it is important not to overlook NHCs as intermediates in chemical reactions [11], more in keeping with the role of "traditional" carbenes in synthesis and Schrock type carbenes in catalysis.

13.1.2 *Reductive Elimination from Alkyl–Metal–NHC Complexes*

Following the synthesis of the first methyl-palladium NHC complexes it was subsequently found that the complexes undergo a facile thermal decomposition process in which the NHC is lost as 2-methylimidazolium salt and the Pd is reduced to Pd(0) (Scheme 13.1) [15–17]. In ensuing studies investigating the reaction behaviour of a range of hydrocarbyl Pd and Ni carbene complexes, it was found that the decomposition reaction is ubiquitous. It occurs with varying ease, for mono-NHC, bis-NHC and donor functionalised-NHC complexes [16–23].

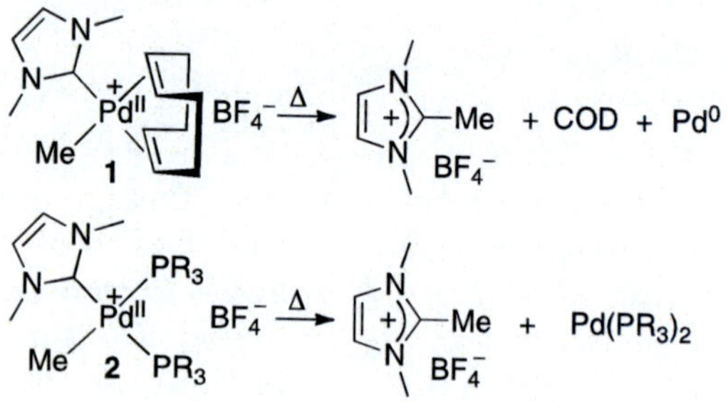

Scheme 13.1 Reductive elimination from methyl–Pd–NHC complexes

A comprehensive experimental and theoretical study was undertaken on the reaction, which was shown to be a concerted reductive elimination process; kinetic studies were consistent with reductive elimination, and DFT calculations on complex **3** (Fig. 13.1) supported an "associative" reductive elimination pathway with a small activation barrier (14.1 kcal mol^{-1}) and overall exothermic reaction (–9.2 kcal·mol^{-1}) [19]. In the transition structure (TS) the P–Pd–P "bite" angle opens out, and the associated angle between the *cis* methyl group and NHC closes down, allowing effective orbital overlap and mixing of the C_{NHC} $p(\pi)$, the C_{Me} and the Pd d_{xy} orbitals, facilitating C_{NHC}–C_{Me} bond formation. In addition, the plane of the NHC ligand is at a substantial angle to the coordination plane (commonly 60–80°) and hence the C_{NHC} $p(\pi)$, which is perpendicular to the NHC plane, is correctly orientated to interact with the Me group (Fig. 13.1). In the encounter complex (EC) the C_{NHC}–C_{Me} bond is fully occupied with no *d* involvement and the N $p(\pi) \rightarrow C_{NHC}$ $p(\pi)$ is purely *p* in character, as in the starting complex. The $p(\pi)$ occupancy of the C_{NHC} has increased due to greater delocalisation of the imidazolium cation. Consistent with these results, when a chelating di-phosphine spectator ligand is used instead of two monodentate ligands there is a much higher barrier to reductive elimination. The chelating ligand prevents contraction of the Me–Pd–NHC angle, and hence prevents effective orbital overlap. Experimentally forcing conditions are required to induce hemilability and hence reductive elimination. Similarly, a chelating ligand in which the NHC forms one component will also impede this process (Fig. 13.1).

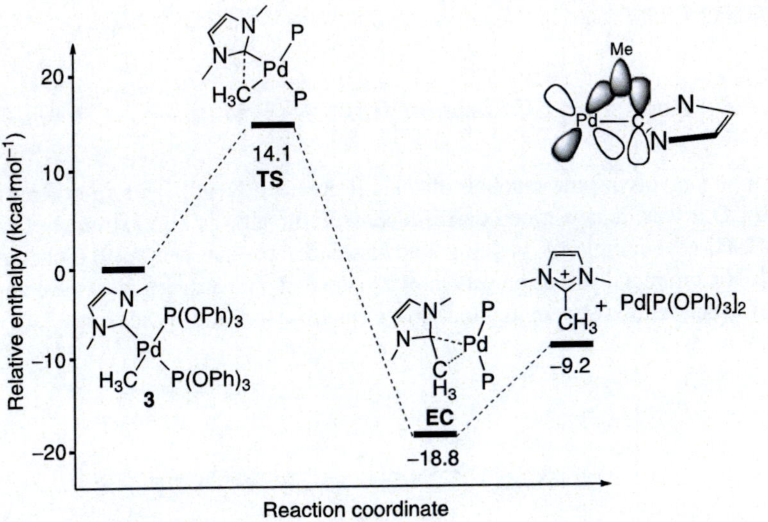

Fig. 13.1 Computational description of the reductive elimination process

In studies on the migratory insertion of ethylene into a phosphine-functionalised NHC–Ni–H complex formed from oxidative addition of the ligand precursor to a Ni(0) complex [24–27], Fryzuk and co-workers observed a modified reductive elimination reaction, leading to the apparently *trans* reductive elimination of 2-ethylimidazolium (Scheme 13.2) [28]. Indeed, labelling studies rule out direct ethyl insertion into the Ni–C_{NHC} bond. However, DFT calculations support a C–C bond-forming step arising *via* an apical agostic ethyl complex facilitated by the steric demands of the coordinated phosphines, thus allowing a *cis* reductive elimination process to occur.

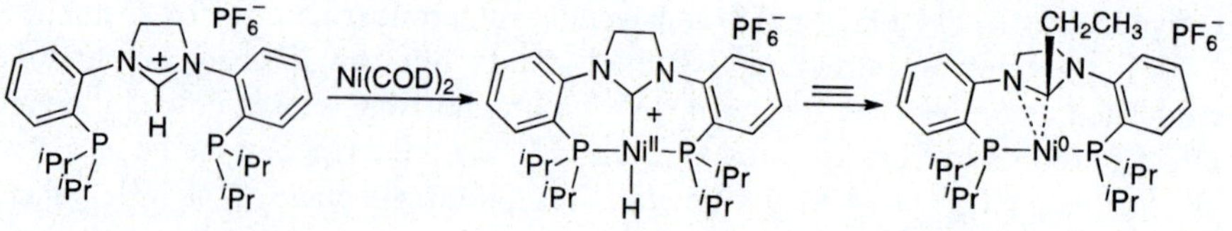

Scheme 13.2 Ethylene insertion followed by elimination of 2-ethylimidazolium

A recent study has also found evidence of reductive elimination of 2-propenylimidazolium salt from a [Pd(π-allyl)(NHC)]$^+$ complex, probably following allyl η^3-η^1 rearrangement (Scheme 13.3) [29]. The elimination reaction generated L–Pd(0) *in situ*, which could then catalyse a novel C–C coupling process.

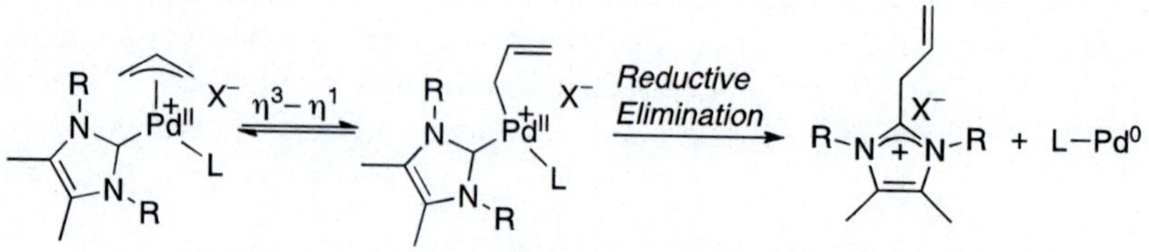

Scheme 13.3 Elimination of 2-propenylimidazolium from a Pd–π–allyl complex

13.1.3 *Reductive Elimination from Acyl–Metal–NHC Complexes*

In studies involving the reaction of a *cis*-[PdCl(Me)(NHC)]$_2$ chloro-bridged dimer with CO it was demonstrated that reductive elimination is also extremely facile for NHC–Pd-acyl complexes, yielding 2-acylimidazolium salts and Pd(0) (Scheme 13.4) [22]. The product distribution was shown to depend on the structure of the complexes from which reductive coupling took place (pathways **A** and **B**, Scheme 13.4).

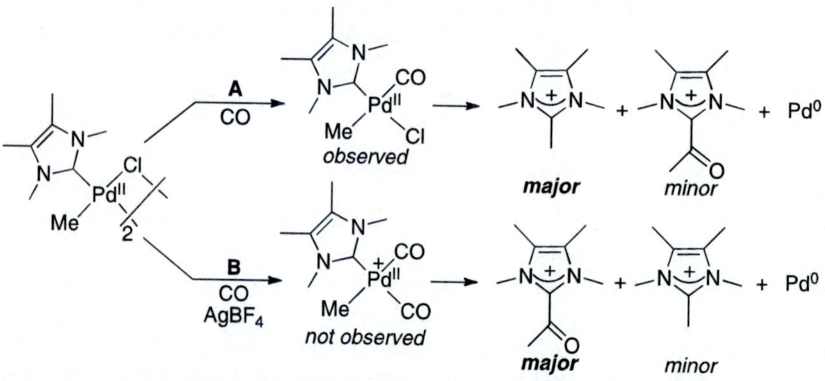

Scheme 13.4 Reductive elimination from acyl–Pd–NHC complexes

Elsevier and co-workers have found that complexes of the type **4** (Scheme 13.5) were surprisingly resistant to reductive elimination [30]. However, this behaviour is not inconsistent with previous observations; the *N*-substituent on the NHC is the bulky mesitylene, and whilst bulky substituents do not prevent decomposition they restrict approach of the *cis*-methyl group, resulting in less effective interaction between the methyl and the NHC [5, 19]. The authors did, however, find that addition of CO to a dichloromethane solution of the complex led to rapid decomposition (R = Me) *via* reductive elimination to give the 2-acylimidazolium salt **5**.

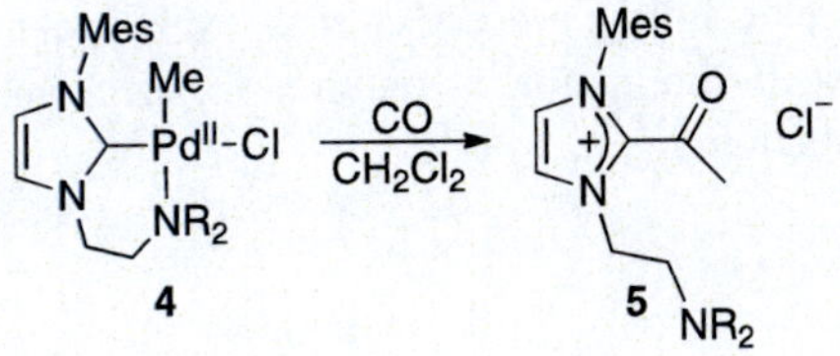

4 **5**

Scheme 13.5 CO-promoted reductive elimination

13.1.4 Reductive Elimination and Catalysis

Most catalytic reactions involve the formation of hydrocarbyl-, acyl-, and/or hydrido-M intermediates at some point in the catalytic cycle, consequently, the reductive elimination reaction can be a serious drawback for the application of NHC–M complexes in catalysis, and must be given due consideration when designing catalyst systems. Key studies on several catalytic reactions demonstrate the pervasive nature of this decomposition process. In a fundamental study on the Heck reaction, oxidative addition of phenyl iodide to [Pd(MAH)(tmiy)$_2$], (tmiy = 1,3,4,5-tetramethylimidazol-2-ylidene; MAH = maleic anhydride) at 60°C yielded the expected Pd$^{(II)}$ complex, [PdI(Ph)(tmiy)$_2$], but also the reductive elimination product, 2-phenyl-1,3,4,5-tetramethylimidazolium iodide. Moreover, on warming, the complex [PdI(*p*-$C_6H_4NO_2$) (tmiy)$_2$] (formed from the room temperature oxidative addition of aryl iodide to [Pd(tmiy)$_2$]) decomposed *via* reductive elimination to give 2-(4-nitrophenyl)-1,3,4,5-tetramethylimidazolium ion as the major product [20]. In another important study, Caddick, Cloke and co-workers observed similar behaviour when investigating catalytic Buchwald-Hartwig amination reactions using Pd-NHC complexes as pre-catalysts; the oxidative addition of aryl chlorides to PdL$_2$ (where L = SIPr) gave only the 2-arylimidazolinium reductive elimination products [31]. However, in recent studies with extremely bulky, ^{t}Bu substituted, imidazolin-2-ylidenes as ligands such a decomposition pathway did not appear to be significant [32]. Marshall and Grushin have provided a further example in which reductive elimination has occurred from a Pd(σ-aryl)Cl(IPr) complex, giving 2-arylimidazolium salt [33]. Interestingly, attempts to prepare the analogous iodo-complex gave only the 2-arylimidazolium iodide salt. It had previously been thought that NHC ligands with bulky aromatic substituents would provide protection from the reductive elimination process.

A stoichiometric Heck reaction between the $[Pd(p\text{-}C_6H_4NO_2)(tmiy)_2]^+$ cation and *n*-butyl acrylate at sub-ambient temperatures elegantly illustrates the competition between decomposition and catalysis (Scheme 13.6) [20]. A complex mixture of products is obtained and it is evident that both Heck chemistry and reductive elimination are occurring in parallel, the relative degree of each is dependent on reaction conditions; at low temperatures (–30°C) reductive elimination predominates and 2-(4-nitrophenyl)-1,3,4,5,-tetramethylimidazolium ion, **6**, (with traces of 2-substituted-imidazolium ion **7**) is the main product. However, at –20°C the Heck product, *n*-butyl(*E*)-4-nitrocinnamate **8**, appears along with small amounts of 1,3,4,5-tetramethylimidazolium salt **9**; at room temperature the main products are Heck product **8,** and salt **9**. Under catalytic conditions Heck coupling proceeds almost exclusively, *i.e.* combined rates of migratory insertion plus β-elimination, as required for Heck catalysis, greatly exceed that of reductive elimination.

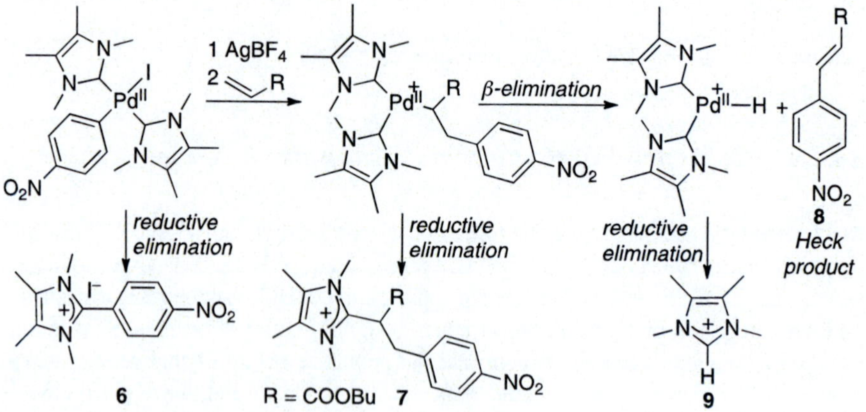

Scheme 13.6 Stoichiometric Heck reaction and reductive elimination products

In a study on the direct intramolecular arylation of a benzene ring with aryl chlorides (Scheme 13.7), Fagnou and co-workers noted that on allowing the reaction to proceed to completion the reductive elimination product **10** was observed, which led to loss of catalyst [34]. Improved catalytic performance was obtained by using the very bulky IPr and by the addition of excess NHC salt to the catalyst mixture. It is apparent that the catalytic and decomposition pathways are occurring in parallel and that catalysis is substantially faster under catalytic conditions [20]. These observations are consistent with previous reports, including the effect of bulky R-substituents on slowing catalyst decomposition, and the effect of added free imidazolium salt on catalyst stability [5, 7]. As noted by the authors, it is possible that free salt oxidatively adds to Pd(0) (formed from reductive elimination) to reform the catalyst – if this is the case there will be a gradual consumption of the salt and build up of reductive elimination product **10** in the product mixture [11].

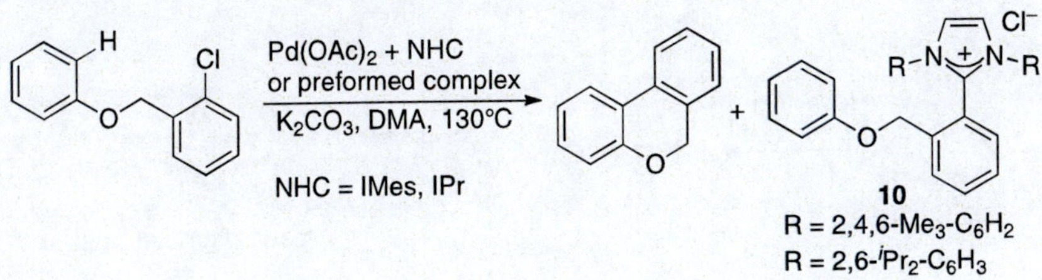

Scheme 13.7 Intramolecular arylation with Pd–NHC complexes

Catalyst systems of the type [NiL_2X_2 + $AlEt_nX_{3-n}$] (where L = PR_3 and X = halide) afford highly active catalysts for olefin dimerisation. However, when complex **11** (Scheme 13.8) is treated with $AlEt_2Cl$ in the presence of 1-butene, in toluene at 20°C the only products observed were decomposition products, **12, 13, 14**; no butene dimers were obtained [22]. At low temperatures (–15°C) and using the complex with 1,3-diisopropylimidazolin-2-ylidene as the NHC ligand, small amounts of butene dimers were observed. It is apparent from these results that Ni–NHC complexes are capable of olefin dimerisation, however, decomposition of the catalyst *via* reductive elimination predominates.

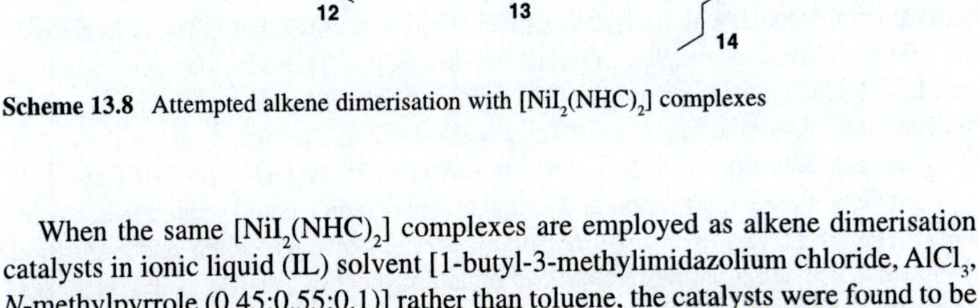

Scheme 13.8 Attempted alkene dimerisation with [$NiI_2(NHC)_2$] complexes

When the same [$NiI_2(NHC)_2$] complexes are employed as alkene dimerisation catalysts in ionic liquid (IL) solvent [1-butyl-3-methylimidazolium chloride, $AlCl_3$, *N*-methylpyrrole (0.45:0.55:0.1)] rather than toluene, the catalysts were found to be highly active, with no evidence of decomposition. Furthermore, product distributions for each of the catalyst systems studied was surprisingly similar, indicating a common active species may have been formed in each case. It was proposed that reductive elimination of the NHC–Ni did indeed occur, as outlined in Scheme 13.8, however, the IL solvent oxidatively adds to the Ni(0) thus formed to yield a new Ni–NHC complex, **15**, stabilised by the IL solvent, and able to effectively catalyse the dimerisation process (Scheme 13.9) [25–27].

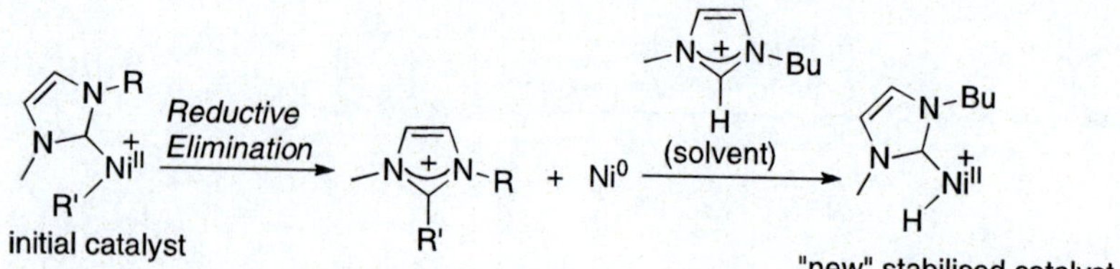

Scheme 13.9 Alkene dimerisation with $(NHC)_2NiI_2$ complexes in IL solvent

In an attempt to avoid the problem of reductive elimination Veige and co-workers synthesised Ir and Rh complexes of semi-rigid, bulky chelating benzimidazolidene-2-ylidene ligands **16** (Fig. 13.2) and investigated the complexes in enantioselective hydrogenation, transfer hydrogenation and hydroformylation [35]. A comprehensive investigation showed that reductive elimination was an ever present problem; under hydrogenation and transfer hydrogenation conditions, the catalysts decompose to M(0), most likely *via* reductive elimination. Attempts at enantioselective hydroformylation also lead to reductive elimination. Similar problems have beset hydroformylation studies using Co–NHC catalysts, and van Rensburg and co-workers have observed reductive elimination of dimesityl-imidazolium salt from $[Co(CO)_3(IMes)]_2$ under various hydroformylation conditions [36, 37].

Reports of reductive elimination from early transition metals are uncommon. However, Bullock and co-workers have reported the elimination of IMes from $[WCp(IMes)(CO)_2][B(C_6F_5)_4]$ to form the 2-H-imidazolium salt, during ketone hydrogenation probably *via* a form of reductive elimination process [38].

13.1.5 *Factors Affecting Reductive Elimination*

Several studies have been undertaken in an attempt to understand factors that affect the susceptibility of a complex to reductive elimination. These include investigations on Me–Pd–CNC pincer complexes [39]; studies on the influence of the geometry of the complexes [40]; the impact of *N*-substituents [41]; and a study on the influence of chelating spectator ligands in complexes of the type $[PdMe(NHC)(P–P)]BF_4$ [42].

Complementary experimental and computational studies demonstrated that pincers **17–19** (Fig. 13.2) are extremely thermally stable; however, when decomposition does occur it is *via* a partial reductive elimination process [39, 43]. The complexes, when operated under catalytic conditions, show long-term stability at high temperatures. A DFT study has been carried out on the influence of spectator ligand bite angle, and NHC twist angle on reductive elimination from $Pd–L_2$ complexes (L = phosphine). Bite angle was found to have a significant impact [40]; increasing the bite angle (across the range 80–130°; **20**) lowered E_{act}, complexes with a large bite angle favoured reductive elimination and those with a smaller bite angle prefer oxidative addition. Interestingly, steric bulk on the spectator ligand(s) appears to be just as important as bite angle in influencing the reaction. Rotation of the NHC with respect to the coordination plane of the complex **21** had little influence on E_{act} for reductive elimination. The instability of the complexes where the NHC twists towards coplanarity with the coordination plane can be attributed to steric strain. A DFT study on the influence of *N*-substituents on both E_{act} and overall thermodynamics of reductive elimination in **22**, demonstrated that electronic factors have a major influence on the activation barrier [41]. The NHC *p*(π) orbital plays a key role in the reductive elimination process and *N*-substituents that remove π-density promote elimination, whereas increased electron donation stabilises against reductive elimination.

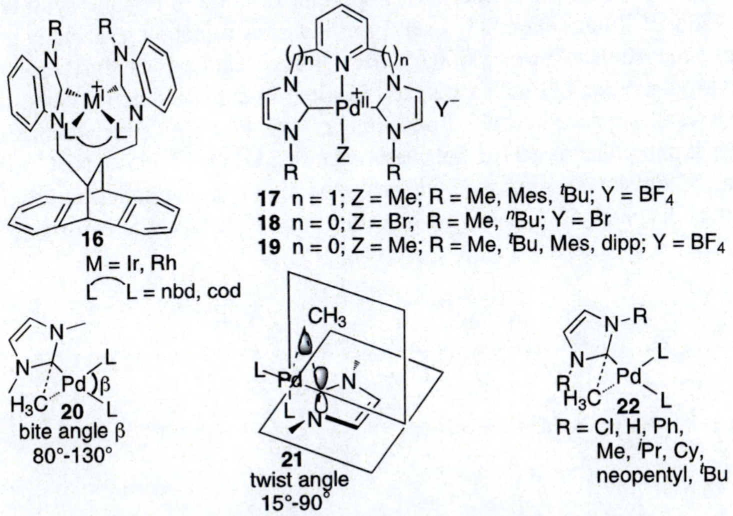

Fig. 13.2 Group 9 and 10 metals NHC complexes prone to reductive elimination

Experimental studies show that chelating spectator ligands impart a degree of stability to complexes of type **23** (Scheme 13.10) [42]. If monodentate phosphine ligands are used decomposition is rapid at 20°C, however, using dppp no decomposition is detected after 24 h [19]. It was found that the rate of decomposition could be linked to the chelate ring size; at 65°C, with dppp decomposition was complete after 6 h, with dppe only a small amount of decomposition occurred after this time [42].

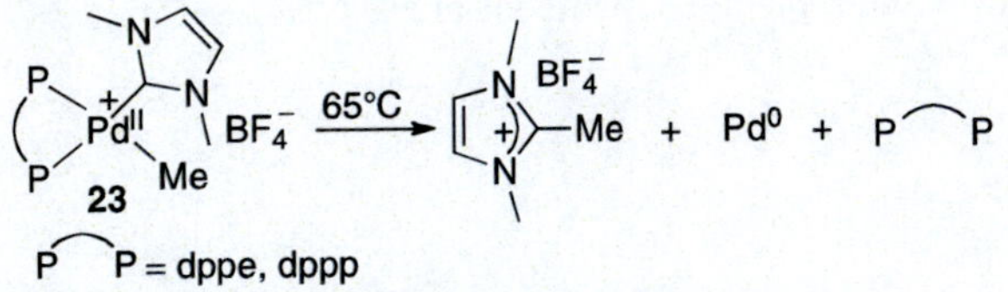

Scheme 13.10 Decomposition of Pd–NHC complexes bearing chelating phosphine ligands

For NHCs to have the overall impact on catalysis originally anticipated it is clear that the problem presented by reductive elimination must be overcome or at least limited. A number of possible solutions have been investigated, and discussed here.

More inventive approaches and hence further research is required. The potential rewards in successfully addressing this problem are substantial. However, it is possible that the use of NHCs may be limited to particular reaction types; not, it could well be that they will never be effective ligands in some catalytic processes.

13.2 Decomposition of Ru Metathesis Catalysts

Olefin metathesis is one of the most important reaction in organic synthesis [44]. Complexes of Ru are extremely useful for this transformation, especially so-called Grubbs catalysts. The introduction of NHCs in Ru metathesis catalysts a decade ago ("second generation" Grubbs catalysts) resulted in enhanced activity and lifetime, hence overall improved catalytic performance [45, 46]. However, compared to the archetypal phosphine-based Ru metathesis catalyst **24** (Fig. 13.3), Ru–NHC complexes such as **25** display specific reactivity patterns and as a consequence, are prone to additional decomposition pathways as well as non NHC-specific pathways [47].

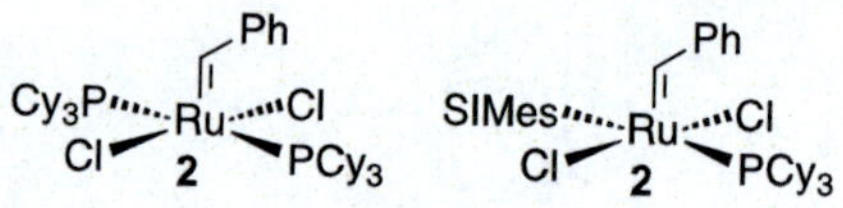

Fig. 13.3 Ru catalysts for olefin metathesis

13.2.1 *C–H Activation in Ru–NHC Catalysts*

In 2007, Grubbs reported complex **26**, a catalyst for ring closing metathesis (RCM) of tetra-substituted olefins. This catalyst was found to decompose into benzylidene-free (*i.e.* inactive in catalysis) complexes **27** and **28** upon heating in C_6H_6 and CD_2Cl_2 (Scheme 13.11) [48]. The authors proposed a sequence involving phosphine dissociation, oxidative addition of the phenyl C–H bond, α-insertion of the resulting hydride into the Ru–benzylidene bond and C–C reductive elimination to give **27**. At this stage, the liberated PCy_3 acts as a base to promote the intramolecular metallation of the other phenyl substituent of the NHC, yielding **28**.

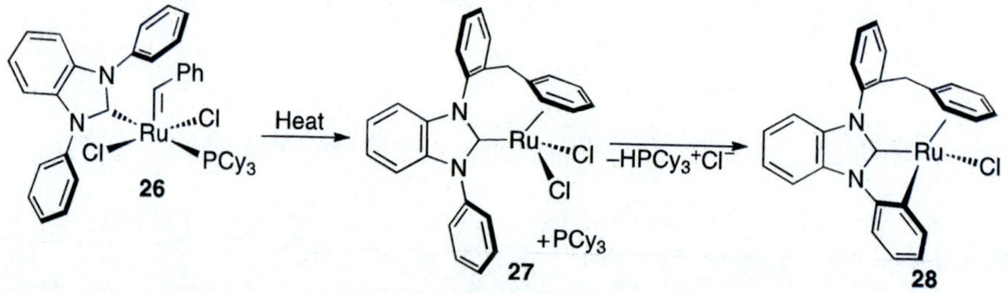

Scheme 13.11 Decomposition of a Ru–NHC catalyst by C–H activation

In another study, Blechert reported the aerobic decomposition of complex **29**, cleanly yielding compound **30** in which the benzylidene is now bound to the aryl ring of the NHC (Scheme 13.12) [49].

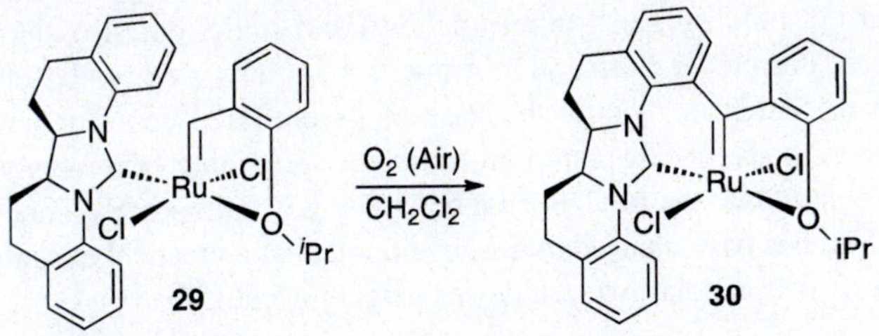

Scheme 13.12 Oxygen promoted C–H activation in a Ru–NHC catalyst

These two studies highlight the high reactivity of C–H bonds close to Ru. Consequently, Grubbs later reported a catalyst in which rotation of the phenyl group is restricted by methyl substituents at the 4-position of the NHC, thus preventing decomposition by C–H activation [50].

13.2.2 Benzylidene Insertion into NHC Substituents

In 2005, Diver and co-workers reported the serendipitous discovery of the CO-promoted insertion of the alkylidene moiety into the NHC aryl ring substituent in **25** [51]. The authors were investigating the use of strongly coordinating CO as a quenching agent in metathesis [52], when they encountered this unusual transformation. It appears that, due to the π-acidic character of CO, its coordination reduces π-backbonding from Ru into the *trans* alkylidene ligand, making the latter more electrophilic. In first generation catalysts the alkylidene undergoes intramolecular nucleophilic attack by the phosphine ligand [53], while in second generation Grubbs catalysts, the enhanced reactivity triggers a so-called Buchner reaction (*i.e.* aromatic ring cyclopropanation by a carbene) between the alkylidene and the NHC, followed by electrocyclisation to give cycloheptatriene NHC complex **31** (Scheme 13.13) [51, 54].

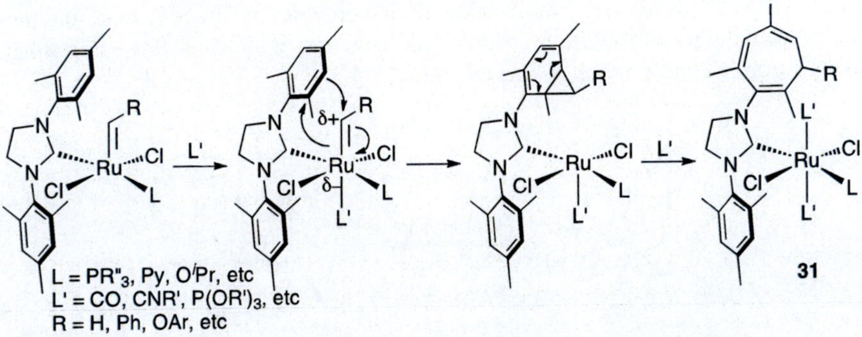

Scheme 13.13 π-Acid ligand-promoted Buchner reaction in metathesis catalysts

In subsequent studies, Diver demonstrated the general character of this transformation: other π-acidic ligands such as isocyanides and phosphites also promote insertion, and even a Fischer carbene complex was found to react, despite the lower electrophilicity of this type of carbenes over alkylidenes [54]. The insertion reaction appears to be controlled not only by electronic (π-acidity) and steric, but also stereoelectronic factors: indeed, despite its π-accepting properties and low steric bulk, ethylene does not promote the Buchner reaction because of its preferred coordination mode (*i.e.* perpendicular to the Ru=CHR plane). In addition to providing extremely valuable and illuminating insight into the fundamental reactivity governing catalyst decomposition, these elegant studies have also led to the development of a practical cleanup procedure for Ru catalysts with a polar isocyanide scavenging agent [55].

13.3 Catalytic Applications of Intramolecular C–H Activation in NHC Complexes

The ability of NHCs to promote intramolecular C–H activation reactions has long been acknowledged [7]. Key to this reactivity are the strong σ-donor properties of NHCs, and the orientation of nitrogen substituents towards the metal, favouring coordinative unsaturation and placing C–H bonds in an ideal position for metallation. Due to their unexpected character, C–H (or C–C [56] and even C–N [57, 58]) activation reactions are usually regarded as mere curiosities, if not a plague. As a consequence, very few studies have looked at catalytic applications of the resulting complexes, which are often considered as dead-end products. However, recent reports have shed a different light on this issue, by showing that the reactivity behind C–H bonds metallation can be exploited in catalysis. These studies highlight the long-term practical importance of investigating decomposition pathways.

13.3.1 C–H Activation at Ru and "Borrowing Hydrogen" Catalysis

In 2004, Whittlesey and Williams demonstrated that the reversible C–H activation of Ru–NHC complexes (e.g. **32a**, Scheme 13.14) provides an effective manifold for tandem dehydrogenation/Wittig reaction/hydrogenation of alcohols, thus generating alkanes from alcohols and phosphorus ylides [56].

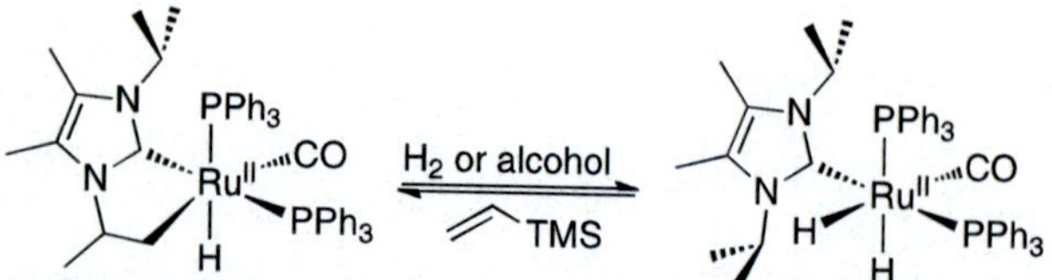

Scheme 13.14 Reversible C–H activation/hydrogenation in Ru–NHC complexes

The concept at the heart of this reaction is the conversion of a hydrogen donor (alcohol) into a hydrogen acceptor (alkene) to close the catalytic cycle (Scheme 13.15).

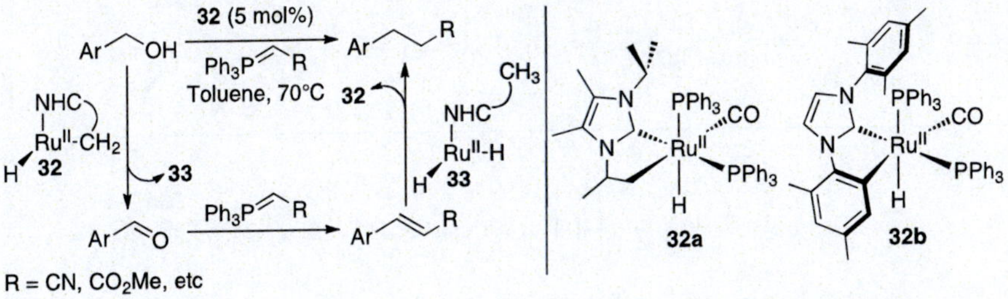

Scheme 13.15 Tandem "Borrowing Hydrogen" Wittig reaction with Ru–NHC complexes

In 2007, the authors showed that **32a** is more active than **32b** [59]. Not surprisingly, these catalysts are also efficient transfer hydrogenation catalysts [60]. Stoichiometric studies indicate that **32b** reacts with benzyl alcohol to yield benzaldehyde and dihydride complex **33b** [59, 61]. After Wittig reaction between the aldehyde and the phosphorus ylide, **33b** hydrogenates the olefin, thus regenerating **32b** (or its *trans* isomer). Finally, a recent report showed that the concept of "Borrowing Hydrogen" in catalysis can be extended to other tandem transformations, such as the Knoevenagel reaction [62].

13.3.2 C–H Activation at Ir and H/D Exchange

Peris and co-workers reported a family of Ir–NHC complexes prone to C–H activation [63, 64]. The authors had initially envisioned the preparation of **34**, which are NHC analogues of H/D exchange catalysts [IrCp*X(L)] (L = phosphine). However, most complexes of type **34** (e.g. **34a**, Scheme 13.16) undergo spontaneous C–H activation to give highly stable cyclometallated Ir complexes **35** [65]. Depending on the synthetic route (*i.e.* using a silver NHC transfer agent, Route A; or *in situ* deprotonation of the NHC precursor, Route B), intermediate **34** may or may not be observed [63, 64]. Complexes **34a** and **35a–b** are excellent catalysts for H/D exchange of a range of substrates in d_4-Methanol, and **35a–b** undergo deuteration themselves, suggesting that the C–H activation chemistry observed with these complexes might be responsible for their interesting activity [64]. Although **34a** is generally more active, cyclometallated **35a–b** can display useful selectivity, such as in the H/D exchange of styrene (Scheme 13.16): in this reaction, **35a–b** are more active and selective than the benchmark phosphine catalyst [IrCp*Cl_2(PMe_3)], and more selective than **34a**. Thus, complete selectivity for the vinylic moiety over the aromatic ring is observed for **35a–b**.

It is clear that these promising studies warrant further investigation in order to assess the potential role of C–H activation in determining H/D exchange activity.

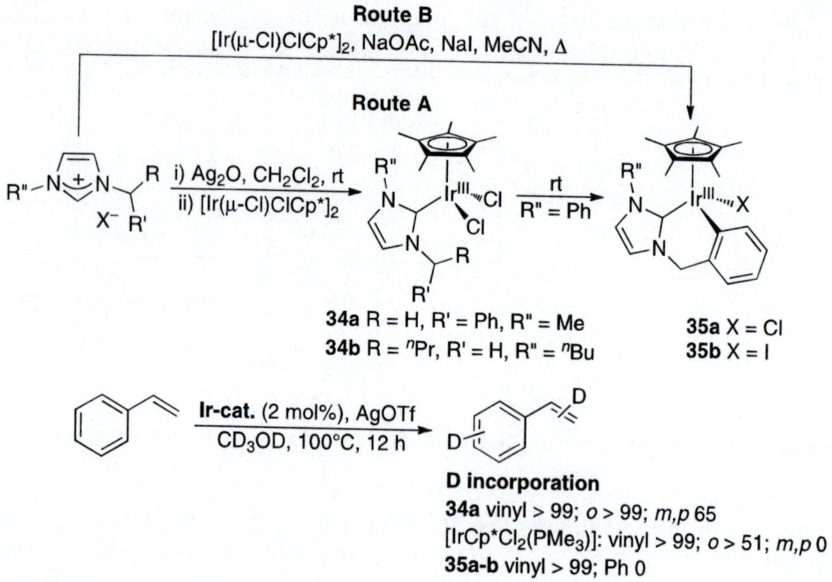

Scheme 13.16 H/D exchange catalysed by monodentate and cyclometallated Ir complexes

13.4 Miscellaneous

Although an increasing number of reports deal with unexpected reactions of NHC complexes [7], there are few studies in which the impact of these reactions on catalysis are even mentioned. Some authors have reported observations pertaining to complex synthesis that might become important for catalysis [66, 67]. More relevant to this chapter, Çetinkaya reported the application of a series of $[PdCl_2(NHC)(L)]$ complexes (L = NHC or phosphine) to Suzuki coupling of aryl bromides and chlorides. Thermogravimetric analyses showed that the catalysts were highly stable and only decomposed above 300°C by ligand dissociation (phosphine then NHC) [68]. Finally, a recent study revealed the decomposition of $[Pd(\pi\text{-allyl})(PR_3)(NHC)]^+$ catalysts in the presence of aryl iodides. The NHC ligand is replaced by an iodide, and 2-arylimidazolium is eliminated. This is the first reported example of NHC ligand loss by transmetallation [69].

References

1. Herrmann WA, Köcher C (1997) Angew Chem Int Ed 36:2162–2187
2. Bourissou D, Guerret O, Gabbai FP, Bertrand G (2000) Chem Rev 100:39–91
3. Herrmann WA (2002) Angew Chem Int Ed 41:1290–1309
4. Zinn FK, Viciu MS, Nolan SP (2004) Annu Rep Prog Chem, Sect B 100:231–249
5. Cavell KJ, McGuinness DS (2004) Coord Chem Rev 248:671–681

6. Peris E, Crabtree RH (2004) Coord Chem Rev 248:2239–2246
7. Crudden CM, Allen DP (2004) Coord Chem Rev 248:2247–2273
8. César V, Bellemin-Laponnaz S, Gade LH (2004) Chem Soc Rev 33:619–636
9. Cavell KJ, McGuinness DS (2007) Palladium complexes with carbonyl, isocyanide and carbene ligands. In: Crabtree RH, Mingos DMP, Canty AJ (eds) Comprehensive organometallic chemistry III. Elsevier, Amsterdam
10. Hahn FE, Jahnke MC (2008) Angew Chem Int Ed 47:3122–3172
11. Cavell KJ (2008) Dalton Trans 6676–6685
12. Praetorius JM, Crudden CM (2008) Dalton Trans 4079–4094
13. Normand AT, Cavell KJ (2008) Eur J Inorg Chem 2781–2800
14. (a) Guest Ed.: Crabtree RH (2007) Coord Chem Rev vol. 251; (b) Guest Ed.: Hahn FE (2009) Dalton Trans themed issue 35; (c) Guest Ed.: Nolan SP (2009) Eur J Inorg Chem special issue 13
15. Green MJ, Cavell KJ, Skelton BW, White AH (1998) J Organomet Chem 554:175–179
16. McGuinness DS, Green MJ, Cavell KJ, Skelton BW, White AH (1998) J Organomet Chem 565:165–178
17. Magill AM, McGuinness DS, Cavell KJ, Britovsek GJP, Gibson VC, White AJP, Williams DJ, White AH, Skelton BW (2001) J Organomet Chem 617–618:546–560
18. McGuinness DS, Cavell KJ (2000) Organometallics 19:741–748
19. McGuinness DS, Saendig N, Yates BF, Cavell KJ (2001) J Am Chem Soc 123:4029–4040
20. McGuinness DS, Cavell KJ, Skelton BW, White AH (1999) Organometallics 18:1596–1605
21. McGuinness DS, Cavell KJ (2000) Organometallics 19:4918–4920
22. McGuinness DS, Mueller W, Wasserscheid P, Cavell KJ, Skelton W, White AH, Englert U (2002) Organometallics 21:175–181
23. McGuinness DS, Cavell KJ, Yates BF (2001) Chem Commun 355–356
24. For similar studies see references [25–27]
25. McGuinness DS, Cavell KJ, Yates BF, Skelton BW, White AH (2001) J Am Chem Soc 123:8317–8328
26. Duin MA, Clement ND, Cavell KJ, Elsevier CJ (2003) Chem Commun 400–401
27. Clement ND, Cavell KJ, Jones C, Elsevier CJ (2004) Angew Chem Int Ed 43:1277–1279
28. Steinke T, Shaw BK, Jong H, Patrick BO, Fryzuk MD, Green JC (2009) J Am Chem Soc 131:10461–10466
29. Normand AT, Stasch A, Ooi L-L, Cavell KJ (2008) Organometallics 27:6507–6520
30. Warsink S, Boer SYd, Jongens LM, Fu C-F, Liu S-T, Chen J-T, Lutz M, Spek AL, Elsevier CJ (2009) Dalton Trans 7080–7086
31. Caddick S, Cloke FGN, Hitchcock PB, Leonard J, Lewis AKdK, McKerrecher D, Titcomb LR (2002) Organometallics 21:4318–4319
32. Esposito O, Gois PMP, de K, Lewis AK, Caddick S, Cloke FGN, Hitchcock PB (2008) Organometallics 27:6411–6418
33. Marshall WJ, Grushin VV (2003) Organometallics 22:1591–1593
34. Campeau L-C, Thansandote P, Fagnou K (2005) Org Lett 7:1857–1860
35. Jeletic MS, Jan MT, Ghiviriga I, Abboud KA, Veige AS (2009) Dalton Trans 2764–2776
36. van Rensburg H, Tooze RP, Foster DF, Otto S (2007) Inorg Chem 46:1963–1965
37. van Rensburg H, Tooze RP, Foster DF, Slawin AMZ (2004) Inorg Chem 43:2468–2470
38. Wu F, Dioumaev VK, Szalda DJ, Hanson J, Bullock RM (2007) Organometallics 26:5079–5090
39. Nielsen DJ, Magill AM, Yates BF, Cavell KJ, Skelton BW, White AH (2002) Chem Commun 2500–2501
40. Graham DC, Cavell KJ, Yates BF (2005) Dalton Trans 1093–1100
41. Graham DC, Cavell KJ, Yates BF (2006) Dalton Trans 1768–1775
42. Magill AM, Yates BF, Cavell KJ, Skelton BW, White AH (2007) Dalton Trans 3398–3406
43. Nielsen DJ, Cavell KJ, Skelton BW, White AH (2002) Inorg Chim Acta 327:116–125
44. Grubbs RH (ed) (2003) Handbook of metathesis, 1st ed. Wiley-VCH, Weinheim
45. Huang J, Stevens ED, Nolan SP, Petersen JL (1999) J Am Chem Soc 121:2674–2678

46. Scholl M, Ding S, Lee CW, Grubbs RH (1999) Org Lett 1:953–956
47. Hong SH, Wenzel AG, Salguero TT, Day MW, Grubbs RH (2007) J Am Chem Soc 129:7961–7968
48. Hong SH, Chlenov A, Day MW, Grubbs RH (2007) Angew Chem Int Ed 46:5148–5151
49. Vehlow K, Gessler S, Blechert S (2007) Angew Chem Int Ed 46:8082–8085
50. Chung CK, Grubbs RH (2008) Org Lett 10:2693–2696
51. Galan BR, Gembicky M, Dominiak PM, Keister JB, Diver ST (2005) J Am Chem Soc 127:15702–15703
52. It is well known that the metathesis activity of Ru catalysts stems from coordinative unsaturation and the generally accepted active species in RCM is a 14 electron $RuCl_2(CH_2)(L)$ fragment, where L = PCy_3 or NHC
53. Galan BR, Pitak M, Keister JB, Diver ST (2008) Organometallics 27:3630–3632
54. Galan BR, Pitak M, Gembicky M, Keister JB, Diver ST (2009) J Am Chem Soc 131:6822–6832
55. Galan BR, Kalbarczyk KP, Szczepankiewicz S, Keister JB, Diver ST (2007) Org Lett 9:1203–1206
56. Jazzar RFR, Macgregor SA, Mahon MF, Richards SP, Whittlesey MK (2002) J Am Chem Soc 124:4944–4945
57. Burling S, Mahon MF, Powell RE, Whittlesey MK, Williams JMJ (2006) J Am Chem Soc 128:13702–13703
58. Caddick S, Cloke GN, Hitchcock PB (2004) Lewis AKdK. Angew Chem Int Ed 43:5824–5827
59. Burling S, Paine BM, Nama D, Brown VS, Mahon MF, Prior TJ, Pregosin PS, Whittlesey MK, Williams JMJ (2007) J Am Chem Soc 129:1987–1995
60. Burling S, Whittlesey MK, Williams JMJ (2005) Adv Synth Cat 347:591–594
61. Diggle RA, Macgregor SA, Whittlesey MK (2004) Organometallics 23:1857–1865
62. Ledger AEW, Mahon MF, Whittlesey MK, Williams JMJ (2009) Dalton Trans 6941–6947
63. Corberan R, Sanau M, Peris E (2006) Organometallics 25:4002–4008
64. Corberan R, Sanau M, Peris E (2006) J Am Chem Soc 128:3974–3979
65. Complexes **35** are isolated by column chromatography on silica gel
66. Fantasia S, Jacobsen H, Cavallo L, Nolan SP (2007) Organometallics 26:3286–3288
67. Waltman AW, Ritter T, Grubbs RH (2006) Organometallics 25:4238–4239
68. Türkmen H, Çetinkaya B (2006) J Organomet Chem 691:3749–3759
69. Normand AT, Nechaev MS, Cavell KJ (2009) Chem Eur J 15:7063–7073

Chapter 14
N-Heterocyclic Carbene Complexes in Industrial Processes

Oliver Briel and Catherine S. J. Cazin

Abstract *N*-Heterocyclic carbene complexes produced on industrial scale are presented in this chapter along with a discussion about their production. Details of processes employing NHC complexes on pilot to industrial scales are discussed. These are frequently oriented towards the synthesis of biologically active molecules, however, examples are given for rubber formation and for 1-octene synthesis, a comonomer for polyethylene synthesis.

14.1 Introduction

The first academic publications introducing the concept of the use of *N*-heterocyclic carbenes as ligands in metal-catalysed applications appeared in the mid 1990s [1]. Since then, an increasing number of scientific groups have explored the scope of potential applications using NHC ligands (see Chapter 1, Fig. 1.1). Similarly to many other catalytic technologies, the time span between the original discovery and the entry of related technology in industrial laboratories is *ca.* 10 years. Currently, there only exists a limited number of publications on NHC complexes

O. Briel (✉)
Umicore AG & Co. KG, Rodenbacher Chaussee 4, P.O. Box 1351,
63403 Hanau-Wolfgang, Germany
e-mail: oliver.briel@eu.umicore.com

C. S. J. Cazin (✉)
EaStCHEM, School of Chemistry, University of St Andrews,
St Andrews, KY16 9ST, UK
e-mail: cc111@st-andrews.ac.uk

C. S. J. Cazin (ed.), *N-Heterocyclic Carbenes in Transition Metal Catalysis and Organocatalysis*, Catalysis by Metal Complexes 32,
DOI 10.1007/978-90-481-2866-2_14, © Springer Science+Business Media B.V. 2011

clearly describing large-scale applications. However, an increasing number of industrial players are filing process patents claiming and protecting the use of such catalysts for their respective applications. This clearly illustrates the great progress achieved by this ligand family that still was, not too long ago, a mere laboratory curiosity. However, NHC-based technologies are still in their early days in terms of industrial uses, and because most companies secure their IP (Intellectual Property) position, process information has not yet been, and might never be publicly disclosed. At this point, gathering information on industrial uses of NHCs proved to be a real challenge. To the best of our knowledge, to date only NHC systems based on ruthenium and palladium [2] have found entries into industrial applications.

In this chapter, we have compiled scientific papers, patent applications and other publicly available information related to large-scale use/commercial applications of ruthenium and palladium NHC complexes. It is not meant to be comprehensive with respect to all applications used to date due to the difficulty met when collecting information. However, this chapter provides a taste of what is currently done on what scale.

14.2 Production of NHC Complexes on Industrial Scale

14.2.1 NHC Complexes Produced on Industrial Scale

As mentioned above, transition metal NHC systems synthesised on industrial scale to date are, to the best of our knowledge, limited to palladium, ruthenium and recently silver [2]. These are listed in Fig. 14.1. For palladium complexes, four types are available: the naphthoquinone-bridged palladium (0) dimers of type **1**, the divinyldisiloxane adducts **2**, the chloride-bridged dimers of type **3**, the monomeric species of type **4** bearing an η^3-alkenic ligand (allyl or cinnamyl) and the 3-chloropyridine-adduct of dimers **3**, the PEPPSI-complexes **5** (PEPPSI: Pyridine-Enhanced Precatalyst Preparation, Stabilization, and Initiation). With respect to ruthenium complexes, there is a wider structural diversity of systems proposed on industrial scale: the benzylidene complexes **6** (Grubbs second generation) and **7**, the butenylidene complex **8**, the indenylidene systems **9** and **10**, the thienylmethylene complexes of type **11**, the ether-boomerang systems **12** (Hoveyda-Grubbs second generation) and the pyridinyl-propylidene complex **13**.

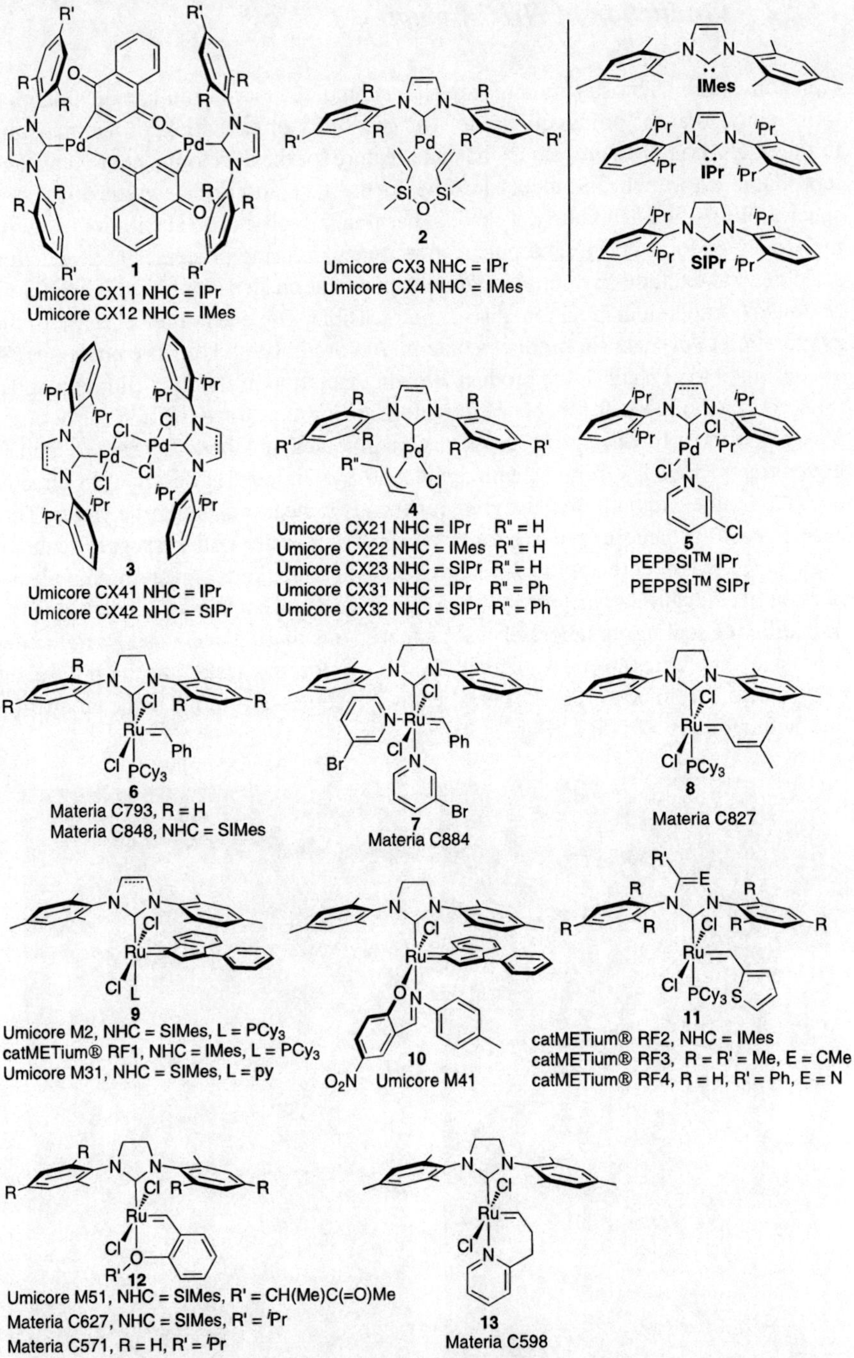

Fig. 14.1 NHC complexes produced on industrial scale

14.2.2 Production of NHC Complexes

Organometallic chemistry requires special techniques not only in chemical laboratories but also at the production scale. The capability of excluding air and moisture throughout a complete process is one main feature for the development of viable and reproducible synthetic protocols leading to the formation of catalytically active species. Whilst in a laboratory, a chemist has much freedom and flexibility in devising strategies for work-up and purification, manufacturing processes at production scale need to exclude any chromatographic purification steps for obvious economic reasons. Another manipulation that is not feasible, for economic reasons, is the evaporation to dryness for product isolation. Accordingly, all Umicore processes [3] are designed to crystallise the product allowing isolation in standard filtration apparatus. Further important aspects of a manufacturing unit for transition metal complexes, particularly when precious metals are involved, are the waste stream/mother liquor treatments. The flow scheme employed by Umicore is designed to directly treat the mother liquid to recycle precious metals in nearly quantitative yields. This ensures cost-efficient procedures for products at any scale. With respect to scale-up strategies, the Umicore approach involves statistical methods permitting the identification of relevant parameters and their optimisation. This allows to determine the possibility of scaling-up laboratory size experiments to 10 L scale using production process technologies. Once validated, the new process is tested on the pilot plant using reactors from 60 to 100 L. Thereafter, the process is transferred to the production unit with reactors >1,000 L (Fig. 14.2).

Fig. 14.2 (**a**) Pilot plant (**b**) Kilo-scale laboratory

While implementing production processes into multi kilogram batch sizes for NHC complexes (Fig. 14.1), a complete quality control of the imidazolium or imidazolidinium starting material was required. Therefore, Umicore has implemented large-scale manufacturing of the salts in-house, resulting in stable and reproducible synthetic protocols for the transition metal complexes.

14.3 Industrial Applications of Ru–NHC Catalysts

Ruthenium–NHC complexes exhibit activity in a very wide field of applications. Due to their unique ability to break and reassemble olefin bonds under reaction conditions very favourable to design simple processes, applications in nearly any chemical discipline can be foreseen. This field may span from manufacturing of specialty polymers and rubbers to pharmaceuticals, pharmaceutical intermediates, agrochemicals, fragrances, dyes, specialty chemicals for electronic applications or fine chemicals from natural feedstock and many more. Below are described Ru–NHC catalysed reactions applied from pilot to full commercial scale.

14.3.1 Ring Closing Metathesis (RCM) Reactions Used in the Pharmaceutical Industry

As a key transformation step, RCM is the most prominent and furthest advanced metathesis reaction technology in the pharmaceutical industry. It has been applied by several organisations on large scale to build up large rings that cannot be synthesised easily on an economically viable pathway using standard organic synthetic protocols. RCM permits these assemblies in fewer steps, thereby rendering a long linear synthetic route much less expensive with minimum waste.

14.3.1.1 Kosan's Epothilone Derivative KOS-1584

In 2002, Danishefsky and co-workers reported that 9,10-dehydro-12,13-desoxyepothilones inhibit the growth of tumour cells, and therefore were promising candidates for novel anticancer agents [4]. In a collaboration between Kosan Biosciences Inc. and F. Hoffmann-La Roche [5], the drug candidate KOS-1584 (R-1645) was developed and moved to clinical phase II. The initial Kosan process employed a Grubbs second generation catalyst, whilst Roche improved reaction yield by using an indenylidene-based ruthenium NHC catalyst (Fig. 14.1). The fact that KOS-1584 is undergoing clinical phase II trials means that large quantities (several kg) of the active substance are produced. The process details have not yet been reported, however, it is likely that KOS-1584 has been synthesised using one of the mentioned catalysts (Fig. 14.3).

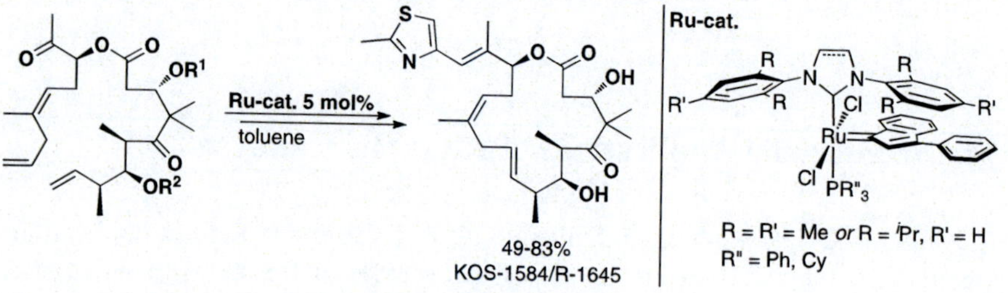

Fig. 14.3 Synthesis of KOS-1584/R1645by RCM

14.3.1.2 Glaxo Smith Kline SB-462795

The Glaxo Smith Kline (GSK) chemical process development group conducted a large-scale RCM reaction leading to the formation of a seven-membered ring. This moiety is a fragment of the molecular architecture of cathepsin K inhibitor SB-462795 [6], a drug candidate for the treatment of osteoporosis. This impressive piece of synthetic and process development work demonstrated the significant influence of the nature of the substrate on yields and on potential side-reactions while conducting RCM reactions. In summary, two synthetic strategies were explored, both involving as key-step a RCM of a chiral diallylic substrate. Both appeared suitable for further scale-up, one was selected and scaled to an 80 kg batch size with a relatively low catalyst loading resulting in nearly quantitative yields of the desired product. In this process, the complex employed is a boomerang-type catalyst (Fig. 14.4).

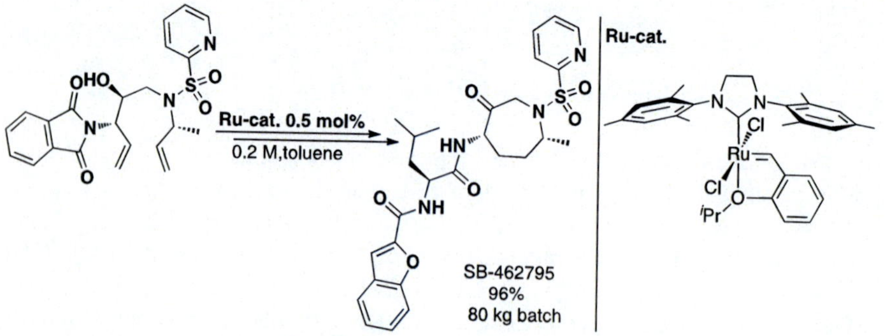

Fig. 14.4 Synthesis of Glaxo Smith Kline SB-462795

However, SB-462795 is no longer in the Glaxo Smith Kline development pipeline – most likely the reason why the GSK researchers were allowed to publish details of this process campaign.

14.3.1.3 Synthesis of HCV Protease Inhibitors

Independently, the Pharmaceutical companies Boehringer Ingelheim (BILN-2091) [7], F. Hoffmann-La Roche (ITMN-191) [8] and Tibotec Pharmaceuticals Ltd., a Johnson & Johnson company (TMC-435) [9] are or have been developing drug candidates for the treatment of Hepatitis C. While Boehringer Ingelheim was the first to successfully scale-up a RCM step to produce >100 kg substrate (using a first generation Hoveyda type catalyst *i.e.* not containing a NHC ligand) both Johnson & Johnson and F. Hoffmann-La Roche have been able to advance their drug development programmes to clinical phase I. We will remind the readers who are not familiar with the pharmaceutical jargon that clinical phase I would require multiple kilogram of active drug substance. There was therefore a need, in these campaigns, to scale-up the RCM step. Unfortunately, as is too often the case in pharmaceutical drug development, the Boehringer Ingelheim drug candidate failed in early clinical testing and resulted in a complete stop of the development of this molecule. Researchers of all three pharmaceutical companies have published, either in scientific publications or in patents, the details of the process chemistry. This chemistry represents a veritable Herculean endeavour, the chemistry evolving from these targeted molecules is simply first rate and teaches much about conformation directing RCM reaction. We strongly encourage the Reader to browse this literature to fully appreciate the intricacies associated with what looks on paper like a simple RCM transformation [10]. In brief, the main achievements for all drug candidates deal with solving the serious initial problem of having to conduct the RCM reaction at high dilutions. Considering the space/volumes requirements in an industrial setting, such high dilutions are costly and impractical, but mostly economically costly. If the reaction was conducted in high concentrations, undesired oligomers/polymers formed which reduced the valuable starting material into waste side-products. High dilution was therefore initially required. The process researchers, after much effort, succeeded in conducting RCM reactions (forming macrocycles possessing ring size of >12) to acceptable concentration levels of up to 0.5 M. This represents a significant advance as the initial Boehringer Ingelheim campaigns required dilutions of 0.01 M. These made use of a first generation metathesis catalyst (non-NHC bearing). All along the aim was and is to conduct the RCM reactions at highest possible concentration to project manageable throughput and reactor capacity use and efficiency. These groups have been able to reduce catalyst loadings significantly to levels of below 0.5 mol% ruthenium loading, another significant achievement, made possible by second generation ruthenium catalysts (NHC-bearing). The Roche and Johnson & Johnson drug candidates have entered clinical phase IIb in 2009 (Fig. 14.5).

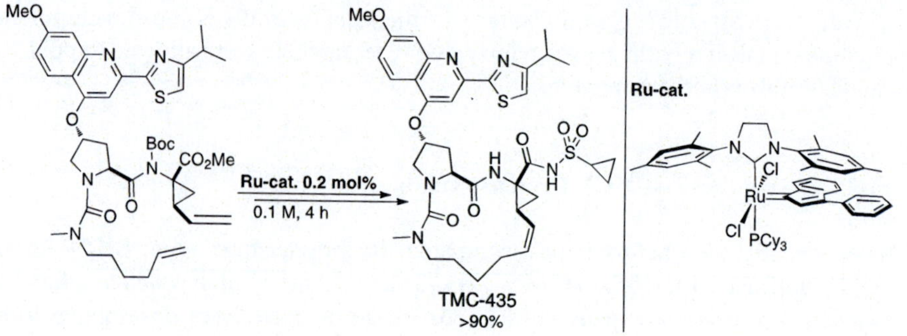

Fig. 14.5 Synthesis of Johnson & Johnson TMC-435

14.3.2 Cross Metathesis of Nitrile Rubber with 1-Hexene

In a series of patents, Lanxess has described its process for manufacturing Hydrogenated Nitrile Butadiene Rubbers (HNBR) with improved properties [11]. By a metathetical degradation of the nitrile rubber in a cross metathesis step with 1-hexene, the resulting HNBR exhibits a lower molecular weight distribution hence also a lower viscosity. This new "Therbane AT" (AT = Advanced Technology) exhibits improved processability in subsequent moulding practices. With different Therbane AT grades, large volume mould with more sophisticated structures can be filled in less time. The patents describe the use of NHC–Ru catalysts.

14.4 Industrial Applications of Pd–NHC Catalysts

Palladium-based homogeneous catalysts are used frequently on large scale in various industries. For instance, the Suzuki-Miyaura, Mizoroki-Heck and Sonogashira coupling reactions are used to synthesise pharmaceutically active ingredients and fine chemicals (see Chapter 6). In the bulk and commodity chemicals sector, there exist two major palladium-based processes, namely the synthesis of methyl methacrylate in the recently introduced Alpha Technology process of Lucite [12], and a process carried out by Dow Chemical for the synthesis of 1-octene. Both processes have an output of > 100,000 metric tons of product annually, both however are performed using a palladium catalyst bearing phosphines as ligands, the *Old Guard*. However, the telomerisation of butadiene involved in the 1-octene process was demonstrated by Oxeno on pilot-scale with Pd–NHC system [13]. On the pilot-scale, more than 25 metric tons of product have successfully been produced with an extremely low catalyst loading, using [Pd(IMes)(dvds)] as catalyst (Fig. 14.6). In spite of these very promising results, the IP (Intellectual Property) owner has not yet decided to implement the technology into a running large-scale process.

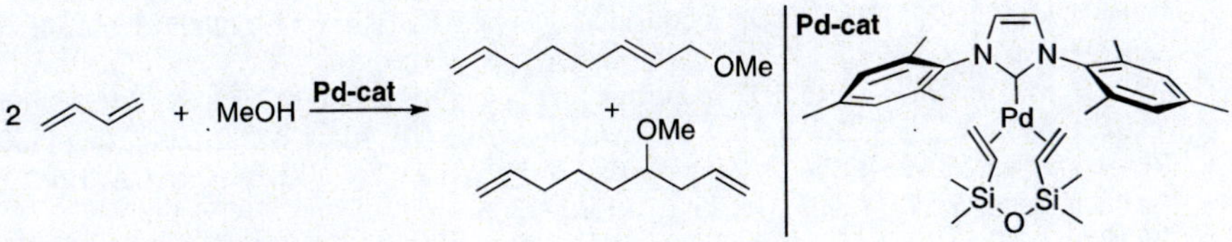

Fig. 14.6 Oxeno telomerisation of butadiene

14.5 Conclusions

The uses of NHC-metal complexes in industrial applications might appear at a very early stage by the relatively few examples provided in this chapter. However, it is clear that much information has not been publicly disclosed for obvious industrial interest reasons. The few examples provided show how efficient the catalysts can be, and how low a catalyst loading can be achieved in large-scale production. Many more larger-scale applications are being carried out using metathesis reactions in their many incarnations, of that we are certain. This initial industrial perspective is hopefully intriguing enough to warrant a further description of the area in a few years. The NHCs have so far caused quite a stir in industry and we can safely say that more will come of these robust ligands.

In view of a future edition, the authors welcome any information that could assist to update the field of use of NHC complexes on industrial scale.

Acknowledgments We are grateful to Christophe Le Ret for his support, Dr Renat Kadyrov (Evonik) and Tom Buel (Materia) for details concerning their product portfolio.

References

1. Herrmann WA, Elison M, Fischer J, Koecher C, Arthus GRJ (1995) Angew Chem Int Ed 34:2371–2373; Herrmann WA, Elison M, Fischer J, Koecher C, Artus GRJ (1996) Chem Eur J 2:772–780
2. Akron Research Commercialization Corp. has recently disclosed a drug candidate based on Ag–NHC complexes (Silvamist™). These NHC complexes were proven efficient for the treatment of respiratory diseases, and the company is currently applying for Investigational New Drug (IND) status with the FDA (US Food and Drug Administration)
3. Due to the lack of available information, this view reflects practices performed at Umicore and may differ from elsewhere
4. Rivkin A, Yoshimura F, Gabarda AE, Chou T-C, Dong H, Tong WP, Danishevsky SJ (2003) J Am Chem Soc 125:2899–2901; WO2004/018478 A2 Sloan-Kettering Institute for Cancer Research: synthesis of Epothilones, Intermediates thereto, analogues and uses thereof
5. WO 2006/111491 A1, F Hoffmann-La Roche AG: Process for the preparation of 9,10-Dehydro-12.13-Desoxyepothilone Derivatives; R1645; KOS1584; Phase II material has been manufactured, http://www.roche.com/home/media/med_div/med_div_pharma/med_pha_2007/med-div-2007-02-27.htm

6. Wang H, Matsuhashi H, Doan BD, Goodman SN, Ouyang X, Clark WM Jr (2009) Tetrahedron 65:6291–6303
7. Wang H, Matsuhashi H, Doan BD, Goodmann SN, Ouang X, Clark WM Jr (2009) Tetrahedron 65:6291–6303; Wang H, Goodman, SN, Dai Q, Stockdale GW, Clark WC Jr (2008) Org Proc Res & Dev 13:250–254; Shu C, Zeng X, Hao M-H, Wei X, Yee NK, Busacca CA, Han Z, Farina V, Senayake CH (2008) Org Lett 10:1303–1306
8. WO 2009/124853 A1, F Hoffmann-La Roche AG: New ruthenium complexes as catalysts for metathesis reactions
9. WO2007/014918 A1, Tibotech Pharmaceuticals Ltd, Medivir AB: Macrocyclic inhibitors of hepatitis C virus; WO 2008/092955 A1, Tibotec Pharmaceuticals Ltd.: Process and intermediates for preparing a macrocyclic protease inhibitor of HCV; Horvath A (2009) Scientific Update Conference: Org Proc Res & Dev, Lisbon, Portugal Sept 9–11
10. (a) Farina V, Shu C, Zeng X, Wei X, Han Z, Yee NK, Senanayake CH (2009) Org Proc Res Dev 13:250–254; (b) Shu C, Zeng X, Hao M-H, Wei X, Yee NK, Busacca CA, Han Z, Farina V, Senanayake CH (2008) Org Lett 10:1303–1306; (c) Nicola T, Brenner M, Donsbach K, Kreye P (2005) Org Proc Res Dev 9:513–515; (d) Faucher A-M, Bailey MD, Beaulieu PL, Brochu C, Duceppe J-S, Ferland J-M, Ghiro E, Gorys V, Halmos T, Kawai SH, Poirier M, Simoneau B, Tsantrizos YS, Llinàs-Brunet M (2004) Org Lett 6:2901–2904
11. WO2005/080456 A1, Lanxess Inc.: Process for the preparation of low molecular weight hydrogenated nitrile rubber; EP2028194 A1, Lanxess Deutschland GmbH: Metathesis of nitrile rubbers in the presence of transition metal complex catalysts; Press release: Lanxess – New York Conference, New York City, USA, Monday August 28, 2006, Address by Dr. Ulrich Koemm "LANXESS: Concepts in Rubber"
12. (a) http://www.luciteinternational.com/newsitem.asp?id=124; (b) Clegg W, Eastham GR, Elsegood MRJ, Tooze RP, Wang XL, Whiston K (1999) Chem Commun 1877–1878; (c) Eastham GR, Tooze RP, Wang XL, Whiston K World Patent 1996, 96/19434
13. Jackstell R, Harkal S, Jiao H, Spannenberg A, Borgmann C, Röttger D, Nierlich F, Elliot M, Niven S, Cavell K, Navarro O, Viciu MS, Nolan SP, Beller M (2004) Chem Eur J 10:3891–3900. Beller M (2006) Scientific Update Conference: Modern Synthetic Methods, Lake Luzern, Switzerland, June 7–8

ERRATUM

N-Heterocyclic Carbenes in Transition Metal Catalysis and Organocatalysis

Catherine S. J. Cazin (Ed.)

C. S. J. Cazin (ed.), *N-Heterocyclic Carbenes in Transition Metal Catalysis and Organocatalysis*, Catalysis by Metal Complexes 32,

DOI 10.1007/978-90-481-2866-2_15

This Erratum has been produced due to a missing co-author.

The co-author of Chapter 5, N-*Heterocyclic Carbene Complexes in Cyclisation Reactions*, was mistakenly not mentioned on page 131. His name is Dr. **Rodrigo Cella,** University of Utah, Department of Chemistry, 315 South 1400 East, Salt Lake City, UT 84112.

The online version of the original chapter can be found at http://dx.doi.org/10.1007/978-90-481-2866-2_5

Index

I

S